Lamb's Problem in Seismology (Volume 1)

Haiming Zhang

Lamb's Problem in Seismology (Volume 1)

Science Press
Beijing

Springer

Haiming Zhang
School of Earth and Space Sciences
Peking University
Beijing, China

ISBN 978-981-95-3169-1 ISBN 978-981-95-3170-7 (eBook)
https://doi.org/10.1007/978-981-95-3170-7

Jointly published with Science Press.
The print edition is not for sale in China mainland. Customers from China mainland please order the print book from: Science Press.
ISBN of the Co-Publisher's edition: 978-7-03-067313-8

This Springer imprint is published by the registered company Springer Nature Singapore Pte Ltd.
The registered company address is: 152 Beach Road, #21-01/04 Gateway East, Singapore 189721, Singapore

Preface

For about 15 years, I have been teaching "Theoretical Seismology" at Peking University. This demanding course, heavily reliant on mathematical tools, lacks beginner-friendly textbooks. Since many students have expressed a desire for notes aligned with the course, I considered writing a thorough textbook but gave up for two reasons: the deep and vast knowledge required for this subject is beyond my current expertise; and with many existing textbooks, like the celebrated *Quantitative Seismology* by Aki and Richards (2002), another textbook seemed unnecessary and challenging to justify.

When I started my graduate studies in seismology, I was intrigued by its vast scope but felt frustrating when reading books and papers. I often resolved to carefully read a book, only to quickly abandon it. Reflecting on this, several factors were at play. Firstly, the content was inherently complex. For instance, Ewing et al. (1957) introduced advanced concepts like the Sommerfeld and Weyl integrals in the very first chapter, followed by complex integrals over a four-sheeted Riemann surface involving hyperbolic secants. Cagniard (1962) was also elusive and confusing to me. Similarly, early chapters of Aki and Richards (2002) were manageable, but by Chap. 6, it became very challenging. Secondly, the textbooks often lacked the detail necessary for clear understanding, which made reading more difficult. Lastly, the information seemed isolated to me; even when I grasped a concept with much effort, it didn't connect to a broader understanding, leaving me unmotivated to continue and quit prematurely.

Since then, I've dreamed of a book that, while not broad, provides detailed enough explanations to facilitate understanding of more complex seismological textbooks and literature. Unfortunately, it is hard to find such a book. Inspired by my research, I chose the classic Lamb's problem in seismology as the central subject to gradually introduce theoretical seismology, exploring the topic from various angles. Although not comprehensive, this book may offer rich insights. This thought motivated me to write, culminating in this completed work, now presented to readers.

The following aspects have been paid special attention in writing.

(1) Positioning and content selection

This book is difficult to categorize. Is it a reference textbook, a specialized monograph, or a work of popular science? It addresses important aspects of theoretical seismology and prominently features the Lamb's problem, marking it as a focused monograph. However, its style and objectives align with popular science, as it aims to elucidate complex theoretical issues in seismology for students and professionals. This unique combination suggests the book could be best described as *a foundational, advanced-level popular science monograph on seismology*. "Advanced" refers to its extensive use of mathematical tools beyond the basics taught in undergraduate courses, and "foundational" indicates it focuses on well-established topics rather than emerging trends. Its detailed discussion on a classic seismological problem justifies it as a monograph.

In terms of content selection, the guiding principle of this book is "less but deeper", focusing solely on topics pertinent to the theme and excluding all irrelevant details. For instance, although the analysis based on the "plane wave assumption" or "high-frequency approximation" is a vital component of seismology, this book does not address these aspects. Instead, it concentrates on *the precise calculation of seismic waves*.

(2) Organization of content

When organizing content, it's crucial to connect ideas to form a cohesive whole, helping readers grasp and structure their understanding. In contrast, disjointed, rigid knowledge fragments the content, making it challenging to build a comprehensive framework. Effective organization infuses knowledge with vitality and coherence.

The history of a discipline often does not unfold in the logical sequence one might expect,[1] and seismology is no exception. Since Lamb (1904), the study of seismic waves has been a focal point for seismologists. The debates over whether the seismic source could be represented by a single or a double couple began following the 1923 Kanto earthquake in Japan. It wasn't until the 1960s that the seismic source representation theorems established by Maruyama (1963) and Burridge and Knopoff (1964) became the theoretical backbone of seismology, settling these disputes. Similar to Aki and Richards (2002), this book does not follow the chronological development of seismology. Instead, it adopts the seismic source representation theorem as its

[1] The development of analysis serves as a key example. Modern higher mathematics textbooks usually begin with the $\varepsilon - \delta$ definition of limits, but this wasn't always the historical approach. In 1666, I. Newton and G. Leibniz independently introduced calculus, displaying significant potential despite its initial logical flaws. It faced criticism from contemporary mathematicians for its foundational issues. It took nearly two centuries for French mathematician A. L. Cauchy and German mathematicians K. Weierstrass and R. Dedekind to rigorously formalize these concepts, solidifying the theoretical underpinnings of calculus.

framework and centers on deriving Green's functions for various media to explore seismic wave problems, including those posed by Lamb.

(3) Manner of expression

The presentation of a book's content significantly influences its fate: whether it is quickly glanced over and forgotten or read intensely and held dear. This book features several unique aspects in its expression.

(a) On language style. The book presents its content in a conversational and approachable style, avoiding a didactic or stiff tone. It features numerous footnotes that enrich the text with detailed explanations, further knowledge, and the author's own observations. This approach aims to expand readers' understanding and increase their interest in the subject.
(b) On narrative style. In line with recent trends in reference books, this book aims to present significant findings in the field in a clear and accessible manner. The goal is for readers with a moderate to advanced undergraduate background in science and mathematics to be able to teach themselves effectively with some effort.
(c) On typesetting and graphic display. The book is typeset using LaTeX and features mostly diagrams drawn with Asymptote, many of which are presented in three dimensions to enhance the reader's intuitive understanding and improve the reading experience. Due to the numerous formulas, the equations for important definitions and results are highlighted by enclosing them in boxes with a gray background, making them easier for readers to refer to.
(d) On references. The book incorporates references, yet it is crafted to be a self-contained work, making extensive reference consultation unnecessary for understanding. References are included primarily to support the discussions and facilitate additional reading opportunities. Curious readers can explore these references for a deeper dive into the subjects, but ignoring them won't compromise the basic understanding of the content.[2] This approach ensures the book can be understood on its own merits, enhancing its accessibility and educational value.

This book appearing in the reader's hands suggests that the reader falls into one of the following categories:

(1) Graduate students, advanced undergraduates, and researchers in seismology who aim to thoroughly understand most or all of the content. While the target audience may be small, the book caters specifically to these dedicated learners. However, mastering the material will not be an easy journey. Deep understanding of the concepts requires hands-on work, including deriving equations and programming. As later sections often reference earlier formulas, frequent cross-referencing within the book is essential. A true measure of mastery is

[2] This choice is based on the author's own reading and learning experiences; if a book frequently requires consulting external references to progress, it inherently diminishes the reader's enjoyment and engagement.

the ability to replicate the derivations and programming independently. Chapters 6 and 7 are crucial as they detail the derivation processes and provide extensive examples and the necessary techniques for numerical implementation. Achieving this level of competence demands significant effort but is achievable with dedication.

(2) Readers who are keen to acquire a foundational understanding of theoretical seismology and gain some insight into Lamb's problem, which may describe the majority of its audience. The detailed exposition provided in the book aims to facilitate this initial learning. However, progressing from a basic to a deep comprehension, and even achieving full mastery of the subject, necessarily involves engaging with mathematical derivations and programming. Some readers might feel intimidated by equations and may have never ventured into programming or derivations on their own. It is encouraged for these readers to attempt these aspects during their study of the book; they might find themselves pleasantly surprised by their capabilities.

(3) Casual readers who pick up this book, flipping to Sects. 1.2, 3.1, and the historical narrative in Chap. 5 can be quite rewarding. Treat these parts like short stories, and you might find them informative and enjoyable.

(4) Seasoned experts in the seismology community, may view this book as an informative resource on popular science, and valuable comments and criticisms are welcome.

This book delves into the significant advancements made in the fields of elastodynamics and theoretical seismology over the past two centuries. The pioneering efforts of past mathematicians, mechanics, and seismologists have lent profound meaning to the popularization of this science, to whom I express deep respect. During the preparation of this manuscript, I consulted numerous books and documents. The contributions from these works have been instrumental in the successful completion of this book, and I extend my gratitude and respect to the authors of these referenced materials.

My ability to teach theoretical seismology today stems largely from the foundational education I received early on from Prof. Liangbao Zhu during my undergraduate studies, and later from Prof. Xiaofei Chen during my graduate years. Under Prof. Chen's guidance, I dedicated my graduate and postdoctoral research to studying the dynamics of seismic sources and theoretical seismology, which formed the bedrock of my later independent research and teaching career. I am deeply grateful to both mentors for their invaluable teaching and guidance.

The research on the Lamb's problem presented in this book draws partly from the efforts of former students in my research group. Notably, the work on the zeros of Rayleigh functions by Dr. Tianshi Liu and the analysis of Rayleigh waves in the frequency domain by Dr. Jie Zhou have enriched the content of this book. I am deeply grateful for their contributions.

In writing this book, I employed LaTeX for typesetting and extensively used the Asymptote software for illustrations, for which I owe a heartfelt thanks to the developers of these tools. The research discussed in this book, as well as its publication, were supported by the National Natural Science Foundation of China under grants 41674050 and 41874047. I am deeply appreciative of this support.

If you find any oversights or inaccuracies in this book, please kindly point them out and feedback to me (zhanghm@pku.edu.cn).

Beijing, China Haiming Zhang

Contents

Chapter 1
Introduction

In the opening chapter of this book, we aim to clarify several key questions: what is the Lamb's problem, and why is it significant? We explore its importance to theoretical seismology and discuss what this field entails and its relevance. Understanding the connection between the Lamb's problem and theoretical seismology is crucial as it sets the stage for a deeper investigation into the subject throughout the book.

1.1 Definition and Significance of Lamb's Problem

To put it simply, if we strip down the seismological issue of ground movements caused by fault displacement to its basics, we could describe it as follows: Imagine the Earth as a homogeneous, isotropic semi-infinite elastic medium. Applying a force to this medium induces motion at its free surface. In essence, this scenario captures what is known as the Lamb's problem.

The Lamb's problem is highly regarded in both seismology and civil engineering, earning its status as a captivating classic. It effectively demonstrates the primary characteristics of seismic waves using a straightforward model, which enables extensive theoretical analysis through mathematical approaches. Nevertheless, despite its deceptively simple appearance, achieving a complete solution to the Lamb's problem is notably difficult.

H. Zhang, *Lamb's Problem in Seismology (Volume 1)*,
https://doi.org/10.1007/978-981-95-3170-7_1

Rayleigh (1885) on what would become known as *Rayleigh waves* laid the groundwork for seismology based on strict elasticity theory.[1] However, it was Lamb's landmark 1904 paper that truly ignited theoretical seismology. Celebrated as one of the most classic papers in the history of seismology, it remains widely cited over a century later. Following Lamb's pioneering work, numerous mathematicians, mechanicians, and seismologists focused on the Lamb's problem for over fifty years, making it a focal point of seismological research in the first half of the 20th century. Their collective efforts deepened the understanding of the fundamental nature of seismic wave motion, establishing a solid foundation for further, more extensive studies in seismology.

1.1.1 What is the Lamb's Problem?

The Lamb's problem, though not strictly defined, is a widely recognized term in seismology. It encompasses two distinct concepts. The first originates from Lamb (1904), in which Lamb explored the effects of a point force applied at the surface to determine displacements at distant surface points. The second, an extension of the first, addresses a broader spectrum of scenarios: the source and observation point may be situated on the surface or beneath it, the assumption of axial symmetry is discarded, the source time function is variable, there are no constraints on the distance of the field, and the source could be a shear dislocation source with seismic implications, rather than a unidirectional point force. In our discussions, we refer to the latter interpretation of the Lamb's problem, and it is crucial to clarify these concepts before delving deeper into the topic.

1.1.1.1 Forms of Analytical Solutions

Analytical solutions, unlike numerical ones obtained through methods like finite element or finite difference, clearly define functional relationships for variables. Numerical methods may fail to predict outcomes for certain variables. Analytical solutions allow precise calculations based on these relationships, even for results expressed through integrals that resist simplification into basic functions and require numerical evaluation. Yet, their well-defined functions still categorize them as analytical solutions.

---

[1] Although the Stokes' solution in an infinite elastic medium remains a staple in seismology textbooks, it lacks practical seismic relevance as it fails to address surface effects, which are crucial for understanding seismic phenomena. Rayleigh (1885), based on the assumption of harmonic oscillations in plane waves, unveiled properties of waves that have universal significance. From a seismological perspective, Rayleigh's research is profoundly meaningful, making it a fitting starting point for seismological studies grounded in rigorous theory of elasticity.

In this book, we discuss the simplicity of function-based solutions, varying from the most straightforward forms to more complex representations. The simplest are expressed through elementary functions like power, exponential, logarithmic, and trigonometric functions, combined using basic arithmetic operations–what we term *closed-form* analytical solutions. Some problems cannot be neatly fit into this strict closed form but can be expressed using a combination of elementary functions and standard elliptic integrals.[2] These are referred to as *generalized closed-form* analytical solutions. In the subsequent volume, we will focus on deriving these generalized closed-form solutions in time-domain for Lamb's problem.

1.1.1.2 Classification of Lamb's Problem

The complexity of solving Lamb's problem depends directly on how the source and field points are distributed, particularly when aiming for closed or generalized closed-form analytical solutions. To facilitate discussion, Lamb's problem is divided into three categories based on the positions of these points:

(1) The first type of Lamb's problem (abbreviated as $\mathcal{LP}_1$), both the source and field point are located on the Earth's surface;
(2) The second type of Lamb's problem (abbreviated as $\mathcal{LP}_2$), the source point is located underground while the field point is on the Earth's surface, or vice versa according to the reciprocity theorem described in Sect. 2.3;
(3) The third type of Lamb's problem (abbreviated as $\mathcal{LP}_3$), both the source point and the field point are located underground.

According to this classification, the issue studied in Lamb (1904) falls under a special category in seismology, known as the $\mathcal{LP}_1$, such as the ground motions caused by nuclear explosions. More commonly relevant in seismology is the $\mathcal{LP}_2$, which aligns with the fact that most observation stations are located on the surface while the sources of seismic activity are within the Earth's medium. Although the models described for solving these problems are not complex, fully resolving them poses significant challenges.

1.1.2 Theoretical Seismological Significance of Lamb's Problem

As previously mentioned, Lamb's problem can be seen as the true beginning of theoretical seismology. This stems from Lamb (1904) efforts to derive the seismic

[2] Standard elliptic integrals, once extensively researched and tabulated by the French mathematician A. M. Legendre, can be expressed as infinite series. Today, mathematical software like Maple, Matlab, and Mathematica offer built-in functions that easily compute these integrals, making them readily accessible for users.

wave field by rigorously solving the equations of elastic motion, marking the first attempt to do so. Over a century ago, any analysis of seismic wave fields required significant simplification. Lamb's research tackled this by considering the problem in its most simplified form, though at the time, solutions were only expressible as integrals. To derive practically meaningful results, Lamb had to focus on the far-field scenario, using asymptotic solutions to produce the first theoretical seismogram in human history. While today this might not seem very difficult, it was undoubtedly a significant historical milestone in seismology at the time.

Choosing Lamb's problem as the entry point for theoretical seismology is meaningful because it focuses on a simplified model. Analyzing this issue from various angles helps further our understanding of the complex interactions between media and seismic sources that influence seismic waves. So, what exactly is theoretical seismology, and why is it important to study? Before addressing these questions, it's useful to briefly review the history of human understanding of earthquakes.

1.2 Humanity's Knowledge of Earthquakes

Earthquakes, now widely understood to result from dislocation in the Earth's subterranean rocks, were not always recognized as such. About 260 years ago, like many natural phenomena, they were believed to be manifestations of divine wrath by our ancient ancestors. The natural philosophers of ancient Greece were among the first to reject supernatural explanations and seek rational explanations for seismic activities. For instance, Aristotle, around 340 B.C., believed that earthquakes were caused by the movement of gases–air that got trapped inside the Earth and caused tremors in its attempt to escape. Similar theories were also explored in ancient China, where earthquakes were thought to be caused by the mysterious accumulation of "Qi". Aristotle's theory influenced European thought for two thousand years until the 1755 Lisbon earthquake.[3] A review of the history of seismology over the past three hundred years can broaden our perspective and inspire us.

[3] The Lisbon earthquake was a significant event in the history of seismology, not merely because of its destructive power. Following the quake, scientists from across Europe, including geology professor J. Michell from Cambridge University, collaborated to study the event. In 1761, Michell concluded from his extensive research that earthquakes originate from within the Earth and that seismic waves are related to the propagation of elastic waves in rocks. While this might seem obvious today, it was the first time such a conclusion was drawn from scientific research. Consequently, this earthquake is regarded as one of the most impactful in the study of earthquakes (Agnew 2002).

The development of seismology can be categorized in different ways depending on the perspective taken.[4] We focus on the theoretical aspects of its evolution, dividing it into three periods: the early seismology period from 1821 to 1903, the classical seismology period from 1904 to 1949, and the modern seismology period starting from 1950 onwards.

1.2.1 Early Period of Seismology (from 1821 to 1903)

Quantitative seismological research dates back to the early 19th century, nearly two centuries ago. Initially, the focus of seismology overlapped significantly with the development of elastic mechanics, as both studied wave propagation in solids.[5] During this period, although there were few theories dedicated solely to seismology, the advancements in general elastic mechanics laid the groundwork for the classical era of seismological research that followed. The progress in seismology during this time was primarily concentrated on earthquake observation.

In 1821, L. Navier introduced the differential equations of elastic mechanics in a paper presented to the French Academy of Sciences, marking a significant advancement. Following this, in 1829, S. D. Poisson theoretically predicted the existence of compressional and shear waves, known in seismology as P-wave and S-wave, respectively, and also studied the vibrations of an elastic sphere. In 1849, G. G. Stokes developed the first mathematical model of an earthquake source, deriving an exact solution for the displacement field caused by a single force in an infinite elastic medium, which will be discussed in detail in Chap. 4 of this book. By 1863, L. Kelvin had made the first quantitative estimate of Earth's vibrational eigenfrequencies. In 1872, E. Betti introduced the reciprocity relation, leading directly to the displacement integral representation theorem in elastodynamics, see Chap. 2 of the text. In 1882, Lamb conducted an in-depth study of the free oscillations of a gravity-free sphere, clearly distinguishing between toroidal and spheroidal oscillations and laying the groundwork for the study of Earth's free oscillations and global seismology. Lastly,

[4] For example, Agnew (2002) segments the evolution of seismology into four distinct phases: the early era from the beginning of the 19th century to 1880, the new seismology era from 1881 to 1920, the classical seismology era from 1921 to 1960, and the modern era starting in 1960. Ben-Menahem (1995) propose a different three-part division: the pre-seismometer era from 1821 to 1891, the pre-computer era from 1892 to 1950, and the computer era beginning in 1951. A comprehensive discussion on the history of seismology exceeds the scope of this book, hence only selected milestone works are briefly mentioned here. Moreover, as this section aims to provide an overview of humanity's understanding of earthquakes rather than a detailed textual research, most research mentioned does not list references. For a detailed historical review and further readings, interested readers can refer to Ben-Menahem (1995), Dahlen and Tromp (1998, p. 1), and Agnew (2002).

[5] The history of elastic mechanics can be traced back to 1660 when R. Hooke established Hooke's Law, but its substantial development began in 1821 with L. Navier's differential equations for elasticity.

in 1885, L. Rayleigh theoretically discovered a surface wave associated with surface movement, known as the Rayleigh wave.[6]

In the early days of seismology, the development of elasticity theory provided a solid foundation for the field. A key figure in this period was R. Mallet, considered the first true seismologist. He not only aimed to systematically develop the study of earthquake waves into a distinct discipline but also coined the term *Seismology*. There was significant progress in earthquake observation in the late 19th century, notably in 1889 when an earthquake in Japan was detected in Potsdam, Germany. This event, highly praised by Ben-Menahem (1995) as the "first revolution" in seismology, marked a pivotal moment as the emergence of teleseismic seismograms transformed seismology into a global science.

1.2.2 Classical Period of Seismology (from 1904 to 1949)

Entering the 20th century, the first half witnessed rapid advancements in the theoretical study of seismology, especially in the propagation of seismic waves, beginning with Lamb (1904). Thanks to the cumulative efforts of generations of mathematicians and seismologists, a magnificent edifice of theoretical seismology was constructed. During this period, seismologists developed detailed models of the Earth's internal structure, demonstrating the power of seismology as the most effective tool for probing the deep structure of the planet.

The year 1904 marked a pivotal moment in the development of seismology, highlighted by Lamb's creation of the first mathematical model of seismic activity in a semi-infinite space. This model addressed the response of an elastic surface to forces applied to it, using Fourier-Bessel integrals to represent solutions. Lacking numerical computation tools, Lamb resorted to far-field approximations, producing the first synthetic seismogram and confirming the existence of Rayleigh wave. In the same year, A. E. H. Love[7] extended the foundational work of G. G. Stokes and L. M. Lorenz on the point source theory in an infinite elastic medium, covering arbitrary initial disturbances and a wide range of body forces, such as dipoles and force couples. This laid the groundwork for mathematical models of seismic sources. Love also expanded on V. Voltarra's work to derive a three-dimensional integral representation

[6] The discovery of Rayleigh waves serves as a classic example of a theoretical prediction preceding empirical confirmation in the history of seismology. Shortly after Rayleigh's theoretical forecast, R. D. Oldham identified the previously predicted P-wave and S-wave by Poisson, as well as the Rayleigh surface wave, in his 1897 study of an earthquake in India. This phenomenon of theory guiding practice is not unique in the history of science, with the discovery of Neptune being a notable example driven by Newtonian mechanics.

[7] A synthesizer in 19th-century elasticity theory, authored *A Treatise on the Mathematical Theory of Elasticity* and pioneered the application of these theories to geophysics, particularly in his specialized work *Some Problems of Geodynamics*. Love's most notable contribution to geophysics was the discovery of the surface wave named after him, the Love wave. He also authored two other educational texts, *Theoretical Mechanics* and *Elementary Calculus*.

theorem, foundational for what became known as the "first principle" of modern seismology. The work of Lamb and Love has been of immeasurable importance to the theoretical development of seismology, and Lamb's problem remained a central topic in theoretical seismology for the next fifty years. By 1907, V. Volterra had developed dislocation theory,[8] and in 1911, Love identified a surface wave not covered by Rayleigh and Lamb's theories, later named the Love wave. Love's insights spurred extensive mathematical research and yielded significant information about the crusts of continents and oceans. In 1923, H. Nakano demonstrated that observed initial motion patterns could be explained using combinations of forces in the Stokes-Love solution, including double-couples, among others.

The period from around 1939 to 1959, referred to by Ben-Menahem (1995) as the "transitional period", marked the decline of the classic era of the *Oxford-Cambridge analytical school*.[9] During this time, numerous mathematicians and seismologists attempted to solve the Lamb's problem using integral transforms. The main focus of theoretical seismology for about fifty years after Lamb (1904) was to find analytical solutions to the Lamb's problem. Notably, French mathematician L. Cagniard published a monograph in 1939 detailing an innovative method based on the Laplace transform that he had developed over the previous decade. This approach, later refined by A. T. de Hoop into the *Cagniard-de Hoop method*,[10] found wide applications

[8] The concept of "dislocation" originally referred to microscopic defects within crystalline materials, where atoms are irregularly arranged locally. In the early 20th century, Italian elasticians like Volterra and Somigliana introduced this idea to the field of elasticity in continuous media. This incorporation marked an evolution in classical elasticity theory, which Love discussed in his seminal work *The Mathematical Theory of Elasticity*.

[9] In the history of mathematics and physics, there have been several distinguished schools of thought named after their locations. For instance, the renowned *Göttingen school*, initiated by the "Prince of Mathematicians" C. F. Gauss, became an international hub for mathematics research and education. Similarly, the *Copenhagen school*, founded by the famous physicist N. Bohr, was once the world's most formidable school in physics. These schools shared several characteristics: they hosted multiple renowned scientists, exhibited similar research styles and philosophies, and significantly impacted their respective fields. Seismology followed this pattern as well. Over a century beginning in the mid-19th century, many mathematicians and physicists who made significant contributions to seismology were associated with Oxford and Cambridge universities, such as Stokes, Kelvin, Rayleigh, Lamb, Love, H. Jefferys, and K. E. Bullen. This cluster of talent made the UK the epicenter of theoretical seismology research for over half a century. Known for using mathematical methods to find analytical solutions, this group was aptly named the "Oxford-Cambridge analytical school" in Ben-Menahem (1995).

[10] The Cagniard method, though groundbreaking, was initially overlooked for 20 years due to its complexity and the fact that the seminal 1939 monograph was written in French, making it less accessible to non-French speakers. However, the situation changed in 1960 when de Hoop published a short paper that introduced the de Hoop transformation. This adaptation presented the essence of the Cagniard method in a more understandable manner, leading to its widespread recognition and subsequent naming as the *Cagniard-de Hoop method* in honor of de Hoop's significant contributions to its popularization. In 1962, further efforts to disseminate this method were made by E. A. Flinn and C. H. Dix, who translated and annotated Cagniard's original monograph. Aki and Richards (2002) later endorsed the Cagniard method as the optimal approach for solving the Lamb's problem.

in various fields of physics. In 1949, British mathematician E. R. Lapwood conducted extensive research on the Lamb's problem, using the saddle point method in asymptotic approaches to study several cylindrical wave reflections at flat interfaces. Ben-Menahem (1995) described this as the "final attack" by the analytical school on Lamb's problem using integral transform techniques.

In the half-century following Lamb (1904), theoretical seismology flourished under the influence of the analytical school, leading to a more profound understanding of Earth's internal structure, propelled by an increase in observational data. Notable advancements include the discovery of the *Moho discontinuity* by A. Mohorovičić in 1909, subsequently named after him, and B. Gutenberg's determination in 1914 of the depth of the boundary between the Earth's mantle and core, later called the *Gutenberg discontinuity*. In 1939, H. Jefferys and his student K. E. Bullen compiled the seismic travel time tables, known as the "J-B travel time tables", which are still in use at basic seismological stations today. These developments significantly enriched our knowledge of the planet's inner workings.

In the 1930 and 1940s, as the analytical era was drawing to a close, seismologists were increasingly aware that limitations in computational power were becoming a major obstacle to progress in their field. The advent of computers in the late 1940s was a timely breakthrough, invigorating nearly all fields that relied on scientific computing, including seismology.

1.2.3 Modern Period of Seismology (After 1950)

In the era of classical seismology, the most significant contributions were made almost exclusively by mathematicians and physicists, primarily due to the lack of computing resources which made exploring complex elastic wave theories nearly impossible without advanced mathematical skills. The introduction of electronic computers marked a revolutionary shift, drastically changing this scenario. Ben-Menahem (1995) referred to this as the "second revolution in seismology." Computers greatly enhanced the capacity to solve problems mathematically, enriching seismological research by integrating theoretical, experimental, and computational methods. This combination of theoretical analysis with computational power not only empowered those without extensive mathematical skills to conduct theoretical research but also led to the development of new computational techniques and the discovery of new phenomena, phasing out outdated methods that were not suited to the computing era.

Over the past three decades, the field of seismology has fully embraced the digital age, marked by the comprehensive digitalization of seismic data and its dramatic increase in volume. Coupled with rapid advancements in computer hardware, these developments have made parallel computing readily accessible. Since the late 1980s, seismology has evolved into a discipline that is predominantly "computer-driven," as noted in Ben-Menahem (1995). If the initial two stages of seismology were primarily driven by individual mathematicians pioneering theoretical developments, the modern era of seismology is now largely propelled by advances in observation and computation. For instance, significant strides in free oscillation and global seismology were initially sparked by observations from the 1960 Chile earthquake.[11] The vigorous growth of computational seismology,[12] driven by rapid technological advancements, now allows for sophisticated simulations of seismic wave propagation through complex three-dimensional structures, detailed reconstructions of subsurface media, and precise temporal and spatial analyses of fault ruptures. However, despite these technological advances, breakthroughs in theoretical seismology have become less frequent, with applied research increasingly dominating the field.[13]

In the early 1950s, W. T. Thomson and N. A. Haskell were pioneers in integrating computers into seismology, calculating the dispersion of surface waves in multilayered media. By 1955, C. L. Pekeris produced the first computer-generated theoretical seismogram for point forces in a Poisson solid (with Poisson's ratio $\nu = 0.25$), marking about fifty years since Lamb's creation of the first synthetic seismogram. In Steketee (1958), J. A. Steketee demonstrated the equivalence theorem by comparing the Stokes-Love solution for a double couple with Volterra's solution for corresponding dislocations in static cases. Dynamic scenarios were later addressed by Burridge and Knopoff (1964). The 1960 Chile earthquake, the largest recorded to date, generated hours of free oscillations worldwide, spurring research that became a highlight of 1960s seismology. This research, initially focused on Earth's free oscillations, essentially the planet's natural modes of vibration, provided a new method for synthesizing theoretical seismograms and a novel tool for understanding Earth's internal structure. From the early 1970s, mathematicians and seismologists like F. Gilbert, J. H. Woodhouse, and F. A. Dahlen developed a comprehensive approach

---

[11] In considering the impact of individual earthquakes on the field of seismology, perhaps none can match the significance of the 1960 Chile earthquake. Dahlen and Tromp (1998) noted that the timing of this earthquake was particularly fortuitous for the study of seismology's normal modes because several critical developments had just occurred: newly available instruments capable of recording long-period free oscillations, the emergence of computers that could handle lengthy time series through Fourier analysis, the recent programming of numerical spectral analysis techniques, and the fresh calculations of Earth's theoretical eigenfrequencies that could be used to identify peaks in observational spectra. Research on the free oscillation recordings from this earthquake continued well into the early 1980s.

[12] *Computational seismology* refers to seismological research that primarily relies on computations rather than theoretical analysis. Emerging during the "dawn of the computer era", as described in Ben-Menahem (1995), this branch of seismology has evolved significantly since the 1950s. Starting in the 1970s, a variety of numerical methods, such as the finite difference method, the finite element method, the pseudospectral method and the spectral element methods, became widely used in seismology.

[13] In the early 20th century, Love already expressed a similar sentiment in his seminal work *A Treatise on the Mathematical Theory of Elasticity* that as a field matures, theorems become scarcer while computations grow more frequent. This trend is not surprising as the foundational principles underlying a discipline are finite and many are uncovered early on. Subsequently, as the volume of observational data increases and computational technologies advance, the scope and depth of practical applications significantly expand.

known as *Global Seismology*, also referred to by Woodhouse as *Long-period Seismology*. Dahlen and Tromp (1998) later detailed this theory extensively in their work spanning over a thousand pages.

The 1960s marked a golden era for the advancement of source physics in seismology, highlighted by significant breakthroughs since Love's 1904 displacement integral theorem. In 1963, T. Maruyama extended Steketee's work to dynamic cases, proving the equivalence of body forces in uniform isotropic media. The following year, Burridge and Knopoff (1964) expanded this proof to non-uniform anisotropic media and formulated the source representation theorem, cementing their work as a milestone in seismology akin to Lamb's in Lamb (1904). In 1966, K. Aki introduced the concept of *seismic moment* (M_0), which was further refined by H. Kanamori in 1977 when he proposed the moment magnitude scale, now the most scientific standard for measuring earthquake energy release. Additionally, in 1970, Gilbert and Kostrov enriched source theory by introducing the concept of the *seismic moment tensor*. This period also saw notable advances in the kinematics and dynamics of sources,[14] with the introduction of quantitative parameters (fault length L, width W, rupture velocity v, final slip D, and rise time T) to characterize sources precisely. N. A. Haskell conducted detailed studies on the near-field problems of seismic energy, spectral density, and source parameter models, known as the Haskell model. Meanwhile, B. V. Kostrov pioneered the study of source dynamics in seismology in 1966 by integrating fracture dynamics theory with Griffith's fracture criterion to analyze spontaneous fracture propagation in anti-plane problems.

From the late 1960s to the 1980s, significant advancements in the study of theoretical seismograms, foundational to seismological research, were driven by developments in computing technology. Before 1950, the study of theoretical seismograms was largely limited due to the lack of computational resources and was based on simplistic Earth models. However, research that began in the 1960s significantly expanded the scope of application. Broadly speaking, there are several types of studies (Xie et al. 1992):

(1) Various integral transform methods developed based on the concepts of the analytical era, using models such as horizontally stratified homogeneous media or spherically symmetric media. Notable among these are the propagation matrix method by F. Gilbert and G. Backus in 1966, the reflectivity method by K. Fuchs and G. Müller in 1971, the discrete wavenumber method by Bouchon (1977), and various versions of the generalized reflection/transmisson coefficient method by researchers such as B. L. N. Kennett, J. E. Luco, R. J. Apsel, Z. Yao, and D. G. Harkrider in the early 1980s. These semi-analytical and semi-numerical approaches leverage matrix properties to connect boundary conditions at interfaces, with solutions typically expressed in integral form that require

[14] *Kinematics* and *dynamics* differ in their focus within physics. Kinematics describes the motion of objects without considering the forces that cause this motion. On the other hand, dynamics delves into the reasons behind the motion, specifically addressing the stresses that provoke it. Simply put, kinematics explores how an earthquake source moves, while dynamics explains why it moves that way.

numerical computation.[15] For the layered spherical Earth model, due to its finite size, vibrations induced by seismic sources can be represented as a sum of normal modes, the coefficients of which are determined by the nature of the source.[16]

(2) Ray-based methods for media models with gradual variations, based on high-frequency approximations and ray series expansions. The complexity of the media often precludes exact analytical solutions, leading researchers to explore approximate solutions at high frequencies. Notably, V. Červený developed the *Asymptotic Ray Theory*,[17] which is improved through the gaussian beam method. Additionally, C. H. Chapman's WKBJ-Maslov method represents another significant contribution to this field.

(3) Numerical methods applicable to any heterogeneous medium theoretically, based on discretization techniques, such as the finite difference method, finite element method, and pseudospectral method, etc. These methods are characterized by their ability to handle intricate calculations, although they are constrained by computational conditions and require addressing technical issues related to their specific implementations. For instance, the outcomes of dynamic simulations of branching fault systems using the finite element method can be influenced by the artificial handling at the junctions of multiple faults.

Advancements in theoretical seismology have, in turn, spurred developments in the study of earthquake source physics. Beginning with J. C. Savage's 1966 study on far-field waveforms, which led to the naming of the "stopping phase" based on frequency-dependent decay behavior of displacement spectra, researchers have built on the Haskell model to explore more realistic processes of nucleation, growth, and arrest of finite-sized faults. This research evolved through the 1970s with studies by P. Molnar and F. A. Dahlen in 1973 and 1974, respectively, who investigated kinematic models of circular and elliptical ruptures. In 1975, J. G. Anderson and P. G. Richards conducted extensive research on near-field motions using the Haskell model and various fault kinematic models. The latter part of the decade saw the

[15] Since these methods incorporate a significant analytical component, they lead to diverse approaches depending on the techniques used. Many studies utilize matrix properties and rely on Fourier transform methods. However, it's important to highlight another category of methods based on the Laplace transform and the Cagniard-de Hoop method. Notable examples include the works of D. V. Helmberger in 1968 and L. R. Johnson in Johnson (1974). These methods are unique in that they ultimately express results as the superposition of potential "seismic rays", an approach known as the *generalized ray method*.

[16] In principle, this analytical approach can compute the complete wavefield for a spherical Earth model. However, when dealing with high frequencies (or short periods), the number of modes needed for calculation increases significantly, which in turn raises the computational costs. Consequently, the method of summing normal modes is practically applied only to low-frequency seismology.

[17] This can be seen as the seismological equivalent of geometric optics. It is premised on the idea that solutions in the frequency domain can be represented as a series of frequencies, where only the most significant term is retained at high frequencies. This approach simplifies the complex behaviors of seismic waves into two main types of equations: one for travel time, known as the *eikonal equation*, and one for amplitude, known as the *transport equation*. This method offers simplicity and generally provides a good description of body wave behavior. However, it falls short in handling specific scenarios, such as areas with velocity jumps or geometric discontinuities.

emergence of numerical simulations of earthquake source dynamics using methods like finite difference by Andrew and Madariaga in 1976, finite element by Archuleta in 1978, and boundary integral equations by Das and Aki in 1977. These studies not only facilitated the dynamical understanding of the causes behind fault ruptures but also corrected erroneous conclusions drawn from simpler models. A classic example includes the discovery of supershear rupture speeds on earthquake faults, correcting earlier scientific misconceptions.[18]

Since the 1980s, seismological research has significantly expanded in both scope and depth, thanks to an increase in digital seismic recordings, the emergence of new observational technologies like GPS and InSar in the 1990s, and rapid advancements in computational capabilities. This period has been characterized by a focus on observation and computation-driven studies, with a proliferation of applied research targeting the subsurface structures of specific regions and particular earthquakes. Theoretical research, while less mainstream, has increasingly integrated with computational technology, moving beyond the constraints of half-space model in the Lamb's problem. With various numerical methods, researchers can now perform parallel computations of seismic wavefields in realistic three-dimensional models. Additionally, inversion techniques have matured, allowing the use of seismic and geodetic data to study the medium structures of regions worldwide. Current research is dominated by inversion techniques that reveal detailed characteristics of fault plane motions based on seismic wave propagation theories. Moreover, the field of source dynamics is burgeoning, with significant advances made in the past two decades in examining how complex factors–such as different frictional properties on the fault, initial stress distributions, and the geometrical complexity of fault systems–affect the dynamics of fault rupture.

In summary, much like in other fields, the Information Age has had a profound impact on seismology.[19] The rapid increase in digital recordings and computational capabilities act as twin guides, continuously propelling humanity forward in our exploration of the unknown realms of earthquakes.

[18] For example, N. F. Mott, a Nobel Prize laureate in Physics, once concluded that the maximum speed of Type I fractures is half the shear wave speed of the material. During the 1960s, due to theoretical and observational limitations, it was commonly believed that the propagation speed of fractures in solids did not exceed 3 km/s, roughly equivalent to the shear wave speed in rocks. However, subsequent numerical simulations and laboratory studies have proven that fracture speeds can exceed shear wave speeds and even reach the speeds of compression waves.

[19] The pioneers of classical seismology could hardly have imagined the rapid advancements in their field over just a few decades. In his book *History of Mechanics* (in Chinese, 2010), the Chinese mechanician J. Wu noted that throughout history, all tools invented by humans have served to extend our physical capabilities. Telescopes and microscopes extend our eyesight, while computers, often referred to as "electronic brains", extend our mental capacity.

1.3 Content and Significance of Theoretical Seismology

In Sect. 1.2, we briefly reviewed the evolution of seismology, which also reflects humanity's growing understanding of earthquakes. We've witnessed an acceleration in this process, delving deeper into the science with each step. Especially in the last fifty years, thanks to richer observational data and enhanced computational power, our understanding of earthquakes has significantly evolved. From H. Reid's elastic rebound theory proposed over a century ago, we now have a comprehensive grasp of various aspects related to earthquakes, particularly in understanding the Earth's media and the sources of quakes. Seismological research today is incredibly diverse, ranging from traditional tasks like earthquake location and travel time analysis to cutting-edge studies like seismic wave simulation in complex Earth media using high-performance computers. In this context, what role does "theoretical seismology" play? What are its research contents and significance?

1.3.1 What is Theoretical Seismology?

Currently, a clear definition of *Theoretical Seismology* is absent from textbooks. Instead of providing a textual definition, we find it more effective to describe its characteristics to better grasp its essence. Research in seismology that meets all three of the following criteria can be classified as theoretical seismology.

(1) Theoretical seismology examines *abstract* and *universally significant* issues within the field of seismology. For instance, the Lamb's problem, abstracted from specific physical phenomena, holds universal significance as it forms the basis for synthesizing theoretical seismograms. Another example is the source representation theorem, which pertains to abstracted source models, and its broad applicability is self-evident.

(2) The research approach in theoretical seismology primarily employs *logical deduction*, which contrasts with the inductive methods predominant in other branches of seismology. Most seismological research begins with, or relies heavily on, induction, aligning with common cognitive patterns. However, theoretical seismology starts with fundamental principles of physics, such as Newton's laws. It builds on these to derive the equations of motion for elastic bodies, combined with specific boundary and initial conditions to solve defined problems. The hallmark of theoretical seismology is directly tackling issues by solving equations, enabling it to *explain* and *predict* phenomena related to earthquakes. A notable example includes the theoretical prediction and subsequent observational confirmation of Rayleigh wave.

(3) In theoretical seismology, the research methods are primarily mathematical due to the field's reliance on logical deduction and theoretical analysis.[20] Historically, before the advent of computers, mathematics was the sole tool for theoretical seismology, leading to significant contributions from mathematicians. A notable example is the Oxford-Cambridge analytical school. With the introduction of computers, the methods in theoretical seismology have evolved to integrate theoretical (analytical) analysis with computational techniques.[21] For instance, the modern method of synthesizing seismograms involves deriving analytical solutions represented by frequency-wavenumber integrals, which are then calculated numerically with computer assistance, exemplifying a semi-analytical and semi-numerical approach.

1.3.2 Significance of Studying and Researching Theoretical Seismology

A natural question arises from the fact that the extensive use of mathematics in certain academic fields can be daunting for some readers: why is it worthwhile to study or research such a challenging subject?

Firstly, the study of theoretical seismology provides foundational insights for further research in various aspects of seismology. One key area within this discipline involves calculating Green's functions, or theoretical seismograms, for different media. Seismology aims to investigate the subsurface structures and earthquake sources that cannot be observed directly, using seismic data. The ultimate goal is to predict and mitigate earthquakes. The primary method for analyzing this data is inversion, which relies on forward modeling–assuming specific conditions for media and sources to compute theoretical seismograms. In this sense, theoretical seismology serves as the groundwork upon which the broader field of seismology is built, much like a foundation supports a building.

Secondly, the study of theoretical seismology also offers a *fundamental* understanding of earthquake dynamics. By analyzing collected seismic records, we can identify certain patterns. These insights are based on observed phenomena. However, when we delve into the essential question of "why" these patterns occur, inductive

[20] The significance of mathematics within any discipline is poignantly underscored by German philosopher I. Kant's bold assertion: "In any given theory, only the part that involves mathematics can be considered true science." This statement, though seemingly extreme, captures the essential role that mathematical principles play in establishing rigorous scientific inquiry.

[21] The essence of theoretical seismology inherently involves analytical components; without them, it would not qualify as theoretical seismology. For example, studying the equations of elastic motion using purely numerical methods does not fall under theoretical seismology but rather *Computational Seismology*.

methods fall short.[22] Theoretical seismology, which primarily uses deductive reasoning, can answer these core questions. The secrets lie within the solutions to the elastic wave equations, and it is only through mathematical approaches that we can uncover these deep insights.[23]

Finally, Theoretical seismology provides *clear functional relationships* between physical quantities for a wide range of problems encountered in the field. This characteristic is a common feature of theoretical analysis methods, helping to deepen our understanding of physical laws.[24] For instance, in addressing the Lamb's problem, theoretical analysis yields an integral representation of displacement. Although complex, this represents a direct correlation between displacement and body forces, fundamentally differing from the discrete solutions obtained through numerical methods.

In the digital age, advancements in computing technology have had an undeniable impact on society, transforming the structure of various academic disciplines. Seismology has also evolved significantly, becoming heavily reliant on computers in ways that would have been unimaginable 60 years ago. The rapid improvement in computational power has not only made it feasible but increasingly easier to solve determinate problems using purely numerical methods. Since the 1970s, *computational seismology* has emerged, leveraging purely numerical techniques to solve traditional equations of elastic motion. Today, we can simulate seismic wave fields in complex models that closely mimic the real Earth. With ongoing advancements in computer hardware, the future of computational seismology looks even more promising.

On one hand, theoretical seismology, which primarily relies on mathematical analysis, is constrained by the complexity of mathematics and limited to simpler theoretical models that do not fully meet the needs for studying more complex and realistic motions. On the other hand, computational seismology, driven by improvements in computational capabilities, can model scenarios that are much closer to reality and is rapidly advancing. This stark contrast raises the question: is there still a need to study theoretical seismology today? The answer is a resounding yes.

---

[22] A classic example in astronomy involves J. Kepler, who meticulously studied the extensive planetary observations recorded by his predecessor, T. Brahe. After years of diligent calculations, Kepler derived the three laws that bear his name. These laws represented a significant advance in knowledge, yet Kepler was unable to explain the reasons behind these laws. It was not until Isaac Newton applied mathematical techniques that a satisfactory explanation was provided.

[23] The renowned mathematician M. Klein once described the 18th-century mathematicians' view of mathematics as a divine mission bestowed upon humanity to use its capabilities to decipher the laws of nature. "The book of nature is open before us, written in a language initially incomprehensible. Only through perseverance, passion, resilience, and deep study can we understand this language, which is, indeed, mathematics." (M. Klein, *Mathematics: The Loss of Certainty*, 1980).

[24] "The defining trait of theoretical research methods lies in their scientific abstraction (approximation), which facilitates the use of mathematical techniques to derive theoretical results. This process distinctly and universally unveils the inherent laws of material motion." (W. Wu, *Fluid Mechanics* (vol.1, in Chinese), 1982).

Firstly, in seismology, our fundamental goal is to understand the physical phenomena related to earthquakes better, with the aim of predicting and, to some extent, "controlling" them through human intervention to reduce the damage caused by disasters. If numerical calculations based on complex models consistently provided better results for this purpose than those derived from analytical or semi-analytical methods, it would make sense to move away from what might be considered "outdated" theories. However, this isn't always the case. In the complex and messy reality of our world, the wise choice often involves simplifying issues to grasp their essence. This strategic focus on main contradictions rather than getting lost in minor details allows us to gain valuable insights. Indeed, while the abundance of complex factors can make it difficult to find clear answers, simpler model-based analyses often reveal significant truths by focusing on the core issues.[25]

Secondly, from a methodological standpoint, both theoretical analysis and numerical computation are crucial methods in scientific research, each with its own strengths and weaknesses. Theoretical analysis excels where it applies to simpler models, as previously discussed, but its limitations become apparent with more complex models. This gap is effectively bridged by numerical computation, which can handle complex models. However, without the insights gained from theoretical analysis, it would be difficult to correctly interpret the results of numerical computations in a detailed manner. Therefore, concluding that one method could replace the other based solely on their individual merits would be shortsighted, akin to the proverbial blind men touching the elephant. The wise approach combines these methods, utilizing each to its fullest potential. If numerical computation is considered as practice, then practice guided by sound theoretical analysis is essential to avoid missteps.[26]

---

[25] In the history of elasticity theory, there is an intriguing fact about the theoretical approaches used to study this field. Research can be conducted from two perspectives: the microscopic, considering molecular kinematics, and the macroscopic, based on the principles of continuum mechanics. Intuitively, one might think that the microscopic approach, which considers the structure of matter at a molecular level, would be more realistic. The macroscopic approach, on the other hand, might seem like a simplified abstraction because it overlooks the microscopic structure and employs differential operations that appear artificially imposed when compared to molecular details. Yet, history shows that the macroscopic approach has proven more successful. For example, in 1821, Navier developed the first differential equations of elasticity using a molecular motion model that included only one elastic parameter. A year later, Cauchy approached elasticity from a macroscopic perspective and formulated the correct equations. Further, regarding the number of elastic constants, in 1828, Cauchy and Poisson, using molecular interactions, erroneously concluded that the most general case involved 15 independent constants, with isotropic bodies having only one–an agreement with Navier's findings. However, in 1838, Green correctly identified from a macroscopic standpoint using the conservation of energy that the most general anisotropic case should involve 21 elastic constants, a result later confirmed by Young in 1855 through thermodynamic laws.

[26] The idea originally expressed by Stalin states that "theory without the backing of revolutionary practice is hollow, and practice unguided by revolutionary theory is blind" (*Stalin Selected Works* (vol. 1, in Chinese), 1979). Cheng-dao Lee also highlighted the interdependence between theory and experimentation in physics with two laws: "My first law of physics: Without experimentalists, theorists tend to drift; my second law: Without theorists, experimentalists tend to waver".

Finally, in reviewing the practical history of disciplinary development, the resilience of theoretical analysis methods is surprisingly robust, finding significant application with the aid of numerical computing techniques. Before computers, mathematicians had to factor in the complexity of calculations when tackling problems. If the complexity exceeded human capabilities, such methods were deemed impractical. However, the advent of computers, especially with increasing computational power, has removed these barriers. From this perspective, not only are theoretical analysis methods far from being obsolete, but they actually invigorate numerical computations with renewed vitality.[27]

Thus, integrating numerical computation with theoretical analysis is the long-term strategy for sustainable development. Given that we are in the information age, where research based on various numerical computations is thriving, theoretical analysis tends to be overlooked. Therefore, it is crucial to emphasize the study and research of theoretical analysis to better serve the objectives of seismological research.

1.4 Content of this Book

As mentioned in the preface, this book aims to introduce the classic Lamb's problem thoroughly to give readers a preliminary understanding of theoretical seismology, laying a solid foundation for advanced studies. Additionally, we adopt a research perspective in discussing the Lamb's problem, illustrating that even a small subject can provide comprehensive insights. Thus, studying the Lamb's problem also helps young students transition into engaging in scientific research.

In preparation for introducing the Lamb's problem, Vol. 1 of this book primarily focuses on explaining the fundamentals of theoretical seismology. Building on this foundation, it then provides a detailed explanation of one approach to solving the Lamb's problem. Specifically, it covers the following topics:

Chapter 2 primarily focuses on establishing the integral representation theorem for displacement, which will serve as the foundation for the source representation theorem in Chap. 3. To achieve this, we take several steps. First, we introduce essential preliminary knowledge, including field theory based on index notation, a mathematical tool necessary for all subsequent discussions. Next, we delve into the basic concepts and formulas of elastodynamics, including the notions of stress and strain, along with geometric equations, motion equations, and constitutive equations, leading to the equations of elastodynamics for isotropic homogeneous media represented

[27] For instance, the matrix-based methods used in creating synthetic seismograms, which cleverly employ the properties of matrices to link boundary conditions at interfaces. This represents a brilliant theoretical analysis that would not have been feasible before the era of computers, as the complex matrix calculations exceeded human computational limits. Only with the aid of computers can the true value of these methods be fully realized. Similarly, the discrete wavenumber method introduced by Bouchon in the late 1970s, as discussed in Sect. 7.1.1, encapsulates profound mathematical and physical insights, yet its value too is only revealed with computer assistance.

through displacement–our starting point. Then, we explore the reciprocal theorems in elastodynamics, including the first reciprocal theorem without time integration and the second with time integration. Fourth, we introduce the Green's functions for the elastodynamics system, setting the stage for the final preparation of the displacement integral representation theorem. Lastly, we derive the displacement integral representation theorem.

Chapter 3 centers on the source representation theorem, discussing related seismic source theory, including the theorem itself, the equivalent body forces of fault sources, and the seismic moment tensor. The chapter is divided into three main parts. First, building on the displacement integral representation theorem established in Chap. 2, we derive the source representation theorem for scenarios involving faults, discussing its significance and applications. Second, we examine the equivalent body forces associated with shear rupture, proving that these forces act as double couples. Third, using the source representation theorem as a basis, we introduce the concept of the seismic moment tensor, which is essential for calculating the seismic waves generated by seismic sources.

Chapter 4 delves into Green's functions in an infinite, homogeneous medium–the simplest elastic medium model and the only scenario that yields closed-form analytical solutions. The chapter is structured in several sections: First, we introduce Lame's theorem, which lays the groundwork for our solutions. Next, we discuss solutions to the wave equation, a critical step when applying Lame's theorem to derive Green's functions. Then, we solve for and analyze Green's functions. Following this, we further derive solutions corresponding to double couples based on Green's functions and analyze these results. Finally, we explore seismic waves generated by dislocation point sources and finite-sized sources within an epicentral coordinate system.

Chapter 5 serves as a transition to the main topic of this book: the study of the Lamb's problem. Before delving into the specifics, it comprehensively reviews the history of research on the Lamb's problem, beginning with Lamb's pioneering work in 1904. Research on the Lamb's problem is generally divided into two approaches: those based on Fourier synthesis, discussed in the last two chapters of this volume, and those employing the Cagniard method, a time-domain solution, which will be the focus of the next volume.

Chapter 6 presents the theoretical formulation of frequency-domain solutions for the Lamb's problem. It begins by clearly defining the boundary value problem, followed by a detailed explanation of the introduction and properties of basis functions. It then discusses the system of ordinary differential equations satisfied by the coefficients derived from the expansion of displacement in terms of these basis functions, along with their general solutions. Using the approach by Zhang (2004), specific forms of these general solutions are derived through boundary conditions, leading to the expression for the frequency-domain Green's function. The properties of the Green's function are analyzed on this basis. Specifically, the contribution of the free surface is addressed by examining complex integrals, resulting in the excitation formula for Rayleigh waves (Zhou and Zhang 2020). Utilizing the Green's function for the Lamb's problem and combining it with the seismic moment tensor, the displacement field generated by shear dislocation point sources is derived. Lastly, the chapter examines special cases in the dynamic solution process–specifically, the eigenvalue degeneracy–and studies the static field problem, using the $\mathcal{LP}_2$ as an example to derive the static Green's function and the surface static solution induced by shear dislocation point sources.

Chapter 7 builds on the theoretical formulations established in Chap. 6 to delve into numerical techniques for integral calculations and their corresponding results. It is organized into four main sections. The first section introduces numerical methods developed for calculating wave number integrals, which are crucial for expressing displacement as a function of both wave number and frequency. This includes adaptive Filon integration methods (Chen and Zhang 2001) and peak-and-trough averaging techniques (Zhang et al. 2001, 2003), alongside discussions on the discrete Fourier transform and essential concepts like the source time function and its spectrum, foundational for computing theoretical seismograms. The second section compares the computed results for three types of Lamb's problems–Green's functions, spatial derivatives of Green's functions, and displacement fields generated by shear dislocation point sources–with previously published works, validating the accuracy of the formulas and procedures developed in this book. The third section examines the characteristics of displacement fields from various perspectives including different time functions, point dislocation sources, and finite-sized dislocation sources, and explores static displacement fields from point forces, point dislocations, and finite-sized dislocation sources. Finally, the fourth section studies the properties of Rayleigh waves generated by single-force and dislocation point sources, in both frequency and time domains.

The next volume of this book will focus on solving the Lamb's problem in the time domain, ultimately presenting a generalized closed-form solution. From the perspective of seeking solutions to the Lamb's problem, a generalized closed-form solution represents the simplest form of an analytical solution that can be achieved.

Chapter 2
Fundamental Theorems of Elastodynamics

The central objective of this chapter is to establish the integral representation theorem of displacement, as the basis of the source representation theorem introduced in Chap. 3. The displacement representation theorem can be seen as a direct consequence of Betti's reciprocity theorem, provided one set of forces takes a specific form that leads to the emergence of Green's functions. Betti's reciprocity theorem itself is derived from the equations of elastic motion–also known as the equations of elastodynamics–using two different sets of forces. Therefore, the equations of elastodynamics form the foundation. To introduce these equations, it is necessary to discuss concepts like stress and strain, as well as related field theory knowledge. This reversal of order is the approach taken in this chapter.

2.1 Prerequisite Knowledge

In this section, we primarily discuss the foundational concepts of field theory that will be essential later on. We'll cover the algebraic definitions and representations of vectors and tensors, along with their algebraic and analytical operations. To facilitate the representation of vectors and tensors, it is necessary to introduce index notation and coordinate transformations.

In this book, we use certain conventions to represent vectors and tensors. Vectors are denoted by bold italic letters, such as $\boldsymbol{a}$, while tensors and matrices are indicated by bold upright letters, like $\mathbf{A}$. Unit basis vectors, which correspond to the axes of the coordinate system $Ox_1x_2x_3$, are represented by bold italic letters topped with a hat and an subscript, such as $\hat{\boldsymbol{e}}_i$. Furthermore, Latin letters such as i, j and k, are used as subscripts and typically take the values 1, 2 and 3, unless otherwise specified. Greek letters, except for α and β which often refer to P-wave and S-wave respectively, such as ζ and η, are used as subscripts with values of 1 and 2. These conventions will not be repeated later in the text.

H. Zhang, *Lamb's Problem in Seismology (Volume 1)*,
https://doi.org/10.1007/978-981-95-3170-7_2

2.1.1 Indicial Notation

This book focuses on solving the equations of elastodynamics within a uniform half-space model, which will later be revealed to involve vector equations. To obtain specific quantitative results, it's necessary to express the vectors in terms of their components within a chosen coordinate system. Naturally, the components of a vector will vary depending on the coordinate system used. For the Lamb's problem we are addressing, it's practical to establish a Cartesian coordinate system.[1] In this system, vector components are represented using indicial notation, where a Latin letter serves as the subscript to indicate the component.

2.1.1.1 Summation Convention

Using the familiar vector $\boldsymbol{a}$ as an example, consider it placed in the Cartesian coordinate system $Ox_1x_2x_3$ with the origin O at the starting point of $\boldsymbol{a}$. In this system, $\boldsymbol{a}$ can be specifically represented as

$$\boldsymbol{a} = \sum_{i=1}^{3} a_i \hat{\boldsymbol{e}}_i = a_1 \hat{\boldsymbol{e}}_1 + a_2 \hat{\boldsymbol{e}}_2 + a_3 \hat{\boldsymbol{e}}_3 \tag{2.1.1}$$

where a_i are the projection values of $\boldsymbol{a}$ on the three coordinate axes. The *summation convention* is introduced to omit the summation symbol in expression (2.1.1) while retaining the same meaning. In other words, if two indices are repeated, it implies summation over that index. Thus, the vector $\boldsymbol{a}$ can be represented as $\boldsymbol{a} = a_i \hat{\boldsymbol{e}}_i$. Indices that are repeated are called *dummy indices*, while those nonrepeated are referred to as *free indices*.

The advantage of adopting the summation convention lies in its simplicity and convenience for notation. Similarly to the integration variable in calculus, the subscript in summation can be substituted with different letters, as seen in the expressions $\boldsymbol{a} = a_i \hat{\boldsymbol{e}}_i = a_j \hat{\boldsymbol{e}}_j$. However, it is important to adhere to two specific rules:

(1) In any term, the same subscript should not appear more than twice[2]; in other words, the summation convention applies only to pairs of identical dummy indices. For example, $\boldsymbol{a} \times \boldsymbol{b} = a_i \hat{\boldsymbol{e}}_i \times b_j \hat{\boldsymbol{e}}_j$, which cannot be rewritten as $\boldsymbol{a} \times \boldsymbol{b} = a_i \hat{\boldsymbol{e}}_i \times b_i \hat{\boldsymbol{e}}_i$, where four identical subscripts i appear.

---

[1] In the context of the Lamb's problem, the surface of the semi-infinite space is flat, making Cartesian coordinates a natural fit. However, for certain types of problems, such as those involving global-scale seismology, it's essential to consider the Earth's curvature. In such cases, using a spherical coordinate system is more appropriate, as exemplified by the studies in Dahlen and Tromp (1998).

[2] If an index appears more than twice in the same term, the summation convention automatically becomes invalid. If, for certain special problems, it is indeed necessary to sum over more than two identical indices, then the summation symbol must be written explicitly and cannot be omitted.

(2) If a subscript appears only once in a term, it must appear only once in each term; in other words, if a free index is replaced with another letter, free indices in all terms must be also be replaced consistently. For example, $f_j = a_i b_i c_j + d_j$ cannot be rewritten as $f_j = a_i b_i c_k + d_m$, but can be expressed as $f_m = a_i b_i c_m + d_m$.

Adhering to these two rules significantly simplifies the representation of vectors and tensors using the summation convention.

2.1.1.2 The Kronecker Delta δ_{ij}

The definition of the Kronecker delta is very straightforward:

$$\delta_{ij} = \begin{cases} 1, & \text{if } i = j \\ 0, & \text{if } i \neq j \end{cases} \tag{2.1.2}$$

Using the Kronecker delta allows for a convenient expression of certain results. For instance, the orthogonality of the basis vectors in the Cartesian coordinate system we introduced can be represented as

$$\hat{\boldsymbol{e}}_i \cdot \hat{\boldsymbol{e}}_j = \delta_{ij} \tag{2.1.3}$$

Another example is the dot product of two vectors,[3]

$$\boldsymbol{a} \cdot \boldsymbol{b} = a_i \hat{\boldsymbol{e}}_i \cdot b_j \hat{\boldsymbol{e}}_j = a_i b_j \left(\hat{\boldsymbol{e}}_i \cdot \hat{\boldsymbol{e}}_j\right) \overset{(2.1.3)}{=\!=\!=} a_i b_j \delta_{ij} \overset{(2.1.2)}{=\!=\!=} a_i b_i \tag{2.1.4}$$

Pay attention to both sides of the final equals sign,

$$b_j \delta_{ij} = b_i \tag{2.1.5}$$

The result on the right side of the equation is equivalent to substituting the index j in b_j on the left side of the equation with i and removing δ_{ij}. Similarly, one can replace the index i in a_i from Eq. (2.1.4) with j, which transforms the expression on the right side of the equation into $a_j b_j$, achieving the same effect. Thus, the Kronecker delta is also known as the *substitution tensor*.[4]

---

[3] In this book, the equation number appearing above the long equals sign indicates that substituting the formula into the left side of the equation yields the right side.

[4] In the upcoming Sect. 2.1.4, we will explore the criteria that determine whether a quantity with two indices qualifies as a second-order tensor. Not all quantities that formally have two indices meet this criterion.

2.1.1.3 The Levi-Civita Symbol ε_{ijk}

The Levi-Civita symbol is defined as

$$\varepsilon_{ijk} = \begin{cases} 1, & \text{if } (i,\ j,\ k) = (1,\ 2,\ 3),\ (2,\ 3,\ 1),\ \text{or } (3,\ 1,\ 2) \\ -1, & \text{if } (i,\ j,\ k) = (2,\ 1,\ 3),\ (3,\ 2,\ 1),\ \text{or } (1,\ 3,\ 2) \\ 0, & \text{if two of } i,\ j, \text{ and } k \text{ are equal} \end{cases} \tag{2.1.6}$$

This means that a non-zero Levi-Civita symbol requires that the three subscripts be distinct, and the specific permutations of i, j, and k determine whether its value is 1 or -1. This definition might seem odd, but its value lies in how it conveniently represents certain results. For example, the cross product of the basis vectors in a Cartesian coordinate system can be expressed as

$$\hat{\boldsymbol{e}}_i \times \hat{\boldsymbol{e}}_j = \varepsilon_{ijk}\hat{\boldsymbol{e}}_k \tag{2.1.7}$$

By varying the indices i and j on the left side of the equation to different combinations and writing out the specific results, and by considering the definition of the Levi-Civita symbol (2.1.6), it's not difficult to verify that this outcome holds true. Effectively, two vectors indexed by i and j are represented by another vector indexed by k through the linkage of the Levi-Civita symbol, which facilitates permutation. Hence, the Levi-Civita symbol is also known as the *permutation symbol* or *alternating symbol*. For instance, the cross product of two vectors can be expressed using the Levi-Civita symbol as

$$\boldsymbol{a} \times \boldsymbol{b} = a_i\hat{\boldsymbol{e}}_i \times b_j\hat{\boldsymbol{e}}_j = a_i b_j \left(\hat{\boldsymbol{e}}_i \times \hat{\boldsymbol{e}}_j\right) \overset{(2.1.7)}{=\!=\!=} a_i b_j \varepsilon_{ijk}\hat{\boldsymbol{e}}_k \tag{2.1.8}$$

2.1.1.4 Relation Between ε_{ijk} and δ_{ij}: the ε–δ Identity

The relationship between the Levi-Civita symbol ε_{ijk} and the Kronecker delta δ_{ij} can be described as follows:

$$\varepsilon_{pij}\varepsilon_{pks} = \delta_{ik}\delta_{js} - \delta_{is}\delta_{jk} \tag{2.1.9}$$

This relationship is known as the ε–δ *identity*, which can be verified by enumeration. To confirm its validity, consider all possible values for the free indices i, j, k, and s and verify each case individually using the definitions of ε_{ijk} from Eq. (2.1.6) and δ_{ij} from Eq. (2.1.2).

Using the ε–δ identity allows for straightforward proofs of certain results. For instance,

$$(\boldsymbol{a} \times \boldsymbol{b}) \cdot (\boldsymbol{c} \times \boldsymbol{d}) = (\boldsymbol{a} \cdot \boldsymbol{c})(\boldsymbol{b} \cdot \boldsymbol{d}) - (\boldsymbol{a} \cdot \boldsymbol{d})(\boldsymbol{b} \cdot \boldsymbol{c})$$

Proof

$$\begin{aligned}(\boldsymbol{a}\times\boldsymbol{b})\cdot(\boldsymbol{c}\times\boldsymbol{d}) &\overset{(2.1.8)}{=\!=\!=} \varepsilon_{ijk}a_ib_j\varepsilon_{pqk}c_pd_q = \left(\varepsilon_{ijk}\varepsilon_{pqk}\right)a_ib_jc_pd_q \\ &\overset{(2.1.9)}{=\!=\!=} \left(\delta_{ip}\delta_{jq}-\delta_{iq}\delta_{jp}\right)a_ib_jc_pd_q \\ &\overset{(2.1.5)}{=\!=\!=} a_ib_jc_id_j - a_ib_jc_jd_i = a_ic_ib_jd_j - a_id_ib_jc_j \\ &\overset{(2.1.4)}{=\!=\!=} (\boldsymbol{a}\cdot\boldsymbol{c})(\boldsymbol{b}\cdot\boldsymbol{d}) - (\boldsymbol{a}\cdot\boldsymbol{d})(\boldsymbol{b}\cdot\boldsymbol{c})\end{aligned}$$

2.1.2 Coordinate Transformation

With a background in indicial notation, we can now introduce coordinate transformations. This step is crucial because understanding the relationships between the components of vectors and tensors in different coordinate systems is essential for accurately representing these quantities and the physical laws they obey. Consider a Cartesian coordinate system $Ox_1x_2x_3$ with basis vectors $\hat{\boldsymbol{e}}_i$. By rotating this system around the origin O, we obtain a new coordinate system $Ox_1'x_2'x_3'$ with basis vectors $\hat{\boldsymbol{e}}_i'$ (see Fig. 2.1).

In the original coordinate system, the new basis vectors $\hat{\boldsymbol{e}}_i'$ can be treated as ordinary vectors that can be expressed in terms of the original basis vectors $\hat{\boldsymbol{e}}_i$,

$$\hat{\boldsymbol{e}}_i' = c_{i1}\hat{\boldsymbol{e}}_1 + c_{i2}\hat{\boldsymbol{e}}_2 + c_{i3}\hat{\boldsymbol{e}}_3 = c_{ij}\hat{\boldsymbol{e}}_j \tag{2.1.10}$$

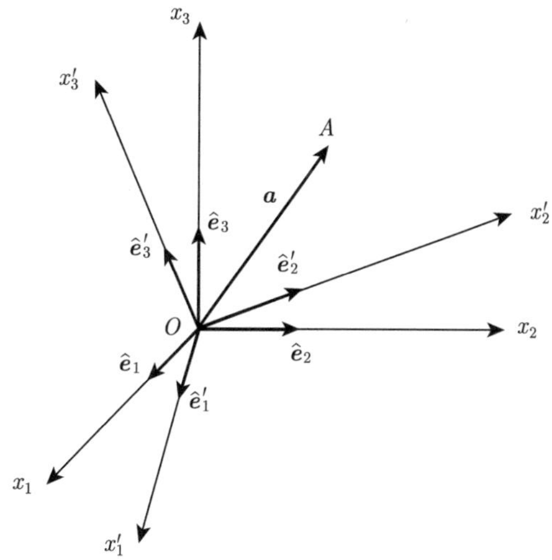

Fig. 2.1 Schematic depiction of coordinate transformation. After the coordinate system $Ox_1x_2x_3$ is rotated around its origin O, it transforms into the coordinate system $Ox_1'x_2'x_3'$. The basis vectors for these coordinate systems are denoted as $\hat{\boldsymbol{e}}_i$ and $\hat{\boldsymbol{e}}_i'$ ($i = 1, 2, 3$), respectively. The vector $\boldsymbol{a}$ originates at O and terminates at point A

where

$$c_{ij} = \hat{\boldsymbol{e}}'_i \cdot \hat{\boldsymbol{e}}_j = \cos\left(\hat{\boldsymbol{e}}'_i,\, \hat{\boldsymbol{e}}_j\right) \tag{2.1.11}$$

is the projection of the new basis vector $\hat{\boldsymbol{e}}'_i$ onto the original basis vector $\hat{\boldsymbol{e}}_j$, and $\left(\hat{\boldsymbol{e}}'_i,\, \hat{\boldsymbol{e}}_j\right)$ is the angle between the two basis vectors. Note that the indices i and j in c_{ij} have different meanings: i refers to the index of the basis vector in the new coordinate system, while j pertains to the basis vector in the original coordinate system. Similarly, the original basis vector $\hat{\boldsymbol{e}}_i$ can also be expressed using the basis vectors of the new coordinate system $\hat{\boldsymbol{e}}'_j$,

$$\hat{\boldsymbol{e}}_i = c_{ji}\hat{\boldsymbol{e}}'_j \tag{2.1.12}$$

The coefficients c_{ij} link the basis vectors of two coordinate systems, thus reflecting the transformation relationship between them. Note that

$$\hat{\boldsymbol{e}}'_i \cdot \hat{\boldsymbol{e}}'_j \overset{(2.1.10)}{=\!=\!=} c_{im}\hat{\boldsymbol{e}}_m \cdot c_{jn}\hat{\boldsymbol{e}}_n \overset{(2.1.3)}{=\!=\!=} c_{im}c_{jn}\delta_{mn} \overset{(2.1.5)}{=\!=\!=} c_{im}c_{jm} = \delta_{ij}$$

similarly,

$$\hat{\boldsymbol{e}}_i \cdot \hat{\boldsymbol{e}}_j \overset{(2.1.12)}{=\!=\!=} c_{mi}\hat{\boldsymbol{e}}'_m \cdot c_{nj}\hat{\boldsymbol{e}}'_n \overset{(2.1.3)}{=\!=\!=} c_{mi}c_{nj}\delta_{mn} \overset{(2.1.5)}{=\!=\!=} c_{mi}c_{mj} = \delta_{ij} \tag{2.1.13}$$

If we consider c_{ij} as an element of the matrix c, then we have:

$$c_{mi}c_{mj} = \left[\mathsf{c}^{\mathrm{T}}\right]_{im}[\mathsf{c}]_{mj} = \left[\mathsf{c}^{\mathrm{T}}\mathsf{c}\right]_{ij} = \delta_{ij} = \left[\mathsf{I}\right]_{ij}$$

in which, I is the identity matrix, and c^{T} denotes the transpose of the matrix c. This implies that $\mathsf{c}^{\mathrm{T}}\mathsf{c} = \mathsf{I}$. Therefore, we have

$$\mathsf{c}^{\mathrm{T}} = \mathsf{c}^{-1}$$

This indicates that c is an *orthogonal matrix*, which represents the rotational transformation between coordinate systems.

2.1.3 Vectors

In mathematics, we recognize vectors as quantities with both magnitude and direction. This is a geometric perspective. However, the algebraic definition of vectors, which is the focus here, offers significant insights. It explains the relationships that should exist between the components of vectors or tensors representing the same physical quantity across different coordinate systems. Physical quantities like temperature (a scalar), displacement (a vector), and stress (a tensor) maintain their intrinsic properties regardless of the coordinate system used. Nevertheless, the numerical values of vector and tensor components vary depending on the chosen coordinate system. What is the relationship between components across different systems? The algebraic definition of vectors addresses this question.

Consider a vector $\boldsymbol{a}$ with starting point O and endpoint A as shown in Fig. 2.1. Two coordinate systems, $Ox_1x_2x_3$ and $Ox_1'x_2'x_3'$, are established with O as the origin, where the latter is derived from the former by a rotation around the origin O.[5] We analyze the components of vector $\boldsymbol{a}$ in these systems, represented as a_i and a_i', respectively, and explore the relationship between them.

First, note that both a_i' and a_i can be viewed as the projections of vector $\boldsymbol{a}$ onto the respective coordinate basis vectors $\hat{\mathbf{e}}_i'$ and $\hat{\mathbf{e}}_i$. Drawing from vector algebra, these projections mathematically correspond to the dot product of the two vectors, therefore

$$a_i' = \boldsymbol{a} \cdot \hat{\boldsymbol{e}}_i', \quad a_i = \boldsymbol{a} \cdot \hat{\boldsymbol{e}}_i$$

Second, by expressing vector $\boldsymbol{a}$ in both coordinate systems $Ox_1x_2x_3$ and $Ox_1'x_2'x_3'$, we can derive the respective formulas for these systems as

$$a_i' = \boldsymbol{a} \cdot \hat{\boldsymbol{e}}_i' \overset{(2.1.10)}{=\!=\!=} a_k\hat{\boldsymbol{e}}_k \cdot c_{ij}\hat{\boldsymbol{e}}_j' \overset{(2.1.3)}{=\!=\!=} c_{ij}a_k\delta_{jk} \overset{(2.1.5)}{=\!=\!=} c_{ij}a_j \tag{2.1.14a}$$

$$a_i = \boldsymbol{a} \cdot \hat{\boldsymbol{e}}_i \overset{(2.1.12)}{=\!=\!=} a_k'\hat{\boldsymbol{e}}_k' \cdot c_{ji}\hat{\boldsymbol{e}}_j \overset{(2.1.3)}{=\!=\!=} c_{ji}a_k'\delta_{jk} \overset{(2.1.5)}{=\!=\!=} c_{ji}a_j' \tag{2.1.14b}$$

Equation (2.1.14) essentially establishes a transformation relationship between the components of two coordinate systems. From this, we can introduce the *algebraic definition of a vector*: an ordered array (a_1, a_2, a_3) constitutes a vector if it satisfies Eq. (2.1.14) under the coordinate transformation c_{ij}. This definition has three key points:

(1) The quantity under consideration is not examined in isolation within one coordinate system; it must also be defined in another coordinate system.
(2) The relationship between the two coordinate systems, represented as c_{ij}, must be understood.
(3) The components in the two coordinate systems must satisfy a specific relationship, as described in Eq. (2.1.14).

This shows that, unlike the geometric definition mentioned at the beginning of this section, not just any three numbers grouped together can form a vector under the

[5] In addition to rotational differences, two coordinate systems may also differ by a translation. This latter relationship is relatively straightforward, involving only a difference in translation between the components in each system, which does not require detailed discussion.

algebraic definition. In fact, it's more accurate to describe this as a criterion provided from an algebraic perspective. This criterion helps determine whether an ordered array represents a vector correctly in different coordinate systems.

With this definition, the problem of representing vectors is effectively resolved. For instance, the vector $\boldsymbol{a}$ in Fig. 2.1, although represented by different components in two coordinate systems, inherently satisfies the relationship described in Eq. (2.1.14) when the rotational relationship between these coordinate systems is established. This ensures that the abstract vector $\boldsymbol{a}$ is accurately represented in each distinct coordinate system.[6]

One notable advantage of the algebraic definition is its ease of extension to higher dimensions. For example, it can be conveniently expanded to include the definition of second-order tensors.

2.1.4 Second-Order Tensors

Consider a high-dimensional space where the coordinate system uses *dyadic basis vectors*, denoted as $\hat{\boldsymbol{e}}_i\hat{\boldsymbol{e}}_j$. In this space, the quantity corresponding to the vector $\boldsymbol{a}$ in three-dimensional space is represented as A and expressed using the dyadic basis as $\mathsf{A} = A_{ij}\hat{\boldsymbol{e}}_i\hat{\boldsymbol{e}}_j$. Similarly, rotating the coordinate system results in a new set of dyadic basis vectors $\hat{\boldsymbol{e}}'_i\hat{\boldsymbol{e}}'_j$. The transformation relationship between the original basis vector $\hat{\boldsymbol{e}}_i$ and the rotated basis vector $\hat{\boldsymbol{e}}'_i$ still adheres to Eqs. (2.1.10) or (2.1.12). Therefore

$$\mathsf{A} = A_{ij}\hat{\boldsymbol{e}}_i\hat{\boldsymbol{e}}_j \xlongequal{(2.1.12)} A_{ij}c_{ki}c_{sj}\hat{\boldsymbol{e}}'_k\hat{\boldsymbol{e}}'_s$$

In the new coordinate system, the tensor A is represented as $\mathsf{A} = A'_{ks}\hat{\boldsymbol{e}}'_k\hat{\boldsymbol{e}}'_s$. Leveraging the orthogonality of the dyadic basis vectors, we find

$$A'_{ks} = c_{ki}c_{sj}A_{ij} \tag{2.1.15}$$

[6] In physics, certain quantities remain unchanged despite transformations in coordinate systems; these are referred to as *invariants*. For example, in elasticity theory, the three stress invariants do not vary. In the context of this discussion, the length of a vector $\boldsymbol{a}$ remains constant, implying that the length does not change with the transformation of the coordinate system. This constancy is demonstrated through the equation

$$a'_i a'_i \xlongequal{(2.1.14)} c_{ij}a_j c_{im}a_m \xlongequal{(2.1.13)} \delta_{jm}a_j a_m \xlongequal{(2.1.2)} a_j a_j$$

Note that although the magnitude of vector a remains constant, its direction varies in different coordinate systems.

Following the algebraic definition of vectors, it's straightforward to define a *second-order tensor*: an ordered array A_{ij} qualifies as a second-order tensor if it satisfies Eq. (2.1.15) under a coordinate transformation c_{ij}.[7]

2.1.5 Operations on Vectors and Tensors

Introducing the vector differential operator, known as the *Hamiltonian operator*:

$$\nabla = \hat{\boldsymbol{e}}_i \frac{\partial}{\partial x_i} \tag{2.1.16}$$

the characteristic of this operator is that it possesses both vectorial and differential properties. This means that it can function as a vector while also performing differential operations on physical quantities.[8] The vector nature of the Hamiltonian operator allows it to engage in various vector and tensor operations such as dot and cross products, corresponding to gradient, divergence, and curl operations, respectively. Each of these functionalities is illustrated through examples below.

2.1.5.1 Dyadic Operation: Gradient

When the Hamilton operator performs a dyadic operation on a vector, it effectively computes the gradient of that vector, resulting in a second-order tensor. For example, taking the gradient of the displacement vector $\boldsymbol{u}$ produces the second-order tensor $\nabla \boldsymbol{u}$, with components $\partial u_i / \partial x_j$. This process is explained through the algebraic definition of a second-order tensor.

---

[7] Just as with vectors, the algebraic definition of a second-order tensor essentially provides a criterion for identifying tensors. Section 2.2 introduces the concepts of stress and strain in elastic mechanics, both of which are second-order tensors. Similar to vectors, simply arranging nine numbers does not necessarily form a second-order tensor. The term "tensor" originates from "*-tens-*", related to "tension," indicating its roots in mechanics. The concept of the stress tensor was introduced by A. L. Cauchy in the early 19th century while studying problems in elastic mechanics. Later, mathematicians like W. R. Hamilton and H. Grassmann enriched the theory of tensors from a mathematical perspective. In physics, tensors are usually defined through coordinate transformations, while mathematicians define them on a broader and more fundamental basis as multilinear mappings within the framework of linear spaces, making tensor theory a part of multilinear algebra in modern mathematics. Tensor theory is extensively applied in fields such as general relativity and continuum mechanics. Although the general theory of tensors is quite complex, this book focuses solely on second-order tensors in Cartesian coordinates, involving mostly differential operations.

[8] In the general theory of continuum mechanics, this operator is capable of differentiating both the physical quantity that follows it and the one that precedes it. However, for the purposes of this book, such complex operations are not required. Therefore, we will only use this operator to differentiate the physical quantities that come after it.

If u_i and u'_i are the components of the vector $\boldsymbol{u}$ in two different coordinate systems, and their gradients are $\partial u_i/\partial x_j$ and $\partial u'_i/\partial x'_j$, respectively, then

$$\frac{\partial u'_i}{\partial x'_j} = \frac{\partial u'_i}{\partial x_k}\frac{\partial x_k}{\partial x'_j} \overset{(2.1.14)}{=\!=\!=} \frac{\partial}{\partial x_k}(c_{il}u_l)\, c_{jk} = c_{jk}c_{il}\frac{\partial u_l}{\partial x_k} \tag{2.1.17}$$

If we define $A_{ji} = \partial u_i/\partial x_j$, then Eq. (2.1.17) can be rewritten as follows

$$A'_{ji} = c_{jk}c_{il}A_{kl}$$

This is in accordance with Eq. (2.1.15), which implies that the gradient of a vector is a second-order tensor. For instance, considering the gradient of a position vector, since $\partial x_i/\partial x_j = \delta_{ij}$, the Kronecker delta δ_{ij} is therefore a second-order tensor.[9]

2.1.5.2 Scalar Product Operation: Divergence

The scalar product of the Hamiltonian operator with a second-order tensor, such as the stress tensor $\boldsymbol{\tau}$, represents taking the divergence of this tensor to yield a vector. As we will see later

$$\nabla\cdot\boldsymbol{\tau} \overset{(2.1.16)}{=\!=\!=} \hat{\boldsymbol{e}}_i\frac{\partial}{\partial x_i}\cdot\tau_{jk}\hat{\boldsymbol{e}}_j\hat{\boldsymbol{e}}_k \overset{(2.1.3)}{=\!=\!=} \delta_{ij}\frac{\partial\tau_{jk}}{\partial x_i}\hat{\boldsymbol{e}}_k \overset{(2.1.5)}{=\!=\!=} \tau_{ik,i}\hat{\boldsymbol{e}}_k$$

where $\tau_{ik,i}$ serves as a shorthand for $\partial\tau_{ik}/\partial x_i$, a notation frequently used to denote the derivative of a physical quantity with respect to spatial coordinates. The above equation indicates that the divergence of a second-order tensor is a vector.

2.1.5.3 Vector Product Operation: Curl

The cross product of the Hamiltonian operator with a vector represents taking the curl of that vector, resulting in another vector. For instance, considering the curl of a displacement vector

$$\nabla\times\boldsymbol{u} \overset{(2.1.16)}{=\!=\!=} \hat{\boldsymbol{e}}_i\frac{\partial}{\partial x_i}\times u_j\hat{\boldsymbol{e}}_j \overset{(2.1.7)}{=\!=\!=} u_{j,i}\varepsilon_{ijk}\hat{\boldsymbol{e}}_k$$

---

[9] Not only that, δ_{ij} is also an *isotropic second-order tensor*, meaning that its values are the same in two different coordinate systems, that is $\delta_{ij} = \delta'_{ij}$. This is because:

$$\delta'_{ij} = c_{im}c_{jn}\delta_{mn} = c_{im}c_{jm} = \delta_{ij}.$$

which implies that the curl of a vector is still a vector. A natural corollary is that taking the curl of the curl of a vector still results in a vector. There is a certain relationship between this new vector and the divergence of the original vector.

2.1.5.4 A Comprehensive Example

Here's a useful conclusion that synthesizes the three operations discussed above:

$$\nabla^2 \boldsymbol{a} = \nabla (\nabla \cdot \boldsymbol{a}) - \nabla \times (\nabla \times \boldsymbol{a}) \tag{2.1.18}$$

Drawing upon the concepts discussed in Sects. 2.1.5.1 to 2.1.5.3, it's not difficult to demonstrate this conclusion:

$$\begin{aligned}\nabla (\nabla \cdot \boldsymbol{a}) - \nabla \times (\nabla \times \boldsymbol{a}) &\xlongequal{(2.1.16),\,(2.1.4),\,(2.1.8)} \hat{\boldsymbol{e}}_i \frac{\partial}{\partial x_i} \frac{\partial a_j}{\partial x_j} - \varepsilon_{ijk} \frac{\partial}{\partial x_j} \varepsilon_{kmn} \frac{\partial}{\partial x_m} a_n \hat{\boldsymbol{e}}_i \\ &\xlongequal{(2.1.9)} \left[a_{j,ji} - \left(\delta_{im}\delta_{jn} - \delta_{in}\delta_{jm}\right) a_{n,jm}\right] \hat{\boldsymbol{e}}_i \\ &\xlongequal{(2.1.5)} \left[a_{j,ji} - \left(a_{j,ji} - a_{i,jj}\right)\right] \hat{\boldsymbol{e}}_i = a_{i,jj} \hat{\boldsymbol{e}}_i = \nabla^2 \boldsymbol{a}.\end{aligned}$$

2.1.5.5 Gaussian Formula

Finally, a formula we'll need later should be mentioned, known as Gauss's formula.[10] This formula, applicable to the vector function $\boldsymbol{a}(x, y, z)$, can be extended to second-order tensors and even higher-order tensors[11]:

$$\oiint_S \mathsf{A} \cdot \hat{\boldsymbol{n}}\, \mathrm{d}S = \iiint_V \nabla \cdot \mathsf{A}\, \mathrm{d}V \tag{2.1.19}$$

where $\mathsf{A} = \mathsf{A}(x, y, z)$ represents a second-order tensor function whose components possess continuous first-order partial derivatives both in the region V and along its boundary S. The symbol $\hat{\boldsymbol{n}}$ denotes the unit normal vector pointing outward from the region.

---

[10] In the context of "Line Integrals and Surface Integrals" in advanced mathematics, the established conclusion is that for a vector function $\boldsymbol{a}(x, y, z)$, if its components have continuous first-order partial derivatives within a region V and on its boundary S, the following relationship holds:

$$\oiint_S \boldsymbol{a} \cdot \hat{\boldsymbol{n}}\, \mathrm{d}S = \iiint_V \nabla \cdot \boldsymbol{a}\, \mathrm{d}V$$

where $\hat{\boldsymbol{n}}$ represents the unit normal vector pointing outward from the region.

[11] The proof method is similar to that used for vector functions. Readers interested in a general proof for tensors of any order can refer to Chou and Pagano (1992, pp. 186–188).

2.2 Basic Concepts and Formulas of Elastodynamics

In theoretical seismology, the fundamental mathematical model is derived from elastodynamics, primarily based on the *equations of elastodynamics*.[12] These equations serve as the starting point for the field. However, this starting point rests on even more fundamental concepts: stress and strain, alongside Newton's laws. In this section, we will introduce the basic ideas of stress and strain, leading to the development of geometric equations, equations of elastic motion, and constitutive relations. Together, these form the complete set of equations in elastic mechanics.

Before delving deeper, it is essential to clarify some fundamental concepts and assumptions in elastodynamics, the study of idealized bodies known as elastic bodies. These bodies are unique in that they possess only elastic properties, unlike most real-world objects, which also exhibit plasticity or viscosity. *Elasticity* refers to a body's ability to return to its original shape once external forces are removed. Elastodynamics investigates how these bodies react internally–specifically, how they displace and stress–under external forces, primarily based on Newton's laws. A critical assumption here is the continuum hypothesis, which disregards the microscopic structure of particles and assumes the elastic body uniformly fills the study area. This assumption allows the use of calculus, differential equations, and differential geometry as primary tools in research. Of course, the validity of conclusions drawn from this assumption depends on their practical applicability. Beyond these basics, elastodynamics is distinguished from other disciplines of continuum mechanics by its unique constitutive relations, specifically the stress-strain relationship defined by the generalized Hooke's law. Simplifying assumptions such as small deformations, isotropy, homogeneity, and no initial stress are often made to ease problem-solving, though these are not fundamental but rather introduced for simplification.

2.2.1 The Concept of Strain and Geometric Equations

The response of elastic bodies to external factors can be analyzed from two perspectives: geometric and physical. Geometrically, we observe that elastic bodies deform, and the physical quantity that measures this deformation is *strain*. Initially, we introduce the concept of strain through a two-dimensional problem from an intuitive

[12] These equations also referred to as the *equations of elastic motion* essentially merge geometric equations and constitutive relations into motion equations expressed in terms of displacement, which are crucial for addressing seismological problems related to the propagation of seismic waves, with time inherently involved. Consequently, the inclusion of time transforms the static equilibrium equations of elastic statics into the dynamic equations of elastodynamics. This transformation is similar to the difference between describing force balance as $\boldsymbol{F} = \boldsymbol{0}$ and using Newton's second law to relate force and acceleration as $\boldsymbol{F} = m\boldsymbol{a}$.

physical standpoint, which leads us to the related geometric relationships. Subsequently, we approach the general three-dimensional problem from an abstract mathematical perspective.

2.2.1.1 Two-Dimensional Case: Intuitive Physical Concepts

Assume the displacement components are $u = u(x, , y)$, $v = v(x, , y)$, and $w = 0$, indicating that displacement occurs only within the xy plane. This situation is characteristic of a *plane strain problem*, where the strain in the z direction is zero, yet stress is not zero. Consider the motion of an infinitesimal element $ABCD$ as shown in Fig. 2.2. For clarity in illustration, both the size of the element and its deformation are greatly exaggerated in the figure; in reality, under the assumption of infinitesimal deformation, the actual deformation is much smaller. The coordinate system is Oxy, with point A located at (x, y). The sides AB and AD are parallel to the x and y axes, respectively, with lengths $\mathrm{d}x$ and $\mathrm{d}y$. Influenced by external factors, this element moves and transforms into $A'B'C'D'$: It is assumed the element first translates to the position of the dashed element, then undergoes deformation with A' fixed. The deformation involves two types: changes in the lengths of the sides of the element, and changes in the angles between the sides, corresponding to *normal* and *shear strain*, respectively.

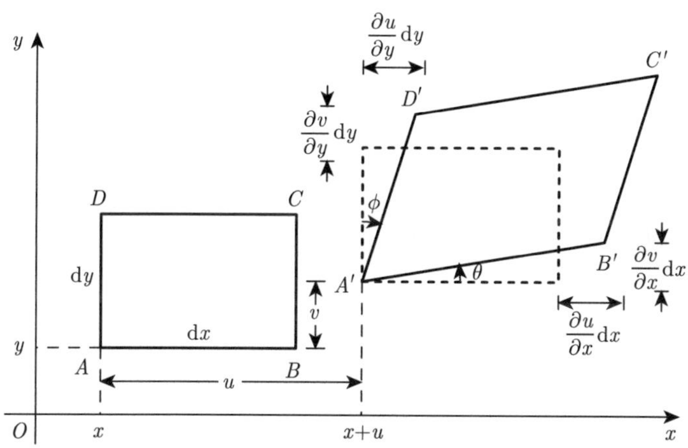

Fig. 2.2 Strain in the two-dimensional case. Under the influence of external factors, the infinitesimal element $ABCD$ changes into $A'B'C'D'$. Point A, initially at coordinates (x, y), moves with displacement components (u, v). The lengths and orientations of the edges AB and AD–now $A'B'$ and $A'D'$ respectively–also change. This results in variations in the dimensions and the angles of the element, indicative of normal and shear strain. For example, AB and AD transform and rotate, with AB rotating counterclockwise by θ and AD clockwise by ϕ

Normal strain is defined as the change in unit length along the coordinate axes of a specified system, represented by ε_x and ε_y. And it is defined that, an increase in length is positive, while a reduction is negative.[13] By this definition,

$$\varepsilon_x = \frac{A'B' - AB}{AB} = \frac{A'B'}{\mathrm{d}x} - 1 \tag{2.2.1a}$$

$$\varepsilon_y = \frac{A'D' - AD}{AD} = \frac{A'D'}{\mathrm{d}y} - 1 \tag{2.2.1b}$$

Shear strain is defined as the change in the angle between the positive directions of the x and y axes of a chosen coordinate system after deformation, measured in radians and denoted by γ_{xy}. As for the sign of shear stress, a decrease in the angle is defined positive, while an increase is negative. According to this definition,

$$\gamma_{xy} = \theta + \phi \tag{2.2.2}$$

The specific expressions for Eqs. (2.2.1) and (2.2.2) can be derived directly from analyzing the deformation shown in Fig. 2.2. The displacement at point A is given by (u, v). Since point B has the same y-coordinate as point A but its x-coordinate is $x + \mathrm{d}x$, the displacement at point B is $(u + (\partial u/\partial x)\ \mathrm{d}x,\ v + (\partial v/\partial x)\ \mathrm{d}x)$. This is obtained by analyzing the changes depicted in Fig. 2.2. Therefore we have

$$\left(A'B'\right)^2 = \left(\mathrm{d}x + \frac{\partial u}{\partial x}\,\mathrm{d}x\right)^2 + \left(\frac{\partial v}{\partial x}\,\mathrm{d}x\right)^2 \xlongequal{(2.2.1a)} \left[\mathrm{d}x\left(1 + \varepsilon_x\right)\right]^2$$

that is

$$1 + 2\frac{\partial u}{\partial x} + \left(\frac{\partial u}{\partial x}\right)^2 + \left(\frac{\partial v}{\partial x}\right)^2 = 1 + 2\varepsilon_x + \varepsilon_x^2$$

In the case of small deformations, the spatial derivatives of the displacement components and the strain components are "very small", making their squares second-order small quantities in comparison to the magnitudes themselves. Consequently, the square terms are neglected and it is found that

$$\varepsilon_x = \frac{\partial u}{\partial x} \tag{2.2.3}$$

[13] In defining normal strain, the focus is solely on changes in length, disregarding any changes in direction. For instance, in Fig. 2.2, even though the direction of segment AB changes to $A'B'$ after deformation, normal strain, represented as ε_x, is still calculated by comparing the difference in length between $A'B'$ and AB directly to the original length of AB, without the need to project $A'B'$ onto the x axis.

Similarly, it can be concluded from examining $\left(A'D'\right)^2$ that

$$\varepsilon_y = \frac{\partial v}{\partial y} \tag{2.2.4}$$

In addition, according to Fig. 2.2,

$$\tan\theta = \frac{\dfrac{\partial v}{\partial x}\mathrm{d}x}{\mathrm{d}x + \dfrac{\partial u}{\partial x}\mathrm{d}x} = \frac{\partial v}{\partial x}\left[1 + \frac{\partial u}{\partial x}\right]^{-1} = \frac{\partial v}{\partial x}\left[1 - \frac{\partial u}{\partial x} + \left(\frac{\partial u}{\partial x}\right)^2 + \cdots\right] \approx \frac{\partial v}{\partial x}$$

In the case of small deformations, the final approximation holds true. Here, the tangent of the angle θ is approximately equal to θ itself, that is, $\tan\theta \approx \theta$. Thus,

$$\theta = \frac{\partial v}{\partial x}$$

Similarly, we have

$$\phi = \frac{\partial u}{\partial y}$$

and finally,

$$\gamma_{xy} = \phi + \theta = \frac{\partial u}{\partial y} + \frac{\partial v}{\partial x} \tag{2.2.5}$$

Equations (2.2.3), (2.2.4), and (2.2.5) together formulate the *geometric equations* for two-dimensional problems, describing the relationship between the spatial rate of change of strain components and displacement components. It is important to note that these geometric equations are valid under the assumption of small deformations. If the deformations are not small, the scenario becomes more complex.

2.2.1.2 Three-Dimensional Case: Abstract Mathematical Concepts

Analyzing two-dimensional problems offers some insights. Since strain components reflect the spatial rate of change of displacement components, it is sensible to begin with the general form of the spatial change rate of displacement when considering three-dimensional scenarios.

In the chosen coordinate system $Ox_1x_2x_3$, the spatial rate of change of displacement, $u_{i,j}$, can be represented as follows,

$$u_{i,j} = \frac{1}{2}\left(u_{i,j} + u_{j,i}\right) + \frac{1}{2}\left(u_{i,j} - u_{j,i}\right)$$

in which $u_{i,j}$ is expressed as the sum of a symmetric tensor and an antisymmetric tensor. As previously proven in Eq. (2.1.17), $u_{i,j}$ is a second-order tensor, which means $u_{j,i}$ is also a second-order tensor. A tensor is termed *symmetric* if it remains unchanged when its indices are swapped, and *antisymmetric* if its value changes sign upon swapping its indices. This forms a general principle where any second-order tensor can be decomposed into the sum of a symmetric and an antisymmetric tensor. The symmetric part is denoted as follows

$$\boxed{\varepsilon_{ij} = \frac{1}{2}\left(u_{i,j} + u_{j,i}\right)} \tag{2.2.6}$$

and this symmetric tensor is referred to as the *strain tensor*.[14] Equation (2.2.6) conveys several key pieces of information. It defines the components of strain in three-dimensional contexts and explicitly details the relationship between these strain components and displacement components. Essentially, this equation serves as a geometric equation. Furthermore, it indicates that the organized array of strain components constitutes a second-order tensor.[15]

---

[14] Accordingly, the antisymmetric component is denoted by

$$\omega_{ij} = \frac{1}{2}\left(u_{i,j} - u_{j,i}\right)$$

and is referred to as the *rotation tensor*. This tensor essentially represents the rotational elements in the motion of an elastic body. It's important to note that this represents the rotational component associated with deformation. In any small element of an elastic body, if there is a deformation component, there is inevitably a rotational component as well. However, the entire elastic body might not necessarily exhibit macroscopic rotational movement.

[15] Based on the preceding discussion, readers may have two questions:

(1) In the analysis of two-dimensional situations, only specific directions of normal and shear strains are defined. For example, normal strains are considered only in the x or y axis directions, and shear strains only in the change of the angle between these axes. How are deformations in other directions, such as along a diagonal in the Oxy coordinate system, measured? What is the normal strain along such a diagonal line?

To address this first question, consider a shift in perspective. Since strain is a second-order tensor, it adheres to the transformation relations of second-order tensors as shown in Eq. (2.1.15). By establishing another coordinate system along this diagonal, where the diagonal direction aligns with the x' axis, the strain in this new coordinate system can be calculated using the original definitions.

(2) According to Eq. (2.2.6), when $i = 1$, $j = 2$, $\varepsilon_{12} = \left(u_{1,2} + u_{2,1}\right)/2$, but Eq. (2.2.5) in indexed form gives $\gamma_{12} = u_{1,2} + u_{2,1}$. This suggests a factor difference between the two. Does this imply two definitions for shear strain?

Indeed, there are two ways to define shear strain: *engineering strain* (denoted as γ_{ij}) and *tensor strain* (denoted as ε_{ij}). Engineering strain, the earlier concept, has a clear physical meaning, whereas tensor strain, which emerged later, has a less intuitive physical interpretation. Its advantage lies in its definition under Eq. (2.2.6), as the components form a second-order tensor and thus exhibit the properties of second-order tensors. The combinations of normal and shear strains defined via engineering strain do not form a second-order tensor, which is particularly noteworthy.

2.2.2 *Stress and Elastic Wave Equations*

As previously discussed, the response of an elastic body to external factors can be examined from both geometric and physical perspectives. The geometric approach introduces the concept of strain and the associated geometric equations. On the physical side, the concept of stress and the equations of elastic motion come into play. Stress is slightly more complex than strain. For example, the complete description of the stress state composed of stress components is a second-order tensor, which is not as obvious as the strain tensor. Before delving into specifics, it is useful to outline our approach to understanding a new physical quantity, which can be broken down into three steps: first, its precise definition; second, its representation, which is particularly challenging with quantities like stress that present unique representational issues; and third, quantitatively determining it, which involves the formulation of equations. Clarifying this seemingly natural sequence is important for a clear and thorough understanding of stress. We will follow this structure to discuss each aspect in turn.

2.2.2.1 Definition of Stress

In understanding the response of an elastic body, it is essential to differentiate between external and internal forces, which are inherently relative concepts. *External forces* refer to those applied from outside a specific region of an elastic body. These can be further divided into body forces and surface forces based on their area of influence and mode of action. *Body forces*, such as gravity and the Coriolis force due to Earth's rotation, act at a distance and do not require direct contact. In contrast, *surface forces* act directly on the surface of an object and involve direct contact, such as the pressure and friction between solids. The distinctive characteristics of these forces manifest differently in problem-solving: body forces enter equations as non-homogeneous terms influencing the field equations directly, whereas surface forces appear in boundary conditions, specifically forming the second type of boundary conditions.

In contrast to external forces, *internal forces* describe the interactions within a body itself.[16] These forces come into play when external forces disturb the equilibrium between the molecules or atoms within an elastic body, resulting in what is known as additional internal forces. For example, consider a hypothetical surface element, ΔS, within the elastic body. The application of external forces generates additional internal forces, causing the parts on either side of ΔS to exert forces on each other. Stress is defined as the additional internal force per unit area within the body. Mathematically, this is expressed as

[16] Internal forces originate within an object and are present even in the absence of external forces. From a microscopic perspective, these forces manifest as the interactions between molecules or atoms. Under the influence of internal forces, the molecules or atoms of an object not subjected to external forces move around their equilibrium positions, ultimately achieving a state of balance.

$$\boldsymbol{T}^{\hat{n}} = \lim_{\Delta S \to 0} \frac{\Delta \boldsymbol{F}}{\Delta S} \tag{2.2.7}$$

where $\hat{\boldsymbol{n}}$ represents the normal direction of the chosen surface element ΔS, while $\Delta \boldsymbol{F}$ is the additional internal force on that element. There are two important aspects to Eq. (2.2.7) to consider: first, the defined stress $\boldsymbol{T}$ is a vector representing the force per unit area, having dimensions of pressure and measured in pascals (Pa). Second, this is not an ordinary vector since its magnitude and direction are dependent on the normal direction $\hat{\boldsymbol{n}}$ of the surface element ΔS, and it is denoted by $\boldsymbol{T}^{\hat{n}}$. This stress vector is also referred to as the *traction*.

2.2.2.2 Representation of Stress

In physics, quantities are typically represented as scalars or vectors, which is straightforward. However, representing stress is more complex because its definition depends on the orientation of the surface element. Stress vectors can be represented in two main ways: One method uses components of the vector as defined in Eq. (2.2.7), such as $T_x^{\hat{n}}$, $T_y^{\hat{n}}$, and $T_z^{\hat{n}}$, which correspond to the projections of the vector $\boldsymbol{T}^{\hat{n}}$ along the three coordinate axes. The other method divides the stress vector for a surface element ΔS into components perpendicular and parallel to the element–known respectively as *normal stress* and *shear stress*, traditionally denoted by σ and τ.[17] The latter representation is more commonly used in practice. Particularly, when the normal direction of the surface element aligns with the coordinate axes, there are specific conventions for these symbols.

First, let's define the orientation of a surface. When selecting a surface element ΔS with its normal aligned along a coordinate axis, such as the x axis, the direction of the normal–whether positive or negative along the x axis–must be specified. Here, the normal is referred to as the "outward normal," implying an inward counterpart. Thus, the selection of this small surface is related to the *body* it is part of,[18] with the outward normal directed from inside to outside of the body. If the outward normal of the surface element points in the positive direction of the x axis, the surface is called the "positive x surface". Conversely, if it points in the negative direction, it is known as the "negative x surface".

After defining the orientation of a surface, stress components can be examined in two steps. Firstly, taking the x surface as an example, the stress vector on this surface is decomposed along the normal (i.e., the x direction) and tangential directions,

---

[17] Due to historical reasons, there are various notational conventions in elastic mechanics, but here we focus on the most commonly used ones. When dealing with stress tensors, both normal and shear stresses are often represented by the same symbol, either τ or σ. Since this usage does not lead to ambiguity, they are typically not distinguished in such contexts.

[18] In continuum mechanics, the fundamental unit is the "material particle", which is characterized by being macroscopically small but microscopically large enough to maintain continuity. The continuum assumption posits that these material particles fill the medium without any gaps. Consequently, any "surfaces" selected within this medium are inherently connected to the body.

resulting in the normal stress σ_x[19] and shear stress. However, the shear stress may not necessarily align with the y or z axes. Thus, it is further decomposed along these axes, yielding two shear stress components, τ_{xy} and τ_{xz}. The first subscript in the shear stress represents the direction of the normal to the face, while the second denotes the direction of the force application.

Like vectors, stress components also have specific directional conventions, though these are more complex. In the second step, the directions of stress components are established by convention: on a "positive surface" (where the outward normal is in the positive direction of the coordinate axis), any stress component along the positive coordinate axis is considered positive; on a "negative face" (where the outward loose is in the negative direction), any stress component along the negative coordinate axis is deemed positive.[20] Figure 2.3 illustrates these conventions using a hexahedron to show all positive stress components on six surfaces of the hexahedron. Representing these with a hexahedron is merely a convenient method to distinguish positive and negative surfaces clearly and to visually articulate the relationship between the body and its surfaces. Although the pairs of parallel faces in the diagram are geometrically the same, they are differentiated by their directions.

The above conventions for the notation and sign of stress components apply to faces with normal directions aligned with the coordinate axes. But what about a face with a normal direction in any arbitrary direction? In this case, we need to consider the stress on an inclined plane. We will derive an important equation known as the stress transformation formula, or Cauchy's formula. Furthermore, based on this formula, we will show that the ordered set of stress components defined for the special faces forms a second-order tensor.

Consider the infinitesimal tetrahedral element $OABC$ shown in Fig. 2.4, where OA, OB, and OC lie along the x_1, x_2, and x_3 axes, respectively. The face $\triangle ABC$ forms an inclined plane within the coordinate system $Ox_1x_2x_3$. Assume the outward normal to this inclined plane is in the direction of x_1', which is an axis in another

---

[19] Normal stress is typically denoted by a single subscript because its direction aligns with the normal direction of the surface. Sometimes, it is represented with a repeated subscript, such as σ_{xx}, to emphasize this alignment.

[20] Under the established conventions, a stress component is considered negative if its actual direction is opposite to the defined positive direction. For instance, a value of $\tau_{xy} = -50$ MPa indicates that the magnitude of this stress component is 50 MPa, and its direction opposes the prescribed positive direction. This convention leads to a deeper inquiry: Why is it defined this way? While the conventions for the positive surface might intuitively make sense, those for the negative surface might seem counterintuitive. Could the conventions be reversed? It is important to note that the sign convention for stress components must be consistent with that for vectors. Consider the example of the positive/negative x surfaces in Fig. 2.3. The stress vectors acting on the positive x surface and the negative x surface form a pair of action and reaction forces, which are equal in magnitude and opposite in direction. Note the Cauchy formula in Eq. (2.2.14) and the definition of vector signs (positive along the positive x axis, and negative along the negative x axis), only by adhering to the current stress sign convention can it match the sign definition of the vector.

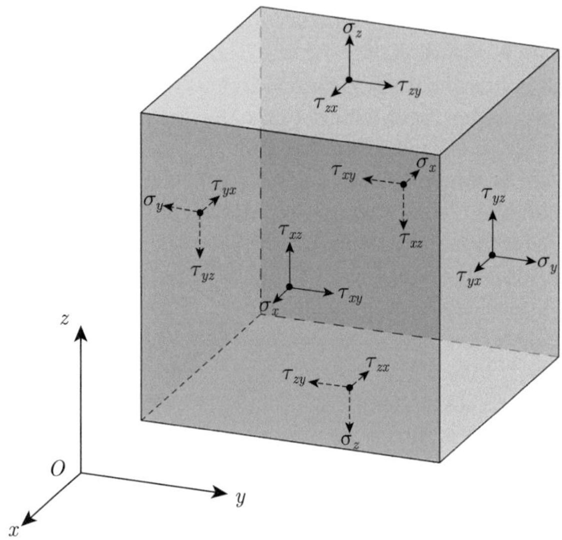

Fig. 2.3 Stress components represented on a hexahedron. The positive directions of stress components are indicated on each surface of the hexahedron. The normal stresses, σ_x, σ_y, and σ_z are positive when they act outward, perpendicular to their respective faces. The shear stresses, τ_{xy} and τ_{xz} on the x surface, point in the positive y and z axes on the positive x surface, and in the negative y and z axes on the negative x surface. The other shear stress components follow a similar pattern. Stresses on the positive and negative surfaces are represented with solid and dashed lines, respectively

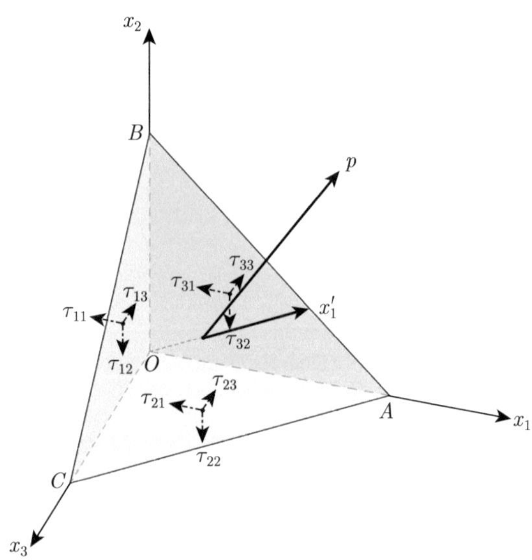

Fig. 2.4 The force equilibrium on the infinitesimal tetrahedral element. The infinitesimal tetrahedral element $OABC$ has the outward normal of the $\triangle ABC$ surface in the x_1' direction, with the stress vector on this surface denoted by $\boldsymbol{p}$. The stress components are indicated on the other triangular faces

coordinate system. The stress vector on this inclined plane is $\boldsymbol{p}$. To analyze the equilibrium of the infinitesimal tetrahedral element in the x_1 direction, we have:

$$p_1 S_{\triangle ABC} = \tau_{11} S_{\triangle BOC} + \tau_{21} S_{\triangle AOC} + \tau_{31} S_{\triangle AOB} \tag{2.2.8}$$

where $S_{\triangle ABC}$ represents the area of $\triangle ABC$, and similar notation applies to the other triangular areas. Note the relationships between the areas of these triangles

$$S_{\triangle AOC} = S_{\triangle ABC} \cos\left(x_1', x_2\right) \overset{(2.1.11)}{=\!=\!=} c_{12} S_{\triangle ABC}$$

and similarly,

$$S_{\triangle BOC} = c_{11} S_{\triangle ABC}, \quad S_{\triangle AOB} = c_{13} S_{\triangle ABC}$$

Thus, by dividing both sides of Eq. (2.2.8) by $S_{\triangle ABC}$, we obtain

$$p_1 = c_{11}\tau_{11} + c_{12}\tau_{21} + c_{13}\tau_{31} = c_{1i}\tau_{i1} \tag{2.2.9}$$

similarly,

$$p_2 = c_{1i}\tau_{i2}, \quad p_3 = c_{1i}\tau_{i3} \tag{2.2.10}$$

Combining the results of Eqs. (2.2.9) and (2.2.10), we have

$$p_k = c_{1i}\tau_{ik} \tag{2.2.11}$$

Projecting p_k in the direction of x_1', we obtain τ_{11}'

$$\tau_{11}' = c_{11}p_1 + c_{12}p_2 + c_{13}p_3 = c_{1k}p_k \overset{(2.2.11)}{=\!=\!=} c_{1i}c_{1k}\tau_{ik} \tag{2.2.12}$$

and similarly

$$\tau_{12}' = c_{1i}c_{2k}\tau_{ik}, \quad \tau_{13}' = c_{1i}c_{3k}\tau_{ik} \tag{2.2.13}$$

Combining the results of Eqs. (2.2.12) and (2.2.13), it is found that

$$\tau_{1n}' = c_{1i}c_{nk}\tau_{ik}$$

If the outward normal direction of the inclined plane is chosen as x_2' or x_3', then the subscript 1 in the left-hand term and the first term on the right-hand side of the above equation should be replaced by 2 and 3, respectively. This implies that

$$\tau_{jn}' = c_{ji}c_{nk}\tau_{ik}$$

This is consistent with Eq. (2.2.15), meaning that all the stress components τ_{ik} together form a second-order tensor.

Given that the unit normal vector to $\triangle ABC$ is denoted as $\hat{\boldsymbol{n}}$ with components $n_i = c_{1i}$, indicating that the stress on this plane is $\boldsymbol{T}^{\hat{n}}$, we can derive from Eq. (2.2.11) that

$$\boxed{T_k^{\hat{n}} = n_i \tau_{ik} \quad \text{or} \quad \boldsymbol{T}^{\hat{n}} = \hat{\boldsymbol{n}} \cdot \boldsymbol{\tau}} \tag{2.2.14}$$

where the symbol $\boldsymbol{\tau}$ represents a second-order tensor composed of stress components τ_{ik}. This formula is known as the *inclined plane stress formula* or *Cauchy's formula*. It establishes the relationship between the stress tensor and the stress vector on a specific plane. If the stress tensor is known, then by projecting the stress tensor in the direction of the normal to a given plane, one can determine the stress vector on that plane. Cauchy's formula also indicates that the stress tensor fully characterizes the stress state at a point.[21]

2.2.2.3 Elastic Motion Equations

Cauchy's formula (2.2.14) demonstrates how to determine the stress components on any plane passing through a point within an elastic body, given a certain stress distribution. However, this relationship alone is not sufficient to determine the overall stress distribution within the material. To achieve this, one must develop equations that describe the spatial variations of the stress components throughout the body.[22]

Consider a closed surface S within an elastic body, enclosing a volume V. Let the body force distribution be $\boldsymbol{f}$ and the surface force on S be $\boldsymbol{T}^{\hat{n}}$.[23] According to Newton's second law, it is easy to write

[21] According to the definition of the stress vector in Eq. (2.2.7), the stress vector on a plane passing through a point depends on the direction of that plane. To fully understand the stress state at that point, one would theoretically need to know the stress vectors on all possible planes through that point. However, Cauchy's formula simplifies this by showing that it's sufficient to know the stress vectors on three mutually perpendicular planes (formed by three pairs of mutually perpendicular coordinate axes). Knowing the stress vectors on these three planes allows us to form a second-order tensor, the stress tensor. Given this tensor, Cauchy's formula can be used to determine the stress vector on any plane through the point by projecting the tensor in the direction of the plane's normal.

[22] In elasticity theory, the study of static problems involves the equilibrium equations. These equations connect the spatial derivatives of different stress components, describing how these components vary throughout the material, and can be used to determine their magnitudes. For dynamic problems that involve time, the equations must also include second-order spatial derivatives of displacement, forming the equations of motion.

[23] According to the definition of $\boldsymbol{T}^{\hat{n}}$, it represents the internal forces within the body, whereas surface forces act on the exterior of the body. Saying "surface force is $\boldsymbol{T}^{\hat{n}}$" might seem conceptually contradictory. However, this can be understood through the relativity of external and internal forces. Relative to the selected volume, the forces acting on S are indeed external forces, but with respect to the entire elastic body being studied, S is within the body, and the forces acting on it are still internal forces.

$$\iiint_V \boldsymbol{f}\,\mathrm{d}V + \oiint_S \boldsymbol{T}^{\hat{\boldsymbol{n}}}\,\mathrm{d}S = \iiint_V \rho\ddot{\boldsymbol{u}}\,\mathrm{d}V \tag{2.2.15}$$

The left side of the equation represents the net external force, where the first term denotes the body force and the second term denotes the surface force. The right side of the equation describes the motion, with $\ddot{\boldsymbol{u}}$ representing the second time derivative of the displacement vector, which is the acceleration vector. For the surface force term,

$$\oiint_S \boldsymbol{T}^{\hat{\boldsymbol{n}}}\,\mathrm{d}S \xlongequal{(2.2.14)} \oiint_S \hat{\boldsymbol{n}} \cdot \boldsymbol{\tau}\,\mathrm{d}S \xlongequal{(2.1.19)} \iiint_V \nabla \cdot \boldsymbol{\tau}\,\mathrm{d}V$$

by substituting into Eq. (2.2.15), we obtain

$$\iiint_V (\nabla \cdot \boldsymbol{\tau} + \boldsymbol{f} - \rho\ddot{\boldsymbol{u}})\,\mathrm{d}V = 0 \tag{2.2.16}$$

Note that V is the volume enclosed by any chosen closed surface S. The equation holds true if and only if the integrand is zero,[24] therefore

$$\boxed{\nabla \cdot \boldsymbol{\tau} + \boldsymbol{f} = \rho\ddot{\boldsymbol{u}}} \tag{2.2.17}$$

This is the elastic motion equation expressed in terms of the stress tensor and the displacement vector.

2.2.3 Constitutive Relation: Generalized Hooke's Law

Previously, the concepts of strain and stress were introduced from the geometric and physical perspectives of how elastic bodies respond to loads. The geometric equations and the equations of elastic motion have been established. The geometric equations include the displacement components and strain components, while the motion equations include the stress components and displacement components. From the perspective of solving the system of equations, this system is not closed, meaning

[24] We can also derive Eq. (2.2.17) in an equivalent manner, as seen in Wang et al. (2002, p. 57). Let point P be a point in the region V. According to the mean value theorem for integrals, Eq. (2.2.16) can be written as:

$$(\nabla \cdot \boldsymbol{\tau} + \boldsymbol{f} - \rho\ddot{\boldsymbol{u}})_i\Big|_{P_i^*} \iiint_V \mathrm{d}V = 0$$

where P_i^* is a point in V, and for different i, P_i^* is not the same. By canceling the integral term and letting the volume V shrink to point P, we have $P_i^* \to P$, thereby reaching the same conclusion as Eq. (2.2.17).

the number of equations is fewer than the number of unknown functions.[25] Therefore, additional equations are needed to link the stress components and strain components, which is where the constitutive relation comes in.

A constitutive relation characterizes the properties of a material, such as the stress-strain relationship, which describes its mechanical behavior. There are other types of constitutive relations as well, such as those that describe thermal conductivity and mass transfer properties. The previously discussed concepts of strain and stress, along with the related equations, apply to all branches of continuum mechanics that are based on the continuum assumption. It is through constitutive relations that different models of continuum motion are distinguished.[26] In the context of elastodynamics, the constitutive relation is given by the generalized Hooke's law,[27] mathematically expressed as:

$$\boxed{\tau_{ij} = C_{ijpq}\varepsilon_{pq}} \tag{2.2.18}$$

where C_{ijpq} is the elastic coefficient, which characterizes the inherent mechanical properties of a material. With four subscripts, C_{ijpq} can be proven to be a fourth-order tensor.[28] A general fourth-order tensor has $3^4 = 81$ components, but C_{ijpq} exhibits certain symmetries, reducing the number of independent components. For instance, due to the symmetries of the stress and strain components, it is evident that $C_{ijpq} = C_{jipq} = C_{ijqp}$. Furthermore, considering the strain energy density, we obtain $C_{ijpq} = C_{pqij}$ (Wang et al. 2002, p. 76). A fourth-order tensor with such

---

[25] Specifically, the geometric Eq. (2.2.6) comprise 6 independent equations, since the strain tensor is symmetric, resulting in only 6 independent components. Combined with the 3 displacement components, there are a total of 9 unknown functions. However, the equations of motion (2.2.17) consist of only 3 equations but involve 9 unknown functions. Besides the 3 displacement components, there are 6 independent stress components. Typically, by applying the conditions of moment equilibrium, the equality of shear stresses is derived. For more details, refer to Wang et al. (2002, pp. 55, 56).

[26] Generally, the constitutive equations of a material should be developed through both theoretical and experimental approaches. Initially, a simplified model is proposed based on experimental observations. Then, the general form of the constitutive equation for this model is derived theoretically. Finally, the parameters appearing in these equations are verified and determined through further experiments.

[27] It is based on the Hooke's law discovered by the English physicist R. Hooke in 1660. Hooke's law states that the force is proportional to the extension in the form of a function. In the context of elasticity theory, this corresponds to the relationship between stress and strain. However, the relationship is much more complex because both stress and strain each have six independent components. A natural assumption is that each stress component is a linear function of all the strain components. This is the generalization of Hooke's law in the field of elasticity theory, known as the *generalized Hooke's law*. Rather than being derived from theoretical analysis, this assumption is based on simple experimental facts, and its validity is judged by experimental results. Indeed, this stress-strain relationship can efectively describe the mechanical properties of materials within a certain range.

[28] Previously, we introduced the algebraic definitions of vectors and second-order tensors. It can be seen that the extension from lower to higher-order tensors is straightforward. For example, the transformation formula for a second-order tensor, as given by Eq. (2.1.15), can naturally be extended to a fourth-order tensor: $A'_{nstv} = c_{ni}c_{sj}c_{tp}c_{vq}A_{ijpq}$. As an exercise, readers can prove that C_{ijpq} satisfies this equation, confirming that it is indeed a fourth-order tensor.

symmetry is known as a *Voigt symmetric fourth-order tensor.* It can be verified that it has 21 independent components. In other words, a general anisotropic elastic material has 21 independent elastic parameters.[29] Several elastic parameters describe the properties of an isotropic elastic material, such as Young's modulus E, Poisson's ratio ν, shear modulus or rigidity modulus μ, Lamé's first parameter λ, and bulk modulus K. However, only two of these parameters are independent. Using Lamé's first parameter λ and the shear modulus μ, for an isotropic elastic material, C_{ijpq} can be expressed as:

$$\boxed{C_{ijpq} = \lambda\delta_{ij}\delta_{pq} + \mu\left(\delta_{ip}\delta_{jq} + \delta_{iq}\delta_{jp}\right)} \tag{2.2.19}$$

Generally speaking, the elastic parameters λ and μ can be functions of position. However, if the medium is homogeneous, these parameters are constant throughout the elastic body. This type of *isotropic homogeneous medium* represents the simplest case and is the primary focus of our study.[30]

For a isotropic homogeneous medium, the form of the generalized Hooke's law can be greatly simplified. Substituting Eq. (2.2.19) into Eq. (2.2.18) yields

$$\begin{aligned}\tau_{ij} &= \left[\lambda\delta_{ij}\delta_{pq} + \mu\left(\delta_{ip}\delta_{jq} + \delta_{iq}\delta_{jp}\right)\right]\varepsilon_{pq} \\ &= \lambda\varepsilon_{kk}\delta_{ij} + \mu\left(\varepsilon_{ij} + \varepsilon_{ji}\right) = \lambda\varepsilon_{kk}\delta_{ij} + 2\mu\varepsilon_{ij}.\end{aligned}$$

2.2.4 Equations for Isotropic Homogeneous Elastic Body

So far, we have derived the geometric equations, the equations of motion, and the constitutive equations (i.e., the generalized Hooke's law). For the sake of convenience in analysis, we now rewrite the equations for an isotropic homogeneous elastic body in component form as follows:

$$\begin{cases}\varepsilon_{ij} = \dfrac{1}{2}\left(u_{i,j} + u_{j,i}\right), & \text{(Geometric equations)} \\ \tau_{ji,j} + f_i = \rho\ddot{u}_i, & \text{(Equations of motion)} \\ \tau_{ij} = \lambda\varepsilon_{kk}\delta_{ij} + 2\mu\varepsilon_{ij}, & \text{(Constitutive equations)}\end{cases} \tag{2.2.20}$$

---

[29] When an anisotropic elastic material has certain symmetries, the number of independent elastic parameters further decreases. For instance, a material with one plane of symmetry has 13 elastic parameters. A material with two planes of symmetry (orthotropic material) has 9 elastic parameters, while a material with one axis of symmetry (transversely isotropic material) has 5 elastic parameters. The simplest case is isotropic materials, which have only two elastic parameters. For a detailed analysis, refer to Wang et al. (2002, pp. 84–87).

[30] In many cases, as a first-order approximation, this simple model can serve as a simplified representation of the medium. However, in certain special situations, such as when anisotropy is significant or the properties of the medium vary sharply, it is necessary to account for the anisotropic or heterogeneous nature of the medium.

in which the unknown functions are u_i (3 functions), τ_{ji} (6 functions), and ε_{ij} (6 functions), making a total of 15 functions.[31] The number of equations is: 6 geometric equations, 3 equations of motion, and 6 constitutive equations. The number of equations matches the number of unknown functions, so Eq. (2.2.20) forms a closed system of equations.

This is a complex system of equations. To simplify solving the system, based on the characteristics of the boundary conditions, there are two possible approaches.

(1) Stress method: In this approach, displacement and strain components are eliminated from the system of equations, leaving only the stress components. Since the equilibrium equations involve only stress components, they remain unchanged. The geometric equations are used to derive the strain compatibility equations, which, using the constitutive stress-strain relationships, are converted into stress compatibility equations. Thus, the entire system is expressed solely in terms of stress components. However, the system remains complex. To further simplify it, an intermediate function, the stress function, is introduced. This transforms the equations of motion into relationships between all stress components and the stress function, and the stress compatibility equations become equations that the stress function must satisfy. For two-dimensional problem, the number of functions to be solved for is reduced to just one, the stress function. Once the stress function is determined, all stress components can be found from their relationships with this function. The trade-off is that the order of the equations increases. For instance, in the special case where the body force is constant, the stress compatibility equation in terms of the stress function becomes a fourth-order biharmonic equation.

(2) Displacement method: In this approach, stress and strain components are eliminated from the system of equations, leaving only the displacement components. On closer inspection of the equation system (2.2.20), it is not as complex as it might initially seem. The geometric equations express strain components in terms of displacement components, the constitutive equations express stress components in terms of strain components, and the equations of motion combine stress and displacement components. A simplification can be achieved by substituting the geometric equations into the constitutive equations, and then substituting the resulting constitutive equations into the equations of motion. This yields the equations of motion expressed solely in terms of displacement components, resulting in three equations for three unknown functions, significantly simplifying the system.

---

[31] According to our convention, the indices i and j take values 1, 2, and 3. Therefore, there are 9 stress components τ_{ji} and 9 strain components ε_{ij}. Based on the geometric equations, ε_{ij} is a symmetric tensor component, meaning it satisfies $\varepsilon_{ij} = \varepsilon_{ji}$, so there are only 6 independent components. Additionally, under the condition of moment equilibrium, the shear stresses are equal, i.e., $\tau_{ij} = \tau_{ji}$, leaving only 6 independent stress components (refer to Wang et al. (2002, pp. 55–56).

Following the displacement method, we derive the equations of motion expressed in terms of displacement components. By substituting the geometric equations into the constitutive equations, we obtain:

$$\tau_{ij} = \lambda u_{k,k}\delta_{ij} + \mu\left(u_{i,j} + u_{j,i}\right)$$

and take its spatial derivative,

$$\begin{aligned}
\tau_{ji,j} &= \lambda u_{k,kj}\delta_{ij} + \mu\left(u_{i,jj} + u_{j,ij}\right) \xlongequal{(2.1.5)} \lambda u_{k,ki} + \mu\left(u_{i,jj} + u_{j,ij}\right) \\
&= (\lambda + 2\mu)u_{j,ji} - \mu\left(u_{j,ji} - u_{i,jj}\right) \\
&= (\lambda + 2\mu)u_{j,ji} - \mu\left(\delta_{im}\delta_{jn} - \delta_{in}\delta_{jm}\right)u_{n,mj} \\
&\xlongequal{(2.1.9)} (\lambda + 2\mu)u_{j,ji} - \mu\varepsilon_{ijk}\frac{\partial}{\partial x_j}\left(\varepsilon_{kmn}\frac{\partial}{\partial x_m}u_n\right)
\end{aligned}$$

then substitute it into the equations of motion and express it in vector form,

$$\boxed{\rho\ddot{\boldsymbol{u}} = (\lambda + 2\mu)\nabla\left(\nabla\cdot\boldsymbol{u}\right) - \mu\nabla\times\left(\nabla\times\boldsymbol{u}\right) + \boldsymbol{f}} \tag{2.2.21}$$

The equation of elastodynamics, expressed in terms of displacement components, serves as the foundation for all subsequent problems we will address.[32]

2.3 Reciprocal Theorem of Elastodynamics

The Betti reciprocal theorem is a crucial principle in elastodynamics.[33] It serves as the foundation for deriving both the displacement representation theorem and the source representation theorem. This theorem establishes the relationship between two different sets of forces applied to an elastic body V and the displacements they cause.

Consider two sets of body forces, $\boldsymbol{f}(\boldsymbol{x}, t)$ and $\boldsymbol{g}(\boldsymbol{x}, t)$, which, under their respective boundary and initial conditions, generate displacement fields $\boldsymbol{u}(\boldsymbol{x}, t)$ and $\boldsymbol{v}(\boldsymbol{x}, t)$

---

32 Generally, when using the displacement method, it is more convenient to apply the first type of boundary condition, where displacements are specified on the boundary. For the Lamb's problem we will study later, the second type of boundary condition, where traction is specified on the free surface, is satisfied. This requires using constitutive and geometric relations to transform it into a displacement representation.

33 It is also known as the *Maxwell-Betti reciprocal work theorem*, or the *Betti-Rayleigh reciprocal work theorem*. In 1864, J. C. Maxwell first proposed a reciprocal work theorem for a specific truss problem. Then in 1872, Betti generalized it to the reciprocal work theorem in elasticity theory. In 1873, Rayleigh extended this theorem to the frequency domain description of elastodynamic problems. In 1887, Lamb conducted a general study of the reciprocal work theorem in dynamics, and in 1936, N. N. Andleev provided several applications of the theorem.

within an elastic body V with boundary S. Correspondingly, the stress fields are $\boldsymbol{\tau}(\boldsymbol{u}(\boldsymbol{x},\ t))$ and $\boldsymbol{\tau}(\boldsymbol{v}(\boldsymbol{x},\ t))$.[34] According to the equations of motion introduced in Sect. 2.2 (specifically, Eq. (2.2.20)), we have

$$\tau_{ji,j}(\boldsymbol{u}(\boldsymbol{x},\ t)) + f_i(\boldsymbol{x},\ t) = \rho\ddot{u}_i(\boldsymbol{x},\ t) \tag{2.3.1a}$$

$$\tau_{ji,j}(\boldsymbol{v}(\boldsymbol{x},\ t)) + g_i(\boldsymbol{x},\ t) = \rho\ddot{v}_i(\boldsymbol{x},\ t) \quad (\boldsymbol{x} \in V) \tag{2.3.1b}$$

Based on this, E. Betti derived two sets of relationships, known as Betti's first reciprocity theorem and Betti's second reciprocity theorem (also referred to as Betti's first and second relations, respectively).

2.3.1 Betti's First Reciprocity Theorem

The mathematical expression of Betti's first reciprocity theorem is

$$\boxed{\begin{aligned}&\iiint_V \left[\boldsymbol{f}(\boldsymbol{x},\ t) - \rho\ddot{\boldsymbol{u}}(\boldsymbol{x},\ t)\right]\cdot\boldsymbol{v}(\boldsymbol{x},\ t)\,\mathrm{d}V + \oiint_S \boldsymbol{T}\left(\boldsymbol{u}(\boldsymbol{x},\ t),\hat{\boldsymbol{n}}\right)\cdot\boldsymbol{v}(\boldsymbol{x},\ t)\,\mathrm{d}S\\ =&\iiint_V \left[\boldsymbol{g}(\boldsymbol{x},\ t) - \rho\ddot{\boldsymbol{v}}(\boldsymbol{x},\ t)\right]\cdot\boldsymbol{u}(\boldsymbol{x},\ t)\,\mathrm{d}V + \oiint_S \boldsymbol{T}\left(\boldsymbol{v}(\boldsymbol{x},\ t),\hat{\boldsymbol{n}}\right)\cdot\boldsymbol{u}(\boldsymbol{x},\ t)\,\mathrm{d}S\end{aligned}} \tag{2.3.2}$$

where $\hat{\boldsymbol{n}}$ denotes the unit normal vector to the boundary surface element, and $\boldsymbol{T}$ represents the stress vector on the surface element with $\hat{\boldsymbol{n}}$ as the normal direction. According to the inclined plane stress formula (2.2.14),

$$\boldsymbol{T}\left(\boldsymbol{u}(\boldsymbol{x},\ t),\hat{\boldsymbol{n}}\right) = \hat{\boldsymbol{n}}\cdot\boldsymbol{\tau}\left(\boldsymbol{u}(\boldsymbol{x},\ t)\right),\quad \boldsymbol{T}\left(\boldsymbol{v}(\boldsymbol{x},\ t),\hat{\boldsymbol{n}}\right) = \hat{\boldsymbol{n}}\cdot\boldsymbol{\tau}\left(\boldsymbol{v}(\boldsymbol{x},\ t)\right) \tag{2.3.3}$$

Alternatively, by substituting Eq. (2.3.3) into Eq. (2.3.2) and omitting the variables, it can be expressed in component form as

$$\iiint_V (f_i - \rho\ddot{u}_i)\,v_i\,\mathrm{d}V + \oiint_S n_j\tau_{ji}v_i\,\mathrm{d}S = \iiint_V (g_i - \rho\ddot{v}_i)\,u_i\,\mathrm{d}V + \oiint_S n_j\tau'_{ji}u_i\,\mathrm{d}S \tag{2.3.4}$$

[34] The stress field, like the displacement field, is a function of spatial coordinates $\boldsymbol{x}$ and temporal coordinates t. However, it is important to note that the stress components can be expressed as linear combinations of the strain components, which in turn can be represented as combinations of the spatial derivatives of the displacement (refer to Eq. (2.2.20)). Hence, stress components can actually be considered as functions of the displacement field. To clearly distinguish which set of body forces induces the stress field, we represent the stress field as a function of the displacement field.

in which, $\tau_{ji} = \tau_{ji}(\boldsymbol{u})$ and $\tau'_{ji} = \tau_{ji}(\boldsymbol{v})$. This equation can be proved as follows:

$$\text{Left side} \overset{(2.3.1),\,(2.2.20)}{=\!=\!=} \iiint_V \left(-\tau_{ji,j}\right) v_i \,\mathrm{d}V + \oint\!\!\!\oint_S n_j \left(C_{jiks} u_{k,s}\right) v_i \,\mathrm{d}S$$

$$\overset{(2.2.20),\,(2.1.19)}{=\!=\!=} - \iiint_V \left(C_{jiks} u_{k,s}\right)_{,j} v_i \,\mathrm{d}V + \iiint_V \left(C_{jiks} u_{k,s} v_i\right)_{,j} \mathrm{d}V$$

$$= \iiint_V C_{jiks} u_{k,s} v_{i,j} \,\mathrm{d}V$$

Similarly, it can be obtained

$$\text{Right side} = \iiint_V C_{jiks} v_{k,s} u_{i,j} \,\mathrm{d}V = \iiint_V C_{skij} u_{k,s} v_{i,j} \,\mathrm{d}V = \iiint_V C_{jiks} u_{k,s} v_{i,j} \,\mathrm{d}V$$

Both sides are equal, hence Eq. (2.3.4) holds. If we consider $-\rho\ddot{u}_i$ and $-\rho\ddot{v}_i$ as inertia forces, then Eqs. (2.3.2) or (2.3.4) indicate that the work done by the first set of forces on the displacement caused by the second set of forces is equal to the work done by the second set of forces on the displacement caused by the first set of forces. This work actually consists of three parts: the work done by body forces, the work done by inertia forces, and the work done by surface forces. The first two correspond to the volume integral terms in Eqs. (2.3.2) or (2.3.4), while the work done by surface forces corresponds to the surface integral term.[35]

Regarding Betti's first reciprocity theorem, two points need to be clarified:

(1) The theorem only mentions the situation where the first set of forces and the second set of forces act within the elastic region V, but it does not state that the first set of forces and the second set of forces *only* act within region V. Therefore, the condition for the application space of Betti's reciprocity theorem can be relaxed: the two sets of body forces $\boldsymbol{f}(\boldsymbol{x}, t)$ and $\boldsymbol{g}(\boldsymbol{x}, t)$ can act respectively in the elastic regions V_1 and V_2, provided they have a common acting region V.[36] In this common acting region, the elastic body's density is ρ, and the elastic parameters are C_{ijkl}. Figure 2.5 shows two typical situations. One is where V_1 and V_2 intersect in region V, but each also has areas outside the other's influence, as shown in Fig. 2.5a; the other is where V_1 is entirely contained within V_2, making V_1 the common acting region, as shown in Fig. 2.5b. Additionally, it should be

[35] This theorem might seem peculiar at first: the first set of forces does not act on the displacement caused by the second set of forces, so what does this "work" mean? For now, let's consider it as a formal "work". In Sects. 2.4 and 2.5, we will see the immense power of this theorem, and in fact, the reciprocity theorem forms the theoretical foundation for what is now known as the boundary element method. It is precisely because of this interchangeability that the theorem is called the *reciprocity theorem*.

[36] Relaxing the conditions on the acting regions of the space is crucial for understanding the choice and continuity conditions of the Green's function in the source representation theorem discussed in Chap. 3. This point will be further elaborated upon later.

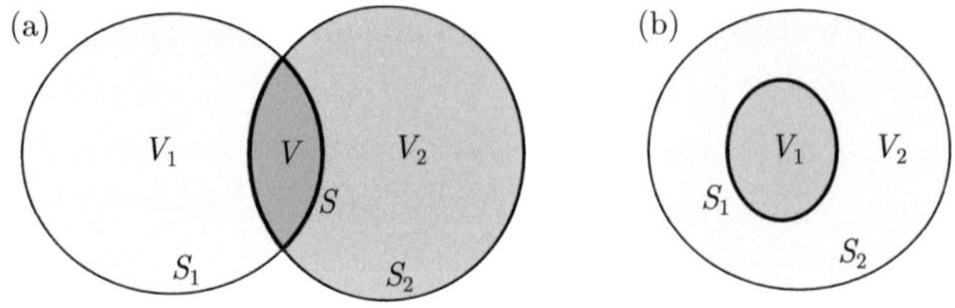

Fig. 2.5 Explanation of the application space in Betti's reciprocity theorem. **a** The elastic region V_1 with boundary S_1 and the elastic region V_2 with boundary S_2 intersect to form the elastic region V with boundary S, where $V = V_1 \cap V_2$. **b** The elastic region V_1 with boundary S_1 is entirely contained within the elastic region V_2 with boundary S_2, in which case $V = V_1$

noted that the boundary conditions corresponding to the two sets of forces on S are not required to be identical.

(2) Even though the body forces, displacement field, and stress field are functions of time, the theorem does not emphasize the time factor. This means that the time instance of the first set of forces (and their resulting displacement and stress fields) does not need to be the same as the time instance of the second set of forces (and their resulting displacement and stress fields). For example, if the time for the first set is t and the time for the second set is $t' = \tau - t$, then Eq. (2.3.4) can be written as:

$$\begin{aligned}&\iiint_V \Big[f_i(\boldsymbol{x},\, t) - \rho\ddot{u}_i(\boldsymbol{x},\, t)\Big] v_i(\boldsymbol{x},\, t')\, \mathrm{d}V + \oiint_S T_i\big(\boldsymbol{u}(\boldsymbol{x},\, t),\, \hat{\boldsymbol{n}}\big) v_i(\boldsymbol{x},\, t')\, \mathrm{d}S \\ =& \iiint_V \Big[g_i(\boldsymbol{x},\, t') - \rho\ddot{v}_i(\boldsymbol{x},\, t')\Big] u_i(\boldsymbol{x},\, t)\, \mathrm{d}V + \oiint_S T_i\big(\boldsymbol{v}(\boldsymbol{x},\, t'),\, \hat{\boldsymbol{n}}\big) u_i(\boldsymbol{x},\, t)\, \mathrm{d}S\end{aligned} \tag{2.3.5}$$

2.3.2 *Betti's Second Reciprocity Theorem*

Integrating both sides of Eq. (2.3.5) with respect to t from $-\infty$ to $+\infty$, we obtain Betti's second reciprocity theorem, which can be mathematically expressed as:

$$\boxed{\begin{aligned}&\int_{-\infty}^{+\infty} \mathrm{d}t \iiint_V f_i(\boldsymbol{x},\, t) v_i(\boldsymbol{x},\, t')\, \mathrm{d}V + \int_{-\infty}^{+\infty} \mathrm{d}t \oiint_S T_i\big(\boldsymbol{u}(\boldsymbol{x},\, t),\, \hat{\boldsymbol{n}}\big) v_i(\boldsymbol{x},\, t')\, \mathrm{d}S \\ =& \int_{-\infty}^{+\infty} \mathrm{d}t \iiint_V g_i(\boldsymbol{x},\, t') u_i(\boldsymbol{x},\, t)\, \mathrm{d}V + \int_{-\infty}^{+\infty} \mathrm{d}t \oiint_S T_i\big(\boldsymbol{v}(\boldsymbol{x},\, t'),\, \hat{\boldsymbol{n}}\big) u_i(\boldsymbol{x},\, t)\, \mathrm{d}S\end{aligned}} \tag{2.3.6}$$

where $t' = \tau - t$. By comparing the integrals of Eqs. (2.3.5) and (2.3.6) with respect to t, it becomes clear that the only difference is the absence of the terms related to inertial forces. Therefore, to prove Eq. (2.3.6), it is sufficient to demonstrate that

$$\int_{-\infty}^{+\infty} \mathrm{d}t \iiint_V \rho \ddot{u}_i(\boldsymbol{x}, t) v_i(\boldsymbol{x}, \tau - t)\,\mathrm{d}V = \int_{-\infty}^{+\infty} \mathrm{d}t \iiint_V \rho \ddot{v}_i(\boldsymbol{x}, \tau - t) u_i(\boldsymbol{x}, t)\,\mathrm{d}V \tag{2.3.7}$$

Assuming that the order of integration with respect to V and t can be interchanged,[37] we have

$$\begin{aligned}
&\iiint_V \mathrm{d}V \int_{-\infty}^{+\infty} \rho \ddot{u}_i(\boldsymbol{x}, t) v_i(\boldsymbol{x}, \tau - t)\,\mathrm{d}t \\
&= \iiint_V \mathrm{d}V \left\{ \left[\rho \dot{u}_i(\boldsymbol{x}, t) v_i(\boldsymbol{x}, \tau - t)\right]\Big|_{-\infty}^{+\infty} + \int_{-\infty}^{+\infty} \rho \dot{u}_i(\boldsymbol{x}, t) \dot{v}_i(\boldsymbol{x}, \tau - t)\,\mathrm{d}t \right\} \\
&= \iiint_V \mathrm{d}V \int_{-\infty}^{+\infty} \rho \dot{u}_i(\boldsymbol{x}, t) \dot{v}_i(\boldsymbol{x}, \tau - t)\,\mathrm{d}t
\end{aligned}$$

The first equality is obtained by using integration by parts, noting that

$$\dot{v}_i(\boldsymbol{x}, \tau - t) = \frac{\partial}{\partial(\tau - t)} v_i(\boldsymbol{x}, \tau - t) = -\frac{\partial}{\partial t} v_i(\boldsymbol{x}, \tau - t)$$

the second equality uses the fact that $\dot{u}_i(\boldsymbol{x}, -\infty) = v_i(\boldsymbol{x}, -\infty) = 0$. Similarly, we can obtain

$$\iiint_V \mathrm{d}V \int_{-\infty}^{+\infty} \rho \ddot{v}_i(\boldsymbol{x}, \tau - t) u_i(\boldsymbol{x}, t)\,\mathrm{d}t = \iiint_V \mathrm{d}V \int_{-\infty}^{+\infty} \rho \dot{v}_i(\boldsymbol{x}, \tau - t) \dot{u}_i(\boldsymbol{x}, t)\,\mathrm{d}t$$

Therefore, Eq. (2.3.7) holds, and consequently, Eq. (2.3.6) holds.

Betti's second reciprocity theorem is very important because, based on it, we can derive the displacement integral representation theorem and the reciprocity property of the Green's function in elastodynamics systems. Specifically, for Betti's second reciprocity theorem (2.3.6), we can consider the following scenario: What happens if we take two sets of external forces, where one set is the actual external forces in a physical problem, and the other set is a special external force? For instance, consider the first volume integral on the left side of Eq. (2.3.6). If the body force $\boldsymbol{f}(\boldsymbol{x}, t)$ is taken as a Dirac δ function in both space and time, we can directly express the result of this integral using the integral properties of the Dirac δ function, which allows us to formally express the displacement. This is precisely what we are about to investigate next. However, before we do so, let's examine the displacement field generated by this special external force, i.e., the Green's function of the elastodynamics equation.

---

[37] Strictly speaking, for the order of integration to be interchangeable, certain continuity conditions must be satisfied. In the physical problems we typically consider, we assume that these continuity conditions are always met.

2.4 Green's Function of Elastodynamics Equations

The Green's function is a significant concept in mathematical physics, first introduced by the British mathematical physicist G. Green in the early 19th century.[38] Different types of equations have their corresponding Green's functions; for example, the Laplace equation, the Poisson equation, the Helmholtz equation, and the wave equation all have their own Green's functions. Moreover, Green's functions are influenced by boundary conditions and/or initial conditions. Once a Green's function is determined, it can be used to construct the field generated by any distribution of sources, both in space and/or time. For elastodynamic problems, if we consider the elastodynamics equation along with its boundary conditions and/or initial conditions as a system, then, simply put, the Green's function for this elastodynamics system is the displacement field produced by a unit concentrated impulsive force.

2.4.1 *Introducing Green's Function for Elastodynamics Equations*

Consider a special type of body force, f_i, which is a unit concentrated impulsive force. *Concentrated* refers to its spatial distribution, meaning this force acts only at a single point $\boldsymbol{\xi}$ in space, along a specific coordinate axis. *Impulsive* refers to its temporal distribution, indicating that this force acts only at a single moment τ. Mathematically, it can be represented as

$$f_i(\boldsymbol{x}, t) = \delta_{in}\delta(\boldsymbol{x} - \boldsymbol{\xi})\,\delta(t - \tau) \tag{2.4.1}$$

The first term on the right side of the equation is the Kronecker delta, representing the direction of the force.[39] The second and third terms are the Dirac δ

[38] In his 1828 work, *An Essay on the Application of Mathematical Analysis to the Theories of Electricity and Magnetism*, he introduced the concept of Green's functions. As Guo explained, "The significance of this concept lies in the following reason: from a physical standpoint, a mathematical equation represents the relationship between a specific field and the source generating that field. The Green's function represents the field produced by a point source. Knowing the field of a point source allows us to calculate the field of any source using the superposition method." (Guo 1991, p. 348).

[39] At first glance, Eq. (2.4.1) seems problematic: the left side has only one subscript i, while the right side has two subscripts, i and n. In this context, n is not considered a free index but rather a predetermined number, taking values 1, 2, or 3, corresponding to the positive direction of a specific coordinate axis x_i. When the free index i equals this predetermined number, the term evaluates to 1; otherwise, it is zero. It is important to note that this interpretation is specific to this equation. For instances like the upcoming Eq. (2.4.2), where both sides have the subscript n, which is treated as a free index.

functions,[40] which indicate that the force acts at a single point $\boldsymbol{\xi}$ in space and at a single moment τ in time. The displacement field produced by the body force described in Eq. (2.4.1) is denoted as $G_{in}(\boldsymbol{x}, t; , \boldsymbol{\xi}, \tau)$, which is the Green's function for the elastodynamics equations. Here, the symbol G represents the Green's function; the first subscript i indicates the component of the displacement, and the second subscript n denotes the direction of the concentrated impulsive force. The variables inside the parentheses are divided into two groups: those before the semicolon represent the spatial and temporal coordinates of the field point, while those after the semicolon represent the spatial and temporal coordinates of the source point.

For a general elastic system, the elastodynamic equations that the Green's functions satisfies are (refer to Eq. (2.2.17))

$$\rho \ddot{G}_{in}(\boldsymbol{x}, t; \boldsymbol{\xi}, \tau) = \frac{\partial}{\partial x_j}\left[C_{ijpq}\frac{\partial}{\partial x_q}G_{pn}(\boldsymbol{x}, t; \boldsymbol{\xi}, \tau)\right] + \delta_{in}\delta(\boldsymbol{x}-\boldsymbol{\xi})\,\delta(t-\tau) \tag{2.4.2}$$

It is important to note that because the Green's function has two sets of spatial coordinates as variables, namely the coordinates of the field point $\boldsymbol{x}$ and the source point $\boldsymbol{\xi}$, special care must be taken when expressing the spatial derivatives of the Green's function, especially when using concise notations such as $G_{pn,q}(\boldsymbol{x}, t; , \boldsymbol{\xi}, \tau)$.[41] If not explicitly stated, this can lead to confusion. In Eq. (2.4.2), the δ_{in} on the right-hand side is a second-order tensor, hence the Green's function $\mathsf{G}(\boldsymbol{x}, t; \boldsymbol{\xi}, \tau)$ is also a second-order tensor, often referred to as the *Green's tensor*.[42]

---

[40] The δ function is a useful mathematical tool for describing quantities that are concentrated at a single point. It was initially introduced by the renowned British physicist P. Dirac in 1930 in his famous book *The Principles of Quantum Mechanics* as a "convenient notation". The δ function has widespread applications in modern physics. The traditional definition of the δ function is a function that satisfies the following conditions:

$$\delta(x) = \begin{cases} 0, & x \neq 0 \\ \infty, & x = 0 \end{cases}, \qquad \int_{-\infty}^{+\infty} \delta(x)\,\mathrm{d}x = 1$$

Many physical quantities, such as point mass, point charge, and instantaneous point source, can be conveniently and clearly described using the δ function. However, this singular function, which is infinite at one point and finite under integration, faced criticism from pure mathematicians since its inception. Soon after (in the 1930s), mathematicians developed the theory of *generalized functions* (or *distribution theory*). This theoretical framework allows for various algebraic and analytical operations on the δ function, much like ordinary functions.

[41] To avoid confusion, in this book, we use $G_{pn,q}$ and $G_{pn,q'}$ to respectively represent the derivatives of the Green's function with respect to the field point coordinates x_i and source point coordinates ξ_i (or x_i').

[42] Specifically, for the static case, it is called the Somigliana function or *Somigliana tensor*, named after the Italian mathematician Somigliana, who lived from the late 19th to the early 20th century.

2.4.2 Reciprocity of Green's Function

As previously mentioned, the Green's function depends on the boundary and initial conditions of the problem. The initial conditions are generally assumed to be

$$\mathsf{G}(\boldsymbol{x},\ t;\ \boldsymbol{\xi},\ \tau)\big|_{t=\tau} = \dot{\mathsf{G}}(\boldsymbol{x},\ t;\ \boldsymbol{\xi},\ \tau)\big|_{t=\tau} = 0$$

This indicates that at the moment of the concentrated impulse force at time τ and prior, both displacement and velocity are zero.[43] For typical problems, boundary conditions often remain constant over time. For instance, in the Lamb's problem we will explore later, the boundary conditions on the free surface always satisfy the condition of zero traction. Notably, the body force term in Eq. (2.4.2) includes the time-dependent term $\delta(t-\tau)$, which means the Green's function is solely dependent on the difference between the reception time at the field point and the action time at the source point, $t-\tau$, and is independent of the actual values of t and τ. Consequently, the origin of time can be arbitrarily shifted. Specifically, there is

$$G_{in}(\boldsymbol{x},\ t;\ \boldsymbol{\xi},\ \tau) = G_{in}(\boldsymbol{x},\ t-\tau;\ \boldsymbol{\xi},\ 0) = G_{in}(\boldsymbol{x},\ -\tau;\ \boldsymbol{\xi},\ -t)$$

This demonstrates the temporal reciprocity of the Green's function. As long as the boundary conditions are unchanged, the displacement field observed at time t resulting from an action at time τ, is identical to the displacement field observed at time $t-\tau$ resulting from an action at time 0, as well as the displacement field observed at time $-\tau$ resulting from an action at time $-t$.

The reciprocal nature of the Green's function extends to spatial relationships as well. Section 2.3 introduces a reciprocity theorem that connects two sets of external forces and their respective displacement fields. By considering two sets of body forces characterized by different points of application, different moments of application, and different directions of impulse forces, one can establish connections between two distinct Green's functions using the reciprocity theorem. If the two sets of body force are taken as

$$f_i(\boldsymbol{x},\ t) = \delta_{im}\delta\left(\boldsymbol{x}-\boldsymbol{\xi}_1\right)\delta\left(t-\tau_1\right),\quad g_i(\boldsymbol{x},\ t) = \delta_{in}\delta\left(\boldsymbol{x}-\boldsymbol{\xi}_2\right)\delta\left(t+\tau_2\right) \tag{2.4.3}$$

then the displacement fields they produce are respectively

$$u_i(\boldsymbol{x},\ t) = G_{im}(\boldsymbol{x},\ t;\ \boldsymbol{\xi}_1,\ \tau_1),\quad v_i(\boldsymbol{x},\ t) = G_{in}(\boldsymbol{x},\ t;\ \boldsymbol{\xi}_2,\ -\tau_2) \tag{2.4.4}$$

[43] The initial conditions, as outlined in Aki and Richards (2002), also include the condition $\boldsymbol{x} \neq \boldsymbol{\xi}$. This is necessary to avoid physical singularities, because the Green's function becomes undefined when $\boldsymbol{x} = \boldsymbol{\xi}$.

Substituting Eqs. (2.4.3) and (2.4.4) into Eq. (2.3.6), there is

$$\begin{aligned}
&\int_{-\infty}^{+\infty} \mathrm{d}t \iiint_V \delta_{im}\delta\left(\boldsymbol{x}-\boldsymbol{\xi}_1\right)\delta\left(t-\tau_1\right)G_{in}(\boldsymbol{x},\ \tau-t;\ \boldsymbol{\xi}_2,\ -\tau_2)\,\mathrm{d}V \\
&+\int_{-\infty}^{+\infty} \mathrm{d}t \oiint_S n_j C_{jipq}\frac{\partial}{\partial x_q}G_{pm}(\boldsymbol{x},\ t;\ \boldsymbol{\xi}_1,\ \tau_1)G_{in}(\boldsymbol{x},\ \tau-t;\ \boldsymbol{\xi}_2,\ -\tau_2)\,\mathrm{d}S \\
&=\int_{-\infty}^{+\infty} \mathrm{d}t \iiint_V \delta_{in}\delta\left(\boldsymbol{x}-\boldsymbol{\xi}_2\right)\delta\left(\tau-t+\tau_2\right)G_{im}(\boldsymbol{x},\ t;\ \boldsymbol{\xi}_1,\ \tau_1)\,\mathrm{d}V \\
&+\int_{-\infty}^{+\infty} \mathrm{d}t \oiint_S n_j C_{jipq}\frac{\partial}{\partial x_q}G_{pn}(\boldsymbol{x},\ \tau-t;\ \boldsymbol{\xi}_2,\ -\tau_2)G_{im}(\boldsymbol{x},\ t;\ \boldsymbol{\xi}_1,\ \tau_1)\,\mathrm{d}S
\end{aligned} \tag{2.4.5}$$

In the case where the boundary conditions of the problem are *homogeneous*, that is, the Green's function or the corresponding tractions are zero at the boundary, the integrands in the surface integral terms on both sides of the equation are identically zero. Consequently, the surface integrals themselves are zero, leaving only the volume integral terms. According to the properties of the Kronecker delta and the Dirac delta function as follows

$$\int_{-\infty}^{+\infty} \delta(t-\tau)f(t)\,\mathrm{d}t = f(\tau), \quad \iiint_V \delta\left(\boldsymbol{x}-\boldsymbol{\xi}\right)f(\boldsymbol{x})\,\mathrm{d}V(\boldsymbol{x}) = f(\boldsymbol{\xi}) \quad (\xi \in V)$$

we obtain

$$G_{mn}(\boldsymbol{\xi}_1,\ \tau-\tau_1;\ \boldsymbol{\xi}_2,\ -\tau_2) = G_{nm}(\boldsymbol{\xi}_2,\ \tau+\tau_2;\ \boldsymbol{\xi}_1,\ \tau_1) \tag{2.4.6}$$

Consider two special cases:

(1) When both τ_1 and τ_2 are zero, the relationship $G_{mn}(\boldsymbol{\xi}_1,\ \tau;\ \boldsymbol{\xi}_2,\ 0) = G_{nm}(\boldsymbol{\xi}_2,\ \tau;\ \boldsymbol{\xi}_1,\ 0)$ holds. This reflects that the subscripts are interchanged, swapping the spatial coordinates of the field and source points without changing the time component. It illustrates that the displacement field in the m-direction observed at point $\boldsymbol{\xi}_1$, caused by a concentrated impulse force in the n-direction at $\boldsymbol{\xi}_2$, is identical to the displacement field in the n-direction observed at $\boldsymbol{\xi}_2$, caused by a concentrated impulse force in the m-direction at $\boldsymbol{\xi}_1$. This demonstrates the spatial reciprocity of the Green's function.
(2) When $\tau = 0$, the equation $G_{mn}(\boldsymbol{\xi}_1,\ -\tau_1;\ \boldsymbol{\xi}_2,\ -\tau_2) = G_{nm}(\boldsymbol{\xi}_2,\ \tau_2;\ \boldsymbol{\xi}_1,\ \tau_1)$ establishes that the subscripts, spatial coordinates, and time components are all swapped. This denotes that the displacement field in the m-direction at point $\boldsymbol{\xi}_1$ observed at time $-\tau_1$, resulting from a concentrated impulse force in the n-direction at point $\boldsymbol{\xi}_2$ at time $-\tau_2$, is equivalent to the displacement field in the n-direction at point $\boldsymbol{\xi}_2$ observed at time τ_2, caused by a concentrated impulse force in the m-direction at point $\boldsymbol{\xi}_1$ at time τ_1. This equality exemplifies the spatiotemporal reciprocity of the Green's function.
The results of the above discussion are listed in Table 2.1.

Table 2.1 Reciprocity of Green's function

Category	Reciprocal relation	Boundary condition
Temporal reciprocity	$G_{in}(\boldsymbol{x},\ t;\ \boldsymbol{\xi},\ \tau) = G_{in}(\boldsymbol{x},\ t-\tau;\ \boldsymbol{\xi},\ 0)$	Time-invariant
Temporal reciprocity	$G_{in}(\boldsymbol{x},\ t;\ \boldsymbol{\xi},\ \tau) = G_{in}(\boldsymbol{x},\ -\tau;\ \boldsymbol{\xi},\ -t)$	Time-invariant
Spatial reciprocity	$G_{mn}(\boldsymbol{\xi}_1,\ \tau;\ \boldsymbol{\xi}_2,\ 0) = G_{nm}(\boldsymbol{\xi}_2,\ \tau;\ \boldsymbol{\xi}_1,\ 0)$	Homogeneous
Spatiotemporal reciprocity	$G_{mn}(\boldsymbol{\xi}_1,\ -\tau_1;\ \boldsymbol{\xi}_2,\ -\tau_2) = G_{nm}(\boldsymbol{\xi}_2,\ \tau_2;\ \boldsymbol{\xi}_1,\ \tau_1)$	Homogeneous

2.5 Integral Representation Theorem of Displacement

Building on the foundation of Betti's reciprocity theorem and the reciprocal properties of the Green's function, this chapter concludes with a significant principle in elastodynamics: the integral representation theorem of displacement.[44] This section specifically aims to calculate the displacement field at a point within an elastic body, $\boldsymbol{u}$, given the distribution of body forces, $\boldsymbol{f}$, within the elastic region V, and the boundary conditions on the surface S.[45] To achieve this, we will utilize the Green's function in conjunction with Betti's reciprocity theorem.

In the context of elastodynamics, Betti's second reciprocity theorem, expressed in Eq. (2.3.6), links two sets of external forces and the resulting displacement fields they create. Analyzing the reciprocal properties of Green's functions in Sect. 2.4 provides an insight: using one set of body forces as a concentrated impulse allows for the explicit expression of volume integral terms involving the Dirac delta function. Indeed, if the other set of body forces represents the actual forces in a problem, the

[44] In the introduction, we revisited the history of seismology and mentioned that the foundational integral representation theorem of displacement in elastodynamics was initially formulated by Volterra in 1894 for two-dimensional cases. This theorem was later extended to three-dimensional contexts by Love in 1904. The discovery of the displacement representation theorem laid the groundwork for the eventual development of the source representation theorem, which emerged six decades later.

[45] At first glance, this objective might seem ambitious since the study of elastodynamics is concerned with the characteristics of displacement and stress fields within an elastic body under external loads. However, the key term here is "representation"; we are merely formally expressing the integral representation of the displacement field. The specifics of calculating this integral are not our current focus.

concentrated impulse force will convolve with the real displacements.[46] The explicitly expressed terms precisely correspond to the displacement field of the problem.

Based on the above approach, we take the first set of body forces and displacements as $f_i(\boldsymbol{x}, t)$ and $u_i(\boldsymbol{x}, t)$, while the second set of body forces and displacements are respectively

$$g_i(\boldsymbol{x}, t) = \delta_{in}\delta\,(\boldsymbol{x} - \boldsymbol{\xi})\,\delta(t), \quad v_i(\boldsymbol{x}, t) = G_{in}(\boldsymbol{x}, t; \boldsymbol{\xi}, 0)$$

Since the time origin can be chosen arbitrarily without loss of generality, the moment of action for the concentrated impulse force is set at $\tau = 0$. By substituting the expressions for the two sets of body forces and displacements into Eq. (2.3.6), we obtain

$$\begin{aligned}
&\int_{-\infty}^{+\infty} \mathrm{d}t \iiint_V f_i(\boldsymbol{x}, t) G_{in}(\boldsymbol{x}, \tau - t; \boldsymbol{\xi}, 0)\,\mathrm{d}V(\boldsymbol{x}) \\
&\quad + \int_{-\infty}^{+\infty} \mathrm{d}t \oiint_S T_i\big(\boldsymbol{u}(\boldsymbol{x}, t), \hat{\boldsymbol{n}}\big) G_{in}(\boldsymbol{x}, \tau - t; \boldsymbol{\xi}, 0)\,\mathrm{d}S(\boldsymbol{x}) \\
&= \int_{-\infty}^{+\infty} \mathrm{d}t \iiint_V \delta_{in}\delta\,(\boldsymbol{x} - \boldsymbol{\xi})\,\delta(\tau - t) u_i(\boldsymbol{x}, t)\,\mathrm{d}V(\boldsymbol{x}) \\
&\quad + \int_{-\infty}^{+\infty} \mathrm{d}t \oiint_S n_j C_{jipq} \frac{\partial}{\partial x_q} G_{pn}(\boldsymbol{x}, \tau - t; \boldsymbol{\xi}, 0) u_i(\boldsymbol{x}, t)\,\mathrm{d}S(\boldsymbol{x})
\end{aligned}$$

This simplifies to

$$\begin{aligned}
u_n(\boldsymbol{\xi}, \tau) = &\int_{-\infty}^{+\infty} \mathrm{d}t \iiint_V f_i(\boldsymbol{x}, t) G_{in}(\boldsymbol{x}, \tau - t; \boldsymbol{\xi}, 0)\,\mathrm{d}V(\boldsymbol{x}) \\
&+ \int_{-\infty}^{+\infty} \mathrm{d}t \oiint_S \Big\{ T_i\big(\boldsymbol{u}(\boldsymbol{x}, t), \hat{\boldsymbol{n}}\big) G_{in}(\boldsymbol{x}, \tau - t; \boldsymbol{\xi}, 0) \\
&- n_j C_{jipq} \frac{\partial}{\partial x_q} G_{pn}(\boldsymbol{x}, \tau - t; \boldsymbol{\xi}, 0) u_i(\boldsymbol{x}, t) \Big\}\,\mathrm{d}S(\boldsymbol{x})
\end{aligned} \tag{2.5.1}$$

In this approach, we have explicitly represented the displacement field in integral form. However, we encounter a challenge in the physical interpretation: the variables of the displacement components on the left side of the equation are $\boldsymbol{\xi}$ and τ, which

[46] The definition of convolution is

$$f(\tau) * g(\tau) = \int_{-\infty}^{+\infty} f(t) g(\tau - t)\,\mathrm{d}t$$

By examining the integrals in Eq. (2.3.6), it is apparent that each term effectively involves convolution.

do not correspond to the spatial and temporal coordinates of the field point, which are actually $\boldsymbol{x}$ and t, respectively. To make this physically interpretable, we exchange $\boldsymbol{x}$ with $\boldsymbol{\xi}$ and τ with t in Eq. (2.5.1),[47] resulting in

$$
\boxed{\begin{aligned} u_n(\boldsymbol{x}, t) = & \int_{-\infty}^{+\infty} \mathrm{d}\tau \iiint_V f_i(\boldsymbol{\xi}, \tau) G_{in}(\boldsymbol{\xi}, t-\tau; \boldsymbol{x}, 0)\, \mathrm{d}V(\boldsymbol{\xi}) \\ & + \int_{-\infty}^{+\infty} \mathrm{d}\tau \oiint_S \Big\{ T_i\big(\boldsymbol{u}(\boldsymbol{\xi}, \tau), \hat{\boldsymbol{n}}\big) G_{in}(\boldsymbol{\xi}, t-\tau; \boldsymbol{x}, 0) \\ & - n_j C_{jipq} \frac{\partial}{\partial \xi_q} G_{pn}(\boldsymbol{\xi}, t-\tau; \boldsymbol{x}, 0) u_i(\boldsymbol{\xi}, \tau) \Big\}\, \mathrm{d}S(\boldsymbol{\xi}) \end{aligned}} \tag{2.5.2}
$$

Equation (2.5.2) illustrates that the displacement at a point $\boldsymbol{x}$ within an elastic body at time t is determined by several contributions: firstly, the contribution of body forces throughout the elastic region V (represented as a volume integral); secondly, the influence of stress vectors on the boundary S of the region (the first term in the area integral); and thirdly, the contribution of the displacement vectors themselves at the boundary S (the second term in the area integral). The term for body forces is the nonhomogeneous term in the elastodynamics equations, while stress and displacement vectors are dictated by boundary conditions depending on the specifics of the problem.[48] Notably, each contribution is weighted differently: the first two by the Green's function itself, and the last by the stress vector corresponding to the Green's function.[49] Typically, finding a Green's function for a problem is as complex as solving the problem itself, so in this sense, Eq. (2.5.2) merely offers a form of representation rather than new information for solving the displacement field. However, the value of Eq. (2.5.2) lies in transforming the problem of solving complex body force distributions in elastodynamics into a problem involving a specific body force (a concentrated impulse force). Once the solution–Green's function–is determined, it can be used to construct the solution for the actual problem at hand.

Upon close inspection of the Green's function in Eq. (2.5.2), it becomes apparent that the coordinates for the field point and the source point are reversed relative to the actual problem. For certain types of boundary conditions, this reversal can be addressed using the spatial reciprocity of the Green's function. From our discussion in Sect. 2.4 (see Table 2.1), we understand that the Green's function, under homogeneous boundary conditions, exhibits this reciprocity. There are two

[47] In the operation of exchanging coordinates, these quantities are merely regarded as ordinary mathematical symbols, without considering their physical meanings for the time being.

[48] If the boundary conditions of the problem are of the first type, specifying the displacement field at the boundary, then the stress vector remains unknown at the boundary. Conversely, if the boundary conditions are of the second type, specifying the stress vector at the boundary, then the displacement vector at the boundary is unknown. Generally, Eq. (2.5.2) forms an integral equation where the function under the integral sign is also unknown.

[49] Sometimes referred to as the *stress Green's function*, it is essentially the spatial derivative of the Green's function. It represents both the stress induced by a point force and can also be considered as the displacement caused by a couple, which will be discussed in detail in Sect. 3.3.

scenarios for homogeneous conditions: one where the Green's function itself is zero at the boundary, corresponding to rigid boundary conditions, and another where the traction force associated with the Green's function is zero, corresponding to free boundary conditions.[50] Both scenarios are explored further below.

(1) The Green's function satisfies rigid boundary conditions, namely

$$G_{in}(\boldsymbol{\xi},\, t-\tau;\, \boldsymbol{x},\, 0)\big|_S = 0$$

At this point, given the spatial reciprocity property (2.4.6), we have

$$G_{in}(\boldsymbol{\xi},\, t-\tau;\, \boldsymbol{x},\, 0) = G_{ni}(\boldsymbol{x},\, t-\tau;\, \boldsymbol{\xi},\, 0)$$

Equation (2.5.2) can thus be written as

$$\begin{aligned} u_n(\boldsymbol{x},t) &= \int_{-\infty}^{+\infty} \mathrm{d}\tau \iiint_V f_i(\boldsymbol{\xi},\, \tau) G_{ni}(\boldsymbol{x},\, t-\tau;\, \boldsymbol{\xi},\, 0)\, \mathrm{d}V(\boldsymbol{\xi}) \\ &\quad - \int_{-\infty}^{+\infty} \mathrm{d}\tau \oiint_S n_j C_{jipq} \frac{\partial}{\partial \xi_q} G_{pn}(\boldsymbol{\xi},\, t-\tau;\, \boldsymbol{x},\, 0) u_i(\boldsymbol{\xi},\, \tau)\, \mathrm{d}S(\boldsymbol{\xi}) \end{aligned} \tag{2.5.3}$$

(2) The Green's function satisfies free boundary conditions, namely

$$n_j C_{jipq} \frac{\partial}{\partial \xi_q} G_{pn}(\boldsymbol{\xi},\, t-\tau;\, \boldsymbol{x},\, 0)\big|_S = 0$$

Similarly, based on the spatial reciprocity property, we have

$$G_{in}(\boldsymbol{\xi},\, t-\tau;\, \boldsymbol{x},\, 0) = G_{ni}(\boldsymbol{x},\, t-\tau;\, \boldsymbol{\xi},\, 0)$$

Equation (2.5.2) is now in the form

$$\begin{aligned} u_n(\boldsymbol{x},t) &= \int_{-\infty}^{+\infty} \mathrm{d}\tau \iiint_V f_i(\boldsymbol{\xi},\, \tau) G_{ni}(\boldsymbol{x},\, t-\tau;\, \boldsymbol{\xi},\, 0)\, \mathrm{d}V(\boldsymbol{\xi}) \\ &\quad + \int_{-\infty}^{+\infty} \mathrm{d}\tau \oiint_S T_i\big(\boldsymbol{u}(\boldsymbol{\xi},\, \tau),\, \hat{\boldsymbol{n}}\big) G_{in}(\boldsymbol{\xi},\, t-\tau;\, \boldsymbol{x},\, 0)\, \mathrm{d}S(\boldsymbol{\xi}) \end{aligned} \tag{2.5.4}$$

[50] A typical example is the free surface (i.e., surface of the Earth), which is of primary interest in seismology since most earthquake recordings are obtained at this surface where the traction is zero. In practical applications, it is first necessary to express the stress at the free surface in terms of displacement.

The expression for the displacement integral representation of the Green's function under two different boundary conditions is respectively given by Eqs. (2.5.3) and (2.5.4).[51]

In essence, the integral representation theorem for displacement formally establishes the relationship between the external loading and the displacement as a response in the elastic body, with the Green's function serving as the bridge connecting the two. This representation is a sum of contributions from body forces within the material, as well as those of stress and displacement vectors on the boundary. Typically, the distribution of body forces is specified, while the boundary conditions determine whether stress or displacement vectors are used on the boundary. It is apparent that the key to solving the entire problem lies in determining the Green's function; if it can be calculated through some means, then the displacement field in the elastic body can be conveniently determined using the theorem. This integral representation theorem for displacement also serves as the foundation for the source representation theorem, which provides the framework of entire theoretical seismology.

2.6 Summary

In this chapter, we start with basic mathematical foundations and review key concepts in elastodynamics, such as strain and stress. We then derive the equations of elastodynamics from the laws of Newtonian mechanics, forming a complete set of equations together with geometric equations and constitutive relations. To solve this set, we use the displacement method, first reformulating the equations solely in terms of displacement vectors. Further, by applying Betti's reciprocal theorem and introducing the Green's function of elastic mechanics, we derive the displacement representation theorem. This theorem explicitly expresses the displacement as a weighted integral of body forces inside the elastic body and displacements and tractions on its surface,

---

[51] Aki and Richards (2002, p.29) note a seeming contradiction between two equations as they represent different integral terms: Eq. (2.5.3) suggests the displacement field depends on displacements specified on the boundary, whereas Eq. (2.5.4) indicates it depends on the traction forces there. This "contradiction" can be resolved by recognizing that at a point on the surface S, it is not possible to specify both displacement and traction simultaneously. Specifying the displacement makes the solution unique, with the traction on the boundary determined by calculation and not arbitrarily set–doing so would lead to inconsistencies. Conversely, if traction is specified, then the displacement values are fixed and should not be arbitrarily adjusted. The choice between using Eqs. (2.5.3) or (2.5.4) in practical problems depends on the boundary conditions associated with the body forces generating the Green's function. For rigid boundary conditions, Eq. (2.5.3) is used. If these forces relate to first-type boundary conditions, the problem is straightforward: displacement values on the boundary are directly substituted. However, if they relate to second-type conditions, the situation is more complex as the boundary displacements are unknown, making Eq. (2.5.3) an integral equation. The same analysis applies when the body forces are associated with free boundary conditions.

with the weights being the Green's function or its spatial derivatives. Thus, solving the displacement response of the elastic body is transformed into solving for the Green's function; once we obtain it, the displacement field for any given body forces and boundary conditions can be determined using the displacement representation theorem.

Although this chapter might not seem directly related to seismology, it lays the groundwork for the source representation theorem we will introduce in Chap. 3. In seismological problems, the source often behaves uniquely, approximating two closely adhering surfaces that slip under certain stress conditions. Based on the displacement representation theorem, deriving the integral representation theorem for such specialized sources is the objective of Chap. 3.

Chapter 3
Theory of Seismic Source Representation

In Chap. 2, we have established the displacement representation theorem within the framework of elastodynamics, which universally applies to how elastic bodies, as continuous media, respond to external loads. Building on this foundation, in this chapter, we will further exploring the displacement representation theorem for elastic bodies with dislocations occurring on faults. This is crucial for representing the displacement fields generated by simplified seismic source models. Special consideration is necessary because elastic bodies containing faults no longer meet the conditions of displacement being single-valued and continuous, typical of a continuous medium.

In this chapter, after briefly reviewing the history of seismic source theory, we first delve into the source representation theorem. This includes how it derives from the displacement representation theorem introduced in Chap. 2, its physical significance, and its applications. We then cover two key concepts closely associated with the source representation theorem: equivalent body forces and the seismic moment tensor. An important seismic principle that *the displacement field created by a shear dislocation is equivalent to that generated by a pair of force couples without moment* will be proved. And additionally, the physical meaning and mathematical formulation of the seismic moment tensor will be discussed.

H. Zhang, *Lamb's Problem in Seismology (Volume 1)*,
https://doi.org/10.1007/978-981-95-3170-7_3

3.1 A Brief History of Seismic Source Theory

In contrast to many scientific and engineering problems where direct observation is possible, the source of earthquakes cannot be directly observed[1]; it is understood only through indirect means, leading to a lengthy process of human comprehension. It wasn't until the 1960s that what is now known as the "first principle of modern seismology", the source representation theorem, was established. In Chap. 1, advancements in the physics of earthquake sources since the 1960s are discussed, which are largely based on the source representation theorem developed between 1963 and 1964. To better understand the context in which this theorem was formulated, we begin with a brief review of the history of source theory.

It wouldn't be an exaggeration to say that the history of human development has progressed alongside the history of earthquakes. Thousands of years ago, people began to explore the causes of earthquakes. In ancient cultures, earthquakes were often seen as manifestations of divine anger. However, ancient civilizations like Greece and China offered natural explanations for earthquakes, generally attributing them to the accumulation of gases (Agnew 2002). This understanding persisted until the 1755 Lisbon earthquake in Portugal, which marked the first clear realization that surface tremors are related to waves traveling through the Earth's medium, a topic already mentioned in Chap. 1. However, the advancement in understanding the sources of earthquakes progressed slowly until the early 20th century. A significant breakthrough came with the destructive 1906 San Francisco earthquake, after which observations revealed that the San Andreas Fault had moved horizontally by up to 6.5 m at some locations. This event led H. Reid to propose the *elastic rebound theory* around 1910,[2] recognizing that ground movement can occur both on and far from fault.

The elastic rebound theory explains how fault ruptures cause surface shaking, and since then, seismologists are keen on linking seismic wave records observed during earthquakes to the actual events at the source. The initial movements of P-waves are distinctively identifiable, leading many seismologists to derive characteristics

---

[1] Seismologists face significant challenges in directly observing earthquake sources due to the inaccessibility of the Earth's interior; they cannot delve deeper than about 10 km below the surface. Additionally, from a practical standpoint, the unpredictability of when and where earthquakes will occur makes it difficult to strategically place observation instruments on fault in advance to directly monitor the seismic process.

[2] Initially referred to as the *elastic rebound hypothesis*, this concept was widely accepted by the seismological community as the elastic rebound theory after being validated through practical experiments. This theory posits that rock layers near fault lines deform due to tectonic movements, storing energy in the form of elastic strain within the rocks. This continues until the deformation accumulates beyond the rocks' capacity to withstand it, leading to a fault slip. Because rocks are elastic, when they break, the deformed rocks snap back to their original shape once the stress is relieved. During this process, the stored elastic strain energy is released–some of it overcoming friction on the fault surfaces and converting into heat, some breaking molecular bonds causing the rocks to fracture, and some transforming into the elastic wave energy that causes ground shaking. This is one of the earliest and most widely applied theories regarding the causes of earthquakes.

of the earthquake source from these observations. With its dense network of seismic stations, Japanese seismologists were pioneers in this field. In 1923, after the devastating Kanto earthquake,[3] Nakano linked the elastic rebound theory with the quadrantal distribution of P-wave first motions, introducing the single couple source model.[4] However, this model had a resultant couple moment, indicating an imbalance of torque in the source region. In 1962, Honda proposed the double couple model to address this imbalance. Both models were the dominant theories proposed by seismologists for earthquake sources at that time. Post World War II, under H. Honda's leadership, Japanese research favored the double couple model, while researchers in the USA and USSR advocated for the single couple model. Theoretically, observations should distinguish between these models, but their P-wave radiation patterns were similar. Despite significant differences in their S-wave patterns, the data quality at the time was insufficient for analyzing these patterns, making it impossible to definitively determine the correct model based solely on observations available then.

While seismologists debated between two earthquake source models, significant theoretical advancements were brewing in the background. The 1950s saw the emergence of fault theory[5] as a pivotal tool for addressing earthquake sources. Steketee, in 1958, laid the groundwork for this theory by exploring the relationship between earthquakes and faults under static conditions. She conceptualized a fault Σ created by a cut in an elastic body, with its two sides designated as Σ^+ and Σ^-. Under stress, these faces would shift, creating discontinuities in displacement that serve as the origin for displacement elsewhere in the medium. This concept later facilitated the development of earthquake source representation theorems. Knopoff and Gilbert (1959, 1960) further explored these ideas under dynamic conditions. By the early 1960s, foundational theorems for earthquake source representation were established. Maruyama (1963) and Burridge and Knopoff (1964) investigated the body force equivalence of dynamic seismic dislocations in special cases involving homogeneous, isotropic media and the general cases of inhomogeneous, anisotropic media, respectively. They theoretically demonstrated that the displacement fields generated

[3] This earthquake, with a magnitude of 8.1, ranks among the largest seismic disasters of the 20th century. The loss of life and property damage resulting from the earthquake and its secondary disasters, particularly fires, were unprecedented. Japan learned a harsh lesson from this tragedy, which had a profound impact on the country's disaster prevention efforts. For instance, the renowned Earthquake Research Institute at the University of Tokyo was established in 1925, two years after the Kanto earthquake, as a direct consequence of these events.

[4] A pair of forces of equal magnitude but opposite direction, acting on two closely spaced parallel lines, is known as a *couple*.

[5] Also known as dislocation theory, originating from the work of the Italian mathematician Volterra in 1907, who was the first to use the term "distorsione" for dislocation. The term "fault" was first introduced by Love in 1927.

by shear dislocation sources are equivalent to those produced by double couples.[6] The publication of these two papers brought an end to a 40-year debate over single and double couple models.

The establishment of the earthquake source representation theorems marked the beginning of a golden age in the study of earthquake sources. These theorems posited that if the dislocations (i.e., displacement discontinuities) on a fault surface are known functions of space and time, and the geometry of the fault is also known, then the movement of the entire Earth's medium caused by the fault slip can be determined. This means that the dislocations on the fault surface can explain the observed seismic motions. Stress is not considered here, so this falls within the realm of kinematics. Accordingly, this approach to describing earthquake sources using fault dislocations is known as the *dislocation model*. In this model, to quantitatively describe the earthquake source, various parameters are typically used, such as the fault's length and width, rupture velocity, slip amount, and rise time.[7] Based on this, research can be conducted on the properties of near-field and far-field seismic waves generated by different rupture models, exemplified by Haskell's model (Haskell 1964), which includes unilateral and bilateral rupture of rectangular faults, as well as other models like circular faults. Dislocation models help understand how the finiteness and geometric characteristics of faults affect seismic wave radiation and were widely used in the late 1960s and 1970s to interpret seismic records. However, it is important to note that in the dislocation model, dislocations are often specified in a somewhat arbitrary manner, such as unilateral or bilateral rupture, and in many cases, these specified dislocations may not meet both the governing equations and fracture criteria.

A natural correction to the flaws mentioned above involves replacing arbitrarily specified dislocations with dislocation histories calculated based on the physical conditions (including boundary and initial conditions) on the fault surface. This model, contrasts with the dislocation model, often called the *crack model*, in which the stress field near the fault surface is involved, hence falls within the realm of dynamics. The application of rock fracture dynamics to earthquake source problems was first introduced by Kostrov (1966), who studied the spontaneous propagation of anti-plane

[6] While Burridge and Knopoff's (1964) primary objective was to demonstrate the equivalence of body forces, the significance and implications of their paper extend far beyond this conclusion. It not only established the earthquake source representation theorems, which later became the theoretical foundation for the study of seismic wave propagation and earthquake source physics, but also provided a basis for defining key parameters that characterize earthquake sources. For instance, based on this paper, K. Aki derived the famous seismic moment formula$M_0 = \mu\overline{[u]}S$ in 1966, where M_0 represents the scalar seismic moment, μ is the shear modulus, $\overline{[u]}$ is the average fault displacement, and S is the fault area.

[7] The time it takes for a point on the fault to start and complete its slip.

cracks.[8] In the fracture model, the fault's initial stress and strength distribution, along with the laws governing how stress changes with slip (known as fracture or friction criteria[9]), collectively dictate the rupture behavior on the fault surface, including its initiation, acceleration, and cessation. Due to the complex mathematical handling involved, many studies in the late 1960s and 1970s focused on addressing fixed rupture velocity in crack propagation and simple two-dimensional spontaneous propagation issues. Analytical solutions for three-dimensional spontaneous rupture problems are challenging to obtain and are typically resolved through numerical simulation methods.

Since the late 1970s, significant advances have been made in both earthquake source kinematics and dynamics, driven by improvements in computational capabilities and an increase in digital seismic recordings. In the realm of earthquake source kinematics, seismologists utilize various types of data, including far-field and near-field seismic data, along with GPS records, to employ different inversion techniques. This approach has enabled the detailed mapping of earthquake rupture processes globally (e.g., Kikuchi and Kanamori 1982, 1991), providing valuable insights into the mechanics of earthquake sources. Notably, after major earthquakes, when transportation and communication are often disrupted, rapid kinematic inversions of seismic data can provide essential intensity distributions within disaster zones, critically aiding emergency response efforts. In the field of earthquake source dynamics, researchers explore the physical mechanisms of earthquake sources from multiple perspectives, seeking to explain the complexities of earthquake genesis. For example, in the late 1970s and early 1980s, Das and Aki (1977), and Kanamori and Stewart (1978) introduced the barrier model and the asperity model, respectively, to describe the heterogeneity in fault strength and stress distribution. Furthermore, the complexity of fault geometry and the coupling effects between surface processes and fault ruptures have led to intricate patterns in earthquake rupture imaging, which are central concerns in earthquake source dynamics (Tada and Yamashita 1997; Aochi et al. 2000; Zhang and Chen 2006). A primary goal within this field is to determine the initial stress distribution in study areas through inversion. Since dynamic forward modeling is time-consuming and such inversion studies have only begun in the past two decades. As computational power continues to improve, it is anticipated that deeper and more frequent dynamic inversion studies will enable seismologists

[8] The term *anti-plane* refers to cracks of type Ⅲ in fracture mechanics, where the direction of crack propagation is perpendicular to that of displacement. Another scenario involves *in-plane* cracks, corresponding to type Ⅱ cracks in fracture mechanics, where the crack propagation direction aligns with the displacement direction. Anti-plane issues are relatively simpler to study as they involve displacement in only one direction. In contrast, in-plane problems involve coupled displacement components in two directions, making them significantly more complex. Spontaneous propagation, as opposed to fixed rupture velocity propagation, means that the extension of the crack is not predetermined but determined by various physical conditions.

[9] Typically, this requires determination based on rock mechanics experiments. The common fracture criteria usually used in numerically modelling include the slip-weakening friction criterion, the rate-weakening friction criterion, and the rate-state dependent friction criterion. In studies of coseismic rupture issues, the slip-weakening friction criterion is most often used.

to map stress fields in seismic regions more accurately, significantly contributing to earthquake prediction efforts.

Reflecting on the historical understanding of earthquake sources, it's evident that over the past century, our grasp has evolved significantly. Initially, we relied on the rudimentary elastic rebound model developed shortly after the 1906 San Francisco earthquake to describe seismic sources. Today, we combine diverse seismic and geodetic data through kinematic inversion to obtain detailed processes of earthquake ruptures. Additionally, we use dynamic forward modeling to explore how various physical and geometric factors influence these processes. A pivotal milestone in this journey has been the source representation theorem, established over half a century ago, marking a significant leap in our understanding of earthquake sources.

3.2 Representation Theorem of Earthquake Sources

3.2.1 Models and Simplifying Assumptions

From a practical perspective, earthquake sources are spatially distributed across a region. Under tectonic stress, two sides of a fault move relative to each other. When focusing on the source zone, one may finds a series of complex physical processes. Due to compression and friction, rocks fracture, generating heat from friction and allowing fluid intrusion. A typical size of such area ranges from centimeters to meters.[10] Even disregarding factors like temperature and fluids, since significant deformations exist in the source region, it is clear that the complex phenomena occurring here cannot be fully described by mere elastodynamics equations.

Looking at the earthquake source zone from a slightly broader perspective, it can be viewed as a fault closely bonded through a geometrical plane, where displacement discontinuities occur between the two contacting surfaces. Based on this fault model, further exploration into how earthquake source issues can be analyzed within the realm of continuum mechanics becomes possible.[11] The complex inelastic behaviors within the source zone can effectively be incorporated into a specific physical quantity (the seismic moment tensor), which will be discussed in detail in Sect. 3.4.2. This allows the problem to still be considered under the model of a continuous medium.

[10] The thickness of the fault fracture zone generated by seismic activity is related to the amount of slip, with their logarithms approximately following a linear relationship (Scholz 1987). For a single earthquake, the slip is generally on the order of meters, corresponding to a fracture zone thickness ranging from centimeters to meters. Over longer timescales, repeated earthquakes along the same fault plane accumulate more significant slip, resulting in a thicker fracture zone. For instance, on the San Andreas Fault in North America, accumulated slip over tens of kilometers corresponds to fracture zone thicknesses on the scale of hundreds of meters.

[11] Choosing the right model is crucial; it must not be overly simplistic or too complex. On one hand, the model should capture the primary complexities of the physical problem, which often involves validating whether theoretical predictions based on the model align with actual measurements. On the other hand, the model needs to be simple enough for us to manage effectively.

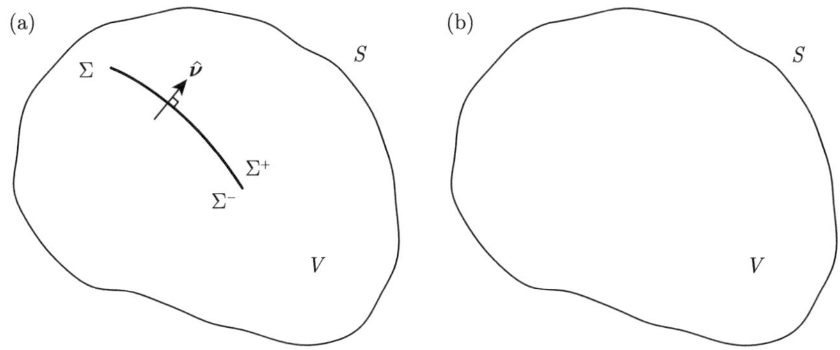

Fig. 3.1 Schematic depiction of fault model. **a** Model I, which addresses real-world issues, Σ represents a fault comprised closely of surfaces Σ^+ and Σ^-, with the normal direction denoted by $\hat{\boldsymbol{\nu}}$, pointing from Σ^- to Σ^+. **b** Model II, related to the Green's function. Both Model I and II feature a same outer boundary S and occupies the same space V

The model we are considering here, referred to as Model I, is illustrated in Fig. 3.1a. The fault plane, Σ, is situated within the elastic body V bordered by S, with Σ^+ and Σ^- representing the upper and lower surfaces of the fault,[12] respectively, and $\hat{\boldsymbol{\nu}}$ indicating the normal direction from the lower to the upper surface. To derive the source representation theorem from the displacement representation theorem, we introduce another model, named Model II, shown in Fig. 3.1b. This corresponds to a space where the Green's function is generated by impulse forces, with the external boundary S and volume V identical to those in Model I, the only difference being the absence of the fault plane.[13] Thus, in Model II, the following equation is applicable,

$$G_{in}\big|_{\Sigma^+} = G_{in}\big|_{\Sigma^-} \tag{3.2.1}$$

In Model II, $G_{in}\big|_{\Sigma^+}$ and $G_{in}\big|_{\Sigma^-}$ represent the values of the Green's functions at two infinitesimally close points, whose positions correspond to the upper and lower surfaces of the fault plane in model I, respectively.

---

[12] In this context, "upper" and "lower" are arbitrary designations typically associated with the boundaries of the problem. For instance, in some scenarios, the free surface forms part of S, and for faults that are not exactly perpendicular to the ground surface, the side closer to the ground surface is defined as "upper", while the opposite side is termed "lower".

[13] The introduction of Model II serves a specific purpose. Recalling the reciprocity theorem discussed in Chap. 2, we dealt with two sets of forces acting within two elastic spaces, V_1 and V_2. Furthermore, our explanation of Betti's first reciprocity theorem emphasized that the spaces where these forces act do not need to be strictly identical; the reciprocity theorem holds only within their intersection. In Model II, the spatial region V bounded by S constitutes a homogeneous body. This condition is crucial for deriving Eq. (3.2.1), which is a necessary requirement for obtaining the source representation theorem. If the Green's function were retained from Model I, this relationship would not hold true due to the discontinuity in displacement across the fault plane.

To facilitate analysis, we also need to make two simplifying assumptions as follows:

(1) Change in body force[14] can be neglected. In seismological studies, the body forces generally refer to gravity. This assumption implies that the effects of changes in gravity before and after an earthquake can be neglected.[15] Consequently, the first volumetric integral term on the right-hand side of the displacement representation theorem (2.5.2) can be omitted.
(2) In Models I and II, the terms u_i and G_{in} each satisfy the homogeneous boundary conditions on the external boundary S. A direct consequence is that the complex area integrals over S in Eq. (2.5.2) can be omitted, which is precisely why this assumption was introduced. However, this is a stringent condition and not just any chosen S will meet this criterion. Rather than viewing this as merely an "assumption", it's more accurate to consider it a deliberate choice of a suitable S for a specific problem, ensuring the condition is met.[16]

3.2.2 Derivation of Representation Theorem

We aim to analyze the displacement field caused by the discontinuity across fault Σ in Model I, as depicted in Fig. 3.1a, starting from the displacement representation theorem (2.5.2). However, a significant challenge arises initially: the elastic medium V, enclosed by the external boundary S in Model I, does not represent a continuous medium due to the discontinuous displacements across Σ, which do not meet the continuity conditions required by continuum mechanics. Consequently, Eq. (2.5.2) is not applicable as it stands.

---

[14] In seismology, the term "body force" actually refers to difference in body force before and after an earthquake. Stress, displacement, and body force all meet the requirements of Eq. (2.2.17) both before and after an earthquake. Subtracting these equations yields the formulas for changes in stress, displacement, and body forces, which also conform to the format of Eq. (2.2.17). The seismic studies we engage with typically derive the relative changes in displacement and corresponding stress changes that occur due to an earthquake, with the body force term in the equation representing the change in these forces. Accordingly, all quantities mentioned in the displacement representation theorem are these changes. In seismic issues, unless specifically dealing with dynamics at the earthquake source, such as friction criteria, the "stress", "displacement", and "body force" discussed are understood as changes, and this will not be reiterated further.

[15] It has been recognized since the late 19th century that gravity's impact is generally negligible in most cases, as noted in Bromwich (1894). However, its effects cannot be ignored when considering long-period seismic waves in global models.

[16] For instance, when addressing problems within an elastic half-space containing faults, we can define $S = S_0 + S_\infty$, where S_0 is the free surface and S_∞ represents the boundary at infinity. On S_0, both u_i and G_{in} meet the homogeneous boundary conditions with zero traction, while on S_∞, they satisfy the conditions of being zero themselves. This selection of S appropriately fulfills the requirements of the assumption.

Fault theory offers a fresh perspective for addressing this issue. By considering the fault surfaces as *internal boundaries*, and focusing on the region bounded by $S + \Sigma^+ + \Sigma^-$, the discontinuities in displacement that violate the conditions of continuum mechanics at the fault's upper and lower surfaces are reinterpreted as boundary conditions of the region. This region is still regarded as a continuous medium. Therefore, Eq. (2.5.2) remains valid; however, the area integral on the right-hand side of the equation now extends over $S + \Sigma^+ + \Sigma^-$, thus we have

$$\begin{aligned} u_n(\boldsymbol{x}, t) &= \int_{-\infty}^{+\infty} \mathrm{d}\tau \iiint_V f_i(\boldsymbol{\xi}, \tau) G_{in}(\boldsymbol{\xi}, t-\tau; \boldsymbol{x}, 0)\, \mathrm{d}V(\boldsymbol{\xi}) \\ &+ \int_{-\infty}^{+\infty} \mathrm{d}\tau \oiint_{S+\Sigma^+ + \Sigma^-} \Big\{ T_i\big(\boldsymbol{u}(\boldsymbol{\xi}, \tau), \hat{\boldsymbol{\nu}}\big) G_{in}(\boldsymbol{\xi}, t-\tau; \boldsymbol{x}, 0) \\ &- u_i(\boldsymbol{\xi}, \tau) \nu_j(\boldsymbol{\xi}) C_{ijpq}(\boldsymbol{\xi}) \frac{\partial}{\partial \xi_q} G_{pn}(\boldsymbol{\xi}, t-\tau; \boldsymbol{x}, 0) \Big\}\, \mathrm{d}S(\boldsymbol{\xi}) \end{aligned} \tag{3.2.2}$$

Under the simplifying assumption (1), the volume integral equals zero. Additionally, it's important to note that the integrands of the two area integrals involve cross-multiplications of displacements (or tractions) from Model I with tractions (or displacements) from Model II. As previously mentioned, homogeneous boundary conditions imply that either displacements or tractions are zero at the boundary. This means that, irrespective of which type of homogeneous boundary condition is met, both area integrals over S equal zero according to simplifying assumption (2). Furthermore, under these boundary conditions, the Green's function maintains spatial reciprocity, as stated in Eq. (2.4.6),

$$G_{in}(\boldsymbol{\xi}, t-\tau; \boldsymbol{x}, 0) = G_{ni}(\boldsymbol{x}, t-\tau; \boldsymbol{\xi}, 0) \tag{3.2.3}$$

Equation (3.2.2) can therefore be simplified as

$$\begin{aligned} u_n(\boldsymbol{x}, t) = \int_{-\infty}^{+\infty} \mathrm{d}\tau \oiint_{\Sigma^+ + \Sigma^-} &\Big\{ T_i\big(\boldsymbol{u}(\boldsymbol{\xi}, \tau), \hat{\boldsymbol{\nu}}\big) G_{ni}(\boldsymbol{x}, t-\tau; \boldsymbol{\xi}, 0) \\ &- u_i(\boldsymbol{\xi}, \tau) \nu_j(\boldsymbol{\xi}) C_{ijpq}(\boldsymbol{\xi}) \frac{\partial}{\partial \xi_q} G_{np}(\boldsymbol{x}, t-\tau; \boldsymbol{\xi}, 0) \Big\}\, \mathrm{d}\Sigma(\boldsymbol{\xi}) \end{aligned}$$

To facilitate the analysis, we separate the integrals over the upper and lower surfaces of the fault, denoting the normal directions of these surfaces with ν_j^+ and ν_j^-, respectively, as illustrated in Fig. 3.2. Consequently, the previous equation is rewritten as follows

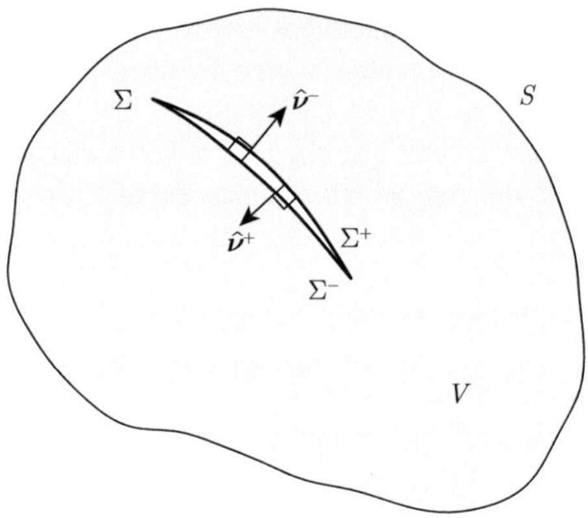

Fig. 3.2 Normal directions of the upper and lower surfaces of the fault. The fault Σ consists of the upper and lower surfaces, Σ^+ and Σ^-, with their normal directions denoted as $\hat{\boldsymbol{\nu}}^+$ and $\hat{\boldsymbol{\nu}}^-$, respectively. To enhance visibility, the upper and lower surfaces of the fault are separated by a certain distance. The external boundary is labeled S, and the gray area represents the region V enclosed by both the internal and external boundaries

$$\begin{aligned}u_n(\boldsymbol{x},t) = &\int_{-\infty}^{+\infty} \mathrm{d}\tau \iint_{\Sigma^+} \Big\{ T_i\big(\boldsymbol{u}(\boldsymbol{\xi},\tau),\hat{\boldsymbol{\nu}}^+\big) G_{ni}(\boldsymbol{x},t-\tau;\boldsymbol{\xi},0) \\ &- u_i(\boldsymbol{\xi},\tau)\nu_j^+(\boldsymbol{\xi}) C_{ijpq}(\boldsymbol{\xi}) \frac{\partial}{\partial \xi_q} G_{np}(\boldsymbol{x},t-\tau;\boldsymbol{\xi},0) \Big\} \mathrm{d}\Sigma(\boldsymbol{\xi}) \\ &+ \int_{-\infty}^{+\infty} \mathrm{d}\tau \iint_{\Sigma^-} \Big\{ T_i\big(\boldsymbol{u}(\boldsymbol{\xi},\tau),\hat{\boldsymbol{\nu}}^-\big) G_{ni}(\boldsymbol{x},t-\tau;\boldsymbol{\xi},0) \\ &- u_i(\boldsymbol{\xi},\tau)\nu_j^-(\boldsymbol{\xi}) C_{ijpq}(\boldsymbol{\xi}) \frac{\partial}{\partial \xi_q} G_{np}(\boldsymbol{x},t-\tau;\boldsymbol{\xi},0) \Big\} \mathrm{d}\Sigma(\boldsymbol{\xi}) \end{aligned} \quad (3.2.4)$$

It's important to recognize that the tractions $T_i\big(\boldsymbol{u}(\boldsymbol{\xi},\tau),\hat{\boldsymbol{\nu}}^+\big)\big|_{\Sigma^+}$ and $T_i\big(\boldsymbol{u}(\boldsymbol{\xi},\tau),\hat{\boldsymbol{\nu}}^-\big)\big|_{\Sigma^-}$ on the upper and lower surfaces respectively act as a pair of reaction forces, thus

$$T_i\big(\boldsymbol{u}(\boldsymbol{\xi},\tau),\hat{\boldsymbol{\nu}}^+\big)\big|_{\Sigma^+} = -T_i\big(\boldsymbol{u}(\boldsymbol{\xi},\tau),\hat{\boldsymbol{\nu}}^-\big)\big|_{\Sigma^-}$$

Additionally, according to Eq. (3.2.1), the first terms in the two area integrals can cancel each other out. Given that the outward normal directions on the upper and lower surfaces of the fault are opposite, if we let

$$\nu_j(\boldsymbol{\xi}) \triangleq \nu_j^-(\boldsymbol{\xi}) = -\nu_j^+(\boldsymbol{\xi}), \quad \big[u_i(\boldsymbol{\xi},\tau)\big] \triangleq u_i(\boldsymbol{\xi},\tau)\big|_{\Sigma^+} - u_i(\boldsymbol{\xi},\tau)\big|_{\Sigma^-}$$

Ultimately, Eq. (3.2.4) can be simplified to

$$\boxed{u_n(\boldsymbol{x}, t) = \int_{-\infty}^{+\infty} \mathrm{d}\tau \iint_{\Sigma} \left[u_i(\boldsymbol{\xi}, \tau)\right] \nu_j(\boldsymbol{\xi}) C_{ijpq}(\boldsymbol{\xi}) G_{np,q'}(\boldsymbol{x}, t-\tau; \boldsymbol{\xi}, 0)\, \mathrm{d}\Sigma(\boldsymbol{\xi})} \tag{3.2.5}$$

where $G_{np,q'} = \partial G_{np}/\partial \xi_q$. Equation (3.2.5) is the *representation theorem of earthquake sources*, which indicates that the displacement at any point within the elastic region V, containing a fault, can be expressed as a weighted integral of the slip function $\left[u_i(\boldsymbol{\xi}, \tau)\right]$ on the fault plane. The weights include the medium parameters C_{ijpq}, the fault's geometric shape ν_j, and the spatial derivatives of the Green's function $G_{np,q'}$. In other words, the seismic wave field is determined by these factors together.

3.2.3 Significance and Applications of Representation Theorem

Based on the representation theorem of seismic sources (3.2.5), assuming we are aware of the medium parameters in the fault zone and understand the geometry of the seismic fault, the radiated seismic wavefield is uniquely determined once the spatiotemporal evolution of the displacement on the fault plane is specified. This connection is made through the spatial derivatives of Green's function. Note that the Green's function belongs to model II, and thus, it is independent of the fault Σ considered in our problem. Theoretically, once the medium model is established, the Green's function is also determined. Of course, how to specifically solve for Green's function is a separate issue and will be discussed in subsequent sections.

Assuming that the Green's function can be calculated by alternative methods, Eq. (3.2.5) essentially establishes a relationship between the seismic wavefield and the displacement on the fault plane. On one hand, by setting the displacement function on the fault plane, we can obtain the resulting displacement field. On the other hand, we can use the various seismic records observed to inversely deduce the spatiotemporal history of the slip on the fault plane. The former process is known as *forward modeling*, while the latter is known as *inversion*. It's important to note that this does not involve the forces that cause the displacements, thus it falls within the realm of *earthquake source kinematics*, as shown in Fig. 3.3.

Based on the displacement on the left side of Eq. (3.2.5), the strain components can be determined through geometric equations. By applying constitutive relations,

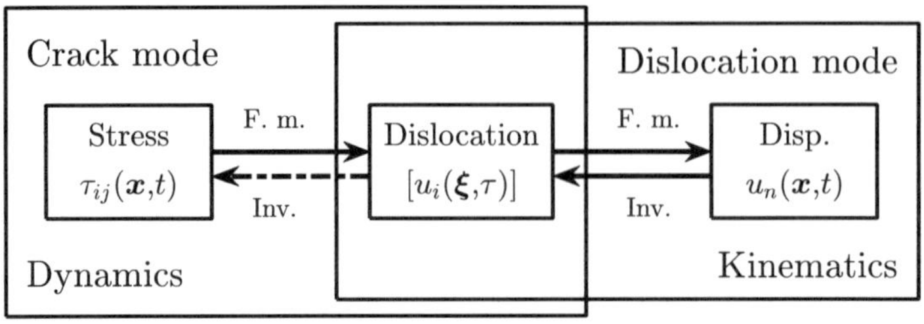

Fig. 3.3 Seismological problems related to seismic sources. The connection of dislocation (i.e., slip) on fault divides seismological studies into dynamics and kinematics. The former relates the stress field to dislocation, while the latter connects dislocation to observed displacement. These correspond to two descriptive models of the seismic source: the crack model and the dislocation model. "Disp.", "F. m." and "Inv." stand for "Displacement", "Forward modeling" and "Inversion", respectively

the stress components are then derived. If the field point is positioned on the fault plane,[17] this method yields the stress distribution on the fault surface. Specifically,

$$\begin{aligned}\tau_{kl}(\boldsymbol{x}_{\mathrm{P}}, t) &= C_{klnm}(\boldsymbol{x}_{\mathrm{P}})u_{n,m}(\boldsymbol{x}_{\mathrm{P}}, t) \\ &= C_{klnm}(\boldsymbol{x}_{\mathrm{P}}) \iint_{\Sigma} m_{pq}(\boldsymbol{\xi}, t) * \left.\frac{\partial^2 G_{np}(\boldsymbol{x},\ t;\ \boldsymbol{\xi},\ 0)}{\partial x_m \partial \xi_q}\right|_{\boldsymbol{x}=\boldsymbol{x}_{\mathrm{P}}} \mathrm{d}\Sigma(\boldsymbol{\xi}) \end{aligned} \quad (3.2.6)$$

in which, $*$ represents convolution operation, $\boldsymbol{x}_{\mathrm{P}}$ is a point in the vicinity of the fault plane, and

$$\boxed{m_{pq}(\boldsymbol{\xi}, t) = \left[u_i(\boldsymbol{\xi},\ t)\right] \nu_j(\boldsymbol{\xi}) C_{ijpq}(\boldsymbol{\xi})} \quad (3.2.7)$$

Equation (3.2.6) can be regarded as another form of the source representation theorem, which is derived directly from Eq. (3.2.5). This involves stress,[18] thus falling within the scope of *earthquake source dynamics*, as shown in Fig. 3.3, which explore issues related to crack propagation. From the perspective of solving mathematical physics equations, this is a problem of moving boundaries, as the boundary between the fractured and unfractured parts shifts during the evolution of the rupture. Near the extending crack front, there is significant stress concentration, demanding high

[17] Strictly speaking, it is not permissible to do so because the field point $\boldsymbol{x}$ in Eq. (3.2.5) should be located *within* the interior of a region, not on its inner boundary. However, we can approximate by allowing $\boldsymbol{x}$ to approach the fault plane indefinitely without being exactly on it to circumvent this issue. Alternatively, we can position $\boldsymbol{x}$ precisely on the fault plane and transform a boundary point into an internal point by removing a small area at the corresponding location on the fault plane and adding an infinitesimally small other surface nearby. Zhang and Chen (2006) provides a detailed account of this process, and interested readers may refer to it for a deeper understanding.

[18] It's important to reiterate that the stress and dislocation components mentioned in Eq. (3.2.6) are relative to changes prior to an earthquake; this equation does not provide information about the absolute stress field. Since both quantities are unknown and the integrand on the right side of the equation includes an unknown function, Eq. (3.2.6) forms an *integral equation*.

resolution in computational methods. Additionally, the friction criteria controlling fracture behavior are often nonlinear. For instance, under a sliding weakening friction criterion, although the stress decreases linearly with the increase in dislocation until a critical sliding weakening displacement is reached, the stages of stress accumulation prior to this point and the subsequent free sliding phase do not adhere to a linear relationship. Due to these complexities, solving source dynamics issues is challenging. Numerical methods are used for computation, requiring high computational costs to achieve certain precision. This makes it difficult to conduct inversion, which includes numerous forward processes. To date, only a few studies have been significantly simplified, and realistic inversion studies in source dynamics have not yet been initiated, which is why the inversion arrow in Fig. 3.3 is depicted with a dashed line.

The discussion above centers on the earthquake source as the subject of study. In seismology, both the earthquake source and the Earth's medium are primary focal points. When the Earth's medium is the research subject, given a known history of dislocation in time and space, the structural properties of the Earth's medium can be determined through the inversion of Eq. (3.2.5) to obtain C_{ijpq}. This indicates that while Eq. (3.2.5) is known as the representation theorem of seismic sources, it also provides a framework for studying the structure of the Earth's medium. It is essential for understanding how seismic waves propagate and studying the source, making it irreplaceable in seismological research.

3.3 Equivalent Body Forces of Dislocation Sources

3.3.1 Why Study Equivalent Body Forces?

Earthquakes result from dislocations along faults, a fact widely accepted for nearly a century. Even if we simplify faults as closely adjacent planes, the spatial distribution of dislocations can still be quite complex. Furthermore, these displacements within an elastic model disrupt the continuity of the medium, necessitating revisions to the equations established in elastic mechanics. In Sect. 3.2, by viewing faults as internal boundaries and modifying them accordingly, we use the displacement representation theorem. Essentially, this transforms internal sources into boundary conditions. Naturally, this leads to the question: Could the dislocations on faults in model I be represented as a type of body force (known as *equivalent body forces*[19]) in model II (see Fig. 3.1)? If so, the medium would be treated as continuous, and the mathematical model would simplify to solving the elastodynamics equations with equivalent body forces as the inhomogeneous terms, under homogeneous boundary conditions on S.

[19] As the name implies, despite different origins, the displacements caused by dislocations on faults in Model I are identical to those produced by the equivalent body force distribution in Model II. In other words, based solely on the displacement at any point, it is impossible to distinguish whether it was caused by dislocations or by equivalent body forces.

Additionally, early seismologists were keenly interested in mechanical models that could explain observed seismic radiation patterns. Thus, the concept of equivalent body forces drew attention as early as the 1920s. As discussed in historical review of source problems in Sect. 3.1, two models were proposed: the single-force couple and the double-force couple. Due to limitations in observational data at the time, which relied solely on P-wave data, distinguishing between these models remained unresolved until the advent of the source representation theorem in the 1960s.[20]

3.3.2 Expression and Properties of Equivalent Body Forces

If the displacement field caused by dislocations can be represented as a distribution of body forces $\boldsymbol{f}^{[u]}$, taking into account simplifying assumption (2) and Eq. (3.2.3), then the displacement representation formula (2.5.2) can be abbreviated as

$$u_n(\boldsymbol{x}, t) = \int_{-\infty}^{+\infty} \mathrm{d}\tau \iiint_V f_p^{[u]}(\boldsymbol{\eta}, \tau) G_{np}(\boldsymbol{x}, t-\tau; \boldsymbol{\eta}, 0)\, \mathrm{d}V(\boldsymbol{\eta}) \tag{3.3.1}$$

If we compare this with Eq. (3.2.5), considering that the displacements caused by equivalent body forces match those due to dislocations, the right-hand side of the equation should be equivalent as well. Notably, there's a distinct difference: the integrand in Eq. (3.3.1) contains the Green's function component G_{np}, while in Eq. (3.2.5), it involves the spatial derivative of this component, $G_{np,q'}$. This indicates that if $G_{np,q'}$ can be represented using G_{np}, then it might be feasible to formulate an expression for $f_p^{[u]}$. According to the properties of the Dirac δ function

$$\int_{-\infty}^{+\infty} \delta'(x-y) f(x)\, \mathrm{d}x = -\int_{-\infty}^{+\infty} \delta(x-y) f'(x)\, \mathrm{d}x = -f'(y) \tag{3.3.2}$$

It is not difficult to obtain[21]

$$\frac{\partial}{\partial \xi_q} G_{np}(\boldsymbol{x}, t-\tau; \boldsymbol{\xi}, 0) = -\iiint_V \frac{\partial}{\partial \eta_q} \delta(\boldsymbol{\eta}-\boldsymbol{\xi}) G_{np}(\boldsymbol{x}, t-\tau; \boldsymbol{\eta}, 0)\, \mathrm{d}V(\boldsymbol{\eta})$$

[20] In fact, Maruyama (1963) and Burridge and Knopoff (1964) centered on the issue of equivalent body forces. In their studies, they incidentally developed the source representation theorem. Our discussion takes the opposite approach, establishing the source representation theorem first and then addressing the equivalent body forces issue.

[21] In this section, we adopt the notation of Aki and Richards (2002), where $\boldsymbol{\xi}$ denotes a point on the fault surface Σ, and $\boldsymbol{\eta}$ indicates a point within the elastic body V.

Substituting the above equation into Eq. (3.2.5) produces

$$
\begin{aligned}
u_n(\boldsymbol{x}, t) &= \int_{-\infty}^{+\infty} \mathrm{d}\tau \iint_{\Sigma} \left[u_i(\boldsymbol{\xi}, \tau)\right] \nu_j(\boldsymbol{\xi}) C_{ijpq}(\boldsymbol{\xi}) \\
&\quad \cdot \left\{ - \iiint_V \frac{\partial}{\partial \eta_q} \delta(\boldsymbol{\eta} - \boldsymbol{\xi}) G_{np}(\boldsymbol{x},\, t - \tau;\, \boldsymbol{\eta},\, 0)\, \mathrm{d}V(\boldsymbol{\eta}) \right\} \mathrm{d}\Sigma(\boldsymbol{\xi}) \\
&= \int_{-\infty}^{+\infty} \mathrm{d}\tau \iiint_V \left\{ - \iint_{\Sigma} \left[u_i(\boldsymbol{\xi}, \tau)\right] \nu_j(\boldsymbol{\xi}) C_{ijpq}(\boldsymbol{\xi}) \frac{\partial}{\partial \eta_q} \delta(\boldsymbol{\eta} - \boldsymbol{\xi})\, \mathrm{d}\Sigma(\boldsymbol{\xi}) \right\} \\
&\quad \cdot G_{np}(\boldsymbol{x},\, t - \tau;\, \boldsymbol{\eta},\, 0)\, \mathrm{d}V(\boldsymbol{\eta})
\end{aligned}
\tag{3.3.3}
$$

The equality in the second equation holds because it involves only the operation of interchanging the order of integration, assuming, of course, that all functions meet the continuity conditions required for such a switch. By comparing the right sides of Eqs. (3.3.3) and (3.3.1), we obtain

$$
\begin{aligned}
f_p^{[u]}(\boldsymbol{\eta},\, \tau) &= - \iint_{\Sigma} \left[u_i(\boldsymbol{\xi}, \tau)\right] \nu_j(\boldsymbol{\xi}) C_{ijpq}(\boldsymbol{\xi}) \frac{\partial}{\partial \eta_q} \delta(\boldsymbol{\eta} - \boldsymbol{\xi})\, \mathrm{d}\Sigma(\boldsymbol{\xi}) \\
&\overset{(3.2.7)}{=\!=\!=} - \iint_{\Sigma} m_{pq}(\boldsymbol{\xi}, \tau) \frac{\partial}{\partial \eta_q} \delta(\boldsymbol{\eta} - \boldsymbol{\xi})\, \mathrm{d}\Sigma(\boldsymbol{\xi})
\end{aligned}
\tag{3.3.4}
$$

This is the equivalent body force of the dislocation function we seek, which holds true for general anisotropic heterogeneous media. The corresponding vector form of Eq. (3.3.4) is

$$
\boxed{\boldsymbol{f}^{[u]}(\boldsymbol{\eta},\, \tau) = - \iint_{\Sigma} \mathbf{m}(\boldsymbol{\xi}, \tau) \cdot \nabla \delta(\boldsymbol{\eta} - \boldsymbol{\xi})\, \mathrm{d}\Sigma(\boldsymbol{\xi})}
$$

Extending the definition of $m_{pq}(\boldsymbol{\xi}, t)$ in Eq. (3.2.7), which is defined on the fault surface, to any point within the elastic body,[22] and using the integral properties of the Dirac δ function derivatives, we can further derive from Eq. (3.3.4) that

$$
f_p^{[u]}(\boldsymbol{\eta},\, \tau) = \iint_{\Sigma} m_{pq}(\boldsymbol{\xi}, \tau) \frac{\partial}{\partial \xi_q} \delta(\boldsymbol{\eta} - \boldsymbol{\xi})\, \mathrm{d}\Sigma(\boldsymbol{\xi}) = - \frac{\partial}{\partial \xi_q} m_{pq}(\boldsymbol{\xi}, \tau) \bigg|_{\boldsymbol{\xi} = \boldsymbol{\eta}}
\tag{3.3.5}
$$

[22] In Sect. 3.4.1, we will see that the $m_{pq}(\boldsymbol{\xi}, t)$ defined in Eq. (3.2.7) actually represents the seismic moment density tensor. When integrated over the fault surface, this yields the seismic moment tensor $M_{pq}(t)$. Equations (3.3.5) and (3.3.6) introduce $m_{pq}(\boldsymbol{\eta}, t)$ defined within the elastic body, which can be seen as an extension of the seismic moment density tensor throughout the volume V of the elastic body, with a zero value outside the fault and the volume integral equating to $M_{pq}(t)$. This extension allows us to use the extended seismic moment density tensor to represent equivalent body forces, as shown in Eq. (3.3.6).

Based on the properties of the elastic coefficients C_{ijpq}, it follows that $m_{pq} = m_{qp}$, thus the above equation can be written as

$$\boldsymbol{f}^{[u]}(\boldsymbol{\eta}, \tau) = -\nabla \cdot \mathbf{m}(\boldsymbol{\eta}, \tau) \overset{(3.2.7)}{=\!=\!=} -\nabla \cdot \left\{ [u_i(\boldsymbol{\eta}, \tau)]\, \nu_j(\boldsymbol{\eta}) C_{ijpq}(\boldsymbol{\eta}) \hat{\boldsymbol{e}}_p \hat{\boldsymbol{e}}_q \right\} \tag{3.3.6}$$

According to the expressions for equivalent body forces given in Eqs. (3.3.4) and (3.3.5), it is easy to find that $\boldsymbol{f}^{[u]}(\boldsymbol{\eta}, \tau)$ exhibits the following properties:

(1) From a temporal perspective, the time-varying behavior of equivalent body forces (as a function of time) matches that of the dislocation function. Spatially, in the model II shown in Fig. 3.1b, the distribution of equivalent body forces is confined to the locations corresponding to the fault Σ in model I, as illustrated in Fig. 3.1a. And in the regions beyond, the equivalent body forces are zero.

(2) The resultant force of the equivalent body forces is zero. This can be readily verified as[23]

$$\begin{aligned}
\iiint_V f_p^{[u]}(\boldsymbol{\eta}, \tau)\, \mathrm{d}V(\boldsymbol{\eta}) &= \iiint_V \left\{ -\iint_\Sigma m_{pq}(\boldsymbol{\xi}, \tau) \frac{\partial}{\partial \eta_q} \delta(\boldsymbol{\eta} - \boldsymbol{\xi})\, \mathrm{d}\Sigma(\boldsymbol{\xi}) \right\} \mathrm{d}V(\boldsymbol{\eta}) \\
&= -\iint_\Sigma m_{pq}(\boldsymbol{\xi}, \tau) \left\{ \iiint_V \frac{\partial}{\partial \eta_q} \delta(\boldsymbol{\eta} - \boldsymbol{\xi})\, \mathrm{d}V(\boldsymbol{\eta}) \right\} \mathrm{d}\Sigma(\boldsymbol{\xi}) \\
&= 0
\end{aligned}$$

the final equality holds due to the result within the large brackets of the integrand being zero, as shown in Eq. (3.3.2) where $f(x) = 1$.

(3) The resultant moment of the equivalent body forces is zero. That is

$$\iiint_V (\boldsymbol{\eta} - \boldsymbol{\eta}_0) \times \boldsymbol{f}^{[u]}(\boldsymbol{\eta}, \tau)\, \mathrm{d}V(\boldsymbol{\eta}) = 0$$

where $\boldsymbol{\eta}_0$ represents the reference point selected for calculating the moment. The proof is left as an exercise for the reader.

The properties (2) and (3) of the equivalent body force mentioned above are expected. Otherwise, it would result in the overall movement or rotation of the source region, which does not align with actual observations. Additionally, the equivalent body force is distributed only in the spatial location corresponding to the fault plane, which matches our intuitive understanding. However, these points only provide a rough understanding of the equivalent body force. The further question is, what is the exact distribution of the equivalent body force? What kind of physical model can we construct based on this?

[23] This is derived using Eq. (3.3.4). As an exercise, readers may start with Eqs. (3.3.5) or (3.3.6) and find that the conclusion remains the same.

As mentioned earlier, Eqs. (3.3.4) and (3.3.5) are valid for general anisotropic and heterogenous media and are relatively complex in form. If we consider the special case of isotropic homogeneous media and establish an appropriate coordinate system for a planar fault, we might be able to answer these questions.

3.3.3 Equivalent Body Force (I): Couple + Single Force

Let's consider a planar shear dislocation source in an isotropic homogeneous medium, as shown in Fig. 3.4. We establish a coordinate system $O\xi_1\xi_2\xi_3$, positioning the fault plane on the $\xi_3 = 0$ plane, with the positive direction of the ξ_1 axis aligned with the direction of the dislocation. This coordinate system is called the *source coordinate system*. Therefore, we have

$$[u_i(\boldsymbol{\xi}, \tau)] = \delta_{i1}[u_1(\xi_1, \xi_2, 0, \tau)], \quad \nu_j = \delta_{j3} \tag{3.3.7}$$

For an isotropic elastic material, the specific expression for the elastic constants C_{ijpq} can be found in Eq. (2.2.19). Substituting them into Eq. (3.3.5), we obtain

$$\begin{aligned} f_p^{[u]}(\boldsymbol{\eta}, \tau) &= -\frac{\partial}{\partial \xi_q}[u_1(\xi_1, \xi_2, 0, \tau)]\Big[\lambda\delta_{13}\delta_{pq} + \mu(\delta_{1p}\delta_{3q} + \delta_{3p}\delta_{1q})\Big]\Big|_{\boldsymbol{\xi}=\boldsymbol{\eta}} \\ &= -\mu\left\{\delta_{1p}\frac{\partial}{\partial \eta_3}\Big([u_1(\eta_1, \eta_2, 0, \tau)]\delta(\eta_3)\Big) + \delta_{3p}\delta(\eta_3)\frac{\partial}{\partial \eta_1}[u_1(\eta_1, \eta_2, 0, \tau)]\right\} \end{aligned}$$

specifically,

$$\begin{cases} f_1^{[u]}(\boldsymbol{\eta}, \tau) = -\mu[u_1(\eta_1, \eta_2, 0, \tau)]\delta'(\eta_3) \\ f_2^{[u]}(\boldsymbol{\eta}, \tau) = 0 \\ f_3^{[u]}(\boldsymbol{\eta}, \tau) = -\mu\dfrac{\partial}{\partial \eta_1}[u_1(\eta_1, \eta_2, 0, \tau)]\delta(\eta_3) \end{cases}$$

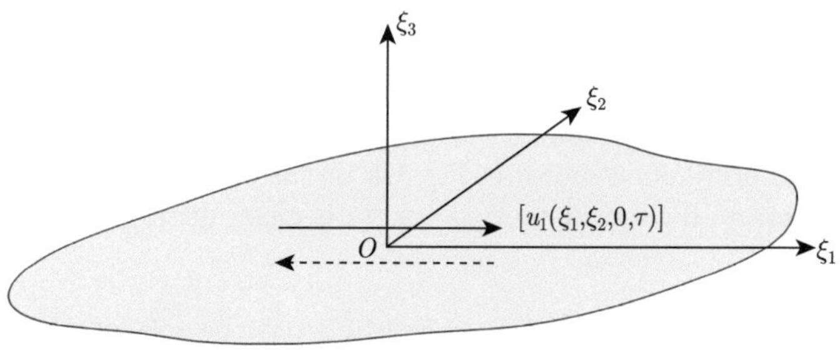

Fig. 3.4 Planar shear dislocation source in the source coordinate system. The fault plane lies within the $\xi_3 = 0$ plane, with the direction of the dislocation along the ξ_1 axis. Solid and dashed arrows represent the displacements occurring on the upper and lower surfaces of the fault, respectively

Thus, we have derived the specific expressions for the three components of the equivalent body force. Next, we will separately consider the properties of $f_1^{[u]}(\boldsymbol{\eta}, \tau)$ and $f_3^{[u]}(\boldsymbol{\eta}, \tau)$.

3.3.3.1 $f_1^{[u]}(\eta, \tau)$

Firstly, note that

$$\iiint_V f_1^{[u]}(\boldsymbol{\eta}, \tau)\,\mathrm{d}V(\boldsymbol{\eta}) = -\mu \iint_\Sigma [u_1(\eta_1, \eta_2, 0, \tau)] \left[\int_{\eta_3^{\min}}^{\eta_3^{\max}} \delta'(\eta_3)\,\mathrm{d}\eta_3\right] \mathrm{d}\eta_1\,\mathrm{d}\eta_2 = 0$$

where $\eta_3^{\min} = \min\{\eta_3(\eta_1, \eta_2)\}$, $\eta_3^{\max} = \max\{\eta_3(\eta_1, \eta_2)\}$. Therefore, the resultant force of $f_1^{[u]}(\boldsymbol{\eta}, \tau)$ is zero. It is important to note that the integral within the brackets with respect to η_3 equals zero. As a result, the resultant force is zero for every surface element $\mathrm{d}\eta_1\,\mathrm{d}\eta_2$ on each fault plane.

Secondly, regarding the resultant moment of $f_1^{[u]}(\boldsymbol{\eta}, \tau)$, there is

$$\begin{aligned}
&\iiint_V (\boldsymbol{\eta} - \boldsymbol{\eta}_0) \times \left[f_1^{[u]}(\boldsymbol{\eta}, \tau)\hat{\boldsymbol{e}}_1\right] \mathrm{d}V(\boldsymbol{\eta}) \xlongequal{\boldsymbol{\eta}_0=\mathbf{0}} \iiint_V \varepsilon_{ij1}\eta_j f_1^{[u]}(\boldsymbol{\eta}, \tau)\hat{\boldsymbol{e}}_i\,\mathrm{d}V(\boldsymbol{\eta})\\
&= \iiint_V \varepsilon_{231}\eta_3 f_1^{[u]}(\boldsymbol{\eta}, \tau)\hat{\boldsymbol{e}}_2\,\mathrm{d}V(\boldsymbol{\eta}) + \iiint_V \varepsilon_{321}\eta_2 f_1^{[u]}(\boldsymbol{\eta}, \tau)\hat{\boldsymbol{e}}_3\,\mathrm{d}V(\boldsymbol{\eta})\\
&= -\mu \iiint_V \left(\eta_3\hat{\boldsymbol{e}}_2 - \eta_2\hat{\boldsymbol{e}}_3\right)[u_1(\eta_1, \eta_2, 0, \tau)]\delta'(\eta_3)\,\mathrm{d}V(\boldsymbol{\eta})\\
&= -\mu \iint_\Sigma [u_1(\eta_1, \eta_2, 0, \tau)] \left[\int_{\eta_3^{\min}}^{\eta_3^{\max}} \delta'(\eta_3)\left(\eta_3\hat{\boldsymbol{e}}_2 - \eta_2\hat{\boldsymbol{e}}_3\right)\mathrm{d}\eta_3\right] \mathrm{d}\eta_1\,\mathrm{d}\eta_2\\
&= \mu\hat{\boldsymbol{e}}_2 \iint_\Sigma [u_1(\eta_1, \eta_2, 0, \tau)]\,\mathrm{d}\eta_1\,\mathrm{d}\eta_2
\end{aligned} \tag{3.3.8}$$

This means that $f_1^{[u]}(\boldsymbol{\eta}, \tau)$ from each fault plane element creates a moment along the ξ_2 axis.

Summarizing the above points, the resultant force of $f_1^{[u]}(\boldsymbol{\eta}, \tau)$ on each fault element is zero, but it generates a moment along the ξ_2 axis. Based on this property, we can use a couple model to describe the distribution of $f_1^{[u]}(\boldsymbol{\eta}, \tau)$[24]: similar to the dislocation distribution shown in Fig. 3.4, on the upper and lower surfaces of each fault element, the direction of $f_1^{[u]}(\boldsymbol{\eta}, \tau)$ is along the positive and negative

[24] When searching for an equivalent body force model, it's neither possible nor necessary to find a "unique" model. This is because the displacement at a field point is a combined result of the entire fault plane, and from the displacement alone, we can't determine the specifics of each fault segment. Therefore, the non-uniqueness of the equivalent body force is inevitable. The criterion for selecting an equivalent body force is that it should have the corresponding properties and be simple and clear in representation.

ξ_1 axis, respectively, with equal magnitude but opposite directions, separated by an infinitesimally small distance, which will generate a resultant moment.

3.3.3.2 $f_3^{[u]}(\boldsymbol{\eta}, \tau)$

Using the same approach as in analyzing $f_1^{[u]}(\boldsymbol{\eta}, \tau)$, we search for the right model by examining the properties of the resultant force and moment.

$$\begin{aligned}
\iiint_V f_3^{[u]}(\boldsymbol{\eta}, \tau)\,\mathrm{d}V(\boldsymbol{\eta}) &= -\mu \iiint_V \frac{\partial}{\partial \eta_1}[u_1(\eta_1, \eta_2, 0, \tau)]\delta(\eta_3)\,\mathrm{d}\eta_1\,\mathrm{d}\eta_2\,\mathrm{d}\eta_3 \\
&= -\mu \iint_\Sigma \frac{\partial}{\partial \eta_1}[u_1(\eta_1, \eta_2, 0, \tau)]\,\mathrm{d}\eta_1\,\mathrm{d}\eta_2 \\
&= -\mu \int_{\eta_2^{\min}}^{\eta_2^{\max}} [u_1(\eta_1, \eta_2, 0, \tau)]\Big|_{\eta_1=\eta_1^{\min}}^{\eta_1=\eta_1^{\max}} \mathrm{d}\eta_2 = 0
\end{aligned}$$

where $\eta_i^{\min} = \min(\eta_i)$ and $\eta_i^{\max} = \max(\eta_i)$ $(i = 1, 2)$. The final equation holds because the dislocation at the fault edge is zero.[25] This means that the resultant force of $f_3^{[u]}(\boldsymbol{\eta}, \tau)$ is also zero. It's important to note that unlike $f_1^{[u]}(\boldsymbol{\eta}, \tau)$, the resultant force for $f_3^{[u]}(\boldsymbol{\eta}, \tau)$ on each element of the fault surface is not zero; however, the integral over the entire fault surface Σ equals zero.[26]

The resultant moment of $f_3^{[u]}(\boldsymbol{\eta}, \tau)$ is

$$\begin{aligned}
&\iiint_V (\boldsymbol{\eta} - \boldsymbol{\eta}_0) \times \left[f_3^{[u]}(\boldsymbol{\eta}, \tau)\hat{\boldsymbol{e}}_3\right] \mathrm{d}V(\boldsymbol{\eta}) \xlongequal{\boldsymbol{\eta}_0=\boldsymbol{0}} \iiint_V \varepsilon_{ij3}\eta_j f_3^{[u]}(\boldsymbol{\eta}, \tau)\hat{\boldsymbol{e}}_i\,\mathrm{d}V(\boldsymbol{\eta}) \\
&= \iiint_V \varepsilon_{123}\eta_2 f_3^{[u]}(\boldsymbol{\eta}, \tau)\hat{\boldsymbol{e}}_1\,\mathrm{d}V(\boldsymbol{\eta}) + \iiint_V \varepsilon_{213}\eta_1 f_3^{[u]}(\boldsymbol{\eta}, \tau)\hat{\boldsymbol{e}}_2\,\mathrm{d}V(\boldsymbol{\eta}) \\
&= -\mu \iiint_V \left(\eta_2\hat{\boldsymbol{e}}_1 - \eta_1\hat{\boldsymbol{e}}_2\right) \frac{\partial}{\partial \eta_1}[u_1(\eta_1, \eta_2, 0, \tau)]\delta(\eta_3)\,\mathrm{d}V(\boldsymbol{\eta}) \\
&= -\mu \int_{\eta_2^{\min}}^{\eta_2^{\max}} \left\{\left(\eta_2\hat{\boldsymbol{e}}_1 - \eta_1\hat{\boldsymbol{e}}_2\right)[u_1(\eta_1, \eta_2, 0, \tau)]\Big|_{\eta_1=\eta_1^{\min}}^{\eta_1=\eta_1^{\max}} \right. \\
&\quad \left. + \int_{\eta_1^{\min}}^{\eta_1^{\max}} [u_1(\eta_1, \eta_2, 0, \tau)]\hat{\boldsymbol{e}}_2\,\mathrm{d}\eta_1 \right\} \mathrm{d}\eta_2 = -\mu\hat{\boldsymbol{e}}_2 \iint_\Sigma [u_1(\eta_1, \eta_2, 0, \tau)]\,\mathrm{d}\eta_1\,\mathrm{d}\eta_2
\end{aligned}$$

---

[25] This assumption is reasonable because our criterion for selecting the fault surface Σ ensures that it is sufficiently large to encompass the area where the rupture occurs. This means that the dislocation at the edges of the fault is zero. If this were not the case, it would indicate that the fault surface Σ is not large enough, as there would still be ruptures occurring beyond it. We would need to enlarge Σ until this condition is met.

[26] This conclusion is apparent from the integrand under the second equality on the right side. Across each fault surface element $\mathrm{d}\eta_1\,\mathrm{d}\eta_2$, the resultant force of $f_3^{[u]}$ is $-\mu\frac{\partial}{\partial \eta_1}[u_1(\eta_1, \eta_2, 0, \tau)]\,\mathrm{d}\eta_1\,\mathrm{d}\eta_2$, which generally is non-zero.

This is evidenced by the fact that it precisely opposes the moment of $f_1^{[u]}(\boldsymbol{\eta}, \tau)$. This indicates that the resultant moment produced by the equivalent body forces is zero, consistent with the general conclusion.

Based on the analysis, the overall properties of $f_3^{[u]}(\boldsymbol{\eta}, \tau)$ are similar to those of $f_1^{[u]}(\boldsymbol{\eta}, \tau)$, both having a resultant force of zero and generating a moment along the ξ_2 axis, which exactly cancels the moment generated by $f_1^{[u]}(\boldsymbol{\eta}, \tau)$. However, a significant difference is that on the fault surface elements, the resultant force of $f_3^{[u]}(\boldsymbol{\eta}, \tau)$ is not zero. Therefore, the couple model is not suitable for $f_3^{[u]}(\boldsymbol{\eta}, \tau)$. It can only be generally stated that across the entire fault surface, $f_3^{[u]}(\boldsymbol{\eta}, \tau)$ represents a distribution of single forces that overall exhibits the described properties.

From the comprehensive analysis of the equivalent body force components, we understand the following: In an isotropic and homogeneous medium, the shear dislocation on a planar fault does not produce any equivalent body force components perpendicular to the dislocation vector within the fault plane. Along the direction parallel to the dislocation within the fault plane, it acts as a couple, while perpendicular to the fault plane, it manifests as a distribution of single forces. Overall, the resultant force and moment of the equivalent body forces are zero.

3.3.4 Equivalent Body Force (II): Double Couple

Initially, we derived the first form of equivalent body forces, couples plus single forces, starting from their expressions. We will now consider this issue directly from the standpoint of the source representation theorem, still within the source coordinate system (see Fig. 3.4). Therefore, Eq. (3.3.7) remains valid. By substituting Eqs. (3.3.7) and (2.2.19) into Eq. (3.2.5), we obtain:

$$\begin{aligned}
u_n(\boldsymbol{x}, t) &= \mu \iint_{\Sigma} \left[u_1(\boldsymbol{\xi}, t)\right] (\delta_{1p}\delta_{3q} + \delta_{1q}\delta_{3p}) * \frac{\partial}{\partial \xi_q} G_{np}(\boldsymbol{x}, t;\ \boldsymbol{\xi}, 0)\,\mathrm{d}\Sigma(\boldsymbol{\xi}) \\
&= \mu \iint_{\Sigma} \left[u_1(\boldsymbol{\xi}, t)\right] * \left\{\frac{\partial}{\partial \xi_3} G_{n1}(\boldsymbol{x}, t;\ \boldsymbol{\xi}, 0) + \frac{\partial}{\partial \xi_1} G_{n3}(\boldsymbol{x}, t;\ \boldsymbol{\xi}, 0)\right\} \mathrm{d}\Sigma(\boldsymbol{\xi}) \\
&= \mu \lim_{\varepsilon \to 0} \iint_{\Sigma} \frac{\left[u_1(\boldsymbol{\xi}, t)\right]}{2\varepsilon} * \Big[G_{n1}(\boldsymbol{x}, t;\ \boldsymbol{\xi} + \varepsilon\hat{\boldsymbol{e}}_3, 0) - G_{n1}(\boldsymbol{x}, t;\ \boldsymbol{\xi} - \varepsilon\hat{\boldsymbol{e}}_3, 0) \\
&\quad + G_{n3}(\boldsymbol{x}, t;\ \boldsymbol{\xi} + \varepsilon\hat{\boldsymbol{e}}_1, 0) - G_{n3}(\boldsymbol{x}, t;\ \boldsymbol{\xi} - \varepsilon\hat{\boldsymbol{e}}_1, 0)\Big]\,\mathrm{d}\Sigma(\boldsymbol{\xi})
\end{aligned} \tag{3.3.9}$$

Based on the explanation of Green's functions in Sect. 2.4.1, the four Green's functions under the integral sign in the equation each correspond to the effects of four forces, $\boldsymbol{f}^{(i)}$ $(i = 1, 2, 3, 4)$, acting near $\boldsymbol{\xi}$ as shown in Fig. 3.5. The forces $\boldsymbol{f}^{(1)}$ and $\boldsymbol{f}^{(2)}$ form a couple separated vertically by 2ε, and $\boldsymbol{f}^{(3)}$ and $\boldsymbol{f}^{(4)}$ form another couple separated horizontally by 2ε. These forces produce moments that cancel each

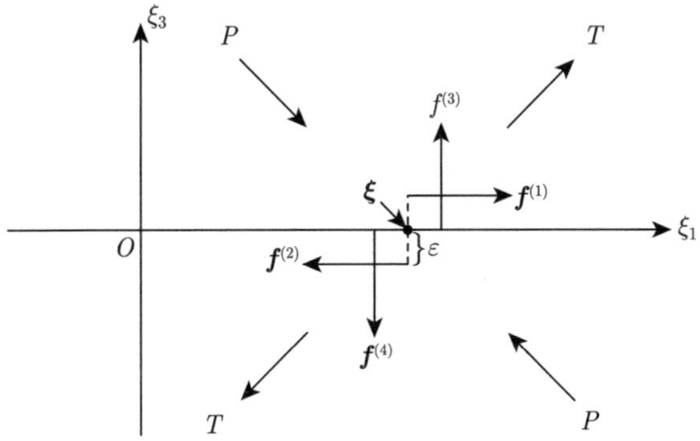

Fig. 3.5 Diagram of the double couple model. Considering the cross-section at $\xi_2 = 0$ in the source coordinate system, the fault plane lies along the ξ_1 axis. At a point $\boldsymbol{\xi}$ on the fault, $\boldsymbol{f}^{(1)}$ and $\boldsymbol{f}^{(2)}$ constitute a couple, separated vertically by 2ε, while $\boldsymbol{f}^{(3)}$ and $\boldsymbol{f}^{(4)}$ form another couple, separated horizontally by 2ε. The directions of the tension axis (T axis) and the compression axis (P axis) are also indicated

other out, resulting in a *double couple* without any resultant moment. Therefore, the displacement field can be viewed as being generated by this moment-free double couple.

To more clearly demonstrate this, in the following we will prove that the displacement field generated by the double couple in Fig. 3.5 is indeed expressed by Eq. (3.3.9).

Consider four concentrated impulse-type forces of the following form

$$\begin{cases} f_p^{(1)}(\boldsymbol{x}, t) = F(\boldsymbol{\xi}, \tau_0)\delta_{p1}\delta\left[\boldsymbol{x} - (\boldsymbol{\xi} + \varepsilon\hat{\boldsymbol{e}}_3)\right]\delta(t - \tau_0) \\ f_p^{(2)}(\boldsymbol{x}, t) = -F(\boldsymbol{\xi}, \tau_0)\delta_{p1}\delta\left[\boldsymbol{x} - (\boldsymbol{\xi} - \varepsilon\hat{\boldsymbol{e}}_3)\right]\delta(t - \tau_0) \\ f_p^{(3)}(\boldsymbol{x}, t) = F(\boldsymbol{\xi}, \tau_0)\delta_{p3}\delta\left[\boldsymbol{x} - (\boldsymbol{\xi} + \varepsilon\hat{\boldsymbol{e}}_1)\right]\delta(t - \tau_0) \\ f_p^{(4)}(\boldsymbol{x}, t) = -F(\boldsymbol{\xi}, \tau_0)\delta_{p3}\delta\left[\boldsymbol{x} - (\boldsymbol{\xi} - \varepsilon\hat{\boldsymbol{e}}_1)\right]\delta(t - \tau_0) \end{cases}$$

where $F(\boldsymbol{\xi}, \tau_0) = \mu[u_1(\boldsymbol{\xi}, \tau_0)]/2\varepsilon$ represents the magnitude of these forces, which depends on the location of the point of application $\boldsymbol{\xi}$ and the time of action τ_0. For example, by substituting $f_p^{(1)}(\boldsymbol{x}, t)$ into Eq. (3.3.1), we have

$$\begin{aligned} u_n^{(1)}(\boldsymbol{x}, t) &= \int_{-\infty}^{+\infty} \mathrm{d}\tau \iiint_V \frac{\mu[u_1(\boldsymbol{\xi}, \tau_0)]}{2\varepsilon}\delta_{p1}\delta\left[\boldsymbol{\eta} - (\boldsymbol{\xi} + \varepsilon\hat{\boldsymbol{e}}_3)\right]\delta(\tau - \tau_0) \\ &\quad \cdot G_{np}(\boldsymbol{x}, t - \tau; \boldsymbol{\eta}, 0)\,\mathrm{d}V(\boldsymbol{\eta}) \\ &= \frac{\mu[u_1(\boldsymbol{\xi}, \tau_0)]}{2\varepsilon}G_{n1}(\boldsymbol{x}, t - \tau_0; \boldsymbol{\xi} + \varepsilon\hat{\boldsymbol{e}}_3, 0) \end{aligned}$$

Similarly, the displacement fields generated by $f_p^{(i)}(\boldsymbol{x}, t)$ $(i = 2, 3, 4)$ are respectively

$$u_n^{(2)}(\boldsymbol{x}, t) = -\frac{\mu[u_1(\boldsymbol{\xi}, \tau_0)]}{2\varepsilon} G_{n1}(\boldsymbol{x}, t - \tau_0;\ \boldsymbol{\xi} - \varepsilon\hat{\boldsymbol{e}}_3, 0)$$

$$u_n^{(3)}(\boldsymbol{x}, t) = \frac{\mu[u_1(\boldsymbol{\xi}, \tau_0)]}{2\varepsilon} G_{n3}(\boldsymbol{x}, t - \tau_0;\ \boldsymbol{\xi} + \varepsilon\hat{\boldsymbol{e}}_1, 0)$$

$$u_n^{(4)}(\boldsymbol{x}, t) = -\frac{\mu[u_1(\boldsymbol{\xi}, \tau_0)]}{2\varepsilon} G_{n3}(\boldsymbol{x}, t - \tau_0;\ \boldsymbol{\xi} - \varepsilon\hat{\boldsymbol{e}}_1, 0)$$

Summing them up, integrating over the fault plane Σ with respect to $\boldsymbol{\xi}$, and integrating over the moment of action τ_0, while taking the limit as ε approaches zero, yields

$$\begin{aligned} u_n(\boldsymbol{x}, t) &= u_n^{(1)}(\boldsymbol{x}, t) + u_n^{(2)}(\boldsymbol{x}, t) + u_n^{(3)}(\boldsymbol{x}, t) + u_n^{(4)}(\boldsymbol{x}, t) \\ &= \mu \lim_{\varepsilon \to 0} \iint_{\Sigma} \frac{[u_1(\boldsymbol{\xi}, t)]}{2\varepsilon} * \Big[G_{n1}(\boldsymbol{x}, t;\ \boldsymbol{\xi} + \varepsilon\hat{\boldsymbol{e}}_3, 0) - G_{n1}(\boldsymbol{x}, t;\ \boldsymbol{\xi} - \varepsilon\hat{\boldsymbol{e}}_3, 0) \\ &\quad + G_{n3}(\boldsymbol{x}, t;\ \boldsymbol{\xi} + \varepsilon\hat{\boldsymbol{e}}_1, 0) - G_{n3}(\boldsymbol{x}, t;\ \boldsymbol{\xi} - \varepsilon\hat{\boldsymbol{e}}_1, 0) \Big] \mathrm{d}\Sigma(\boldsymbol{\xi}) \end{aligned}$$

This aligns with Eq. (3.3.9). Consequently, we can conclude that *the displacement created by a shear dislocation on a fault is effectively equivalent to that produced by a double couple*. The discussions reveal that this conclusion holds true point-by-point along the fault. That is, at each point $\boldsymbol{\xi}$ on the fault, the displacement field generated by the shear dislocation can be equivalently represented by the double couple depicted in Fig. 3.5. It should be noted that when the direction of the dislocation is fixed, the magnitude and direction of the four single forces within the double couple are uniquely determined. Moreover, the stress states caused by $\boldsymbol{f}^{(i)}$ $(i = 1, 2, 3, 4)$ can be intuitively represented by the T axis, the resultant direction of $\boldsymbol{f}^{(1)}$ and $\boldsymbol{f}^{(3)}$, and the P axis, perpendicular to the T axis,[27] as shown in Fig. 3.5. These axes correspond to the directions of the principal tensile and compressive stresses, respectively, which aids in further analysis of the characteristics of the displacement fields generated by shear dislocation sources.

[27] "T" and "P" represent "tension" and "compression", respectively, which referred to here are solely the stress components resulting from dislocations. The absolute stress values are the sum of the stress induced by the dislocations and the initial stress before an earthquake, thus the overall stress field is not merely as described here. Generally, the stress fields underground are in a state of compression. Of course, there are cases of tension caused by local tectonic forces, which are not widespread and beyond the scope of our discussion.

3.3.5 Relationship Between Two Sets of Equivalent Body Forces

In the above text, we have discussed two models of equivalent body forces derived from both the specific forms of equivalent body forces and the direct application of the earthquake source representation theorem. As previously mentioned, for equivalent body forces, it is only required that the displacement field they produce matches the one generated by the dislocation source. However, the displacement field observed at a point is a composite result of the displacement fields produced by various surface sources on the fault plane. Therefore, it is expected that there can be different models of equivalent body forces since the force distribution on each surface element cannot be uniquely determined based solely on this observation.[28]

We now aim to explore the underlying connection between two sets of equivalent body forces, which are essentially equivalent. Drawing from the definition of Green's functions and the analysis in Sect. 3.3.4, it becomes apparent that *a concentrated impulse type of single force, and a couple made up of such single forces, produce displacement fields corresponding to the Green's function and its spatial derivative, respectively*. Thus, from the results on the right side of the second equality in Eq. (3.3.9), it is evident that the forces generating these results are two sets of force couples oriented in the directions of ξ_1 and ξ_3, leading to a model of equivalent body forces represented as a double couple.

Considering the expressions within the curly brackets on the right side of the second equality in Eq. (3.3.9), the derivatives of the Green's functions are taken with respect to ξ_3 and ξ_1. Here, $\mathrm{d}\Sigma(\boldsymbol{\xi}) = \mathrm{d}\xi_1\,\mathrm{d}\xi_2$, where ξ_1 serves as the variable of integration, while ξ_3 does not. This distinction allows for the spatial derivative with respect to ξ_1 to be transferred to the dislocation function through integration by parts, whereas the spatial derivative with respect to ξ_3 cannot be further transformed. Specifically,

$$\begin{aligned}
u_n(\boldsymbol{x}, t) = &\, \mu \iint_\Sigma \big[u_1(\boldsymbol{\xi}, t)\big] * \left\{\frac{\partial}{\partial \xi_3} G_{n1}(\boldsymbol{x}, t;\ \boldsymbol{\xi}, 0) + \frac{\partial}{\partial \xi_1} G_{n3}(\boldsymbol{x}, t;\ \boldsymbol{\xi}, 0)\right\} \mathrm{d}\Sigma(\boldsymbol{\xi}) \\
= &\, \mu \iint_\Sigma \big[u_1(\boldsymbol{\xi}, t)\big] * \frac{\partial}{\partial \xi_3} G_{n1}(\boldsymbol{x}, t;\ \boldsymbol{\xi}, 0)\, \mathrm{d}\Sigma(\boldsymbol{\xi}) \\
&- \mu \iint_\Sigma \frac{\partial}{\partial \xi_1} \big[u_1(\boldsymbol{\xi}, t)\big] * G_{n3}(\boldsymbol{x}, t;\ \boldsymbol{\xi}, 0)\, \mathrm{d}\Sigma(\boldsymbol{\xi})
\end{aligned}$$

---

[28] In fact, the application of the concept of "equivalence" is far more widespread than we imagine. Even the foundational theory of elastodynamics, on which we rely, is itself a product of "equivalence". As a branch of continuum mechanics, the basic assumption of elastodynamics is the continuum hypothesis, which considers the elastic body to be filled with mass elements that are macroscopically small enough and microscopically large enough to allow the use of calculus tools for analysis. Clearly, the continuum hypothesis does not reflect reality, as can be easily verified using an electron microscope. Its existence is justified because the results obtained under this assumption are 'equivalent' to those produced by real objects, which can be directly observed.

The two terms on the right side of the equation include the spatial derivatives of Green's function and Green's function itself. From the analysis above, these are caused by force couples and single forces, respectively. Notably, this conclusion also stems from the earthquake source representation theorem, aligning with findings derived from equivalent body forces. This indicates that both the couple plus single force model and the double couple model are viable as equivalent body forces, with both models being equivalent.

Given this context, we can opt for a simpler and clearer model as the equivalent body force. The specific distribution of the single force is unclear, known only by its resultant force and moment being zero on the fault plane. In contrast, the model of a double couple at every point on the fault is clearer, which is why we choose the double couple as the equivalent body force for shear dislocation sources. This leads us to conclude that *the displacement field produced by shear dislocation on the fault is equivalent to that produced by a moment-free double couple*. This conclusion is valid for every point on the fault plane.

3.4 Seismic Moment Tensor

3.4.1 Definition and Properties of Seismic Moment Tensor

In Eq. (3.2.6), for the sake of simplicity, we introduced m_{pq}, which is defined in Eq. (3.2.7) as follows:

$$m_{pq}(\boldsymbol{\xi}, t) = \left[u_i(\boldsymbol{\xi}, t)\right] \nu_j(\boldsymbol{\xi}) C_{ijpq}(\boldsymbol{\xi}) \tag{3.4.1}$$

In many situations, we study problems in the far field.[29] In such cases, the source representation theorem, as stated in Eq. (3.2.5), can be rewritten as

$$u_n(\boldsymbol{x}, t) = \iint_{\Sigma} m_{pq}(\boldsymbol{\xi}, t)\, \mathrm{d}\Sigma(\boldsymbol{\xi}) * G_{np,q'}(\boldsymbol{x}, t;\ \boldsymbol{\xi}_0, 0) = M_{pq}(t) * G_{np,q'}(\boldsymbol{x}, t;\ \boldsymbol{\xi}_0, 0) \tag{3.4.2}$$

where $\boldsymbol{\xi}_0$ represents a point arbitrarily selected on the fault plane to characterize the position of the source point in the Green's function under far field conditions, while $M_{pq}(t)$ is defined as

[29] As the term implies, the far field refers to the observation point being significantly distant from the source. The distance between the field point and the source is much greater than the characteristic dimensions of the fault, rendering the distribution of the Green's function on the fault surface indistinguishable from the field point. This means that the spatial variations of the Green's function in the fault area can be considered negligible, allowing it to be approximated and moved outside the integral sign.

$$\boxed{M_{pq}(t) = \iint_{\Sigma} m_{pq}(\boldsymbol{\xi}, t)\,\mathrm{d}\Sigma(\boldsymbol{\xi}) = \iint_{\Sigma} \left[u_i(\boldsymbol{\xi},\, t)\right] \nu_j(\boldsymbol{\xi}) C_{ijpq}(\boldsymbol{\xi})\,\mathrm{d}\Sigma(\boldsymbol{\xi})} \tag{3.4.3}$$

The dimension of $M_{pq}(t)$ is [N · m], the same as moment. For a plane fault shear dislocation in an isotropic homogeneous medium, within the source coordinate system and according to Eq. (3.4.3), we have

$$M_{pq}(t) = \mu\left(\delta_{1p}\delta_{3q} + \delta_{3p}\delta_{1q}\right) \iint_{\Sigma} \left[u_1(\boldsymbol{\xi},\, t)\right] \mathrm{d}\Sigma(\boldsymbol{\xi}) \tag{3.4.4}$$

Written in matrix form, it is

$$\mathbf{M}(t) = M_0(t)\begin{bmatrix} 0 & 0 & 1 \\ 0 & 0 & 0 \\ 1 & 0 & 0 \end{bmatrix}, \quad M_0(t) = \mu \iint_{\Sigma} \left[u_1(\boldsymbol{\xi},\, t)\right] \mathrm{d}\Sigma(\boldsymbol{\xi}) = \mu\overline{[u_1(t)]}A$$

in which,

$$\overline{[u_1(t)]} = \frac{1}{A} \iint_{\Sigma} \left[u_1(\boldsymbol{\xi},\, t)\right] \mathrm{d}\Sigma(\boldsymbol{\xi})$$

is the average dislocation, A is the area of the fault surface, and $M_0(t)$ represents the *seismic moment*. From the equation, it is clear that $M_{13} = M_{31} = M_0(t)$, which is identical in magnitude to the moment generated by $f_1^{[u]}$ in Eq. (3.3.8). This reveals that $M_{pq}(t)$ represents the moment of some equivalent body forces. Additionally, the definition of $M_{pq}(t)$ easily confirms its alignment with the criteria for a second-order tensor, as outlined in Eq. (2.1.15). Consequently, $M_{pq}(t)$, defined according to Eq. (3.4.3), is known as the *seismic moment tensor*, while the integrand $m_{pq}(\boldsymbol{\xi}, t)$ in its integral form is referred to as the *seismic moment density tensor*.

According to Eq. (3.4.3), the properties of the elastic coefficients C_{ijpq} (specifically $C_{ijpq} = C_{ijqp}$) indicate that $M_{pq} = M_{qp}$, meaning $M_{pq}(t)$ is a symmetric tensor. In the case of shear dislocations within an isotropic medium, it's straightforward to derive from Eq. (3.4.4) that $M_{pp} = 0$. This shows that for shear dislocations, the trace of the seismic moment tensor is zero.

In a typical Cartesian coordinate system, for plane shear dislocations within an isotropic homogeneous medium, assume the unit vector in the direction of dislocation is $\hat{\boldsymbol{e}}$. Consequently, $[u_i(\boldsymbol{\xi}, t)] = [u(\boldsymbol{\xi}, t)]e_i$. The spatial average of $[u(\boldsymbol{\xi}, t)]$ is denoted as $\overline{[u(t)]}$. Therefore, based on Eq. (3.4.3), the corresponding seismic moment tensor can be expressed as

$$M_{pq}(t) = \mu \iint_{\Sigma} \left[u_i(\boldsymbol{\xi},\, t)\right] \nu_j(\delta_{ip}\delta_{jq} + \delta_{iq}\delta_{jp})\,\mathrm{d}\Sigma(\boldsymbol{\xi}) = M_0(t)(e_p\nu_q + e_q\nu_p) \tag{3.4.5}$$

where $M_0(t) = \mu\overline{[u(t)]}A$. In the expression for $M_{pq}(t)$, the roles of $\hat{\boldsymbol{e}}$ and $\hat{\boldsymbol{\nu}}$ are completely equivalent. In other words, swapping these two vectors does not affect the seismic moment tensor $M_{pq}(t)$. This suggests that for seismic waves emitted from a shear dislocation point source, it is impossible to distinguish between the two mutually perpendicular planes, one being the fault plane and the other, referred to as the *auxiliary plane*.

3.4.2 Physical Significance of Seismic Moment Tensor

In Sect. 3.2, we treat the fault plane as two closely adjacent surfaces with a displacement discontinuity, using these surfaces as internal boundaries. Within the region enclosed by these boundaries, the displacement remains single-valued and continuous. Hence, we derive the source representation theorem based on the displacement representation theorem valid in the context of continuum mechanics. However, if we focus specifically on the fault plane itself, the displacement discontinuities occurring there do not meet the conditions for continuity of displacement. Therefore, the source process is clearly not explainable by elastic theories based on the assumption of a continuous medium; the displacement discontinuities are manifestations of the inelastic effects at the source. It's conceivable that the moment density tensor $m_{pq}(\boldsymbol{\xi}, \tau)$, as defined by dislocations (see Eq. (3.4.1)), also inherently contains information about inelasticity, related to the inelastic processes occurring at the source. Let's consider this issue from a more general perspective (Chen and Gu, 2023).

In Sect. 3.2.1, while discussing the model and assumptions underlying the source representation theorem, it was noted that the seismic process involves complex physical phenomena that cannot be solely described by the elastic constitutive laws following the generalized Hooke's law. Elastic processes alone do not create a seismic source. Therefore, from a broader perspective, seismic sources can be considered as volumetric sources that deviate from elastic strain. This process involves plastic deformation, thermal effects, and phase transitions, all contributing to the occurrence of stress-free strains in certain regions.[30] The total strain ε_{pq} is the sum of the elastic strain ε^{e}_{pq}, resulting from external forces, and the stress-free strain (i.e., the inelastic strain) ε^{*}_{pq} produced by inelastic effects

$$\varepsilon_{pq} = \varepsilon^{e}_{pq} + \varepsilon^{*}_{pq}$$

[30] Let's consider the strain produced by thermal expansion as an example. At a specific temperature, the application of an external force typically results in both stress and strain. However, the scenario changes when temperature variations accompany the external force. Imagine an elastic body that expands during a temperature increase without any external force acting on it. Since strain measures deformation relative to the state before any forces acted, the expansion due to increased temperature in this elastic body results in a positive strain. Yet, because the body is not constrained, no additional internal forces are generated, and thus no stress is produced. Of course, if external forces are present during the temperature increase, these forces will continue to induce both stress and strain. The final stress is caused by the external force, while the total strain is the sum of the part induced by the external force and the part resulting from the temperature change.

Stress and elastic strain adhere to the generalized Hooke's law as stated in Eq. (2.2.18),

$$\tau_{ij} = C_{ijpq}\varepsilon^{e}_{pq} = C_{ijpq}(\varepsilon_{pq} - \varepsilon^{*}_{pq})$$

Inserting it into the elastic motion Eq. (2.2.17), which does not involve external forces ($f_i = 0$), we find that

$$(C_{ijpq}\varepsilon_{pq})_{,j} - (C_{ijpq}\varepsilon^{*}_{pq})_{,j} = \rho\ddot{u}_i$$

If we consider $-(C_{ijpq}\varepsilon^{*}_{pq})_{,j}$ as the equivalent body force $f_i^{[u]}$,

$$f_i^{[u]} = -(C_{ijpq}\varepsilon^{*}_{pq})_{,j}$$

If we derive stress from the total strain using the elastic relationship, it still satisfies the elastic motion equation,

$$(C_{ijpq}\varepsilon_{pq})_{,j} + f_i^{[u]} = \rho\ddot{u}_i$$

If we define the *non-elastic stress* τ^{*}_{ij} as the stress formally derived from the non-elastic strain according to the generalized Hooke's law,[31] i.e.,

$$\tau^{*}_{ij} = C_{ijpq}\varepsilon^{*}_{pq}$$

then we have

$$f_i^{[u]} = -\tau^{*}_{ij,j}, \quad \text{i.e.} \quad \boldsymbol{f}^{[u]} = -\nabla\cdot\boldsymbol{\tau}^{*}$$

Comparing Eq. (3.3.6), it is apparent that

$$m_{ij}(\boldsymbol{\eta}, \tau) = \tau^{*}_{ij}(\boldsymbol{\eta}, \tau) \tag{3.4.6}$$

Equation (3.4.6) indicates that the seismic moment density tensor is equal to the non-elastic stress calculated using the linear elastic constitutive relationship from non-elastic strains. This non-elastic stress, τ^{*}_{ij}, also referred to as the *stress glut* (Backus and Mulcahy 1976), is derived by subtracting the actual stress caused by external forces, τ_{ij}, from the model stress $C_{ijpq}\varepsilon_{pq}$, which is the stress calculated from the true strain ε_{pq} using generalized Hooke's Law.

In summary, the seismic moment density tensor represents the non-elastic stress tensor, or stress glut, in the earthquake source region, which corresponds to the stress associated with deformations caused by non-elastic processes within the source. It is important to note that this analysis views the earthquake source as a physical

[31] This definition may seem unusual. It's important to emphasize that it is merely defined using the form of elastic constitutive relationships and does not imply any connection between non-elastic and elastic stresses.

process occurring within a specific volume. Hence, Eq. (3.4.6) defines the moment density tensor at a point in space. Specifically, for a dislocation source, the moment density tensor is defined on the fault plane as seen in Eq. (3.4.1). The non-elastic strain corresponds to the dislocation occurring on the fault plane.[32] As expressed by the area integral in Eq. (3.4.3), the seismic moment tensor is a cumulative measure of these effects across the fault surface.

3.4.3 Specific Expression of Seismic Moment Tensor

Based on Eq. (3.4.4), in isotropic homogeneous media, the seismic moment tensor $M_{pq}(t)$ for plane shear fractures is simplified such that $M_{13}(t) = M_{31}(t) = M_0(t)$, with all other components being zero. This representation is a streamlined expression of the seismic moment tensor. However, for the ease of analysis in certain situations, it is necessary to express the seismic moment tensor in different coordinate systems. For instance, consider a coordinate system centered at the epicenter (the surface projection of the earthquake's initiation point) with the north and east directions defining the x_1 and x_2 axes, respectively, and the downward direction perpendicular to the surface as the x_3 axis, known as *epicentral coordinate system*, as shown in Fig. 3.6.

The fault plane can appear in any form within this epicentral coordinate system. To quantitatively describe the shear fractures on the fault plane in this system, several critical angles must be introduced.

(1) The strike of a fault, denoted as ϕ_s, refers to the direction along which the line of intersection between the fault plane and a horizontal plane extends. This direction, illustrated by the dashed line in Fig. 3.6, is called the strike of the fault. ϕ_s is measured clockwise from north; when facing in the direction of the strike, the hanging wall of the fault appears on the right-hand side of the observer. The possible range for ϕ_s is $[0°, 360°)$.
(2) The dip of a fault, denoted as δ, is defined by the angle between the line on the fault plane perpendicular to the strike direction, known as the dip line, and its projection on the earth's surface. This angle corresponds to the angle between the normal vector $\hat{\boldsymbol{\nu}}$ of the fault plane and the vertical axis, measured from the $-x_3$ direction as shown in Fig. 3.6. The range for δ is from $[0°, 90°]$.

[32] From a certain perspective, in the theory of faults, we can consider the complex non-elastic processes occurring on the fault plane as equivalent to displacements due to faults. It is important to emphasize again that although the expression for the seismic moment density tensor includes the elastic coefficients C_{ijpq}, this is merely a use of the form of elastic constitutive relationships. It does not imply that the processes occurring in the earthquake source region are based on an elastic model.

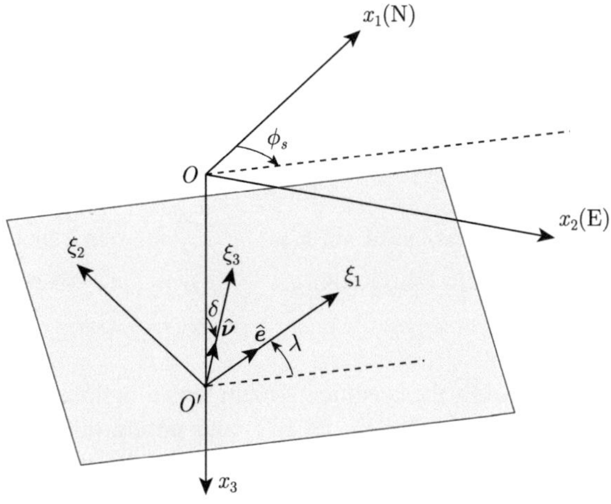

Fig. 3.6 Planar fault in the epicentral coordinate system. The point O marks the epicenter, which is the surface projection of the earthquake's source. The x_1 and x_2 axes point north and east respectively, while the x_3 axis extends downward, perpendicular to the surface. The fault plane is depicted as a gray rectangle. The $O\xi_1\xi_2\xi_3$ coordinate system is centered at the earthquake source, with $\hat{\boldsymbol{e}}$ representing the unit vector in the direction of dislocation, and $\hat{\boldsymbol{\nu}}$ as the normal vector to the fault plane. The dashed line indicates the fault's strike, and the angles δ, ϕ_s and λ represent the dip, strike, and rake of the fault plane respectively

(3) The rake or slip angle of a fault, denoted as λ, is defined by the angle between the direction of slip on the fault plane, represented by $\hat{\boldsymbol{e}}$, and the strike line. This angle is measured counterclockwise from the strike line, with a possible value range $(-180°,\ 180°]$ (though sometimes the range is given as $[0°,\ 360°)$). The vector $\hat{\boldsymbol{e}}$ represents the slip direction on the fault plane. The difference in displacement between the upper and lower fault surfaces, indicated by $[\boldsymbol{u}(\boldsymbol{\xi}, \tau)]$, determines the type of fault: a reverse fault for angles $0° < \lambda < 180°$ and a normal fault for angles $-180° < \lambda < 0°$. Specifically, $\lambda = 0°$ indicates a left-lateral strike-slip fault, where if you stand on one side of the fault, the other side appears to move from right to left. Conversely, $\lambda = 180°$ denotes a right-lateral strike-slip fault. When $\lambda = 90°$, the fault is classified as a dip-slip fault.

After introducing these angles, we can use them to express the components of the seismic moment tensor in the epicenter coordinate system. There are two approaches: one is by utilizing coordinate transformation. Given that the seismic moment tensor is a second-order tensor, it adheres to the coordinate transformation equation for second-order tensors (2.1.15). This allows us to obtain the expression in the epicenter coordinate system by rotating the axes from the source coordinate system. The other approach is based on Eq. (3.4.5), where we directly substitute the expressions of e_i and ν_i in the epicenter coordinate system. In the following, we will illustrate the first method, and leave the solution by the second method as an exercise for the reader.

If we consider the epicenter coordinate system $Ox_1x_2x_3$ in Fig. 3.6 as the new primed coordinate system and the source coordinate system $O'\xi_1\xi_2\xi_3$ as the old unprimed coordinate system in Eq. (2.1.15), then the components of the transformation matrix between the two coordinate systems are $c_{ij} = \cos(\hat{\boldsymbol{x}}_i, \hat{\boldsymbol{\xi}}_j)$. Based on the geometric relationships shown in Fig. 3.6,[33] we have

$$c_{11} = \cos\lambda\cos\phi_s + \cos\delta\sin\lambda\sin\phi_s, \qquad c_{13} = -\sin\delta\sin\phi_s \tag{3.4.7a}$$

$$c_{21} = \cos\lambda\sin\phi_s - \cos\delta\sin\lambda\cos\phi_s, \qquad c_{23} = \sin\delta\cos\phi_s \tag{3.4.7b}$$

$$c_{31} = -\sin\lambda\sin\delta, \qquad c_{33} = -\cos\delta \tag{3.4.7c}$$

According to Eq. (2.1.15), the seismic moment tensor in the epicenter coordinate system is $M'_{ij} = c_{ip}c_{jq}M_{pq}$. With Eq. (3.4.7), and noting that the only non-zero components are $M_{13}(t) = M_{31}(t) = M_0(t)$ (see Eq. (3.4.4)), we obtain

$$\begin{aligned} M'_{11}(t) &= c_{11}c_{13}M_{13}(t) + c_{13}c_{11}M_{31}(t) \\ &= -M_0(t)(\sin\delta\cos\lambda\sin 2\phi_s + \sin 2\delta\sin\lambda\sin^2\phi_s) \end{aligned} \tag{3.4.8a}$$

$$\begin{aligned} M'_{12}(t) &= M'_{21}(t) = c_{11}c_{23}M_{13}(t) + c_{13}c_{21}M_{31}(t) \\ &= M_0(t)\left(\sin\delta\cos\lambda\cos 2\phi_s + \frac{1}{2}\sin 2\delta\sin\lambda\sin 2\phi_s\right) \end{aligned} \tag{3.4.8b}$$

$$\begin{aligned} M'_{13}(t) &= M'_{31}(t) = c_{11}c_{33}M_{13}(t) + c_{13}c_{31}M_{31}(t) \\ &= -M_0(t)\,(\cos\delta\cos\lambda\cos\phi_s + \cos 2\delta\sin\lambda\sin\phi_s) \end{aligned} \tag{3.4.8c}$$

$$\begin{aligned} M'_{22}(t) &= c_{21}c_{23}M_{13}(t) + c_{23}c_{21}M_{31}(t) \\ &= M_0(t)(\sin\delta\cos\lambda\sin 2\phi_s - \sin 2\delta\sin\lambda\cos^2\phi_s) \end{aligned} \tag{3.4.8d}$$

$$\begin{aligned} M'_{23}(t) &= M'_{32}(t) = c_{21}c_{33}M_{13}(t) + c_{23}c_{31}M_{31}(t) \\ &= -M_0(t)\,(\cos\delta\cos\lambda\sin\phi_s - \cos 2\delta\sin\lambda\cos\phi_s) \end{aligned} \tag{3.4.8e}$$

$$M'_{33}(t) = M_0(t)\sin 2\delta\sin\lambda \tag{3.4.8f}$$

---

[33] Take c_{11} as an example. It represents the projection of the unit vector $\hat{\boldsymbol{e}}$ along the ξ_1 axis onto the x_1 axis. First, decompose $\hat{\boldsymbol{e}}$ on the fault plane into components along the dashed line (strike direction) $\hat{\boldsymbol{e}}^{(1)}$ and perpendicular to the dashed line $\hat{\boldsymbol{e}}^{(2)}$. Clearly, $|\hat{\boldsymbol{e}}^{(1)}| = \cos\lambda$ and $|\hat{\boldsymbol{e}}^{(2)}| = \sin\lambda$. Translate $\hat{\boldsymbol{e}}^{(1)}$ to the dashed line in the x_1x_2 plane and project it onto the x_1 direction to get $\cos\lambda\cos\phi_s$. For $\hat{\boldsymbol{e}}^{(2)}$, first project it onto the plane parallel to x_1x_2 to get $\sin\lambda\cos\delta$, which lies on a line perpendicular to the dashed line. Then, further project this onto the x_1 axis to get $\cos\delta\sin\lambda\sin\phi_s$. The sum of these two parts gives c_{11}, which is:

$$c_{11} = \cos\lambda\cos\phi_s + \cos\delta\sin\lambda\sin\phi_s$$

Similarly, all other components can be derived.

The above results provide the seismic moment tensor in the epicenter coordinate system,[34] expressed in terms of the seismic moment $M_0(t)$ and the three angles δ, ϕ_s, and λ. This demonstrates that a pure shear dislocation source can be represented using these four quantities, sometimes referred to as *fault parameters*.

3.5 Summary

The generation of earthquake source and the propagation of seismic waves through the Earth's medium are two key topics in seismology. Building on the displacement representation theorem obtained in Chap. 2 we use dislocation theory to derive the source representation theorem in this chapter. This theorem links the slip on the fault plane with the radiated seismic wave field in the Earth's medium, forming the basis for many research directions in seismology. Utilizing this theorem, we also thoroughly discusses the concept of equivalent body forces, providing both a mathematical expression and a clear, simple model: the double-couple model. Additionally, we delves into the concepts of seismic moment density tensor and seismic moment tensor, explaining that the seismic moment density tensor corresponds to the stress during inelastic processes in the source region, and presents the seismic moment tensor in the epicenter coordinate system expressed using four fault parameters.

The source representation theorem indicates that if the dislocation function on the fault plane is known, the seismic wave field can be determined; conversely, the spatial and temporal distribution of the dislocation can be inferred from seismic records. The link between the dislocation and the seismic wave field is the spatial derivative of the Green's function. In seismology, Green's functions are solutions to the equations of elastic motion under the condition of a concentrated impulsive force, and the equations themselves remain unchanged. However, it is important to note that Green's functions vary with different medium models and boundary conditions. Therefore, the next major task is to determine the Green's functions and their spatial derivatives for various medium models.

Since the mid-19th century, countless mathematicians, mechanicians, and geophysicists have devoted themselves to studying Green's functions, starting with those in infinite homogeneous media. Among the most influential contributions to seismology is Lamb's pioneering work in 1904, which introduced the Lamb's problem. Over the past century, especially with the advent of computers, numerical solutions for seismic wave problems in various complex media have become a major research focus. However, the half-space medium, as the simplest and most essential approximation model of the Earth, continues to attract significant attention. The solution for the half-space model comprises two parts: the contribution from the infinite elastic

[34] As a second-order tensor, the seismic moment tensor naturally possesses all the properties of second-order tensors. For instance, it has eigenvalues and eigenvectors, can be decomposed in various ways, and can be represented using intuitive geometric methods, among other things. These aspects are not highly relevant to the subsequent discussion, so they will be omitted.

medium itself and the contribution from the free surface. Due to the interactions of elastic waves near the free surface, much more complex phenomena occur in a semi-infinite elastic medium than in an infinite one. A typical example is the generation of Rayleigh waves. Before delving into these complex effects, it is necessary to focus on the seismic wave problems associated with an infinite elastic medium without a free surface. In the upcoming Chap. 4, we will first consider the seismic wave problems in an infinite homogeneous medium.

Chapter 4
Seismic Waves in Infinite Homogeneous Medium

Starting from this chapter, our primary focus will be on solving seismic wave problems in two simplified medium models: an isotropic homogeneous infinite medium and an isotropic homogeneous semi-infinite medium. The main topic of this book, the Lamb's problem, pertains to the latter. However, the infinite medium is the only scenario in elastodynamics where simple closed-form solutions exist, serving as the foundation for further consideration of the semi-infinite medium problem. In this chapter, we will first address seismic wave issues within this simple model, with a focus on Green's functions in infinite space and the displacement field caused by a dislocation source.

In various boundary value problems where the field equations are fixed but the boundary conditions differ, the model of an infinite medium provides boundary conditions at infinity. From the perspective of solving mathematical physics equations, this is undoubtedly the simplest case. Therefore, starting with this scenario is a good choice, as it allows us to focus on handling the equations themselves without the need to deal with complex boundary conditions.[1]

In this chapter, we will use the potential function method based on Lamé's theorem. By separately solving the wave equations corresponding to the potential functions for P-wave and S-wave, we will ultimately combine them to form the desired Green's function. The Green's function in infinite space has a straightforward form,

[1] For a partial differential equation, the equation itself is not enough to find a solution; specific boundary conditions must be applied. When the governing equation is fixed, the solution can vary greatly depending on the boundary conditions. In fact, the complexity of some problems often arises not from the equation itself but from the boundary conditions. The Green's function solutions for the two models considered in this book provide an excellent example. In the case of an infinite medium, the solution is quite simple; however, for a semi-infinite medium, the presence of the boundary makes the problem much more complex. In the following chapters of this book, we will discuss and analyze this issue in depth.

H. Zhang, *Lamb's Problem in Seismology (Volume 1)*,
https://doi.org/10.1007/978-981-95-3170-7_4

making the displacement field caused by a dislocation source relatively simple as well. We will analyze its properties from both spatial and temporal perspectives. These results will lay the groundwork for solving the Lamb's problem in subsequent chapters.

4.1 Solution Strategy

At the end of Sect. 2.2, we obtained Eq. (2.2.21), which serves as the starting point for solving the Green's function. Its component form is as follows,

$$\rho \ddot{u}_i = (\lambda + \mu) u_{j,ji} + \mu u_{i,jj} + f_i \tag{4.1.1}$$

This is a very complex partial differential equation, more intricate than the Laplace, Poisson, Helmholtz, heat conduction, and wave equations we are more familiar with. Before we proceed with solving it, it is essential to review the tools we have and select the most suitable one for the task.

To solve a partial differential equation, the general approach follows the principle of "transforming the unknown into the known and the complex into the simple". Various methods are employed to simplify it into ordinary differential equations or even algebraic equations, making it easier to solve. A primary technique is using integral transforms. Common integral transforms like the Laplace transform and Fourier transform possess the property of converting differentiation in the original domain into multiplication in the transform domain, thus simplifying the operations. For problems with multiple variables, partial or complete integral transforms of the independent variables can be used to achieve simplification.

Of course, if the problem itself has some symmetry, utilizing this symmetry to simplify the problem first can often yield highly efficient results. This is the case with our current problem. The medium model we are solving for is an infinite space model, and the force is a unit concentrated impulse. Therefore, only the point of force application is unique within the entire spatial region. If we take this point as the coordinate origin, the wave field is the same in all directions, giving the problem spherical symmetry. This means if we use spherical coordinates (r, θ, ϕ) to solve it, the function we seek depends only on r and not on θ or ϕ. Thus, without needing any integral transforms, the problem automatically reduces to an ordinary differential equation.[2]

Reflecting on the methods for solving stress problems in elastic mechanics, there is an approach known as the potential function method. The basic idea is that for conservative forces, by introducing potential functions, the equilibrium equations

[2] However, it's important to note that this spherical symmetry is a convenience only available in an infinite medium model. If we consider a half-space problem with finite boundaries, such as the Lamb's problem discussed in the following chapters, this symmetry no longer exists, and we will need to find alternative methods.

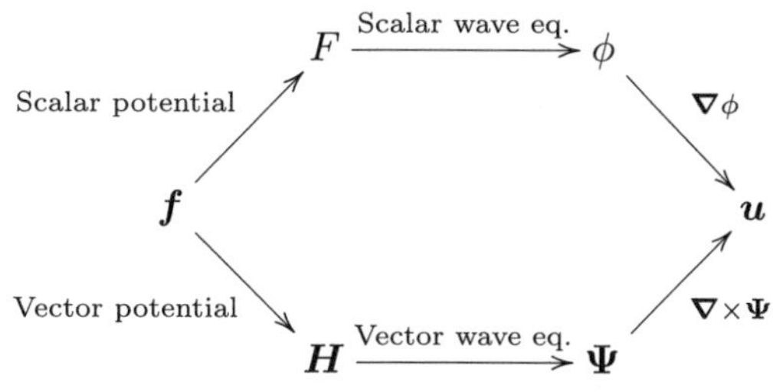

Fig. 4.1 The approach for solving elastodynamic equations based on Lamé's theorem

can be expressed in terms of stress components using these potential functions. This transforms the equilibrium equations into equations that the stress functions must satisfy, reducing the number of equations to be solved. By solving the stress functions, we can then determine the stress components. We can apply a similar approach to the problem at hand. Using the Helmholtz decomposition for vectors, we can express the displacement field in terms of two sets of potential functions. According to the conclusion of Lamé's theorem, these two sets of potential functions satisfy scalar and vector wave equations, respectively. This transforms the solution of complex equations into solving relatively simpler wave equations, thereby solving the entire problem, as shown in Fig. 4.1.

4.2 Lamé's Theorem

The theorem states: For a displacement field that satisfies the elastic motion Eq. (4.1.1), if the body force $\boldsymbol{f}$ can be expressed using Helmholtz potential,[3] then

$$\boldsymbol{f} = \nabla F + \nabla \times \boldsymbol{H} \tag{4.2.1}$$

where $\nabla \cdot \boldsymbol{H} = 0$, then there exist potential functions ϕ and $\boldsymbol{\Psi}$ such that

(1) $\boldsymbol{u} = \nabla\phi + \nabla \times \boldsymbol{\Psi}$, and $\nabla \cdot \boldsymbol{\Psi} = 0$;
(2) $\ddot{\phi} = F/\rho + \alpha^2\nabla^2\phi$, and $\alpha^2 = (\lambda + 2\mu)/\rho$;
(3) $\ddot{\boldsymbol{\Psi}} = \boldsymbol{H}/\rho + \beta^2\nabla^2\boldsymbol{\Psi}$, and $\beta^2 = \mu/\rho$.

[3] According to Helmholtz's theorem, for any vector $\boldsymbol{a}$, there always exist a scalar potential φ and a vector potential $\boldsymbol{b}$ such that $\boldsymbol{a}$ can be expressed as $\boldsymbol{a} = \nabla\varphi + \nabla \times \boldsymbol{b}$, and $\nabla \cdot \boldsymbol{b} = 0$. This can be proven by constructing a Poisson equation. Assume $\boldsymbol{u}$ is a solution to the Poisson equation $\nabla^2\boldsymbol{u} = \boldsymbol{a}$. Noting that, according to Eq. (2.1.18),

$$\nabla^2\boldsymbol{u} = \nabla(\nabla \cdot \boldsymbol{u}) - \nabla \times (\nabla \times \boldsymbol{u}) = \boldsymbol{a}$$

it is evident that taking $\varphi = \nabla \cdot \boldsymbol{u}$ and $\boldsymbol{b} = -\nabla \times \boldsymbol{u}$ suffices.

Proof. Assume that the initial values of displacement and velocity $\boldsymbol{u}(\boldsymbol{x}, 0)$ and $\dot{\boldsymbol{u}}(\boldsymbol{x}, 0)$ are represented by the Helmholtz decomposition as

$$\dot{\boldsymbol{u}}(\boldsymbol{x}, 0) = \nabla A + \nabla \times \boldsymbol{B}, \quad \boldsymbol{u}(\boldsymbol{x}, 0) = \nabla C + \nabla \times \boldsymbol{D}$$

in which, $\nabla \cdot \boldsymbol{B} = \nabla \cdot \boldsymbol{D} = 0$. Let[4]

$$\phi(\boldsymbol{x}, t) = \frac{1}{\rho} \int_0^t (t - \tau)\big[F(\boldsymbol{x}, \tau) + (\lambda + 2\mu)\nabla \cdot \boldsymbol{u}(\boldsymbol{x}, \tau)\big]\mathrm{d}\tau + tA + C$$

$$\boldsymbol{\Psi}(\boldsymbol{x}, t) = \frac{1}{\rho} \int_0^t (t - \tau)\big[\boldsymbol{H}(\boldsymbol{x}, \tau) - \mu\nabla \times \boldsymbol{u}(\boldsymbol{x}, \tau)\big]\mathrm{d}\tau + t\boldsymbol{B} + \boldsymbol{D}$$

Therefore, the following equation holds

$$\begin{aligned}
\nabla\phi + \nabla \times \boldsymbol{\Psi} &= \int_0^t (t - \tau)\ddot{\boldsymbol{u}}(\boldsymbol{x}, \tau)\mathrm{d}\tau + t\dot{\boldsymbol{u}}(\boldsymbol{x}, 0) + \boldsymbol{u}(\boldsymbol{x}, 0) \\
&= [(t - \tau)\dot{\boldsymbol{u}}(\boldsymbol{x}, \tau)]\Big|_0^t + \int_0^t \dot{\boldsymbol{u}}(\boldsymbol{x}, \tau)\mathrm{d}\tau + t\dot{\boldsymbol{u}}(\boldsymbol{x}, 0) + \boldsymbol{u}(\boldsymbol{x}, 0) \\
&= -t\dot{\boldsymbol{u}}(\boldsymbol{x}, 0) + \boldsymbol{u}(\boldsymbol{x}, t) - \boldsymbol{u}(\boldsymbol{x}, 0) + t\dot{\boldsymbol{u}}(\boldsymbol{x}, 0) + \boldsymbol{u}(\boldsymbol{x}, 0) = \boldsymbol{u}(\boldsymbol{x}, t)
\end{aligned}$$

And since $\nabla \cdot \nabla \times \boldsymbol{u} = \nabla \cdot \boldsymbol{H} = \nabla \cdot \boldsymbol{B} = \nabla \cdot \boldsymbol{D} = 0$, it is clear that $\nabla \cdot \boldsymbol{\Psi} = 0$.

Using the derivative relationship of the integral with a parameter

$$\frac{\partial}{\partial t} \int_0^t f(t, \tau)\mathrm{d}\tau = \int_0^t \frac{\partial}{\partial t} f(t, \tau)\mathrm{d}\tau + f(t, t) \quad \text{and} \quad \frac{\partial}{\partial t} \int_0^t f(\tau)\mathrm{d}\tau = f(t)$$

and

$$\nabla \cdot \boldsymbol{u} = \nabla \cdot \nabla\phi + \nabla \cdot \nabla \times \boldsymbol{\Psi} = \nabla^2\phi, \quad \nabla \times \boldsymbol{u} = \nabla \times \nabla\phi + \nabla \times \nabla \times \boldsymbol{\Psi} = -\nabla^2\boldsymbol{\Psi}$$

it can be obtained that

$$\begin{aligned}
\ddot{\phi} &= \frac{1}{\rho}\frac{\partial}{\partial t} \left\{ \int_0^t [F(\boldsymbol{x}, \tau) + (\lambda + 2\mu)\nabla \cdot \boldsymbol{u}(\boldsymbol{x}, \tau)]\, \mathrm{d}\tau + A \right\} \\
&= \frac{F}{\rho} + \alpha^2\nabla \cdot \boldsymbol{u} = \frac{F}{\rho} + \alpha^2\nabla^2\phi
\end{aligned}$$

---

[4] The conclusion of Lamé's theorem is that there exist two potential functions that satisfy certain properties. If the forms of these two potential functions are explicitly given, the theorem can be proven by verifying that they meet all the required properties. The key issue is how to construct these two potential functions. Inspiration can be drawn from the form of the equations of motion that $\boldsymbol{u}$ satisfies, and the result that $\nabla\phi + \nabla \times \boldsymbol{\Psi}$ should equal $\boldsymbol{u}$. Specifically, ϕ should contain $F + (\lambda + 2\mu)\nabla \cdot \boldsymbol{u}$, and $\boldsymbol{\Psi}$ should contain $\boldsymbol{H} - \mu\nabla \times \boldsymbol{u}$. By constructing these two functions as convolutions with t and making slight adjustments using the initial conditions, the proof can be achieved.

Similarly, we have

$$\ddot{\boldsymbol{\Psi}} = \frac{\boldsymbol{H}}{\rho} - \beta^2 \nabla \times \boldsymbol{u} = \frac{\boldsymbol{H}}{\rho} + \beta^2 \nabla^2 \boldsymbol{\Psi}$$

Lamé's theorem states that if we have a scalar potential function ϕ and a vector potential function $\boldsymbol{\Psi}$ that satisfy the scalar and vector wave equations respectively, then the sum of the gradient of ϕ and the curl of $\boldsymbol{\Psi}$ gives us the displacement field $\boldsymbol{u}$ we need to solve for.[5] This provides a method for finding the Green's function in an infinite space: solve the wave equations for the scalar potential ϕ associated with P-wave and the vector potential $\boldsymbol{\Psi}$ associated with S-wave, and the displacement field we are looking for is $\boldsymbol{u} = \nabla\phi + \nabla \times \boldsymbol{\Psi}$. Therefore, obtaining the solutions to the wave equations is the first step.

4.3 Solution of the Wave Equation

In a Cartesian coordinate system, the vector wave equation can be seen as a combination of wave equations satisfied by each component of the vector. Therefore, we only need to consider the scalar wave equation. Our goal is to find the solution of the wave equation of the following form, under the conditions of infinite boundary and zero initial conditions,

$$\ddot{\phi}(\boldsymbol{x}, t) = \frac{F(\boldsymbol{x}, t)}{\rho} + \alpha^2 \nabla^2 \phi \tag{4.3.1}$$

There are various methods to solve the wave equation. In this book, we use the displacement representation theorem to obtain the displacement field, which then requires solving for the Green's function. Essentially, this approach is a Green's function method. To gain a deeper understanding of this method, we will use the Green's function method to solve the wave equation mentioned above. We will first solve for the Green's function of the corresponding wave equation, and then use it to construct the solution to Eq. (4.3.1).

[5] Conversely, the same holds true: if we decompose the displacement field $\boldsymbol{u}$ as in the conclusion (1) of Lamé's theorem, then the two potential functions will satisfy the scalar and vector wave equations respectively. Additionally, because the propagation speeds of these wave equations correspond to the speeds of P-wave and S-wave, this decomposition of $\boldsymbol{u}$ aligns with the decomposition into P-wave and S-wave. Note that $\nabla \times \nabla\phi = 0$ and $\nabla \cdot \nabla \times \boldsymbol{\Psi} = 0$, which from a field theory perspective means that P-wave are irrotational fields and S-wave are solenoidal fields. It is necessary to point out that this is not the only way to decompose displacement. There are other methods, such as decomposing displacement according to the polarization direction into longitudinal and transverse waves. These two methods are not always equivalent, and only under far-field conditions are they so. We will clarify this in Sect. 4.5.1.

4.3.1 Green's Function Solution of the Wave Equation

Let the Green's function of the wave equation be denoted as $g(\boldsymbol{x}, t; \boldsymbol{\xi}, \tau)$, which satisfies the following boundary value problem

$$\ddot{g}(\boldsymbol{x}, t; \boldsymbol{\xi}, \tau) = \delta(\boldsymbol{x} - \boldsymbol{\xi})\delta(t - \tau) + c^2\nabla^2 g(\boldsymbol{x}, t; \boldsymbol{\xi}, \tau) \tag{4.3.2a}$$

$$g(\boldsymbol{x}, t; \boldsymbol{\xi}, \tau)\Big|_{x\to\infty} = 0 \tag{4.3.2b}$$

$$g(\boldsymbol{x}, t; \boldsymbol{\xi}, \tau)\Big|_{t<\tau} = 0, \quad \dot{g}(\boldsymbol{x}, t; \boldsymbol{\xi}, \tau)\Big|_{t<\tau} = 0 \tag{4.3.2c}$$

The wave equation involves three spatial variables x_i and one time variable t. To simplify the analysis, we first apply the Laplace transform, which makes it easier to use the initial conditions, and convert the equation to the Laplace domain. Then, using spatial symmetry, we reduce it to an ordinary differential equation for further handling. Let's denote

$$\bar{g} = \bar{g}(\boldsymbol{x}, p; \boldsymbol{\xi}, \tau) = \mathscr{L}\{g(\boldsymbol{x}, t; \boldsymbol{\xi}, \tau)\} = \int_0^{+\infty} g(\boldsymbol{x}, t; \boldsymbol{\xi}, \tau)\mathrm{e}^{-pt}\mathrm{d}t$$

where $\mathscr{L}\{\cdot\}$ denotes the Laplace transform. According to the properties of the Laplace transform

$$\mathscr{L}\{\ddot{g}(\boldsymbol{x}, t; \boldsymbol{\xi}, \tau)\} = p^2\bar{g}(\boldsymbol{x}, p; \boldsymbol{\xi}, \tau) - pg(\boldsymbol{x}, 0; \boldsymbol{\xi}, \tau) - \dot{g}(\boldsymbol{x}, 0; \boldsymbol{\xi}, \tau)$$

Combining with Eq. (4.3.2c), we have $\mathscr{L}\{\ddot{g}(\boldsymbol{x}, t; \boldsymbol{\xi}, \tau)\} = p^2\bar{g}$. Applying the Laplace transform to both sides of Eq. (4.3.2a), we get

$$c^2\nabla^2\bar{g} - p^2\bar{g} = -\delta(\boldsymbol{x} - \boldsymbol{\xi})\mathrm{e}^{-p\tau} \tag{4.3.3}$$

This is an equation in the Laplace domain that involves derivatives with respect to only three spatial variables. Considering the symmetry of the problem, if we use spherical coordinates with the point of force application $\boldsymbol{\xi}$ as the origin, where the coordinates are (r, θ, φ), $\bar{g}$ depends only on r and is independent of θ and φ. Thus, based on the expression for ∇^2 in spherical coordinates

$$\nabla^2 = \frac{1}{r^2}\frac{\partial}{\partial r}\left(r^2\frac{\partial}{\partial r}\right) + \frac{1}{r^2\sin\theta}\frac{\partial}{\partial\theta}\left(\sin\theta\frac{\partial}{\partial\theta}\right) + \frac{1}{r^2\sin^2\theta}\frac{\partial^2}{\partial\varphi^2} = \frac{1}{r^2}\frac{\mathrm{d}}{\mathrm{d}r}\left(r^2\frac{\mathrm{d}}{\mathrm{d}r}\right)$$

Equation (4.3.3) can be simplified as

$$c^2\frac{1}{r^2}\frac{\mathrm{d}}{\mathrm{d}r}\left(r^2\frac{\mathrm{d}\bar{g}}{\mathrm{d}r}\right) - p^2\bar{g} = -\delta(\boldsymbol{x})\mathrm{e}^{-p\tau}$$

This is a special ordinary differential equation that is homogeneous under normal circumstances (when $r \neq 0$), and it becomes non-homogeneous only when $r = 0$. Note that when $r \neq 0$,

$$c^2 \frac{1}{r^2} \frac{\mathrm{d}}{\mathrm{d}r} \left(r^2 \frac{\mathrm{d}\bar{g}}{\mathrm{d}r} \right) - p^2 \bar{g} = c^2 \frac{\mathrm{d}^2}{\mathrm{d}r^2} (r\bar{g}) - p^2 (r\bar{g}) = 0$$

it is not difficult to obtain

$$\bar{g} = \frac{A}{r} \mathrm{e}^{\pm \frac{p}{c} r} \quad \xrightarrow[\mathrm{Re}(p),c,r>0]{(4.3.2\mathrm{b})} \quad \bar{g} = \frac{A}{r} \mathrm{e}^{-\frac{p}{c} r}$$

where A is an undetermined coefficient that needs to be determined from the non-homogeneous equation at $r = 0$. Given that the non-homogeneous term in Eq. (4.3.3) is the Dirac δ function, we take a small sphere V_ε with its center at the origin (i.e., the point $\boldsymbol{\xi}$) and radius $\varepsilon \to 0$. Then, we integrate both sides of Eq. (4.3.3) over the volume V_ε,

$$\iiint_{V_\varepsilon} c^2 \nabla^2 \bar{g} \mathrm{d}V \xlongequal{(2.1.19)} \iint_{S_\varepsilon} c^2 \frac{\partial \bar{g}}{\partial n} \mathrm{d}S = c^2 \int_0^{2\pi} \mathrm{d}\phi \int_0^{\pi} \left[r^2 \sin\theta \frac{\mathrm{d}\bar{g}}{\mathrm{d}r} \right] \Bigg|_{r=\varepsilon} \mathrm{d}\theta$$
$$= -4\pi c^2 \lim_{\varepsilon \to 0} A \left(1 + \frac{\varepsilon p}{c} \right) \mathrm{e}^{-\frac{p}{c}\varepsilon} = -4A\pi c^2$$
$$\iiint_{V_\varepsilon} p^2 \bar{g} \mathrm{d}V = \lim_{\varepsilon \to 0} \int_0^{2\pi} \mathrm{d}\phi \int_0^{\pi} \sin\theta \mathrm{d}\theta \int_0^{\varepsilon} p^2 A r \mathrm{e}^{-\frac{p}{c} r} \mathrm{d}r$$
$$= -\lim_{\varepsilon \to 0} 4\pi p^2 A \frac{p}{c} \left[\varepsilon \mathrm{e}^{-\frac{p}{c}\varepsilon} + \frac{c}{p} \mathrm{e}^{-\frac{p}{c} r} \Big|_0^{\varepsilon} \right] = 0$$

Therefore, $A = \mathrm{e}^{-p\tau}/(4\pi c^2)$, and thus

$$\bar{g} = \frac{1}{4\pi c^2 r} \mathrm{e}^{-p\left(\tau + \frac{r}{c}\right)}$$

the solution $\bar{g}$ in the transformed domain is obtained.

The next task is to find the solution g in the original domain using the solution $\bar{g}$ from the transformed domain. According to the general inversion formula for the

Laplace transform,[6] let $p = s + \mathrm{i}\sigma$, and taking into account the properties of the Dirac δ function,

$$\delta(t) = \frac{1}{2\pi}\int_{-\infty}^{+\infty} \mathrm{e}^{\mathrm{i}\omega t}\mathrm{d}\omega, \quad \text{and} \quad f(x)\delta(x-a) = f(a)\delta(x-a)$$

we obtain

$$\begin{aligned} g(\boldsymbol{x}, t; \boldsymbol{\xi}, \tau) &= \frac{1}{2\pi \mathrm{i}c^2}\int_{s-\mathrm{i}\infty}^{s+\mathrm{i}\infty} \frac{\mathrm{e}^{\left(t-\tau-\frac{r}{c}\right)p}}{4\pi r}\mathrm{d}p = \frac{\mathrm{e}^{s\left(t-\tau-\frac{r}{c}\right)}}{4\pi c^2 r}\frac{1}{2\pi}\int_{-\infty}^{+\infty} \mathrm{e}^{\mathrm{i}\left(t-\tau-\frac{r}{c}\right)\sigma}\mathrm{d}\sigma \\ &= \frac{\mathrm{e}^{s\left(t-\tau-\frac{r}{c}\right)}}{4\pi c^2 r}\delta\left(t-\tau-\frac{r}{c}\right) = \frac{1}{4\pi c^2 r}\delta\left(t-\tau-\frac{r}{c}\right) \\ &= \frac{1}{4\pi c^2}\frac{\delta\left(t-\tau-\dfrac{|\boldsymbol{x}-\boldsymbol{\xi}|}{c}\right)}{|\boldsymbol{x}-\boldsymbol{\xi}|} \end{aligned} \tag{4.3.4}$$

4.3.2 Solution of the Wave Equation

After obtaining the Green's function solution of the wave equation, it can be used to find the solution of the wave Eq. (4.3.1). By comparing Eq. (4.3.1) with Eq. (4.3.2a), it is clear that the only difference between them lies in the inhomogeneous term on the right-hand side. Note that

$$\frac{F(\boldsymbol{x}, t)}{\rho} = \int_{-\infty}^{+\infty}\mathrm{d}\tau \iiint_V \frac{F(\boldsymbol{\xi}, \tau)}{\rho}\delta(\boldsymbol{x}-\boldsymbol{\xi})\delta(t-\tau)\mathrm{d}V(\boldsymbol{\xi})$$

substituting Eq. (4.3.4) into Eq. (4.3.2a), multiplying both sides by $F(\boldsymbol{\xi}, \tau)/\rho$, and integrating over τ and V, it is easy to see through comparison that

[6] If the function $F(p) = F(s+\mathrm{i}\sigma)$ satisfies the following conditions in the region $\mathrm{Re}(p) = s > s_0$: (1). $F(p)$ is analytic; (2). $F(p)$ uniformly tends to 0 as $|p| \to \infty$; (3). The infinite integral along the line L: $\mathrm{Re}(p) = s$, $\int_{s-\mathrm{i}\infty}^{s+\mathrm{i}\infty} |F(p)|\,\mathrm{d}\sigma$ $(s > s_0)$ converges for all $\mathrm{Re}(p) - s > s_0$; then the inverse function of $F(p)$ for $\mathrm{Re}(p) = s > s_0$ is given by

$$f(t) = \frac{1}{2\pi \mathrm{i}}\int_{s-\mathrm{i}\infty}^{s+\mathrm{i}\infty} F(p)\mathrm{e}^{pt}\mathrm{d}p.$$

$$\begin{aligned}\phi(\boldsymbol{x},t) &= \int_{-\infty}^{+\infty}\mathrm{d}\tau\iiint_V \frac{F(\boldsymbol{\xi},\tau)}{4\pi\rho\alpha^2}\frac{\delta\left(t-\tau-\dfrac{|\boldsymbol{x}-\boldsymbol{\xi}|}{\alpha}\right)}{|\boldsymbol{x}-\boldsymbol{\xi}|}\mathrm{d}V(\boldsymbol{\xi})\\ &= \frac{1}{4\pi\rho\alpha^2}\iiint_V \frac{F\left(\boldsymbol{\xi},t-\dfrac{|\boldsymbol{x}-\boldsymbol{\xi}|}{\alpha}\right)}{|\boldsymbol{x}-\boldsymbol{\xi}|}\mathrm{d}V(\boldsymbol{\xi})\end{aligned} \tag{4.3.5}$$

4.4 Derivation of Green's Function Solution

According to the solution approach shown in Fig. 4.1, we first need to obtain the potential functions F and $\boldsymbol{H}$ for the body force $\boldsymbol{f}$. Then, based on the solution from Sect. 4.3 for the wave function, we take the gradient of the scalar potential function ϕ and the curl of the vector potential function $\boldsymbol{\Psi}$. Adding these two together forms the Green's function.

4.4.1 Expressions for Potential Functions F and H of Body Force

Take the divergence of both sides of Eq. (4.2.1) to obtain

$$\nabla\cdot\boldsymbol{f} = \nabla\cdot\nabla F + \nabla\cdot\nabla\times\boldsymbol{H} = \nabla^2 F$$

Since the body force $\boldsymbol{f}$ is known, this becomes a Poisson equation for F, with the solution given by[7]

$$F(\boldsymbol{x},t) = -\frac{1}{4\pi}\iiint_V \frac{\nabla\cdot\boldsymbol{f}(\boldsymbol{\xi},t)}{|\boldsymbol{x}-\boldsymbol{\xi}|}\mathrm{d}V(\boldsymbol{\xi}) \tag{4.4.1}$$

Similarly, taking the curl of both sides of Eq. (4.2.1), we obtain

$$\nabla\times\boldsymbol{f} = \nabla\times\nabla F + \nabla\times\nabla\times\boldsymbol{H} \overset{(2.1.18)}{=\!=\!=} \nabla(\nabla\cdot\boldsymbol{H}) - \nabla^2\boldsymbol{H} = -\nabla^2\boldsymbol{H}$$

[7] The Poisson equation is given by $\nabla^2 F(\boldsymbol{x}) = G(\boldsymbol{x})$. When compared to the wave equation in Eq. (4.3.1), it can be seen as a degenerate version of the wave equation without the time-dependent derivative terms. Therefore, the analysis process outlined in Sect. 4.3 for the wave equation also applies to solving the Poisson equation, but it is simpler since it doesn't involve the Laplace transform and its inverse. The solution is

$$F(\boldsymbol{x}) = -\frac{1}{4\pi}\iiint \frac{G(\boldsymbol{\xi})}{|\boldsymbol{x}-\boldsymbol{\xi}|}\mathrm{d}V(\boldsymbol{\xi})$$

The detailed derivation is left as an exercise for the reader.

therefore

$$\boldsymbol{H}(\boldsymbol{x},t)=\frac{1}{4\pi}\iiint_V \frac{\nabla\times\boldsymbol{f}(\boldsymbol{\xi},t)}{|\boldsymbol{x}-\boldsymbol{\xi}|}\mathrm{d}V(\boldsymbol{\xi}) \tag{4.4.2}$$

Our goal is to find the Green's function, so $\boldsymbol{f}$ has a specific form, as shown in Eq. (2.4.1). Given that only the point of force application is unique in space, we choose the coordinate origin at the point where the force is applied and assume the force acts at $t=0$. This allows us to represent the concentrated impulsive force source as

$$f_i(\boldsymbol{x},t)=\delta_{ip}\delta(\boldsymbol{x})\delta(t)$$

where p represents a specific direction. Therefore,

$$\nabla\cdot\boldsymbol{f}(\boldsymbol{x},t)=f_{i,i}=\delta_{ip}\frac{\partial\delta(\boldsymbol{x})}{\partial x_i}\delta(t)$$

$$\nabla\times\boldsymbol{f}(\boldsymbol{x},t)=\varepsilon_{ijk}\delta_{kp}\frac{\partial\delta(\boldsymbol{x})}{\partial x_j}\delta(t)\hat{\boldsymbol{e}}_i=-\varepsilon_{ipj}\frac{\partial\delta(\boldsymbol{x})}{\partial x_j}\delta(t)\hat{\boldsymbol{e}}_i$$

Substituting them into Eqs. (4.4.1) and (4.4.2), we obtain the expressions for the potential functions of the body force F and $\boldsymbol{H}$ as follows,

$$F(\boldsymbol{x},t)=-\frac{1}{4\pi}\frac{\partial}{\partial x_p}\frac{1}{|\boldsymbol{x}|}\delta(t),\qquad H_i(\boldsymbol{x},t)=-\frac{1}{4\pi}\varepsilon_{ipj}\frac{\partial}{\partial x_j}\frac{1}{|\boldsymbol{x}|}\delta(t) \tag{4.4.3}$$

4.4.2 *Expressions for Potential Functions ϕ and $\boldsymbol{\Psi}$ of Displacement*

According to Lamé's theorem, the scalar potential function of displacement ϕ and the vector potential function $\boldsymbol{\Psi}$ satisfy the wave equations, respectively,

$$\ddot{\phi}=\frac{F}{\rho}+\alpha^2\nabla^2\phi \quad\text{and}\quad \ddot{\boldsymbol{\Psi}}=\frac{\boldsymbol{H}}{\rho}+\beta^2\nabla^2\boldsymbol{\Psi}$$

Using the results from Eq. (4.3.5) along with the specific expression for the body force potential function in Eq. (4.4.3), it is straightforward to write out

$$\phi(\boldsymbol{x},t)=-\frac{1}{4^2\pi^2\rho\alpha^2}\iiint_V \frac{\delta\left(t-\dfrac{|\boldsymbol{x}-\boldsymbol{\xi}|}{\alpha}\right)}{|\boldsymbol{x}-\boldsymbol{\xi}|}\frac{\partial}{\partial\xi_p}\frac{1}{|\boldsymbol{\xi}|}\mathrm{d}V(\boldsymbol{\xi}) \tag{4.4.4a}$$

$$\boldsymbol{\Psi}(\boldsymbol{x},t)=-\frac{\varepsilon_{ipj}\hat{\boldsymbol{e}}_i}{4^2\pi^2\rho\beta^2}\iiint_V \frac{\delta\left(t-\dfrac{|\boldsymbol{x}-\boldsymbol{\xi}|}{\beta}\right)}{|\boldsymbol{x}-\boldsymbol{\xi}|}\frac{\partial}{\partial\xi_j}\frac{1}{|\boldsymbol{\xi}|}\mathrm{d}V(\boldsymbol{\xi}) \tag{4.4.4b}$$

We aim to obtain the closed-form expressions for ϕ and $\boldsymbol{\Psi}$. By comparing these two equations, we can see that the integral forms on the right-hand side are identical, which can be denoted as I_1,

$$I_1 = \iiint_V \frac{\delta\left(t - \frac{|\boldsymbol{x} - \boldsymbol{\xi}|}{c}\right)}{|\boldsymbol{x} - \boldsymbol{\xi}|} \frac{\partial}{\partial \xi_p} \frac{1}{|\boldsymbol{\xi}|} \mathrm{d}V(\boldsymbol{\xi})$$

Noting that I_1 is the integral over the entire space with respect to the field point $\boldsymbol{\xi}$, and that the integrand includes $|\boldsymbol{x} - \boldsymbol{\xi}|$, it is convenient to choose a spherical coordinate system centered at $\boldsymbol{x}$, therefore

$$I_1 = \int_0^{+\infty} \left[\iint_{|\boldsymbol{\xi}-\boldsymbol{x}|=R} \frac{\delta\left(t - \frac{R}{c}\right)}{R} \frac{\partial}{\partial \xi_p} \frac{1}{|\boldsymbol{\xi}|} \mathrm{d}S(\boldsymbol{\xi}) \right] \mathrm{d}R = \int_0^{+\infty} \frac{\delta\left(t - \frac{R}{c}\right)}{R} I_2(R) \mathrm{d}R \tag{4.4.5}$$

in which,

$$I_2(R) = \iint_{|\boldsymbol{\xi}-\boldsymbol{x}|=R} \frac{\partial}{\partial \xi_p} \frac{1}{|\boldsymbol{\xi}|} \mathrm{d}S(\boldsymbol{\xi}) \tag{4.4.6}$$

The integral expression for $I_2(R)$ is already quite simple, but it is not easy to compute directly. To calculate the integral in Eq. (4.4.6), we introduce an auxiliary vector $\boldsymbol{\eta}$, so that

$$\frac{1}{|\boldsymbol{\xi}|} = \lim_{|\boldsymbol{\eta}| \to 0} \frac{1}{|\boldsymbol{\xi} - \boldsymbol{\eta}|}$$

This allows us to use this relationship to transform the derivative with respect to ξ_q into a derivative with respect to η_q. Since $\boldsymbol{\eta}$ is not an independent variable, the differentiation can be moved outside the integral sign, thereby simplifying the integral calculation. If we note and utilize the following conclusion

$$\frac{\partial}{\partial \eta_p} \frac{1}{|\boldsymbol{\xi} - \boldsymbol{\eta}|} = -\frac{\partial}{\partial \xi_p} \frac{1}{|\boldsymbol{\xi} - \boldsymbol{\eta}|} \tag{4.4.7}$$

the integral of $I_2(R)$ can be transformed as

$$\begin{aligned} I_2(R) &= \lim_{|\boldsymbol{\eta}| \to 0} \iint_{|\boldsymbol{\xi}-\boldsymbol{x}|=R} \frac{\partial}{\partial \xi_p} \frac{1}{|\boldsymbol{\xi} - \boldsymbol{\eta}|} \mathrm{d}S(\boldsymbol{\xi}) \\ &= -\lim_{|\boldsymbol{\eta}| \to 0} \iint_{|\boldsymbol{\xi}-\boldsymbol{x}|=R} \frac{\partial}{\partial \eta_p} \frac{1}{|\boldsymbol{\xi} - \boldsymbol{\eta}|} \mathrm{d}S(\boldsymbol{\xi}) = -\lim_{|\boldsymbol{\eta}| \to 0} \frac{\partial}{\partial \eta_p} I_3(R, \boldsymbol{\eta}) \end{aligned} \tag{4.4.8}$$

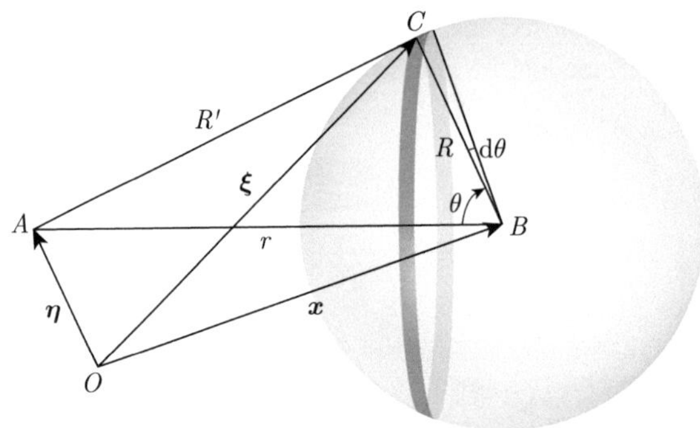

Fig. 4.2 Diagram illustrating the solution of the integral I_3. The integral I_3 represents the integration over the surface of a sphere centered at $\boldsymbol{x}$ with a radius of $|\boldsymbol{\xi} - \boldsymbol{x}|$. O is the source point, and $\boldsymbol{\eta}$ is an auxiliary vector starting from O. The distances are $\overline{AC} = R'$, $\overline{AB} = r$, and $\overline{BC} = R$, with $\angle ABC = \theta$. The dark gray strip represents the surface element $\mathrm{d}S = 2\pi R^2 \sin\theta \mathrm{d}\theta$

where

$$I_3(R, \boldsymbol{\eta}) = \iint_{|\boldsymbol{\xi}-\boldsymbol{x}|=R} \frac{1}{|\boldsymbol{\xi} - \boldsymbol{\eta}|} \mathrm{d}S(\boldsymbol{\xi})$$

Figure 4.2 illustrates the diagram for solving the integral I_3. In the triangle $\triangle ABC$, the lengths are $\overline{AC} = R'$, $\overline{AB} = r$, and $\overline{BC} = R$, with $\angle ABC = \theta$. According to the cosine law:

$$R'^2 = r^2 + R^2 - 2rR\cos\theta \qquad \Longrightarrow \qquad \frac{\sin\theta \mathrm{d}\theta}{R'} = \frac{\mathrm{d}R'}{rR}$$

we have that

$$\begin{aligned} I_3(R, \boldsymbol{\eta}) &= \int_0^{\pi} \frac{1}{R'(\theta)} 2\pi R^2 \sin\theta \mathrm{d}\theta = 2\pi R^2 \int_{\theta=0}^{\theta=\pi} \frac{\mathrm{d}R'}{rR} = \frac{2\pi R}{r} R' \Big|_{\theta=0}^{\theta=\pi} \\ &= \begin{cases} \dfrac{4\pi R^2}{r}, & r > R \\ 4\pi R, & r < R \end{cases} = 4\pi R + 4\pi R^2 \left(\frac{1}{r} - \frac{1}{R} \right) H(r - R) \end{aligned}$$

where $r = |\boldsymbol{x} - \boldsymbol{\eta}|$. Note that in the final equation, a step function is used to piece together the results expressed as a piecewise function. This is necessary because when substituting the result of $I_3(R, \boldsymbol{\eta})$ back into the integral $I_2(R)$, we need to differentiate with respect to η_p. Using a piecewise function would miss the derivative at the points of discontinuity.[8]

Substitute the result of I_3 back into Eq. (4.48), and note that R is independent of η_q. Therefore,

$$
\begin{aligned}
I_2(R) &= -4\pi R^2 \lim_{|\eta|\to 0} \left\{ \frac{\partial}{\partial \eta_p} \frac{1}{r} H(r-R) + \left(\frac{1}{r} - \frac{1}{R}\right) \delta(r-R) \right\} \\
&\overset{(4.4.7)}{=\!=} 4\pi R^2 \lim_{|\eta|\to 0} \frac{\partial}{\partial x_p} \frac{1}{r} H(r-R) \qquad \left(\text{using } x\delta(x)=0\right) \\
&= 4\pi R^2 \frac{\partial}{\partial x_p} \frac{1}{|\boldsymbol{x}|} H(|\boldsymbol{x}| - R)
\end{aligned}
$$

Substituting it into Eq. (4.4.5), and then back into Eq. (4.4.4), we obtain

$$
\begin{aligned}
\phi(\boldsymbol{x}, t) &\overset{\tau=\frac{R}{\alpha}}{=\!=} -\frac{1}{4^2\pi^2\rho\alpha^2} \int_0^{\frac{|\boldsymbol{x}|}{\alpha}} \frac{\delta(t-\tau)}{\tau} 4\pi R^2 \frac{\partial}{\partial x_p} \frac{1}{|\boldsymbol{x}|} \mathrm{d}\tau \\
&= -\frac{1}{4\pi\rho} \frac{\partial}{\partial x_p} \frac{1}{|\boldsymbol{x}|} \int_0^{\frac{|\boldsymbol{x}|}{\alpha}} \tau \delta(t-\tau) \mathrm{d}\tau
\end{aligned} \tag{4.4.9a}
$$

$$
\Psi_i(\boldsymbol{x}, t) \overset{\tau=\frac{R}{\beta}}{=\!=} -\frac{\varepsilon_{ipj}}{4\pi\rho} \frac{\partial}{\partial x_j} \frac{1}{|\boldsymbol{x}|} \int_0^{\frac{|\boldsymbol{x}|}{\beta}} \tau \delta(t-\tau) \mathrm{d}\tau \tag{4.4.9b}
$$

4.4.3 Expression of Green's Function

According to Lamé's theorem, $\boldsymbol{u} = \nabla\phi + \nabla \times \boldsymbol{\Psi}$. When the body force is a concentrated impulse force, $\boldsymbol{u}$ becomes the Green's function. Therefore, substituting Eq. (4.4.9) yields

$$
\begin{aligned}
G_{np}(\boldsymbol{x}, t) &= \frac{\partial \phi}{\partial x_n} + \varepsilon_{nki} \frac{\partial \Psi_i}{\partial x_k} \\
&= -\frac{1}{4\pi\rho} \frac{\partial^2}{\partial x_n \partial x_p} \frac{1}{|\boldsymbol{x}|} J(|\boldsymbol{x}|, \alpha) - \frac{1}{4\pi\rho} \frac{\partial}{\partial x_p} \frac{1}{|\boldsymbol{x}|} \frac{\partial}{\partial x_n} J(|\boldsymbol{x}|, \alpha) \\
&\quad - \frac{1}{4\pi\rho} \varepsilon_{nki}\varepsilon_{ipj} \left\{ \frac{\partial^2}{\partial x_k \partial x_j} \frac{1}{|\boldsymbol{x}|} J(|\boldsymbol{x}|, \beta) + \frac{\partial}{\partial x_j} \frac{1}{|\boldsymbol{x}|} \frac{\partial}{\partial x_k} J(|\boldsymbol{x}|, \beta) \right\}
\end{aligned} \tag{4.4.10}
$$

[8] For example, the step function itself is defined as a piecewise function, which is constant in both intervals $(-\infty, 0)$ and $(0, +\infty)$. If we differentiate it piecewise, the result is zero, which fails to capture the discontinuity at $x = 0$. In the context of generalized functions, the derivative of the step function is the Dirac delta function, $H'(x) = \delta(x)$. The discontinuity of the step function at $x = 0$ is represented by the infinite value of the Dirac δ function.

in which,

$$J(|\boldsymbol{x}|, c) = \int_0^{\frac{|\boldsymbol{x}|}{c}} \tau\delta(t-\tau)\mathrm{d}\tau$$

Note that

$$\frac{\partial}{\partial x_p}|\boldsymbol{x}| = \frac{x_p}{r} \triangleq \gamma_p, \quad \frac{\partial}{\partial x_p}\frac{1}{|\boldsymbol{x}|} = -\frac{1}{r^2}\gamma_p$$

$$\frac{\partial^2}{\partial x_n \partial x_p}\frac{1}{|\boldsymbol{x}|} = \frac{1}{r^3}\left(3\gamma_n\gamma_p - \delta_{np}\right), \quad \frac{\partial}{\partial x_n}J(|\boldsymbol{x}|, \alpha) = \frac{\gamma_n}{\alpha^2}r\delta\left(t-t_{\mathrm{P}}\right)$$

$$\int_{t_{\mathrm{P}}}^{t_{\mathrm{S}}} \tau\delta(t-\tau)\mathrm{d}\tau = t\left[H\left(t-t_{\mathrm{P}}\right) - H\left(t-t_{\mathrm{S}}\right)\right], \quad t_{\mathrm{P}} = \frac{r}{\alpha}, \quad t_{\mathrm{S}} = \frac{r}{\beta}$$

where γ_p varies with the coordinates, reflecting the directional effects and is referred to as the *directional factor*. Substituting the above results into Eq. (4.4.10), the Green's function for infinite space is finally obtained as

$$\boxed{\begin{aligned} G_{np}(\boldsymbol{x}, t; \boldsymbol{0}, 0) = {} & \frac{3\gamma_n\gamma_p - \delta_{np}}{4\pi\rho r^3}t\left[H\left(t-t_{\mathrm{P}}\right) - H\left(t-t_{\mathrm{S}}\right)\right] \\ & + \frac{\gamma_n\gamma_p}{4\pi\rho\alpha^2 r}\delta\left(t-t_{\mathrm{P}}\right) - \frac{\gamma_n\gamma_p - \delta_{np}}{4\pi\rho\beta^2 r}\delta\left(t-t_{\mathrm{S}}\right) \end{aligned}} \tag{4.4.11}$$

4.5 Properties of Green's Function and General Displacement in Infinite Space

4.5.1 Properties of Green's Function

Equation (4.4.11) represents the expression for the Green's function in infinite space. First, it is important to note that this is a closed-form analytical expression, which is relatively simple and facilitates the analysis of its properties. The Green's function is the sum of terms proportional to r^{-2}[9] and r^{-1}, dominating in the near field and far field respectively. Hence, they are referred to as the *near-field term* and *far-field term*. The near-field term is non-zero only between the arrival times of the P-wave and S-wave ($t_{\mathrm{P}} < t < t_{\mathrm{S}}$) and is independent of the velocities of the P-wave or S-wave. In contrast, the far-field term separates the P-wave and S-wave, forming a far-field

[9] At first glance, the first term appears to be proportional to r^{-3}. Note that the numerator includes t, and this term is only non-zero when $r/\alpha < t < r/\beta$. Therefore, $t = r/c$ where $\beta < c < \alpha$, and one r in the denominator is canceled out.

P-wave term dependent only on α, and a far-field S-wave term dependent only on β. Therefore, the Green's function can be written as follows:

$$G_{np}(\boldsymbol{x}, t; \boldsymbol{0}, 0) = G_{np}^{\mathrm{N}}(\boldsymbol{x}, t; \boldsymbol{0}, 0) + G_{np}^{\mathrm{FP}}(\boldsymbol{x}, t; \boldsymbol{0}, 0) + G_{np}^{\mathrm{FS}}(\boldsymbol{x}, t; \boldsymbol{0}, 0)$$

The three terms on the right-hand side of the equation are the near-field term, the far-field P-wave term, and the far-field S-wave term, respectively,

$$G_{np}^{\mathrm{N}}(\boldsymbol{x}, t; \boldsymbol{0}, 0) = \frac{3\gamma_n\gamma_p - \delta_{np}}{4\pi\rho r^3} t\left[H\left(t - t_{\mathrm{P}}\right) - H\left(t - t_{\mathrm{S}}\right)\right] \tag{4.5.1a}$$

$$G_{np}^{\mathrm{FP}}(\boldsymbol{x}, t; \boldsymbol{0}, 0) = \frac{\gamma_n\gamma_p}{4\pi\rho\alpha^2 r}\delta\left(t - t_{\mathrm{P}}\right) \tag{4.5.1b}$$

$$G_{np}^{\mathrm{FS}}(\boldsymbol{x}, t; \boldsymbol{0}, 0) = -\frac{\gamma_n\gamma_p - \delta_{np}}{4\pi\rho\beta^2 r}\delta\left(t - t_{\mathrm{S}}\right) \tag{4.5.1c}$$

4.5.1.1 Time-Varying Properties of Green's Function

If we fix the position of a point in space and examine the behavior of the components of the Green's function as t varies, it is not difficult to see that:

(1) The P-wave term and the S-wave term each have a delay time from the moment of the pulse action, which are the P-wave arrival time t_{P} and the S-wave arrival time t_{S}, respectively.
(2) The far-field P-wave term and the far-field S-wave term each have a pulse only at their respective arrival times.
(3) The near-field term increases linearly with t in the time period $t_{\mathrm{P}} < t < t_{\mathrm{S}}$.

Figure 4.3 shows the Green's function generated by a unit force applied in the x_3 direction.

4.5.1.2 Spatial Variation Properties of Green's Function

In Eq. (4.5.1), both the near-field and far-field terms are modulated by a directional factor, commonly referred to as the *radiation pattern*. This pattern illustrates how the components of the Green's function vary with the observation point at the same moment. This is crucial for deducing the characteristics of the earthquake source based on seismic records from different locations. Denote the radiation pattern as $\mathscr{A}$. The radiation patterns for near-field, far-field P-wave, and far-field S-wave are, respectively,

$$\mathscr{A}_{np}^{\mathrm{N}} = 3\gamma_n\gamma_p - \delta_{np}, \quad \mathscr{A}_{np}^{\mathrm{FP}} = \gamma_n\gamma_p, \quad \mathscr{A}_{np}^{\mathrm{FS}} = -(\gamma_n\gamma_p - \delta_{np})$$

Without loss of generality, let's continue using the example of a force acting along the x_3 axis ($p = 3$). Note that according to the definition of the directional factor,

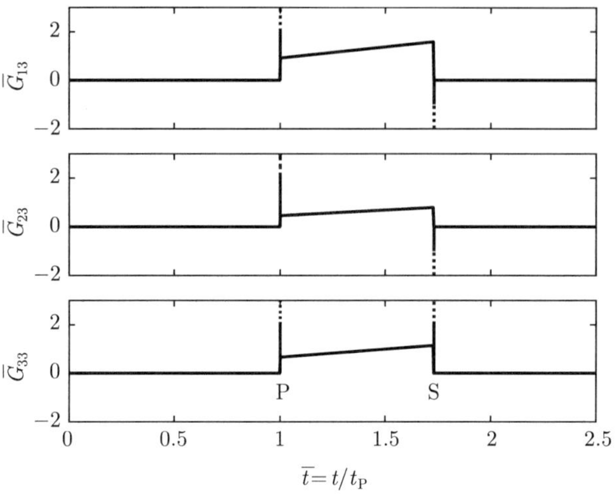

Fig. 4.3 $\bar{G}_{i3} = 4\pi\mu r G_{i3}$ generated by a unit concentrated impulse force in the x_3 direction. The observation point is located at $\boldsymbol{x} = (2, 1, 3)$ km. The dashed lines at the arrival times of P-wave and S-wave (also marked with "P" and "S") represent propagation to infinity either upwards or downwards

$\hat{\boldsymbol{\gamma}}$ represents the unit vector pointing from the source of the force to the field point. Therefore, if we consider a sphere centered at the source of the force, $\hat{\boldsymbol{\gamma}}$ is the direction of the outward normal to the surface of the sphere. Notice that

$$\mathscr{A}_{n3}^{\text{FP}} \propto \gamma_n, \quad \mathscr{A}_{n3}^{\text{FS}}\gamma_n \propto -(\gamma_n\gamma_n\gamma_3 - \gamma_n\delta_{n3}) = 0$$

This indicates that the displacement of the far-field P-wave is parallel to $\hat{\boldsymbol{\gamma}}$ (i.e., perpendicular to the spherical surface), while the displacement of the far-field S-wave is perpendicular to $\hat{\boldsymbol{\gamma}}$ (i.e., tangent to the spherical surface), implying that far-field P-wave and S-wave are longitudinal and transverse waves, respectively.

Figures 4.4 and 4.5 illustrate the distribution of the radiation patterns for far-field P-wave and far-field S-wave on the spherical surface, respectively. For the radiation pattern of the far-field P-wave, the arrows in Fig. 4.4b are perpendicular to the surface at all positions. It is zero in the x_1x_2 plane and increases in magnitude along the surface towards the positive x_3 axis, being positive, and similarly increases towards the negative x_3 axis, being negative. This aligns with intuition, considering the force is applied along the x_3 axis. For the radiation pattern of the far-field S-wave (Fig. 4.5), the pattern is tangent to the surface at all positions. The maximum magnitude appears in the x_1x_2 plane, and the magnitude decreases along the surface towards both the positive and negative x_3 axes, becoming zero at the poles, with the direction pointing along the surface towards the positive x_3 axis.

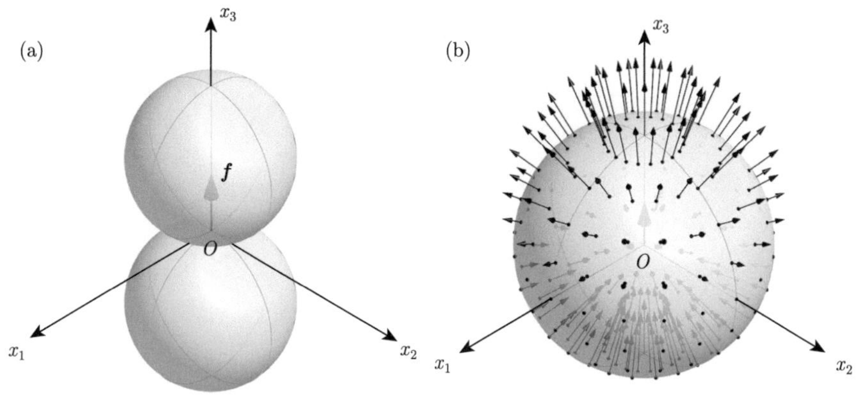

Fig. 4.4 Spatial distribution of the far-field P-wave radiation pattern $\mathscr{A}_{n3}^{\mathrm{FP}}$. **a** The amplitude (absolute value) of the radiation pattern; **b** The vector distribution on the spherical surface, where the length and direction of the arrows represent the magnitude and direction of the radiation pattern. The direction of the applied force $\boldsymbol{f}$ is indicated in the figure

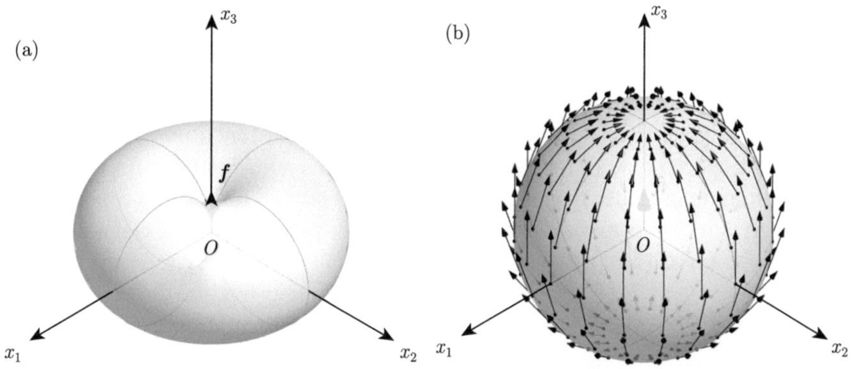

Fig. 4.5 Spatial distribution of the far-field S-wave radiation pattern $\mathscr{A}_{n3}^{\mathrm{FS}}$.The same as Fig. 4.4, except here it is for the far-field S-wave

The near-field radiation pattern does not have the simple spatial distribution seen in the far-field P-wave and far-field S-wave patterns. Figure 4.6 shows the distribution of the near-field radiation pattern on a spherical surface. Compared to Figs. 4.4 and 4.5, it is clearly more complex and can be seen as a combination of both. Overall, the component perpendicular to the sphere dominates near the $\pm x_3$ axis, while the component tangential to the sphere dominates near the x_1x_2 plane.

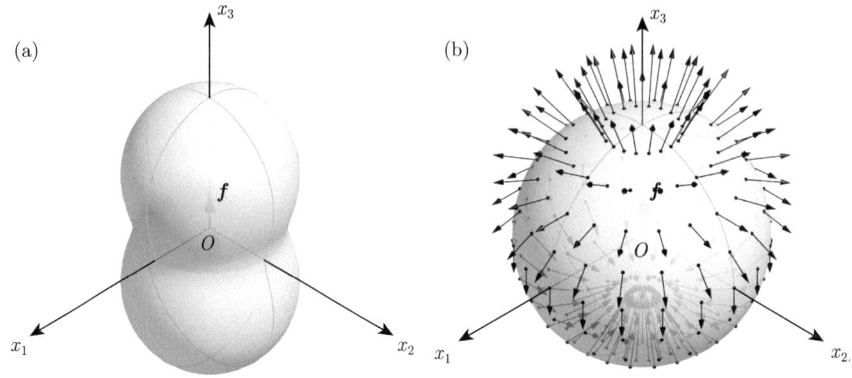

Fig. 4.6 Spatial distribution of the near-field radiation pattern $\mathscr{A}_{n3}^{\mathrm{N}}$. The same as Fig. 4.4, except here it is for the near-field term, which can be regard as a combination of P-wave and S-wave

4.5.2 Displacement Due to Point Source with General Time Function

In the Green's function expression of Eq. (4.4.11), both the far-field P-wave term and the far-field S-wave term are in the form of a delta function. Since the delta function is a special type of function that is infinite at a point but has a finite integral, it is important to consider the case where the concentrated force has a time-varying function $S(t)$ to better explain the nature of the displacement field generated by a point force source. This time-varying function $S(t)$ is commonly referred to as the *source time function* (STF). Correspondingly, to distinguish it, the Green's function is denoted as $G_{np}^{S(t)}$. According to the displacement representation theorem (2.5.3), we have

$$
\begin{aligned}
G_{np}^{S(t)}(\boldsymbol{x}, t; \boldsymbol{0}, 0) &= G_{np}(\boldsymbol{x}, t; \boldsymbol{0}, 0) * S(t) \\
&= \frac{3\gamma_n\gamma_p - \delta_{np}}{4\pi\rho r^3} t\big[H\left(t - t_{\mathrm{P}}\right) - H\left(t - t_{\mathrm{S}}\right)\big] * S(t) \\
&\quad + \frac{\gamma_n\gamma_p}{4\pi\rho\alpha^2 r}\delta\left(t - t_{\mathrm{P}}\right) * S(t) - \frac{\gamma_n\gamma_p - \delta_{np}}{4\pi\rho\beta^2 r}\delta\left(t - t_{\mathrm{S}}\right) * S(t)
\end{aligned}
\tag{4.5.2}
$$

Clearly, the result of $G_{np}^{S(t)}$ depends on the specific form of $S(t)$.

4.5.2.1 Displacement Field Generated by Step Function

To gain a concrete understanding, let's first consider an important special case where $S(t) = H(t)$, with $H(t)$ being the step function, also known as the *Heaviside function*. Note that

$$\delta\left(t - t_{\mathrm{P}}\right) * H(t) = \int_{-\infty}^{+\infty} H(t-\tau)\delta(\tau - t_{\mathrm{P}})\mathrm{d}\tau = H(t - t_{\mathrm{P}})$$

$$\begin{aligned} tH(t - t_{\mathrm{P}}) * H(t) &= \int_{-\infty}^{+\infty} H(t-\tau)\tau H(\tau - t_{\mathrm{P}})\mathrm{d}\tau = H(t - t_{\mathrm{P}}) \int_{t_{\mathrm{P}}}^{t} \tau \mathrm{d}\tau \\ &= \frac{1}{2}(t^2 - t_{\mathrm{P}}^2)H(t - t_{\mathrm{P}}) \end{aligned}$$

thus

$$\begin{aligned} G_{np}^{H(t)}(\boldsymbol{x}, t; \boldsymbol{0}, 0) &= \frac{3\gamma_n\gamma_p - \delta_{np}}{8\pi\rho r^3}\left[(t^2 - t_{\mathrm{P}}^2)H\left(t - t_{\mathrm{P}}\right) - (t^2 - t_{\mathrm{S}}^2)H\left(t - t_{\mathrm{S}}\right)\right] \\ &+ \frac{\gamma_n\gamma_p}{4\pi\rho\alpha^2 r}H\left(t - t_{\mathrm{P}}\right) - \frac{\gamma_n\gamma_p - \delta_{np}}{4\pi\rho\beta^2 r}H\left(t - t_{\mathrm{S}}\right) \end{aligned} \tag{4.5.3}$$

Notice that as $t \to \infty$, $G_{np}^{H(t)}(\boldsymbol{x}, t; \boldsymbol{0}, 0)$ simplifies to the static solution $G_{np}^{H:\mathrm{static}}(\boldsymbol{x}; \boldsymbol{0})$,

$$G_{np}^{H:\mathrm{static}}(\boldsymbol{x}; \boldsymbol{0}) = \frac{1}{8\pi\rho r}\left(\frac{\delta_{np} - \gamma_n\gamma_p}{\alpha^2} + \frac{\delta_{np} + \gamma_n\gamma_p}{\beta^2}\right)$$

This is a very concise result. Continuing with the example of a force applied along x_3, note that this is not an impulsive force but a continuously applied concentrated force, which induces a static displacement field in the space. Figure 4.7 shows the displacement on the spherical surface centered at the point of application of the force. From the figure, it can be seen that due to the force acting along the x_3 axis, all points on the sphere experience an upward displacement. However, in the region where $x_3 > 0$, the horizontal displacement is away from the x_3 axis, while in the region where $x_3 < 0$, the horizontal displacement converges towards the x_3 axis. To clearly illustrate this, Fig. 4.8 shows the distribution of displacement on the plane Π_1 above the x_3 axis and the plane Π_2 below the x_3 axis. The planes Π_1 and Π_2 before deformation are shown in gray, and the planes after deformation are shown in black. The corresponding changes of the selected grid points are indicated by gray arrows. The figure clearly shows that the vertical force produces horizontal separation in the region $x_3 > 0$ and convergence in the region $x_3 < 0$, while in the vertical direction, an upward displacement is generated in both regions. This effect diminishes with distance from the origin.

Comparing Eq. (4.5.3) with Eq. (4.4.11), it is evident that, apart from a constant difference in the near-field term, the far-field radiation patterns are identical. There-

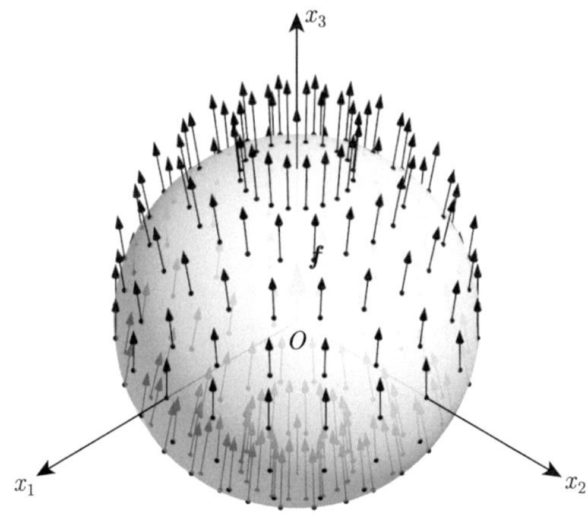

Fig. 4.7 Distribution of $G_{n3}^{H:\text{static}}(\boldsymbol{x}; \boldsymbol{0})$ in infinite space on the spherical surface

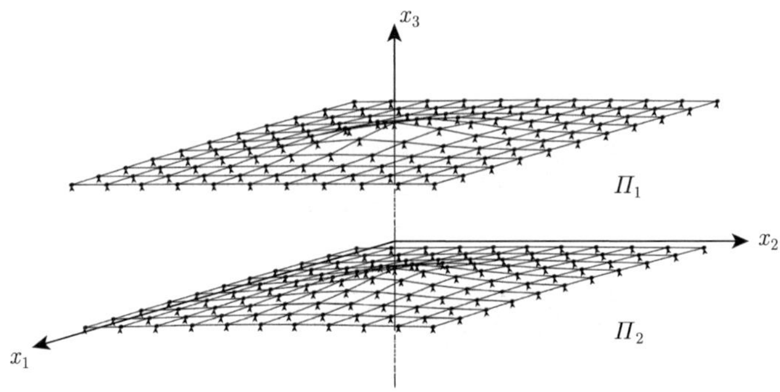

Fig. 4.8 Distribution of $G_{n3}^{H:\text{static}}(\boldsymbol{x}; \boldsymbol{0})$ in infinite space on the planes Π_1 and Π_2. The planes before deformation are shown in gray, and the planes after deformation are shown in black. The gray arrows represent displacement

fore, the difference between the displacement field caused by the step function $H(t)$ and the Green's function lies only in their temporal behavior. For ease of comparison, we continue using the scenario depicted in Fig. 4.3, with all parameters unchanged except for the time function of the concentrated force, which now varies as $H(t)$. Figure 4.9 illustrates the time variation of the displacement components $G_{i3}^{H(t)}$ at two positions in space, $\boldsymbol{x} = (2, 1, 3)$ km and $\boldsymbol{x} = (2, 1, -3)$ km. The horizontal axis represents dimensionless time $\bar{t}$, scaled by the P-wave arrival time t_{P}, and the vertical axis represents dimensionless displacement, scaled by $(8\pi\mu r)^{-1}$. Very steep seismic phases appear at the P-wave arrival time $\bar{t}_{\text{P}} = 1$ and the S-wave arrival time $\bar{t}_{\text{S}} = \sqrt{3}$, corresponding to the two pulses in Fig. 4.3, with the latter being the derivative of the former in time. After $\bar{t}_{\text{S}}$, the displacement no longer changes, representing the static displacement discussed in the previous section.

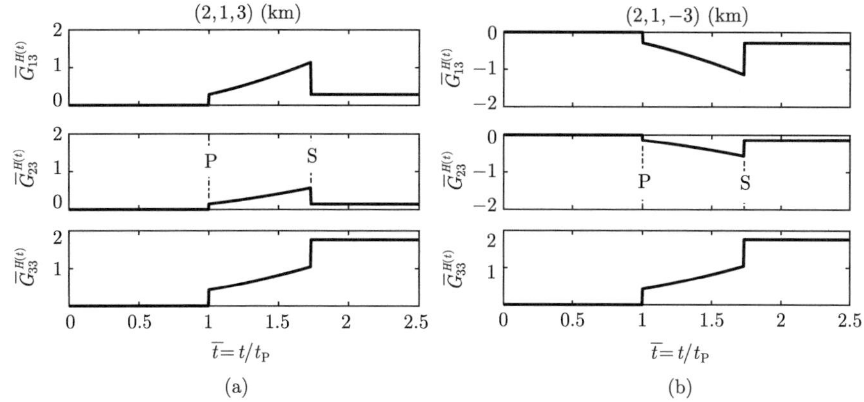

Fig. 4.9 $\bar{G}_{i3}^{H(t)}$ at different $\boldsymbol{x}$ caused by a unit concentrated step function force in the x_3 direction. **a** $\boldsymbol{x} = (2, 1, 3)$ km; **b** $\boldsymbol{x} = (2, 1, -3)$ km. $\bar{G}_{i3}^{H(t)}$ is the dimensionless displacement: $\bar{G}_{i3}^{H(t)} = G_{i3}^{H(t)} \cdot 8\pi\mu r$

Fig. 4.10 Particle trajectories at (2, 1, 3) km and (2, 1, −3) km caused by a step-function force

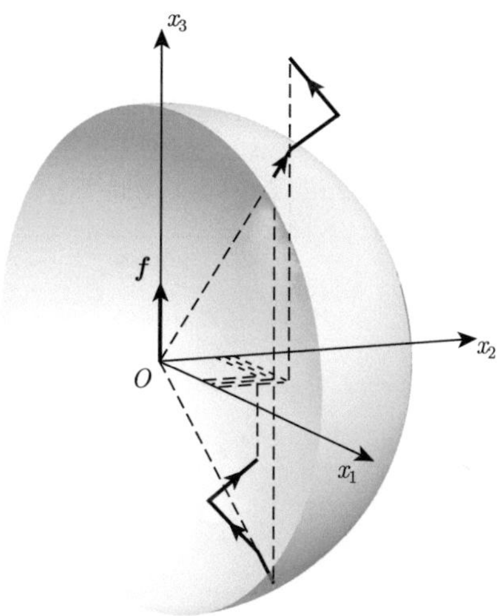

To better visualize the motion of particles within the medium under the force, Fig. 4.10 shows the trajectories of particles at the two aforementioned points. It can be observed that the particles start moving from positions on a sphere, initially moving along the positive and negative radial directions, corresponding to the motion at the P-wave arrival in Fig. 4.9. Then, during the time interval $t_{\mathrm{P}} < t < t_{\mathrm{S}}$, the motion deviates from the radial direction. At the S-wave arrival time t_{S}, the particles abruptly change direction, moving perpendicular to the radial direction, which corresponds to the S-wave phase. Finally, after the S-wave phase arrives, the motion ceases. A close analysis of the particle positions at the cessation of motion reveals that the particle at $\boldsymbol{x} = (2, 1, 3)$ km is displaced horizontally away from the x_3 axis, while the particle at $\boldsymbol{x} = (2, 1, -3)$ km is displaced closer to the x_3 axis. This reflects the fact that their horizontal components are opposites, as shown in Fig. 4.9. Both particles experience a displacement in the positive x_3 direction.

4.5.2.2 Displacement Generated by Ramp Function

Although the step function is closely related to the delta function in Green's functions, being its direct integral, such a time function does not accurately reflect real-world scenarios. In reality, the slip of a seismic source starts from zero and gradually increases to its final value; this implies that the force is applied progressively. This behavior is better represented by the *ramp function* $R(t)$, which can be expressed as follows,

$$R(t) = \frac{t}{t_0}H(t) + \left(1 - \frac{t}{t_0}\right)H(t - t_0) = \begin{cases} 0, & \text{if } \ t < 0 \\ \dfrac{t}{t_0}, & \text{if } \ 0 \leqslant t \leqslant t_0 \\ 1, & \text{if } \ t > t_0 \end{cases} \tag{4.5.4}$$

where t_0 is the *rise time*, which is the duration needed for the force to increase from zero to its constant value.[10] Figure 4.11 illustrates the shape of the ramp function. By substituting Eq. (4.5.4) into Eq. (4.5.2), we note that

$$R(t) * \delta(t - t_{\mathrm{P}}) = \int_{-\infty}^{+\infty} R(t - \tau)\delta(\tau - t_{\mathrm{P}})\mathrm{d}\tau$$

$$R(t) * tH(t - t_{\mathrm{P}}) = \int_{-\infty}^{+\infty} R(\tau)(t - \tau)H(t - \tau - t_{\mathrm{P}})\mathrm{d}\tau = H(t - t_{\mathrm{P}})T(t, t_{\mathrm{P}})$$

in which,

$$T(t, \tau) = \frac{1}{6t_0}\left[H(t_0 + \tau - t)(t - \tau)^2(t + 2\tau) + H(t - t_0 - \tau)t_0(3t^2 + t_0^2 - 3t_0t - 3\tau^2)\right]$$

and thus can obtain the displacement field $G_{np}^{R(t)}(\boldsymbol{x}, t; \boldsymbol{0}, 0)$ generated by the ramp function $R(t)$ as

$$\begin{aligned} G_{np}^{R(t)}(\boldsymbol{x}, t; \boldsymbol{0}, 0) = {} & \frac{3\gamma_n\gamma_p - \delta_{np}}{4\pi\rho r^3}\left[H\left(t - t_{\mathrm{P}}\right)T(t, t_{\mathrm{P}}) - H\left(t - t_{\mathrm{S}}\right)T(t, t_{\mathrm{S}})\right] \\ & + \frac{\gamma_n\gamma_p}{4\pi\rho\alpha^2 r}R\left(t - t_{\mathrm{P}}\right) - \frac{\gamma_n\gamma_p - \delta_{np}}{4\pi\rho\beta^2 r}R\left(t - t_{\mathrm{S}}\right) \end{aligned}$$

---

[10] Note that in the mathematical definition of the ramp function, the rise time t_0 appears in the denominator, so it cannot be zero. Therefore, the step function $H(t)$ cannot be obtained from the ramp function by simply setting t_0 to zero.

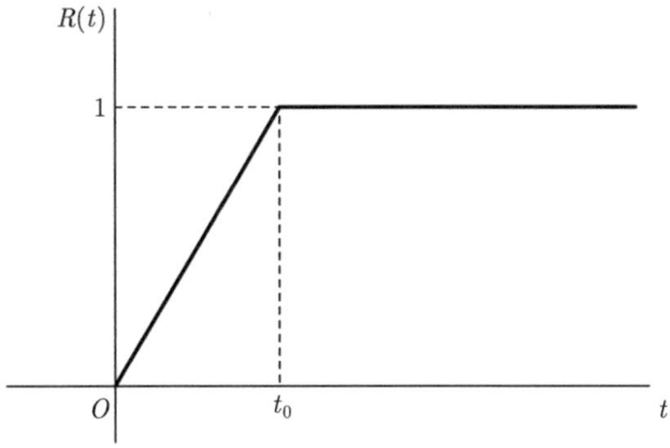

Fig. 4.11 Ramp function $R(t)$ with a rise time of t_0

Similar to the solution for the step function in Eq. (4.5.3), the spatial radiation pattern in the solution for the ramp function matches that of the Green's function, with differences only in the time-varying part. Given that the ramp function includes a crucial parameter t_0, it is expected that this parameter will affect the displacement field $G_{np}^{R(t)}(\boldsymbol{x}, t; \boldsymbol{0}, 0)$. Figure 4.12 illustrates the displacement field at $\boldsymbol{x} = (2, 1, 3)$ km for

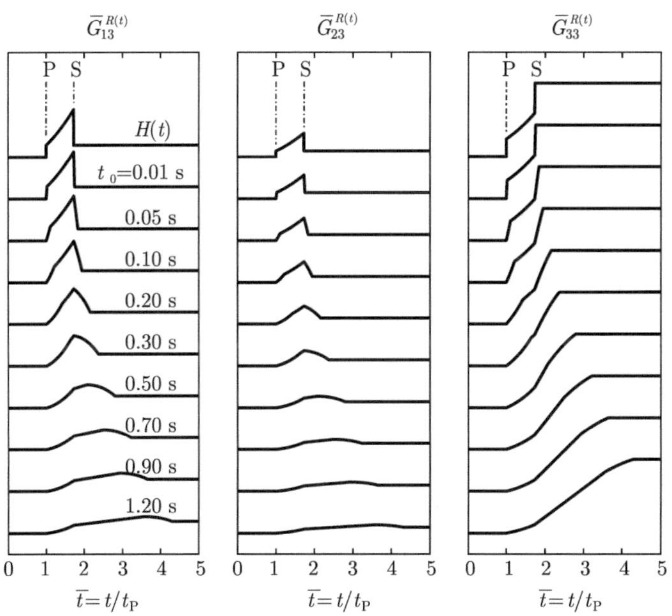

Fig. 4.12 $\bar{G}_{i3}^{R(t)}$ generated by a force in the x_3 direction from $R(t)$ with different t_0. The observation point is at $\boldsymbol{x} = (2, 1, 3)$ km. The horizontal axis represents time made dimensionless using the P-wave arrival time t_P. Dimensionless displacement $\bar{G}_{i3}^{R(t)} = G_{i3}^{R(t)} \cdot 4\pi\mu r$. In the figure, "P" and "S" denote the P-wave and S-wave, with arrival times of $t_P = 0.47$ s and $t_S = 0.81$ s, respectively

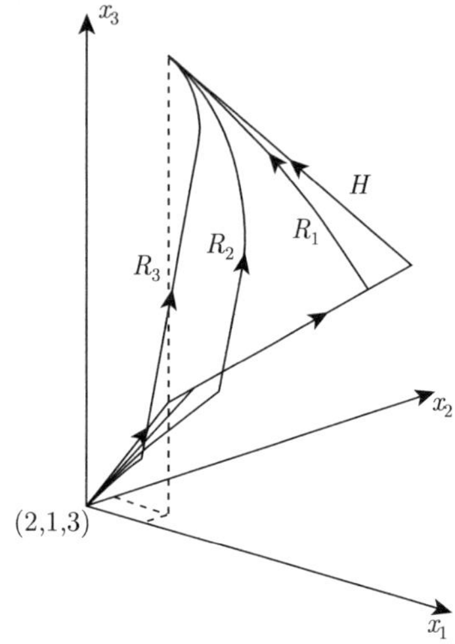

Fig. 4.13 Particle motion trajectories caused by ramp function with different rise times t_0. The observation point is located at $\boldsymbol{x} = (2, 1, 3)$ km. R_1, R_2, and R_3 represent the particle motion trajectories caused by a step function with rise times of $t_0 = 0.1$ s, 0.5 s, and 1.2 s, respectively, while H represents the particle motion trajectory caused by a step function

different values of t_0. For comparison, the result for the step function is also shown. It is very clear that t_0 significantly influences the waveform. As t_0 increases, the displacement waveform becomes smoother. When $t_0 > t_P$, it is almost impossible to distinguish the P-wave and S-wave phases from the waveform.

The differences in displacement waveforms with varying rise times inevitably affect particle motion trajectories as well. Figure 4.13 shows the particle motion trajectories at $\boldsymbol{x} = (2, 1, 3)$ km for several values of t_0. The trajectories R_1, R_2, and R_3 correspond to $t_0 = 0.1$ s, 0.5 s, and 1.2 s, respectively. For comparison, the particle motion trajectory for the step function (labeled as H) is also shown. It is evident that as t_0 increases, the R_i trajectories deviate further from H, making the P-wave and S-wave phases increasingly difficult to distinguish. However, the particles ultimately come to rest at the same point.

4.6 Seismic Waves Generated by Shear Dislocation Point Source

In Sects. 4.4 and 4.5, the Green's function for an infinite homogeneous medium was solved and its spatial and temporal distribution properties were analyzed. However, a concentrated impulse force is not an actual earthquake source model. Real earthquakes are caused by dislocations on fault planes. To understand seismic waves in an infinite homogeneous medium, it is necessary to go further and analyze the seismic wave solutions generated by dislocation sources based on the Green's function and study their properties.

There are different approaches to obtaining the seismic wave field. In Chap. 3, the equivalent body force expression (3.3.4) for a dislocation source was derived.

According to Lamé's theorem, starting from this equivalent body force and performing the Helmholtz potential decomposition, one can obtain the displacement field of the dislocation source following the same steps as solving the Green's function.[11] Another approach is to directly use the source representation theorem, taking the spatial derivatives of the Green's function and substituting them into Eq. (3.2.5). Since we have the closed-form expression of the Green's function, we can directly compute the spatial derivatives. This method will be used in the following analysis.

4.6.1 *Mathematical Solution to Seismic Wave of Shear Dislocation Point Source*

The current focus is on the point source, which is not a geometrically shapeless "point" but rather an approximation in the far-field case. Therefore, according to Eq. (3.4.2), the seismic wave field can be expressed as

$$u_n(\boldsymbol{x},t) = M_{pq}(t) * G_{np,q'}(\boldsymbol{x},t;\boldsymbol{\xi}_0,0) \tag{4.6.1}$$

where $\boldsymbol{\xi}_0$ represents a reference point on the fault plane. According to Chap. 3, the components of the seismic moment tensor M_{pq} characterize the source information, which is known in the current study. Therefore, we only need to determine $G_{np,q'}$ to further obtain the displacement field generated by the dislocation source.

Starting from Eq. (4.4.11), differentiate the right-hand side of the equation. Note the following conclusions,

$$\frac{\partial r}{\partial \xi_q} = -\gamma_q, \quad \frac{\partial \gamma_n}{\partial \xi_q} = \frac{\gamma_n\gamma_q - \delta_{nq}}{r}, \quad \frac{\partial}{\partial \xi_q}\delta\left(t-t_{\text{P}}\right) = \frac{\gamma_q}{c}\dot{\delta}\left(t-t_{\text{P}}\right)$$

$$\frac{\partial r}{\partial \xi_q}\{t[H\left(t-t_{\text{P}}\right) - H\left(t-t_{\text{S}}\right)]\} \overset{x\delta(x)=0}{=\!=\!=} \gamma_q\left[\frac{t_{\text{P}}}{\alpha}\delta\left(t-t_{\text{P}}\right) - \frac{t_{\text{S}}}{\beta}\delta\left(t-t_{\text{S}}\right)\right]$$

it's easy to obtain

$$\begin{aligned} G_{np,q'} = {} & \frac{15\gamma_n\gamma_p\gamma_q - 3\gamma_n\delta_{pq} - 3\gamma_p\delta_{nq} - 3\gamma_q\delta_{np}}{4\pi\rho r^4} t\left[H\left(t-t_{\text{P}}\right) - H\left(t-t_{\text{S}}\right)\right] \\ & + \frac{6\gamma_n\gamma_p\gamma_q - \gamma_n\delta_{pq} - \gamma_p\delta_{nq} - \gamma_q\delta_{np}}{4\pi\rho\alpha^2 r^2}\delta\left(t-t_{\text{P}}\right) + \frac{\gamma_n\gamma_p\gamma_q}{4\pi\rho\alpha^3 r}\dot{\delta}\left(t-t_{\text{P}}\right) \\ & - \frac{6\gamma_n\gamma_p\gamma_q - \gamma_n\delta_{pq} - \gamma_p\delta_{nq} - 2\gamma_q\delta_{np}}{4\pi\rho\beta^2 r^2}\delta\left(t-t_{\text{S}}\right) - \frac{(\gamma_n\gamma_p - \delta_{np})\gamma_q}{4\pi\rho\beta^3 r}\dot{\delta}\left(t-t_{\text{S}}\right) \end{aligned}$$

[11] As an exercise, readers can try to solve it using this method and compare the results with those obtained using the other approach presented later.

Substituting into Eq. (4.6.1), we have

$$\boxed{\begin{aligned} u_n(\boldsymbol{x},t) = &\frac{15\gamma_n\gamma_p\gamma_q - 3\gamma_n\delta_{pq} - 3\gamma_p\delta_{nq} - 3\gamma_q\delta_{np}}{4\pi\rho r^4}\int_{t_\text{P}}^{t_\text{S}} \tau M_{pq}(t-\tau)\mathrm{d}\tau \\ &+\frac{6\gamma_n\gamma_p\gamma_q - \gamma_n\delta_{pq} - \gamma_p\delta_{nq} - \gamma_q\delta_{np}}{4\pi\rho\alpha^2 r^2} M_{pq}\left(t-t_\text{P}\right) \\ &-\frac{6\gamma_n\gamma_p\gamma_q - \gamma_n\delta_{pq} - \gamma_p\delta_{nq} - 2\gamma_q\delta_{np}}{4\pi\rho\beta^2 r^2} M_{pq}\left(t-t_\text{S}\right) \\ &+\frac{\gamma_n\gamma_p\gamma_q}{4\pi\rho\alpha^3 r}\dot{M}_{pq}\left(t-t_\text{P}\right) - \frac{(\gamma_n\gamma_p - \delta_{np})\gamma_q}{4\pi\rho\beta^3 r}\dot{M}_{pq}\left(t-t_\text{S}\right) \end{aligned}} \tag{4.6.2}$$

This describes the displacement field caused by the seismic moment tensor $M_{pq}(t)$. It holds for any form of dislocation function $[u_i(\boldsymbol{\xi},t)]$, any elastic medium $C_{ijpq}(\boldsymbol{\xi})$, and any fault shape $\nu_j(\boldsymbol{\xi})$. If we further limit the problem to considering shear rupture on a planar fault in an isotropic homogeneous medium, then the seismic moment tensor $M_{pq}(t)$ can be expressed as in Eq. (3.4.5). Substituting this into Eq. (4.6.2), we obtain

$$\boxed{\begin{aligned} u_n(\boldsymbol{x},t) = &\frac{30\gamma_p e_p\gamma_q\nu_q\gamma_n - 6e_p\gamma_p\nu_n - 6\gamma_q\nu_q e_n}{4\pi\rho r^4}\int_{t_\text{P}}^{t_\text{S}} \tau M_0(t-\tau)\mathrm{d}\tau \\ &+\frac{12\gamma_p e_p\gamma_q\nu_q\gamma_n - 2e_p\gamma_p\nu_n - 2\gamma_q\nu_q e_n}{4\pi\rho\alpha^2 r^2} M_0\left(t-t_\text{P}\right) \\ &-\frac{12\gamma_p e_p\gamma_q\nu_q\gamma_n - 3e_p\gamma_p\nu_n - 3\gamma_q\nu_q e_n}{4\pi\rho\beta^2 r^2} M_0\left(t-t_\text{S}\right) \\ &+\frac{2\gamma_p e_p\gamma_q\nu_q\gamma_n}{4\pi\rho\alpha^3 r}\dot{M}_0\left(t-t_\text{P}\right) \\ &-\frac{2\gamma_p e_p\gamma_q\nu_q\gamma_n - \gamma_p e_p\nu_n - \gamma_q\nu_q e_n}{4\pi\rho\beta^3 r}\dot{M}_0\left(t-t_\text{S}\right) \end{aligned}} \tag{4.6.3}$$

4.6.2 Properties of Seismic Wave of Shear Dislocation Point Source

Equation (3.4.5) provides the displacement expression for a shear dislocation point source represented by the seismic moment time function $M_0(t)$. The temporal variation of the displacement field is clearly determined by the temporal variation of the seismic moment, because $M_0(t) = \mu[u(t)]A$, which is directly related to the time variation of the slip on the fault plane. Based on the different power decay rates with increasing r, the displacement field $u_n(\boldsymbol{x},t)$ in Eq. (4.6.3) is classified into three categories:

(1) Near-field term, proportional to $\dfrac{1}{r^4}\displaystyle\int_{t_\text{P}}^{t_\text{S}} \tau M_0(t-\tau)\mathrm{d}\tau$, is denoted as $u_n^\text{N}(\boldsymbol{x},t)$.

(2) Intermediate field term,[12] is proportional to $r^{-2}M_0(t-t_\text{P})$ or $r^{-2}M_0(t-t_\text{S})$. Since the terms containing α and β are separate, they are further divided into the

[12] The term "intermediate field" might lead one to mistakenly think it, like the near-field and far-field terms, dominates the displacement in a certain spatial region. However, it never actually dominates. This fact can be clearly seen by comparing the curves of r^{-3}, r^{-2}, and r^{-1}. The name "intermediate

intermediate field P-wave term $u_n^{\text{IP}}(\boldsymbol{x}, t)$ and the intermediate field S-wave term $u_n^{\text{IS}}(\boldsymbol{x}, t)$.

(3) Far-field term is proportional to $r^{-1}\dot{M}_0(t - t_{\text{P}})$ or $r^{-1}\dot{M}_0(t - t_{\text{S}})$.[13] Similar to the intermediate field term, the P-wave and S-wave components are separate, resulting in the far-field P-wave term $u_n^{\text{FP}}(\boldsymbol{x}, t)$ and the far-field S-wave term $u_n^{\text{FS}}(\boldsymbol{x}, t)$.

4.6.2.1 Radiation Pattern of Shear Dislocation Point Source

Based on the above properties, we can explicitly express the displacement field as a sum of individual terms,

$$u_n(\boldsymbol{x}, t) = u_n^{\text{N}}(\boldsymbol{x}, t) + u_n^{\text{IP}}(\boldsymbol{x}, t) + u_n^{\text{IS}}(\boldsymbol{x}, t) + u_n^{\text{FP}}(\boldsymbol{x}, t) + u_n^{\text{FS}}(\boldsymbol{x}, t) \tag{4.6.4}$$

in which,

$$u_n^{\text{N}}(\boldsymbol{x}, t) = \frac{\mathscr{A}_n^{\text{N}}}{4\pi\rho r^4}\int_{t_{\text{P}}}^{t_{\text{S}}} \tau M_0(t-\tau)\mathrm{d}\tau \tag{4.6.5a}$$

$$u_n^{\text{IP}}(\boldsymbol{x}, t) = \frac{\mathscr{A}_n^{\text{IP}}}{4\pi\rho\alpha^2 r^2} M_0\left(t - t_{\text{P}}\right), \quad u_n^{\text{IS}}(\boldsymbol{x}, t) = \frac{\mathscr{A}_n^{\text{IS}}}{4\pi\rho\beta^2 r^2} M_0\left(t - t_{\text{S}}\right) \tag{4.6.5b}$$

$$u_n^{\text{FP}}(\boldsymbol{x}, t) = \frac{\mathscr{A}_n^{\text{FP}}}{4\pi\rho\alpha^3 r} \dot{M}_0\left(t - t_{\text{P}}\right), \quad u_n^{\text{FS}}(\boldsymbol{x}, t) = \frac{\mathscr{A}_n^{\text{FS}}}{4\pi\rho\beta^3 r} \dot{M}_0\left(t - t_{\text{S}}\right) \tag{4.6.5c}$$

where $\mathscr{A}_n^{\text{N}}$, $\mathscr{A}_n^{\text{IP}}$, $\mathscr{A}_n^{\text{IS}}$, $\mathscr{A}_n^{\text{FP}}$, and $\mathscr{A}_n^{\text{FS}}$ represent the radiation patterns of the respective components, defined as follows:

$$\mathscr{A}_n^{\text{N}} \triangleq 30\gamma_p e_p \gamma_q \nu_q \gamma_n - 6e_p\gamma_p\nu_n - 6\gamma_q\nu_q e_n \tag{4.6.6a}$$

$$\mathscr{A}_n^{\text{IP}} \triangleq 12\gamma_p e_p \gamma_q \nu_q \gamma_n - 2e_p\gamma_p\nu_n - 2\gamma_q\nu_q e_n \tag{4.6.6b}$$

$$\mathscr{A}_n^{\text{IS}} \triangleq -(12\gamma_p e_p \gamma_q \nu_q \gamma_n - 3e_p\gamma_p\nu_n - 3\gamma_q\nu_q e_n) \tag{4.6.6c}$$

$$\mathscr{A}_n^{\text{FP}} \triangleq 2\gamma_p e_p \gamma_q \nu_q \gamma_n \tag{4.6.6d}$$

$$\mathscr{A}_n^{\text{FS}} \triangleq -(2\gamma_p e_p \gamma_q \nu_q \gamma_n - \gamma_p e_p\nu_n - \gamma_q\nu_q e_n) \tag{4.6.6e}$$

Noting the above definitions, it is not difficult to see that

$$\mathscr{A}_n^{\text{N}} = 9\mathscr{A}_n^{\text{FP}} - 6\mathscr{A}_n^{\text{FS}}, \quad \mathscr{A}_n^{\text{IP}} = 4\mathscr{A}_n^{\text{FP}} - 2\mathscr{A}_n^{\text{FS}}, \quad \mathscr{A}_n^{\text{IS}} = -3\mathscr{A}_n^{\text{FP}} + 3\mathscr{A}_n^{\text{FS}} \tag{4.6.7}$$

---

field" simply reflects its decay behavior with increasing r, which falls between that of the near-field and far-field terms. Aki and Richards (2002, p.77) also noted this point.

[13] It is important to note that, unlike the far-field term of the Green's function, the far-field term of a dislocation point source is not proportional to the seismic moment M_0 itself, but to its derivative to time $\dot{M}_0$. This means that the far-field P-wave and S-wave terms are controlled by the time derivative of the dislocation function rather than the function itself.

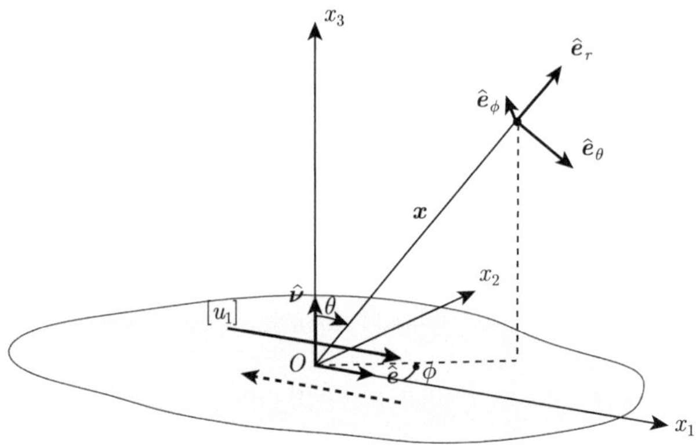

Fig. 4.14 Spherical coordinates in the source coordinate system. The normal direction of the fault plane is $\hat{\boldsymbol{\nu}} = (0, 0, 1)$, and the unit vector of the dislocation direction is $\hat{\boldsymbol{e}} = (1, 0, 0)$. The spherical coordinate basis at the position $\boldsymbol{x} = (r, \theta, \phi)$ is $(\hat{\boldsymbol{e}}_r, \hat{\boldsymbol{e}}_\theta, \hat{\boldsymbol{e}}_\phi)$

This indicates that the radiation patterns of both the near-field and intermediate-field terms can be expressed as combinations of the radiation patterns of the far-field P-wave and far-field S-wave terms. Therefore, by understanding the characteristics of the far-field P-wave and S-wave radiation patterns, the patterns of other terms can be derived through their combinations.

The expression for the radiation pattern in Eq. (4.6.6) involves three unit vectors: $\hat{\boldsymbol{\gamma}}$, $\hat{\boldsymbol{e}}$, and $\hat{\boldsymbol{\nu}}$. Recall from Chap. 3, when we studied the equivalent body force, we introduced the source coordinate system for convenience. In this system, $\hat{\boldsymbol{e}}$ and $\hat{\boldsymbol{\nu}}$ have their simplest forms. Here, we will use the source coordinate system and express the radiation pattern in spherical coordinates to examine its spatial distribution. Figure 4.14 shows the source coordinate system, where the spherical coordinates of a point $\boldsymbol{x}$ are (r, θ, ϕ). It is easy to find that the Cartesian components of the basis vectors in the spherical coordinate system can be expressed in terms of spherical coordinates as

$$\hat{\boldsymbol{e}} = (1,\ 0,\ 0), \quad \hat{\boldsymbol{\nu}} = (0,\ 0,\ 1), \quad \hat{\boldsymbol{\gamma}} = \hat{\boldsymbol{e}}_r = (\sin\theta\cos\phi,\ \sin\theta\sin\phi,\ \cos\theta)$$

Substitute into Eqs. (4.6.6d) and (4.6.6e), and note that

$$\hat{\boldsymbol{e}}_\theta = (\cos\theta\cos\phi,\ \cos\theta\sin\phi,\ -\sin\theta), \quad \hat{\boldsymbol{e}}_\phi = (-\sin\phi,\ \cos\phi,\ 0)$$

we find

$$\vec{\mathscr{A}}^{\mathrm{FP}} = \sin 2\theta\cos\phi(\sin\theta\cos\phi, \sin\theta\sin\phi, \cos\theta) = \sin 2\theta\cos\phi\hat{\boldsymbol{e}}_r \tag{4.6.8a}$$

$$\begin{aligned}\vec{\mathscr{A}}^{\mathrm{FS}} &= \big(\cos\theta - \sin\theta\sin 2\theta\cos^2\phi, -\sin\theta\sin 2\theta\sin\phi\cos\phi, -\sin\theta\cos 2\theta\cos\phi\big)\\ &= \cos 2\theta\cos\phi\hat{\boldsymbol{e}}_\theta - \cos\theta\sin\phi\hat{\boldsymbol{e}}_\phi\end{aligned} \tag{4.6.8b}$$

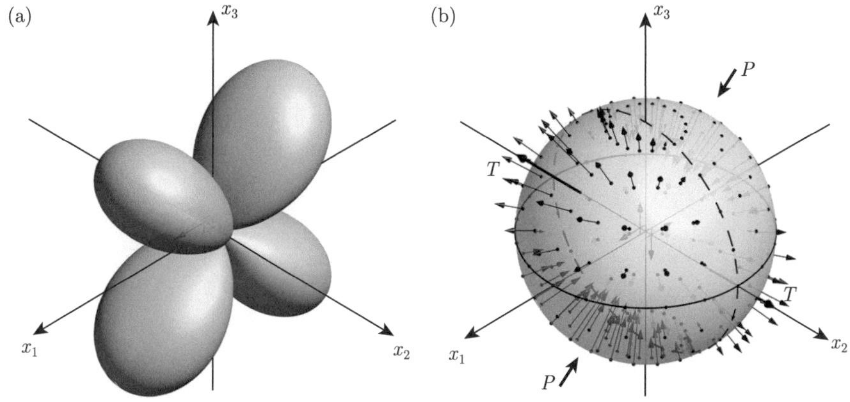

Fig. 4.15 Spatial distribution of $\mathscr{A}_n^{\mathrm{FP}}$ from a dislocation point source. **a** The amplitude (absolute value) of the radiation pattern; **b** The vector distribution on the sphere, where the length and direction of the arrows represent the magnitude and direction of the radiation pattern. The solid and dashed circles on the sphere represent the fault plane and the auxiliary plane. "*T*" and "*P*" denote the tensile axis and the compressional axis, respectively

By substituting Eq. (4.6.8) into Eq. (4.6.7), we obtain the radiation patterns for the near-field and intermediate-field terms,

$$\overrightarrow{\mathscr{A}}^{\mathrm{N}} = 9\sin 2\theta\cos\phi\hat{\boldsymbol{e}}_r - 6\left(\cos 2\theta\cos\phi\hat{\boldsymbol{e}}_\theta - \cos\theta\sin\phi\hat{\boldsymbol{e}}_\phi\right) \tag{4.6.9a}$$

$$\overrightarrow{\mathscr{A}}^{\mathrm{IP}} = 4\sin 2\theta\cos\phi\hat{\boldsymbol{e}}_r - 2\left(\cos 2\theta\cos\phi\hat{\boldsymbol{e}}_\theta - \cos\theta\sin\phi\hat{\boldsymbol{e}}_\phi\right) \tag{4.6.9b}$$

$$\overrightarrow{\mathscr{A}}^{\mathrm{IS}} = -3\sin 2\theta\cos\phi\hat{\boldsymbol{e}}_r + 3\left(\cos 2\theta\cos\phi\hat{\boldsymbol{e}}_\theta - \cos\theta\sin\phi\hat{\boldsymbol{e}}_\phi\right) \tag{4.6.9c}$$

Based on Eq. (4.6.8), the radiation pattern of the far-field P-wave is in the same direction as $\hat{\boldsymbol{e}}_r$. If we still consider a sphere with the source point as its center, it is along the normal direction of the sphere's surface. The radiation pattern of the far-field S-wave is a combination of the components in the $\hat{\boldsymbol{e}}_\theta$ and $\hat{\boldsymbol{e}}_\phi$ directions, so it is tangential to the sphere's surface. The S-wave in the $\hat{\boldsymbol{e}}_\theta$ direction is usually called the SV wave, and the S-wave in the $\hat{\boldsymbol{e}}_\phi$ direction is called the SH wave.[14]

Figures 4.15 and 4.16 illustrate the spatial distribution of the far-field radiation patterns for P-wave and S-wave, respectively. The amplitude of the far-field P-wave radiation pattern (Fig. 4.15a) forms a regular four-lobed shape in space. If we take a cross-section at $x_2 = 0$, it shows the classic "rose petal" pattern seen in seismology textbooks. The vector distribution on the sphere's surface (Fig. 4.15b) provides more detailed information. If we use x_2 as the axis and draw circles in the x_1x_2 and x_2x_3 planes on the sphere, dividing it into four parts, the vectors of the radiation pattern

[14] There is a close relationship between SV waves and P waves; for instance, they can convert into each other at an interface. However, SH waves are not related to the other two. In Chap. 6, we will mathematically explain the reasons behind this.

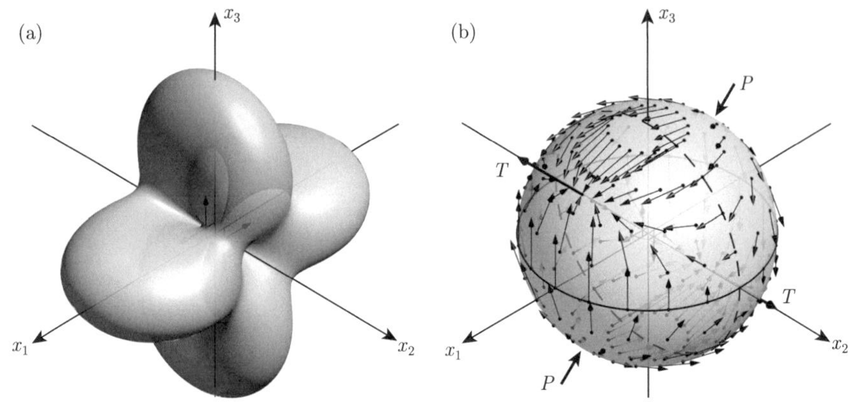

Fig. 4.16 Spatial distribution of $\mathscr{A}_n^{\mathrm{FS}}$ from a dislocation point source. The same as Fig. 4.4, except here it is for the far-field S-wave

alternately point inward and outward on the sphere's surface. The solid line circle in the x_1x_2 plane lies within the fault plane, while the dashed line circle perpendicular to it is known as the auxiliary plane in seismology. The center of the sphere is marked with the equivalent body force (double couple), indicating that the direction of the far-field P-wave radiation pattern closely relates to the four single force directions of the double couple. The outward parts of the pattern correspond to a tensile stress state, while the inward parts correspond to a compressive stress state. The radial directions at the centers of these four parts are the T-axis and P-axis, respectively. This alternating inward and outward distribution of the radiation pattern directly corresponds to the P-wave first motion directions observed at the surface, allowing us to infer the distribution of the earthquake fault plane from the observed P-wave first motion directions in different regions.

The amplitude of the far-field S-wave radiation pattern (Fig. 4.16a) also forms a similar four-lobed shape in space but is rotated by 45° compared to the P-wave pattern. As expected, the vectors of the radiation pattern on the sphere's surface (Fig. 4.16b) are tangential to the sphere. Unlike the longitudinal motion characteristics of far-field P-wave, the transverse motion of far-field S-wave displays interesting features in the corresponding force diagrams. In the four spherical regions defined by the fault plane and the auxiliary plane, the vectors spread out from P-axis and converge at T-axis, as clearly shown in Fig. 4.16b.

Since the far-field S-wave radiation pattern includes components in both the $\hat{\boldsymbol{e}}_\theta$ and $\hat{\boldsymbol{e}}_\phi$ directions, the overall vector distribution on the sphere is relatively complex. To gain a more detailed understanding of the respective distributions along the meridians and latitudes, Figs. 4.17 and 4.18 show the radiation patterns for the far-field SV wave ($\hat{\boldsymbol{e}}_\theta$ component) and far-field SH wave ($\hat{\boldsymbol{e}}_\phi$ component), respectively. These patterns are oriented along meridian and latitude directions on the sphere. However, the amplitude distribution of the radiation patterns is more intricate. The amplitude distribution of the far-field SV wave radiation pattern is relatively regular along the

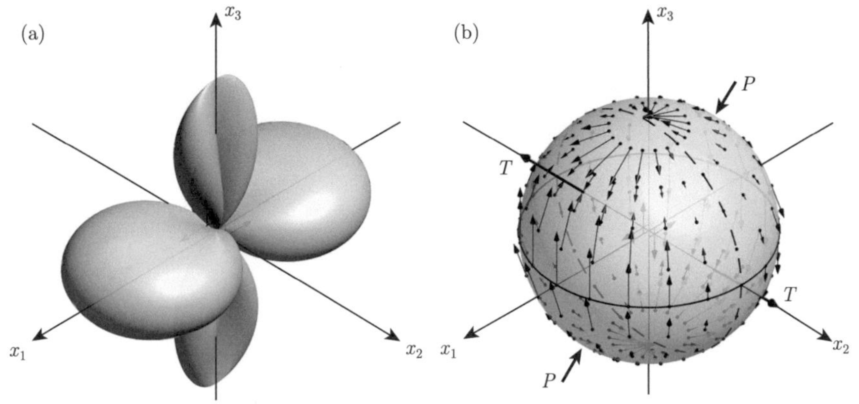

Fig. 4.17 Spatial distribution of $\mathscr{A}_n^{\mathrm{FSV}}$ from a dislocation point source. The same as Fig. 4.4, except here it is for the far-field SV-wave

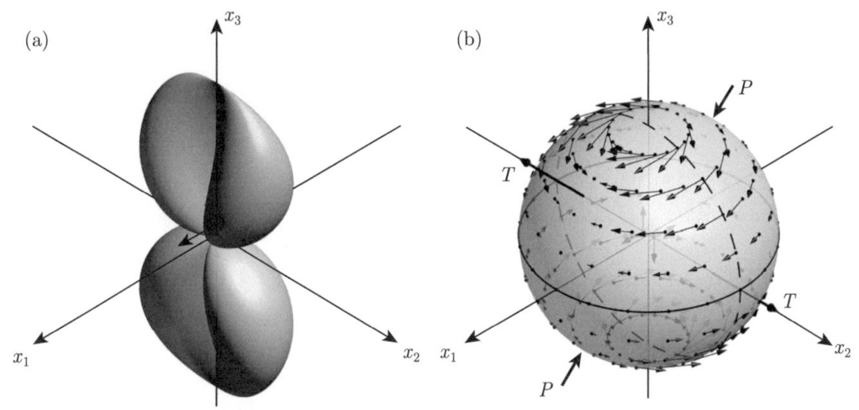

Fig. 4.18 Spatial distribution of $\mathscr{A}_n^{\mathrm{FSH}}$ from a dislocation point source. The same as Fig. 4.4, except here it is for the far-field SH-wave

x_1 direction but irregular along the x_3 direction. Similarly, the amplitude distribution of the far-field SH wave radiation pattern is irregular in the x_3 direction. Despite these irregularities, the combined effect of both components results in a regular amplitude distribution for the far-field S-wave radiation pattern, as shown in Fig. 4.16a.

Unlike the far-field terms, the intermediate-field P-wave and S-wave terms, as well as the near-field terms, include components in the $\hat{\boldsymbol{e}}_r$, $\hat{\boldsymbol{e}}_\theta$, and $\hat{\boldsymbol{e}}_\phi$ directions. This makes their vector distribution on the sphere more complex, as shown in Figs. 4.19, 4.20 and 4.21. However, the amplitude distribution of the radiation patterns is relatively regular.

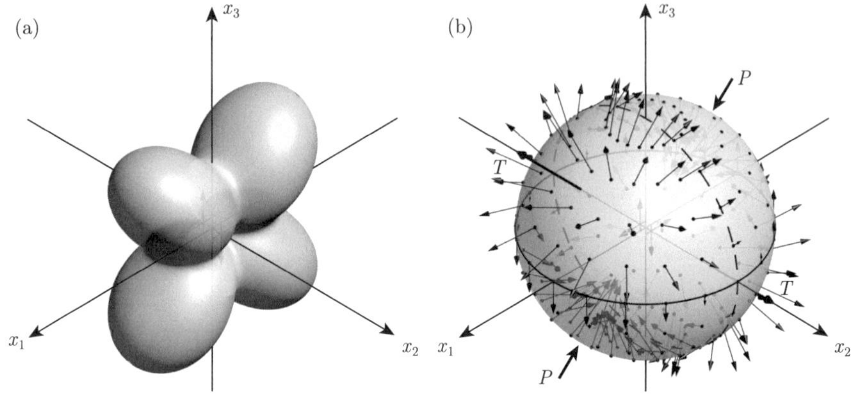

Fig. 4.19 Spatial distribution of $\mathscr{A}_n^{\mathrm{IP}}$ from a dislocation point source. The same as Fig. 4.4, except here it is for the intermediate-field P-wave

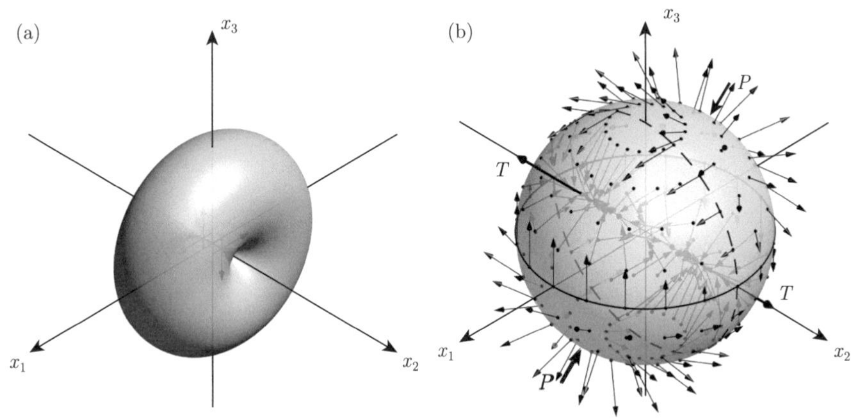

Fig. 4.20 Spatial distribution of $\mathscr{A}_n^{\mathrm{IS}}$ from a dislocation point source. The same as Fig. 4.4, except here it is for the intermediate-field S-wave

From the above analysis, it is evident that, aside from the far-field P-wave and S-wave terms, the radiation patterns of the intermediate-field and near-field terms exhibit more complex distributions. This implies that in regions closer to the seismic source, the seismic wave field shows intricate variations with azimuth. In the far-field situation, the near-field and intermediate-field terms attenuate quickly, while the far-field terms dominate, resulting in a more pronounced overall directional characteristic of the seismic wave field.

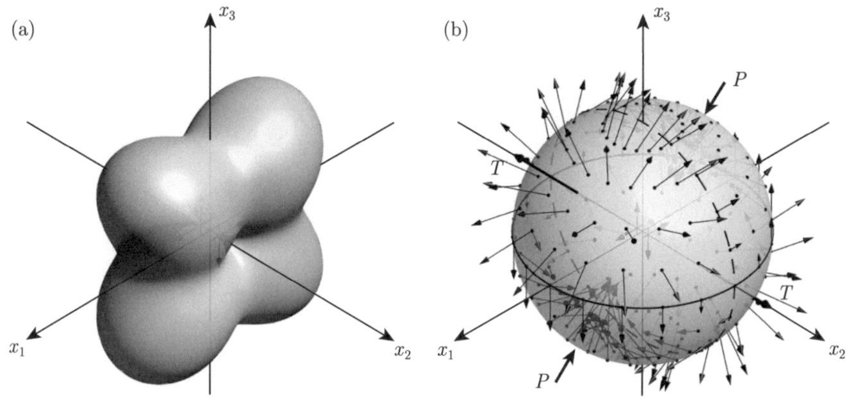

Fig. 4.21 Spatial distribution of $\mathscr{A}_n^{\mathrm{N}}$ from a dislocation point source. The same as Fig. 4.4, except here it is for the near-field wave

4.6.2.2 Displacement Generated by a General Time Function

In Eq. (4.6.3), the time-varying quantity is the seismic moment $M_0(t) = \mu\overline{[u(t)]}A$. The average displacement $\overline{[u(t)]} = \overline{[u]}S(t)$, where $S(t)$ is the time function. Therefore,

$$M_0(t) = \mu\overline{[u]}AS(t) = M_0 S(t), \quad M_0 \triangleq \mu\overline{[u]}A$$

Equation (4.6.3) can be rewritten as

$$u_n(\boldsymbol{x}, t) = \frac{M_0\mathscr{A}_n^{\mathrm{N}}}{4\pi\rho r^4}\int_{t_{\mathrm{P}}}^{t_{\mathrm{S}}} \tau S(t-\tau)\mathrm{d}\tau + \frac{M_0\mathscr{A}_n^{\mathrm{IP}}}{4\pi\rho\alpha^2 r^2}S\left(t - t_{\mathrm{P}}\right) + \frac{M_0\mathscr{A}_n^{\mathrm{IS}}}{4\pi\rho\beta^2 r^2}S\left(t - t_{\mathrm{S}}\right)$$
$$+ \frac{M_0\mathscr{A}_n^{\mathrm{FP}}}{4\pi\rho\alpha^3 r}\dot{S}\left(t - t_{\mathrm{P}}\right) + \frac{M_0\mathscr{A}_n^{\mathrm{FS}}}{4\pi\rho\beta^3 r}\dot{S}\left(t - t_{\mathrm{S}}\right) \tag{4.6.10}$$

If the time function $S(t)$ is taken as a step function $H(t)$, since

$$\int_{t_{\mathrm{P}}}^{t_{\mathrm{S}}} \tau H(t-\tau)\mathrm{d}\tau = \frac{1}{2}(t^2 - t_{\mathrm{P}}^2)H(t - t_{\mathrm{P}}) - \frac{1}{2}(t^2 - t_{\mathrm{S}}^2)H(t - t_{\mathrm{S}}) \triangleq P(t)$$

then Eq. (4.6.10) can be specifically expressed as

$$u_n(\boldsymbol{x}, t) = \frac{M_0\mathscr{A}_n^{\mathrm{N}}}{4\pi\rho r^4}P(t) + \frac{M_0\mathscr{A}_n^{\mathrm{IP}}}{4\pi\rho\alpha^2 r^2}H\left(t - t_{\mathrm{P}}\right) + \frac{M_0\mathscr{A}_n^{\mathrm{IS}}}{4\pi\rho\beta^2 r^2}H\left(t - t_{\mathrm{S}}\right)$$
$$+ \frac{M_0\mathscr{A}_n^{\mathrm{FP}}}{4\pi\rho\alpha^3 r}\delta\left(t - t_{\mathrm{P}}\right) + \frac{M_0\mathscr{A}_n^{\mathrm{FS}}}{4\pi\rho\beta^3 r}\delta\left(t - t_{\mathrm{S}}\right) \tag{4.6.11}$$

Notice that the time function for the far-field terms is $\delta(t - t_P)$ and $\delta(t - t_S)$, which means the waveform shows two pulses at the arrival times of the P-wave t_P and the S-wave t_S. This is similar to the Green's function itself, as illustrated in Fig. 4.3. Since impulsive seismic phases don't have practical significance, we will then consider the simplest time function suitable for a dislocation source–the ramp function.

However, Eq. (4.6.11) can provide useful information about the static displacement field. Considering the case as $t \to \infty$, the static displacement field is obtained as follows:

$$
\begin{aligned}
u_n^{\text{static}}(\boldsymbol{x}) &= \frac{M_0}{8\pi\mu r^2}\left[\mathscr{A}_n^{\text{N}}(1-k^2) + 2k^2\mathscr{A}_n^{\text{IP}} + 2\mathscr{A}_n^{\text{IS}}\right] \qquad \left(k = \frac{\beta}{\alpha}\right) \\
&\overset{(4.6.9)}{=\!=} \frac{M_0}{8\pi\mu r^2}\left[(3-k^2)\mathscr{A}_n^{\text{FP}} + 2k^2\mathscr{A}_n^{\text{FS}}\right] \\
&\overset{(4.6.8)}{=\!=} \frac{M_0}{8\pi\mu r^2}\left[(3-k^2)\sin 2\theta\cos\phi\,\hat{\boldsymbol{e}}_r + 2k^2\left(\cos 2\theta\cos\phi\,\hat{\boldsymbol{e}}_\theta - \cos\theta\sin\phi\,\hat{\boldsymbol{e}}_\phi\right)\right]
\end{aligned}
$$

The far-field P-wave and S-wave pulse signals become zero as $t \to \infty$, so they do not contribute to the static displacement. According to the above results, the static displacement field caused by a dislocation point source has components in all directions. Figures 4.22 and 4.23 show the displacement distribution on a sphere centered at the point source and on planes above and below the x_3 axis, respectively. The static displacement amplitude distribution in Fig. 4.22a is somewhat similar to the radiation pattern of the far-field P-wave. The distribution on the sphere clearly shows that in the tensional region, the displacement is outward from the sphere, while in the compressional region, the displacement is inward, with the maximum

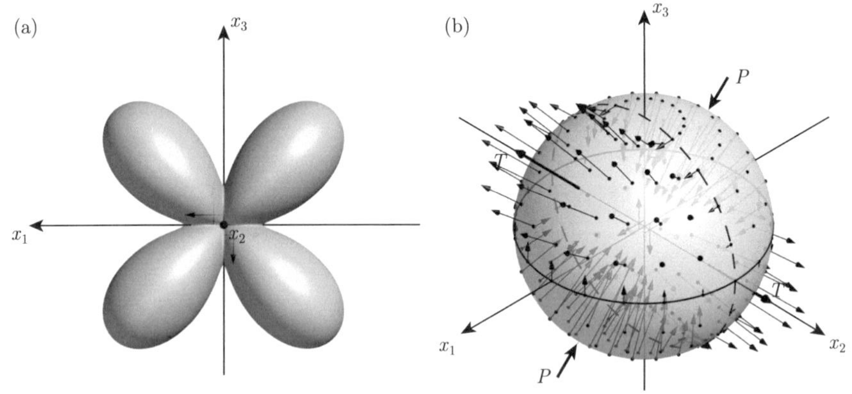

Fig. 4.22 Static displacement field caused by a dislocation point source. The same as Fig. 4.4, except here it is for the static displacement

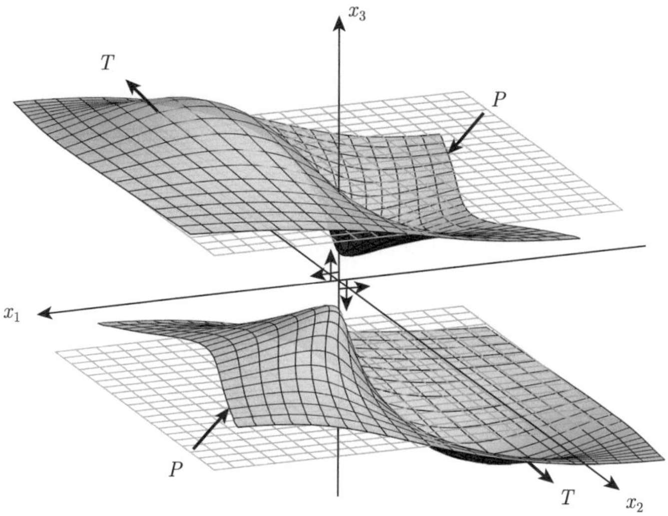

Fig. 4.23 Distribution of the static displacement field on the planes above and below the x_3 axis. "T" and "P" denote the tensional axis and compressional axis, respectively. The plane before deformation is represented by a gray grid

displacement occurring in the x_1x_3 plane. The results of the static displacement field on the two planes in Fig. 4.23 also exhibit this characteristic.

If the time function $S(t)$ in Eq. (4.6.10) is taken as the ramp function $R(t)$, see Eq. (4.5.4), since

$$\begin{aligned}\dot{R}(t) &= \frac{1}{t_0}\big[H(t) + t\delta(t) - H(t - t_0) - (t - t_0)\delta(t - t_0)\big] \\ &\xlongequal{x\delta(x)=0} \frac{1}{t_0}\big[H(t) - H(t - t_0)\big] \triangleq B(t) \qquad (4.6.12)\end{aligned}$$

the time variation behavior of the far-field P-wave and S-wave caused by the ramp function is box-shaped, that is, $\dot{R}(t) = 1$ when $0 < t < t_0$, and zero otherwise. Additionally,

$$\frac{1}{t_0}\int_{t_\text{P}}^{y} \tau(x - \tau)H(x - \tau)\mathrm{d}\tau = P(x, t_\text{P}) - P(x, y)$$

in which,

$$P(x, y) = \frac{1}{6t_0}\big[x^3 - (3x - 2y)y^2\big]H(x - y) \qquad (4.6.13)$$

thus

$$\begin{aligned}\int_{t_\mathrm{P}}^{t_\mathrm{S}} \tau R(t-\tau)\mathrm{d}\tau &= \frac{1}{t_0}\int_{t_\mathrm{P}}^{t_\mathrm{S}} \big[\tau(t-\tau)H(t-\tau) - \tau(t-t_0-\tau)H(t-t_0-\tau)\big]\mathrm{d}\tau \\ &= P(t,t_\mathrm{P}) - P(t,t_\mathrm{S}) - P(t-t_0,t_\mathrm{P}) + P(t-t_0,t_\mathrm{S}) \end{aligned} \tag{4.6.14}$$

Substituting Eqs. (4.6.12) and (4.6.14) into Eq. (4.6.10), we obtain

$$\begin{aligned} u_n(\boldsymbol{x},t) =& \frac{M_0\mathscr{A}_n^{\mathrm{N}}}{4\pi\rho r^4}\big[P(t,t_\mathrm{P}) - P(t,t_\mathrm{S}) - P(t-t_0,t_\mathrm{P}) + P(t-t_0,t_\mathrm{S})\big] \\ &+ \frac{M_0\mathscr{A}_n^{\mathrm{IP}}}{4\pi\rho\alpha^2 r^2} R\left(t-t_\mathrm{P}\right) + \frac{M_0\mathscr{A}_n^{\mathrm{IS}}}{4\pi\rho\beta^2 r^2} R\left(t-t_\mathrm{S}\right) \\ &+ \frac{M_0\mathscr{A}_n^{\mathrm{FP}}}{4\pi\rho\alpha^3 r} B\left(t-t_\mathrm{P}\right) + \frac{M_0\mathscr{A}_n^{\mathrm{FS}}}{4\pi\rho\beta^3 r} B\left(t-t_\mathrm{S}\right) \end{aligned} \tag{4.6.15}$$

The expressions for the radiation pattern can be found in Eqs. (4.6.6), (4.6.8), and (4.6.9). The definitions of $P(t_1, t_2)$, $R(t)$, and $B(t)$ are given in Eqs. (4.5.4), (4.6.12), and (4.6.13), respectively.

Figure 4.24 illustrates the displacement component $\bar{u}_i$ generated by a dislocation point source with a ramp function as the time function, for different rise times t_0. A noticeable feature of the waveforms is the distinct box-shaped phase following the arrivals of the P-wave and S-wave. This is characteristic of the far-field wave terms from a dislocation source with a ramp time function, as analyzed earlier. The width of this box-shaped phase is t_0, so when $t_0 \to 0$, the far-field phases approach a pulse. Conversely, when $t_0 > t_\mathrm{S} - t_\mathrm{P}$ (i.e., $\bar{t}_0 > \sqrt{3} - 1 \approx 0.732$), as seen for $\bar{t}_0 =$ 1.07, 1.50, 1.92, and 2.57 in the figure, the box-shaped phases following the P-wave and S-wave overlap. As t_0 increases, the amplitude of the waveforms decreases.

To clearly understand the contributions of the near-field, intermediate-field, and far-field components to the overall waveform, Fig. 4.25 illustrates the total displacement components $\bar{u}_i$ (shown with thick lines) for $t_0 = 0.2$ s ($\bar{t}_0 = 0.43$), along with their corresponding near-field components ($\bar{u}_i^{\mathrm{N}}$), intermediate-field components ($\bar{u}_i^{\mathrm{IP}}$ and $\bar{u}_i^{\mathrm{IS}}$), and far-field components ($\bar{u}_i^{\mathrm{FP}}$ and $\bar{u}_i^{\mathrm{FS}}$). As shown in Eq. (4.6.15), the near-field, intermediate-field, and far-field correspond to the integral of the ramp function, the ramp function itself, and the boxcar function (the derivative of the ramp function), respectively. Consequently, the waveform exhibits progressively discontinuous features. Overall, for a dislocation point source with a ramp function as the time function, the most prominent feature in the waveform is the boxcar phase of the far-field P-wave and S-wave, with a width equal to the rise time t_0. The presence of the far-field P-wave and S-wave components makes the waveform distinctly different from a step function, highlighting the significant impact of the time function on the seismic waveforms. Through the inversion of seismic wave records, the time history of the dislocation on the earthquake fault can be understood.

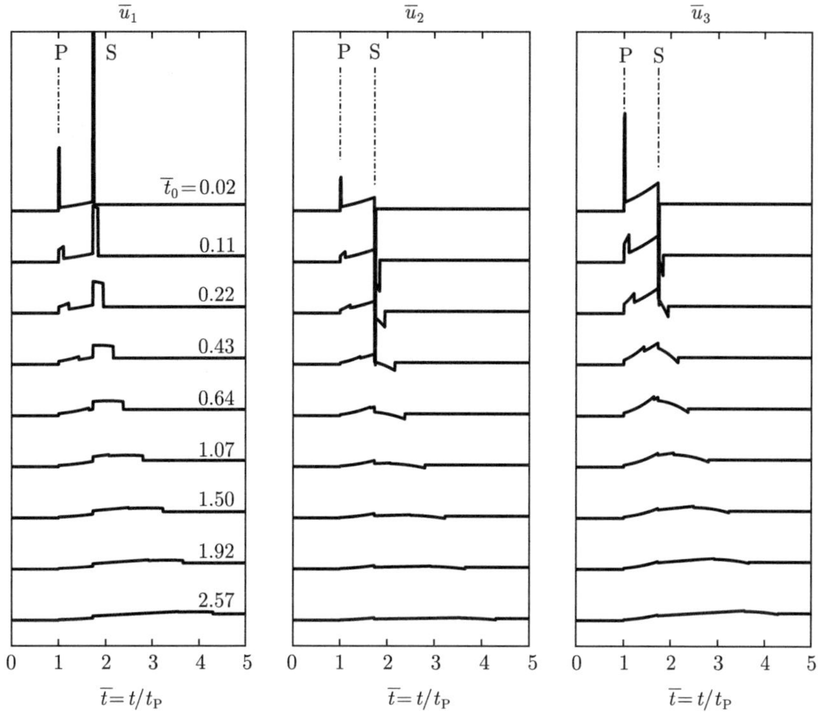

Fig. 4.24 $\bar{u}_i(t)$ generated by dislocation point source with ramp function $R(t)$ for different t_0. The observation point is located at $\boldsymbol{x} = (2, 1, 3)$ km. The horizontal axis represents time nondimensionalized by the P-wave arrival time t_P, with dimensionless displacement $\bar{u}_i(t) = 4\pi\mu r^2 u_i(t)$. In the figure, $\bar{t}_0 = t/t_P$. "P" and "S" represent the P-wave and S-wave, with arrival times of $t_P = 0.47$ s and $t_S = 0.81$ s, respectively

Figure 4.26 shows the particle trajectory generated by a dislocation point source with a ramp function of $t_0 = 0.2$ s ($\bar{t}_0 = 0.43$). The final position of the particle is similar to the case of a step function single force shown in Fig. 4.10, but the intermediate process is much more complex. During the P-wave phase, the particle moves radially away from the sphere, while during the S-wave phase, the particle moves perpendicular to the radial direction. This is evident when comparing to the analysis in Fig. 4.24.

As shown in Fig. 4.24, the amplitude of the waveform decreases significantly as t_0 increases. This characteristic is also evident in the particle motion trajectory, as illustrated in Fig. 4.27. The figure displays the particle motion trajectories for $\bar{t}_0 = 0.22$, 1.07, and 2.57 (refer to Fig. 4.24). Although the motion paths differ significantly, they all eventually terminate at the same point. This also highlights the nature of transient processes: even though the final static displacement is the same, the intermediate stages can involve vastly different motion amplitudes.

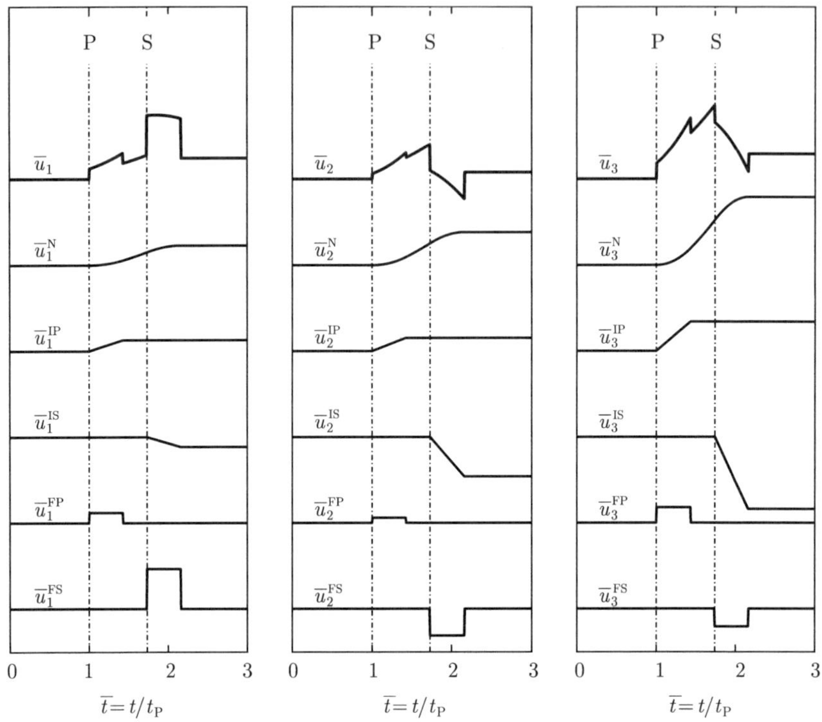

Fig. 4.25 $\bar{u}_i$ and its components generated by dislocation point source with $R(t)$ ($t_0 = 0.2$ s). Including the near-field component ($\bar{u}_i^{\rm N}$), the intermediate-field components ($\bar{u}_i^{\rm IP}$ and $\bar{u}_i^{\rm IS}$), and the far-field components ($\bar{u}_i^{\rm FP}$ and $\bar{u}_i^{\rm FS}$). The total displacement is shown with a thick line. The horizontal axis represents time nondimensionalized by the P-wave arrival time $t_{\rm p}$, with the dimensionless displacement $\bar{u}_i(t) = 4\pi\mu r^2 u_i(t)$. The dashed lines indicate the arrival times of the P-wave and S-wave

Notice that the examples shown above, including the spatial distribution of radiation patterns and the displacement fields caused by the dislocation point source, are all presented in the source coordinate system. While this coordinate system is convenient for illustrating radiation patterns and calculating displacement fields, it is not practical for real-world applications. This is because seismic waves are recorded at the Earth's surface. Therefore, we must calculate the displacements in the surface coordinate system to compare with actual seismic records and obtain information about the source and the medium.

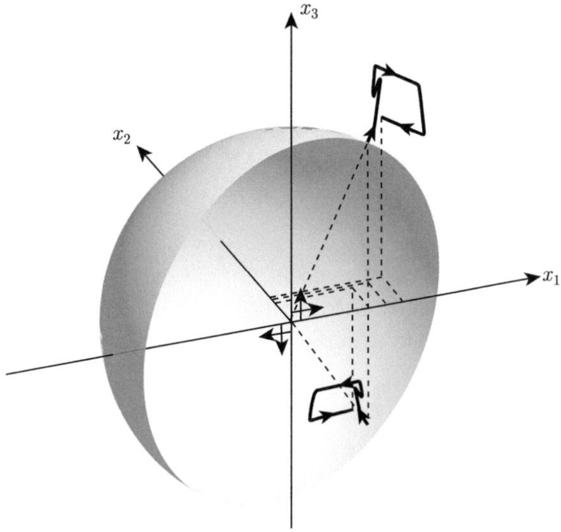

Fig. 4.26 Particle motion trajectory generated by a dislocation point source with a ramp function. Two observation points are located at $\boldsymbol{x} = (2, 1, 3)$ km and $(2, 1, -3)$ km. The rise time of the dislocation function is $t_0 = 0.2$ s. Similar to Fig. 4.10, the particle starts its motion from the surface of the sphere and eventually comes to rest at a point

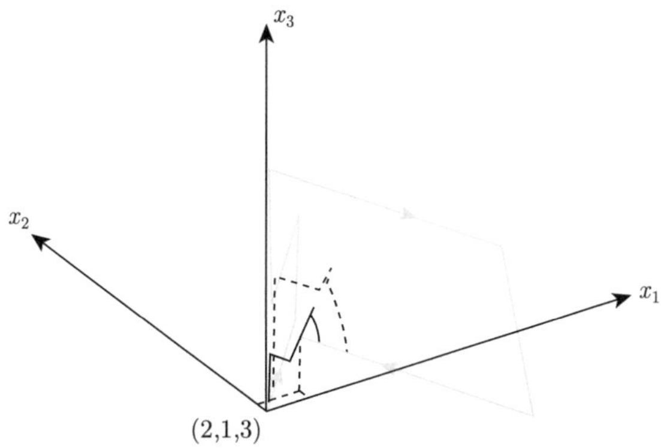

Fig. 4.27 Particle motion trajectories due to dislocation point source with $R(t)$ for different t_0. The observation point is located at $\boldsymbol{x} = (2, 1, 3)$ km. The light gray solid line, black dashed line, and black solid line correspond to $\bar{t}_0 = 0.22$, 1.07, and 2.57, respectively (refer to Fig. 4.24)

4.7 Displacement of Dislocation Point and Finite Source in Epicenter Coordinates

Although the medium model considered in this chapter is an infinite space without a surface, it is necessary to assume an imaginary "surface plane" and establish a coordinate system based on it. This approach helps in calculating the displacement field in this coordinate system and provides a basis for future studies of the Lamb's problem to better understand the effects of a free surface. Note that this "surface plane" does not actually exist; it is an arbitrarily selected plane in the infinite space.

In the coordinate system established on the "surface", it is natural to place two axes (typically the x_1 and x_2 axes) within the surface plane, with the other axis (usually the x_3 axis) perpendicular to the surface and pointing downward. However, there is some flexibility in choosing the origin of the coordinate system. For convenience in addressing the problem, the origin O is usually placed at the epicenter, introducing the epicenter coordinate system described in Sect. 3.4.3 (see Fig. 3.6).

4.7.1 Displacement of Dislocation Point Source in Epicenter Coordinates

According to the displacement field expression (4.6.3) generated by a dislocation point source, we need to use the specific expressions for the three unit vectors $\hat{\boldsymbol{\gamma}}$, $\hat{\boldsymbol{\nu}}$, and $\hat{\boldsymbol{e}}$ in the epicenter coordinate system. These are provided in Fig. 3.6 and Eq. (3.4.7),

$$\begin{aligned}\hat{\boldsymbol{e}} &= (c_{11},\ c_{21},\ c_{31})\\ &= (\cos\lambda\cos\phi_s + \cos\delta\sin\lambda\sin\phi_s,\ \cos\lambda\sin\phi_s - \cos\delta\sin\lambda\cos\phi_s,\ -\sin\lambda\sin\delta)\\ \hat{\boldsymbol{\nu}} &= (c_{13},\ c_{23},\ c_{33}) = (-\sin\delta\sin\phi_s,\ \sin\delta\cos\phi_s,\ -\cos\delta)\end{aligned} \tag{4.7.1}$$

$\hat{\boldsymbol{\gamma}}$ is the unit vector in the direction from the source to the field point. As shown in Fig. 4.28, if we denote the angle between the source-to-field line and the vertical axis as θ, and the angle between the projection of the source-to-field line on the surface and the x_1 axis (the northward axis) as ϕ, then it is straightforward to write according to the geometric relationship as

$$\hat{\boldsymbol{\gamma}} = (\sin\theta\cos\phi,\ \sin\theta\sin\phi,\ -\cos\theta) \tag{4.7.2}$$

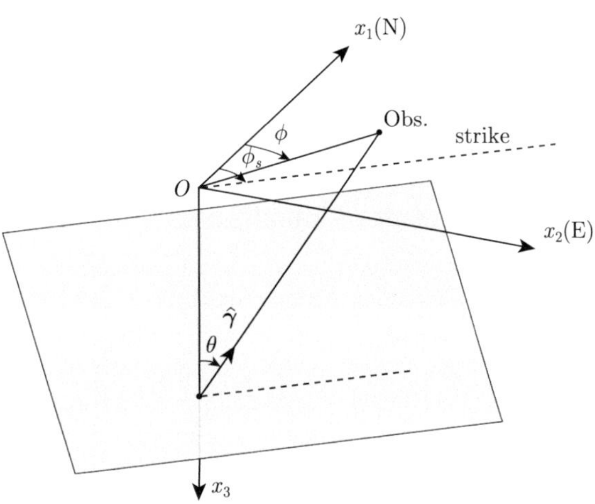

Fig. 4.28 Unit vector $\hat{\boldsymbol{\gamma}}$ in epicenter coordinate system. θ and ϕ are the angles between the source-to-field line and the vertical axis, and between the projection of the source-to-field line on the surface and the x_1 axis (northward axis), respectively. The dashed line represents the fault's strike, which forms an angle ϕ_s with the northward direction. "Obs." indicates the observation point

Table 4.1 Parameters for numerical calculation

Parameter	Value	Parameter	Value
δ	90°	α	8.00 km/s
λ	0°	β	4.62 km/s
ϕ_s	90°	ρ	3.30 g/cm^3
t_0	3.00 s	$\overline{[u]}$	1.00 m
h	10 km	A	1.00 km^2

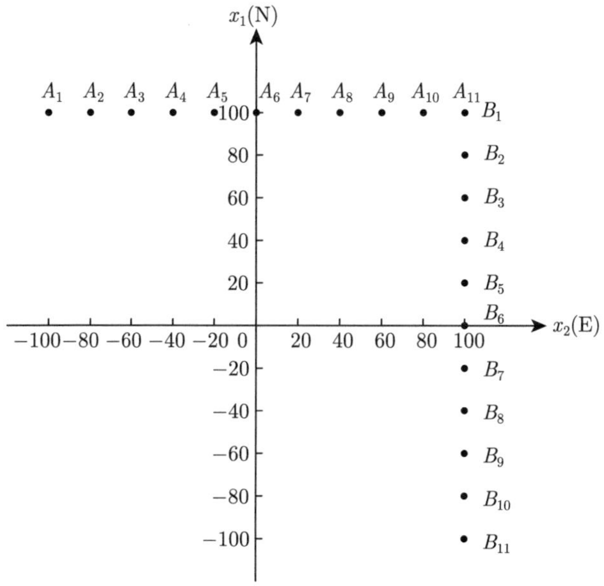

Fig. 4.29 Distribution of observation points (top view). Along the axes parallel to x_2 and x_1, 11 measurement points, denoted as A_i and B_i $(i = 1, 2, \ldots, 11)$, are set respectively, with A_{11} coinciding with B_1

By substituting Eqs. (4.7.1) and (4.7.2) into Eq. (4.6.6), and then substituting the result into Eqs. (4.6.4) and (4.6.5), we can ultimately obtain the displacement field expressed in the epicenter coordinate system.[15]

The following example demonstrates this. Consider the displacement generated at the "surface" by a vertical left-lateral strike-slip fault with a depth of $h = 10$ km and dimensions of 1 km × 1 km. The material properties and calculation parameters are listed in Table 4.1.

To illustrate the orientation effect, we calculated the displacements at points A_i and B_i $(i = 1, 2, \ldots, 11)$ on two lines. These lines are parallel and perpendicular to the fault's strike, as shown in Fig. 4.29. Since the distances from these points to the

[15] Given the complexity of the expressions, we won't list the results here. The substitution process mentioned above is clear and very conducive to program implementation.

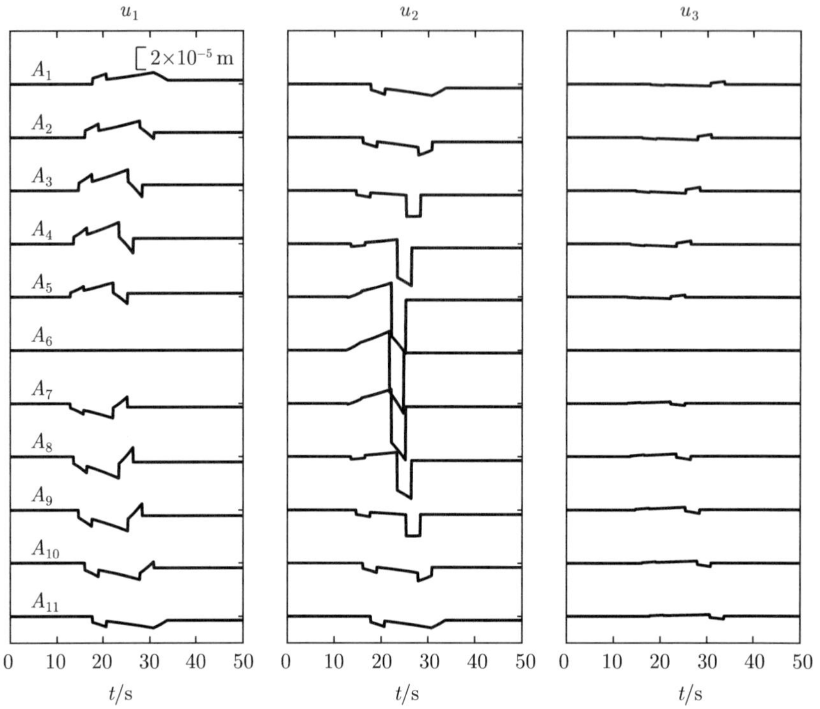

Fig. 4.30 $u_i(t)$ $(i = 1, 2, 3)$ at each A_i $(i = 1, 2, \ldots, 11)$ caused by left-lateral strike-slip rupture

earthquake source (approximately 100 to 140 km) are much greater than the size of the source (1 km), we can treat this as a far-field problem and consider the source as a point source.

Figures 4.30 and 4.31 show the displacement fields at points A_i and B_i $(i = 1, 2, \ldots, 11)$ caused by a vertical left-lateral strike-slip rupture. In Fig. 4.30, the u_1 and u_3 components exhibit symmetry with respect to the x_1 axis, while the u_2 component is continuously distributed along the x_2 axis. Since the measurement points A_i are centered around A_6 and the distance to both sides and to the seismic source increases, the arrival times of the seismic phases at each point are symmetrical around A_6 and increasingly delayed towards both sides. The displacement characteristics at the points B_i, which are perpendicular to the x_2 axis, differ from those at A_i. Figure 4.31 shows that at the points B_i, the u_2 and u_3 components are symmetrical with respect to the x_2 axis, while the u_1 component is continuously distributed along the x_1 axis. Table 4.2 lists the positive and negative static displacement fields for the u_1, u_2, and u_3 components in the northeast, northwest, southwest, and southeast quadrants.[16]

[16] A displacement field is a vector, and the sign of the vector components is defined based on their alignment with the specified coordinate axes. For example, in the northwest and southeast quadrants, the signs of u_1 are positive and negative, respectively. This indicates that u_1 points in the positive x_1

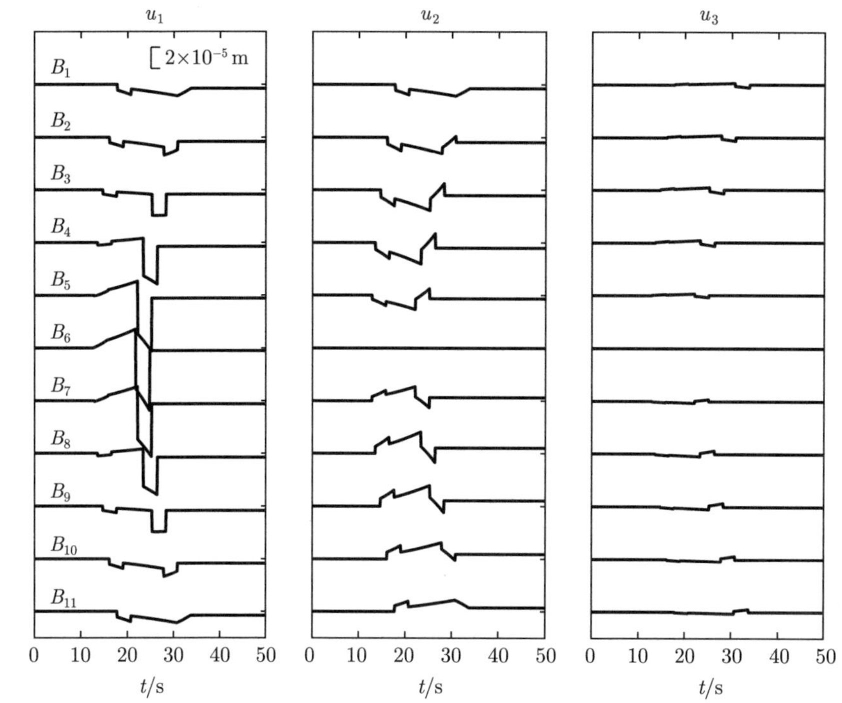

Fig. 4.31 $u_i(t)$ $(i = 1, 2, 3)$ at each B_i $(i = 1, 2, \ldots, 11)$ caused by left-lateral strike-slip rupture

Table 4.2 Signs of the static displacement components in each quadrant

Quadrant	u_1	u_2	u_3
Northwest ($x_1 > 0, x_2 < 0$)	+	−	−
Southeast ($x_1 < 0, x_2 > 0$)	−	+	−
Northeast ($x_1 > 0, x_2 > 0$)	−	−	+
Southwest ($x_1 < 0, x_2 < 0$)	+	+	+

Apart from this, the displacement fields at points A_i and B_i $(i = 1, 2, \ldots, 11)$ have the following characteristics:

(1) Since it can be considered a far-field situation, the far-field wave components dominate, which is reflected in the box-shaped seismic phase that follows immediately after the P-wave and S-wave arrivals.
(2) The amplitude of the instantaneous change is much greater than the final static displacement.

direction in the northwest and the negative x_1 direction in the southeast, reflecting the characteristic of moving away from the fault in the surface projection.

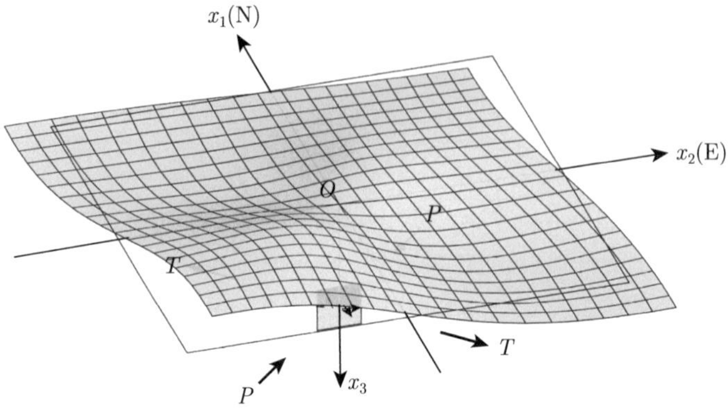

Fig. 4.32 Surface static displacement field generated by vertical strike-slip fault. The region before deformation is rectangular. The equivalent double couple and the P and T axes are shown

(3) Overall, the amplitudes of the lateral displacement components u_1 and u_2 are significantly greater than that of the vertical displacement component u_3.

Understanding the equivalent body forces and the distribution of P and T-axes for a strike-slip fault helps clarify the waveform characteristics mentioned earlier. Figure 4.32 illustrates the surface static displacement field, showing the equivalent double couple and the P and T-axes for a vertical left-lateral strike-slip fault. With a strike of 90°, the northern fault block moves west relative to the southern block. Consequently, the equivalent double force couple lies in a plane parallel to the surface, with the tension axis (T-axis) oriented northwest-southeast and the compression axis (P-axis) oriented northeast-southwest. This stress distribution causes the Earth's medium to stretch in the northwest-southeast direction and compress in the northeast-southwest direction. As a result, horizontal displacement in the northwest and southeast quadrants moves away from the fault, while in the northeast and southwest quadrants, it converges toward the fault. Additionally, due to Poisson's ratio, the surface in the northwest and southeast quadrants subsides, whereas it uplifts in the northeast and southwest quadrants. Given the Poisson's ratio of 0.25 in this example, the magnitude of vertical displacement is smaller than that of horizontal displacement. This illustrates the characteristic of a strike-slip fault, which primarily induces horizontal motion with relatively minor vertical movement. This explanation accounts for the waveform phenomena described earlier.

Finally, it is important to consider the magnitude of the displacement components. Figures 4.30 and 4.31 indicate the displacement amplitudes, providing a rough idea of the values. For a fault with an area of 1 km^2 at a depth of 10 km and an average slip of 1 m, the static and dynamic displacement amplitudes at the "surface" 100 km away are on the order of micrometers and tens to hundreds of micrometers, respectively. The seismic moment M_0 is proportional to the average slip and the fault area, so the larger the average slip and fault area, the larger the displacement field it produces.

For instance, considering the above example, what would be the result if the fault were not 1 km × 1 km, but 50 km × 10 km?

4.7.2 Displacement of Dislocation Finite Source in Epicenter Coordinates

Whether a seismic source can be approximated as a point source depends on the specific situation. In the example considered in Sect. 4.7.1, if the fault dimensions are 50 km × 10 km, this scale is comparable to the distance between the fault and the observation point. Therefore, it cannot be treated as a point source, and the finite size effects of the fault must be taken into account. The common approach is to divide the finite-sized fault plane into several sub-faults. Each sub-fault is small enough to be approximated as a point source, and then the displacement fields from all the sub-faults are summed.

To illustrate this, let's use the example from Sect. 4.7.1 again. Keeping all other parameters unchanged, we replace the fault with one that is 50 km × 10 km. As shown in Fig. 4.33, the upper boundary of the fault is 10 km from the x_2 axis. The fault is divided into 50 × 10 square sub-faults, each 1 km^2 in size, with the position of each sub-fault defined by its center point. Each sub-fault is the same size as in Sect. 4.7.1, so they can be considered point sources.

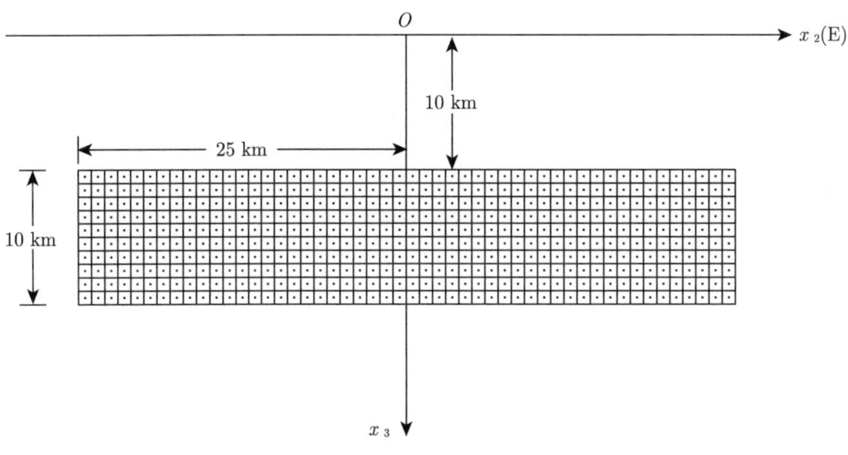

Fig. 4.33 Finite-size vertical fault model. The fault is 50 km × 10 km, with its upper boundary 10 km from the x_2 axis. The fault is divided into 50 × 10 square sub-faults, each 1 km^2 in size, represented by black dots at their centers

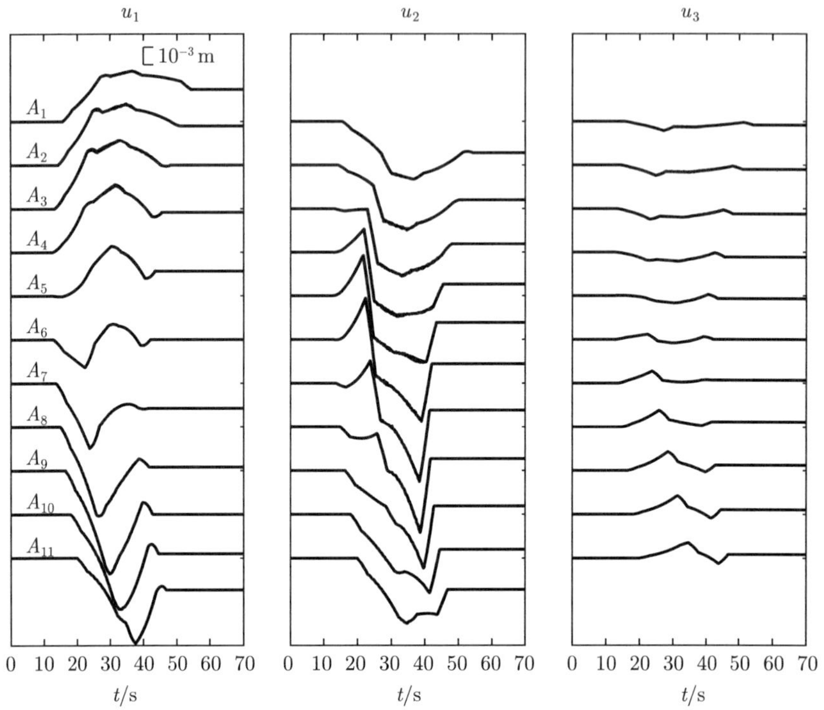

Fig. 4.34 Displacement at A_i $(i = 1, 2, \ldots, 11)$ caused by finite fault consisting of 50×10 sub-faults

Consider a classic unilateral rupture Haskell-model: at time $t = 0$, the fault begins rupturing from the leftmost row of sub-faults, then propagates to the right at a speed of $v_c = 3$ km/s (approximately 0.65β).[17] Figures 4.34 and 4.35 show the displacement field resulting from this finite source. It's evident that the overall waveform is much smoother than that of a single point source, and the arrival times of the seismic phases do not exhibit the sharp changes characteristic of a single point source. This is a typical feature of displacement fields generated by finite-size sources. Although each individual sub-fault can be considered a point source when observed at stations A_i and B_i $(i = 1, 2, \ldots, 11)$, showing distinct far-field characteristics in the waveform with box-shaped seismic phases (Figs. 4.30 and 4.31), the combined effect of the finite fault shown in Fig. 4.33 does not display the box-shaped far-field seismic phases in the waveform due to the delay effect caused by fault propagation. The presence of this delay effect lengthens the overall wave train time.

Comparing Figs. 4.30, 4.31, 4.34 and 4.35 reveals an interesting phenomenon. At the central position A_6 on line A, the u_1 and u_3 components caused by a single sub-fault on the x_3 axis are zero (Fig. 4.30), but these components are not zero when

[17] When superimposing point sources, it's important to account for the delay effect caused by rupture propagation, as not all points rupture at $t = 0$.

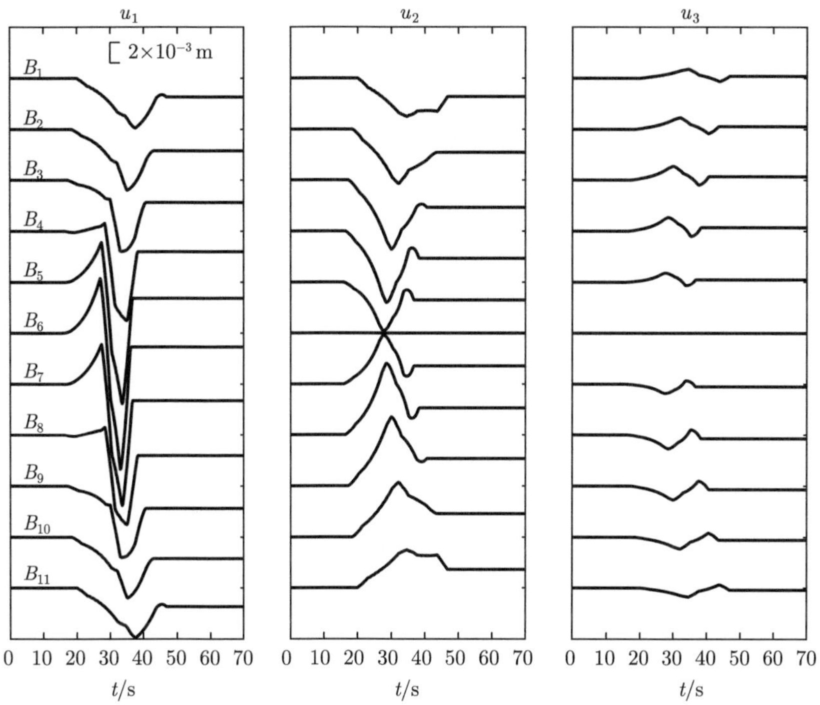

Fig. 4.35 Displacement at B_i ($i = 1, 2, \ldots, 11$) caused by finite fault consisting of 50 × 10 sub-faults

caused by the finite fault (Fig. 4.34). Conversely, at the central position B_6 on line B, the u_2 and u_3 components of the displacement field are zero for both a single sub-fault and the finite fault. Notably, for the finite-scale fault shown in Fig. 4.33, the positions of A_6 and B_6 are asymmetrical: at B_6, the azimuth angle ϕ of all sub-faults is 90°, resulting in zero 2 and 3 components of all radiation patterns (see Eq. (4.6.6)). However, for points on line A, only A_6 can have zero 1 and 3 components of all radiation patterns (resulting in u_1 and u_3 being zero). This relationship does not hold for other sub-faults distributed along the fault, leading to the observed superposition effects.

For finite seismic sources, using the method of dividing into sub-faults treated as point sources and superimposing their displacement fields naturally raises a question: what is the appropriate size for the sub-faults? Larger sub-faults reduce computational cost but may compromise accuracy. Careful examination of Fig. 4.34 reveals waveform "jitters" in some displacement components, such as the u_1 component at A_2 through A_4 and the u_2 component at A_2 through A_6. Is this due to the inherent nature of the physical problem, or is it caused by using too large sub-faults? To investigate this, we divided the fault in Fig. 4.33 into 100 × 20 sub-faults, with each

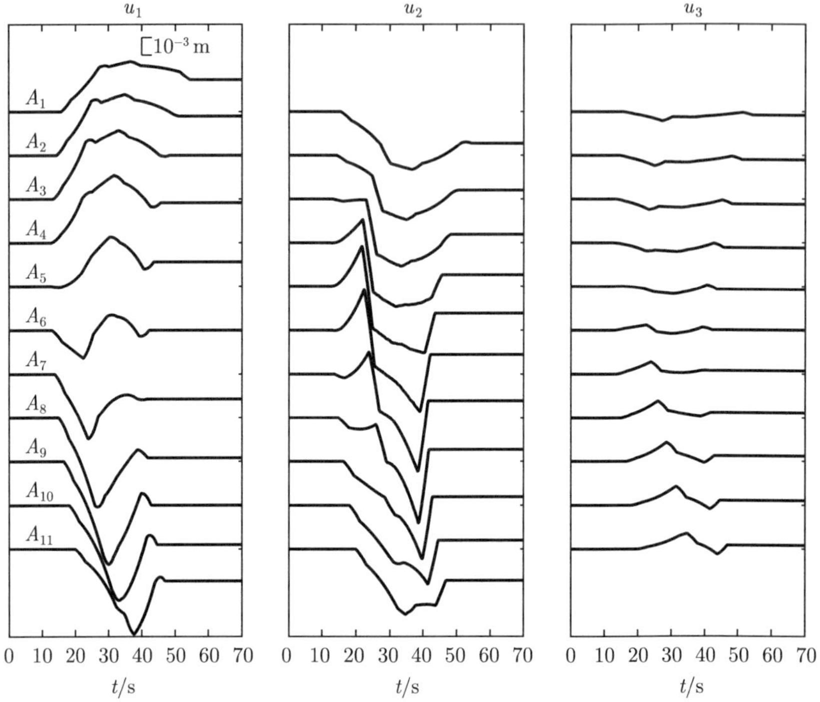

Fig. 4.36 Refined results of Fig. 4.34 with 100 × 20 sub-faults

sub-fault measuring 0.5 km × 0.5 km. Figure 4.36 shows the displacement components at each point along line A after this refinement. Compared to Fig. 4.34, the noticeable "jitter" phenomenon has disappeared.[18]

Finally, let's pay attention to the displacement amplitude. For this finite-scale fault of 50 km × 10 km, assuming an average slip of 1 m, the displacement field at a "surface" location 100 km away can reach nearly a centimeter. This is quite a significant amount. If the fault extends over hundreds of kilometers and is closer to the surface, it can be very destructive. Moreover, it's important to note that an infinite medium is assumed in current model. If a real surface is present, wave behavior becomes more complex at the surface, leading to more severe damage. This will be the focus of our next study.

[18] In practical calculations, it's difficult to have a theoretical quantitative standard for determining the sub-fault size. Generally, we select an appropriate sub-fault size based on the calculation conditions. If the results show unreasonable "jitter" phenomena like those observed here, we refine the sub-fault size and check if the results improve, continuing this process until reasonable results are achieved.

4.8 Summary

As the simplest model of an elastic medium, an isotropic and homogeneous infinite medium is the best starting point for studying seismic wave problems. In this chapter, we first solve the Green's function for the infinite medium model. This is achieved by applying Lamé's theorem, decomposing the displacement into potentials, and solving the wave equations for these potentials. The most notable feature of the Green's function in an infinite medium is its simple closed-form solution, which greatly facilitates further analysis of the wave field's properties. The Green's function can be explicitly written as the sum of near-field and far-field terms, with the radiation pattern representing directional effects and the time function terms being separated. This allows us to conveniently study the spatial distribution and temporal behavior of each wave field component.

The Green's function induced by a concentrated impulsive force forms the foundation for analyzing seismic wave problems caused by shear dislocation sources, which are more significant in seismology. In the second part of this chapter (Sects. 4.6 and 4.7), which is a key focus, we examine the seismic wave field and its properties resulting from shear dislocation sources. Although more complex in form than the Green's function, the displacement field caused by shear dislocation sources can also be clearly expressed in terms of various wave field components. Each component can similarly separate the radiation pattern from the time function. We thoroughly discuss the radiation patterns of different wave field components and use the ramp function as an example to study the temporal behavior of the wave field in detail. According to the equivalent body forces discussed in Chap. 3, a shear dislocation source is equivalent to a double couple without moment. The associated P-axis and T-axis provide powerful tools for analyzing and interpreting waveforms.

The infinite medium model is not suitable for the Earth, as it does not account for the presence of the surface. How do seismic waves behave when the surface is considered? This will be the focus of the subsequent chapters of this book. We will see that introducing the surface greatly increases the complexity of the problem. However, it also reveals more interesting physical phenomena, such as the emergence of Rayleigh surface waves, which do not exist in the infinite medium model.

Chapter 5
Overview of Researches on Lamb's Problem

With the groundwork laid in the previous chapters, we now finally delve into the main topic of this book–the Lamb's problem. As mentioned at the beginning of Chap. 1, the Lamb's problem is a long-standing classic issue in seismology and related fields. Since Lamb's 1904 work, research on this problem has continued to emerge and persists to this day.

Before delving into the specific research on the Lamb's problem, it is essential to review previous studies. Given that research related to the Lamb's problem involves extensive mathematical treatment, especially before the advent of computers, various complex asymptotic analysis techniques, such as the method of steepest descents and the saddle point method, were employed to study integrals. These methods are unfamiliar to most people today. To avoid getting bogged down in mathematical details, we will omit all formulas and instead summarize the methods and ideas. This approach is sufficient for understanding the research history. Additionally, it is important to note that the research history of the Lamb's problem spans over a century and is vast; it is neither possible nor necessary to exhaustively cover all related studies. We will only discuss the most representative and significant works[1] to provide readers with a clear understanding of the background and research history of the Lamb's problem.

[1] Defining "the most representative and significant works" involves a degree of subjectivity, as opinions may vary. The works selected for introduction here are simply those personally deemed most representative. It is important to note that some special researches related to the Lamb's problem, such as moving sources and anisotropic half-spaces, are not included in discussion here.

H. Zhang, *Lamb's Problem in Seismology (Volume 1)*,
https://doi.org/10.1007/978-981-95-3170-7_5

5.1 Lamb's Pioneering Work

5.1.1 Background of the Original Lamb (1904)

The theory of elasticity was established in the 1820s. Before that, seismology lacked a solid theoretical foundation and mainly relied on semi-empirical qualitative studies. After its establishment, seismologists[2] began focusing on explaining seismic phenomena using strict elasticity theory. Shortly after the theory of elasticity emerged, Poisson noted the existence of two types of waves with different velocities in elastic media, later known as P-wave and S-wave. A classic work on the propagation of waves in elastic media is Stokes' solution for elastic waves in an infinite medium in the mid-19th century. Although this was far from practical seismological problems, it was a significant step towards them. Stokes' solution provided the fundamental laws of wave propagation in elastic media, which have been thoroughly examined in Chap. 4.

Before Lamb's work in 1904, the most significant research was done by Rayleigh in 1885. Rayleigh discovered that under the condition of a free surface boundary, a plane wave solution of simple harmonic vibration exists. This wave, traveling at a speed slightly less than the S-wave speed (approximately 0.9194β), exhibited characteristics distinct from the previously known P-wave and S-wave. For instance, it primarily propagated along the Earth's surface; its amplitude rapidly decayed with depth; the motion at the surface traced a retrograde ellipse; and its amplitude decayed more slowly with distance than P-wave and S-wave. Later, in 1897, Oldham identified Poisson-predicted P-wave and S-wave, along with the predicted Rayleigh waves, in his study of an Indian earthquake. This confirmed that the effect of the free surface on ground motion characteristics exceeded intuitive expectations. However, Rayleigh's study was based on the assumption of plane waves, essentially waves excited by a source at an infinite distance. This raised questions about the effects of force application and non-harmonic vibrations, which remained unanswered. Lamb's in-depth research in 1904 aimed to address these questions.

5.1.2 Main Content of Lamb (1904)

Nearly 20 years after Rayleigh (1885), British physicist Horace Lamb published a paper titled "On the Propagation of Tremors over the Surface of an Elastic Solid"

[2] Strictly speaking, those who pioneered some key aspects of seismology were not professional seismologists but rather mathematicians or mechanicians who occasionally became interested in certain issues in seismology and achieved results of significant importance to the field. For instance, Rayleigh and Love, who discovered the two types of surface waves, and Lamb, the originator of the Lamb's problem mentioned here.

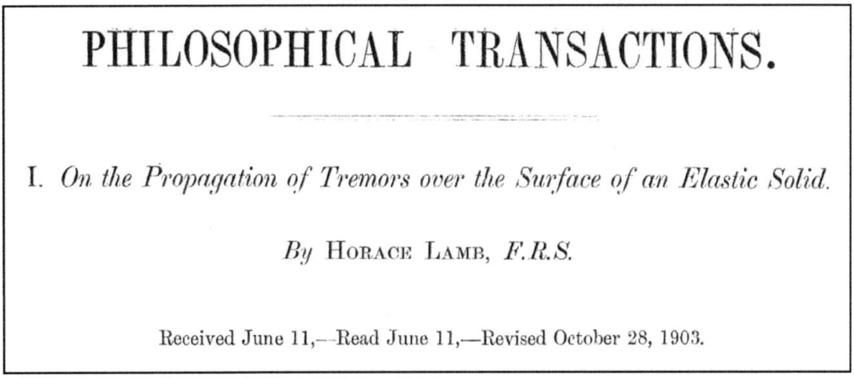
PHILOSOPHICAL TRANSACTIONS.

I. *On the Propagation of Tremors over the Surface of an Elastic Solid.*

By HORACE LAMB, *F.R.S.*

Received June 11,—Read June 11,—Revised October 28, 1903.

Fig. 5.1 Lamb (1904) published in *Philosophical Transactions of the Royal Society of London*

in 1904 in the *Philosophical Transactions of the Royal Society of London*.[3] See Fig. 5.1.

Lamb primarily considered the problem of surface tremors caused by force sources acting on a semi-infinite space. He began with the two-dimensional case, calculating the pressure applied along a line on the surface, starting with harmonic oscillations and then transitioning to arbitrary time functions. Based on this, he addressed the three-dimensional problem with vertical axis symmetry. Lamb also examined the problem of an internal line source of compressional waves, providing the integral solution for it.

In the first part of the paper, titled "Two-Dimensional Problems", Lamb starts from the equations of motion that displacement satisfies. By introducing potential functions, he transforms the coupled partial differential equations for the displacement components into two separate wave equations for the potential functions. Assuming harmonic oscillation, the wave equations reduce to the Helmholtz equations. Lamb first considers a boundary force distributed periodically along the coordinates on a free surface and derives the concentrated force through integration.[4] By ensuring that the general solution to the Helmholtz equations satisfies the boundary conditions, he obtains the integral expression for displacement. For a line source at a certain depth,

[3] This 42-page paper can be quite challenging to understand today because it employs many mathematical techniques that were necessary for addressing the problem at that time, with which, however, the vast majority of today's readers are not familiar.

[4] It should be noted that in the early 20th century, when Lamb published his paper, there were no mathematical tools like the Dirac δ function to represent concentrated forces. As a result, Lamb had to use this integral approach. In fact, the Fourier integral representing the line source is essentially the Fourier inverse transform of the δ function. As mentioned earlier, the δ function was first introduced by Dirac around 1930, and it was much later that mathematicians developed the theory of generalized functions to provide a rigorous mathematical foundation for it.

Lamb also provides the corresponding integral results.[5] The remaining challenge is to handle the integrals, ideally aiming for a closed-form solution. However, due to the complexity, a closed-form solution is not achievable, so he resorts to asymptotic solutions for special cases (far-field conditions). By carefully analyzing branch points and poles on the integration path[6] and using the residue theorem, he expresses the integral as a sum of the pole contribution in closed form and the principal value integral of the remaining part.[7] For far-field conditions, asymptotic expansion yields an approximate estimate of the principal value integral, resulting in an approximate solution for displacement in terms of elementary functions. The final approximate displacement solution can be expressed as a sum of three elementary functions, where one term, representing the contribution of the pole, corresponds to the Rayleigh wave. Considering the harmonic oscillation time factor, it is found that the Rayleigh wave follows an elliptical trajectory. The other two terms, representing the contributions of the path integrals, correspond to the P-wave and S-wave, respectively.

In the second part of the paper, "Three-Dimensional Problems", Lamb continues to investigate the displacement field caused by a point source under the assumption of symmetry about the z axis.[8] Lamb first considers forces that vary periodically over time. The approach to solving the problem is generally similar to the two-dimensional case, but an important difference is the introduction of 0th-order and 1st-order Bessel functions in the expression for the displacement potential function. The appearance of Bessel functions naturally leads to the use of the Hankel transform (also known as the Fourier-Bessel transform) in the expression of forces, a

[5] For a line source at depth, since the source lies within the region to be solved, it is impossible to obtain a solution through homogeneous equations. Lamb's approach involves first extending the space to consider an infinite medium and providing the solution for this infinite medium. He then places a mirror image source at the symmetric point on the free surface, attempting to satisfy the boundary conditions by the sum of the two sources. However, he finds this approach inadequate and introduces a third vertical force acting on the free surface. By the combined effect of these three line sources, the equations and boundary conditions are satisfied, resulting in the displacement caused by the buried source.

[6] The integral expression for displacement is essentially an integral over real variables. However, the first step in handling the integral is to extend the real variable to the complex domain. While this approach might seem to complicate the problem, it actually has significant advantages, allowing for flexible manipulation of the integration path to achieve unexpected results. In the second volume of this book, the Cagniard-de Hoop method, which leverages integration techniques based on complex functions, is used to obtain the time-domain solution for Lamb's problem. After the advent of computers, real integrals can be computed numerically, but in Lamb's time, this was not possible; considering the complex domain was the only feasible method.

[7] The contribution of the poles essentially transforms into an integral around a semicircular arc centered on the poles, which can be expressed as elementary functions. So far, the original integral expression has just been converted into other integrals. However, the significance of this transformation is that these integrals can be approximated more easily.

[8] That is, the z axis is perpendicular to the free surface. This assumption simplifies the problem, as all variations with respect to the azimuthal angle in the cylindrical coordinate system are zero. Displacement can be represented by two quantities: q, the component within the horizontal plane, and w, the component parallel to the z axis. However, the trade-off is that only vertical forces can be considered, and this problem is not truly "three-dimensional".

transformation unfamiliar to most readers today.[9] Similarly, based on the results for spatially periodic forces, the integral expression for the displacement caused by a vertically concentrated force, which includes Bessel functions, is obtained. Lamb uses an integral representation of the Bessel functions to achieve results that can be derived directly from the two-dimensional case through simple differentiation. As a result, an asymptotic expression for the displacement in the far field is obtained, which explicitly includes the contributions of Rayleigh waves, as well as P-wave and S-wave.

The harmonic variation of the applied force inevitably leads to harmonic variation in displacement, which clearly has significant limitations. Lamb noticed this and attempted to generalize it to arbitrary time variations; specifically, he considered a time function that resembles a pulse. Taking the Rayleigh wave term as an example, Lamb replaced the harmonic variation term in the results with an arbitrary time function[10] and demonstrated the corresponding displacement pattern for Rayleigh waves; similar treatment was applied to the other two terms in the displacement results. As the most important and intuitive result of his paper, Lamb presented the displacement curves shown in Fig. 5.2 at the end of the main text. This figure shows the variation of displacement over time at the surface, far from the source of the approximate pulse source he considered, with q_0 and w_0 representing horizontal and vertical displacements, respectively. This was the first theoretical seismogram in history, with the initial waves being P-wave, followed by S-wave, which Lamb referred to as "minor tremors". The large amplitude waves appearing at the end were Rayleigh waves, which Lamb called the "main shock". Although the seismogram Lamb presented differed significantly from actual observed seismograms,[11] the result undoubtedly revealed an important conclusion: far from the source, Rayleigh waves are the most prominent component of the displacement waveform. Additionally, Lamb noted that the P-wave and S-wave phases reflect the time integral of the original pulse action, while the Rayleigh waves reflect the original pulse itself.

---

[9] The direct and inverse transformations are respectively as follows

$$\tilde{f}(\zeta) = \int_0^\infty f(r) J_n(\zeta r) r \, \mathrm{d}r, \quad f(r) = \int_0^\infty \tilde{f}(\zeta) J_n(\zeta r) \zeta \, \mathrm{d}\zeta.$$

[10] In fact, this "replacement" is not strict because if the time function in the force term is arbitrary, the equation cannot be directly reduced to the Helmholtz equation at the initial stage of solving. Instead, an integral transform with respect to the time variable, such as the Laplace transform, must be used to obtain the Helmholtz equation in the transformed domain, followed by an inverse Laplace transform to get the final result. This process not only significantly increases the complexity of the solution but also ensures that the form of the solution differs from that of the harmonic time variation. Perhaps Lamb was just making a rough estimate.

[11] The reason is clear: Lamb considered an overly simplified model that differs significantly from reality. Regarding the medium, the uniform half-space model ignores spatial variations, thus eliminating dispersion phenomena. As for the force source, Lamb only considered the action of vertical force, which obviously does not represent actual earthquake conditions. As introduced in Chap. 3, the shear dislocation on a fault that causes real earthquakes is equivalent to a pair of force couples with zero moment, far from being represented by a simple vertical force.

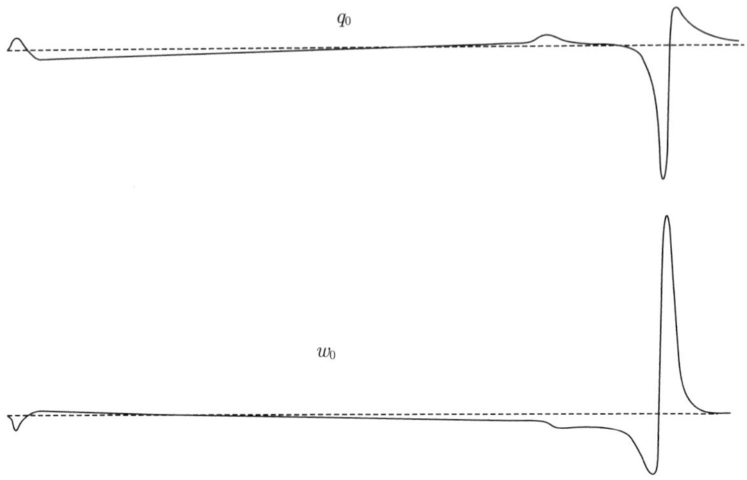

Fig. 5.2 The first theoretical seismogram in history obtained by Lamb (1904). This is the time variation of the horizontal displacement q_0 and vertical displacement w_0 at the Earth's surface far from the source, generated by an approximate pulse source

5.1.3 Some Comments

Lamb's research was a groundbreaking milestone. His classic paper opened up the use of rigorous elasticity theory to study seismic waves, establishing the "Lamb's problem" as a cutting-edge research topic in seismology for over half a century. In his paper, Lamb presented the first theoretical seismogram in history, which had an immeasurable demonstrative significance. Theoretical seismograms form the basis for many seismological studies, and it's no exaggeration to say that they are a key tool for seismologists to understand the Earth's medium and seismic sources. Therefore, the importance of Lamb's work for subsequent seismological research cannot be overstated.

When discussing Lamb (1904), one cannot overlook Rayleigh (1885). Compared to Rayleigh's analysis, which was based on plane waves and completely free surface boundary conditions, Lamb's (1904) analysis was much more complex, particularly in the asymptotic estimation of displacement integral expressions. While Rayleigh (1885) merely demonstrated the existence of Rayleigh waves, Lamb (1904) mathematically revealed the reasons behind their formation. This method of analysis based on elastic theory quickly caught the attention of applied mathematicians and seismologists, leading to an expansion of research within the framework of his fundamental analytical approach.

Of course, from a critical perspective, Lamb's research is not without flaws. Some assumptions made for the sake of analytical convenience limit its applicability.

(1) Only the displacement at the surface is examined.
(2) The scenario of a force acting on the surface was considered. Although Lamb provided an integral expression for a force source located within the medium, he did not analyze this expression further. When the force is at the surface, it can be treated as a boundary force, simplifying the boundary conditions and avoiding the need to deal with inhomogeneous equations, which makes the solution process much easier.
(3) In the three-dimensional case, only the special scenario of vertical forces was considered. This type of problem has symmetry around the axis of the applied force, which allows it to be simplified to a single integral, similar to the two-dimensional case. This greatly eases the solution process.
(4) As for the time variation of the force source, periodic forces is primarily considered. Although non-periodic forces were also discussed, they lacked rigorous analysis.
(5) Only the far-field asymptotic solution is obtained. For non-far-field situations, the integral expressions are difficult to simplify further, so the far-field assumption is used as a compromise.

5.1.4 Supplementary Notes on the Source in Lamb's Problem

In Sect. 1.1.1, we previously classified the Lamb's problem into three categories based on the different spatial positions of the source point and the field point. There are two points regarding the source in the Lamb's problem that need to be explained:

(1) On the direction of the applied force. In Lamb's three-dimensional problem, only forces applied perpendicular to the surface were considered. This is important primarily for convenience, as a vertical force implies axial symmetry along the force direction, significantly simplifying the problem. However, in the Lamb's problem we will study later, this assumption is not made. Forces can be applied in any direction. Removing this restriction is necessary because a double-couple source, which is seismologically significant, cannot be composed solely of vertical forces.
(2) On the time function. When seeking solutions to the elastic motion equation (2.2.21), it is crucial to distinguish between single-frequency motion and transient motion. The former, also known as steady state or harmonic motion, is primarily considered in Lamb's studies. In this case, the time function can be expressed as $e^{i\omega t}$, where ω is the angular frequency. This implies that the problem is in a dynamic equilibrium from the outset, meaning there are no initial conditions. This form of time variation of the force is reflected in the solution. Although this situation is mathematically more convenient to handle, its application is limited. For transient motion, the time function of the point source can

take any form and is assumed to be applied from a certain moment. Therefore, the corresponding boundary value problem requires initial conditions. From a seismological perspective, this is a more realistic scenario because an earthquake source starts rupturing at a specific moment, inherently introducing an initial condition to the problem. Consequently, handling transient motion is more complex than dealing with steady state motion.

The points above outline the main features of the Lamb's problem that we need to solve. While the description of the boundary value problem may seem straightforward, finding a complete solution is quite challenging. Since Lamb's 1904 paper, his analytical method has been adopted by subsequent researchers. At the same time, overcoming the limitations mentioned earlier has become a primary goal for those following in his footsteps.

Subsequent research can be broadly classified into two categories. The first category extends Lamb's approach, starting with solutions for harmonic waves and then using Fourier synthesis to solve the time-domain problem. For harmonic waves (single-frequency), the solution is in the form of an integral, so this method involves the computation of multiple integrals. The second category is based on the Cagniard-de Hoop method, which, through analysis in the complex domain, cleverly avoids the direct calculation of integrals to solve the time-domain problem. This method was initially proposed by the French mathematician Cagniard in the 1930s and later refined in de Hoop (1960). It has since become a powerful tool for addressing problems in various fields, including seismology.

5.2 Methods Based on Fourier Synthesis

Lamb (1904) aimed to obtain the response to a vertical impulse force, equivalent to a partial component of the Green's function. He first derived the single-frequency solution under harmonic force and then used Fourier transform to synthesize it into a transient solution. This can be expressed as a double integral. Lamb did not directly calculate this transient solution; instead, he studied their general properties and based on this, provided an approximate motion history of the free surface. Lamb's approach can be considered straightforward since applying Fourier synthesis is a common technique. However, calculating this double integral was challenging. Without computers at that time, achieving quantitative results required the use of asymptotic analysis techniques.

For a considerable time following Lamb's pioneering work, research extending his approach remained the mainstream in studying Lamb's problem. Since computers were not available in the first half of the 20th century, numerical integration was not feasible. Researchers often had to rely on mathematical techniques like the method of steepest descent and the stationary phase method to obtain asymptotic solutions for integrals. Consequently, applying these methods was fraught with difficulties. Each breakthrough addressing the limitations of Lamb's work faced technical challenges, leading to diverse research focuses. Among these, the work of Nakano (1925) and Lapwood (1949) stands out as particularly representative.

5.2.1 *Nakano (1925)'s Study on Rayleigh Waves*

Nakano (1925) produced the first significant study following Lamb (1904). In his 94-page paper titled "On Rayleigh Waves", Nakano sought to elucidate how Rayleigh waves form and propagate when there are sources of force within a solid. To simplify the analysis, he conducted a detailed discussion of the two-dimensional problem based on elasticity theory.[12] Unlike Lamb (1904), who did not focus specifically on Rayleigh waves, Nakano noted that in earthquake records where the epicentral distance is not much larger than the source depth, Rayleigh waves are difficult to identify. Thus, he examined the conditions under which Rayleigh waves appear.

Similar to Lamb (1904), Nakano also started with the case of periodic line sources and derived the same solutions as Lamb (1904) for the two-dimensional homogeneous equations of motion for displacement components. He first considered the case where no boundary forces are applied to the free surface, which was the simple scenario both Rayleigh and Lamb had examined, naturally arriving at the same conclusions. He then separately considered the effects of the potential functions corresponding to P-wave and S-wave, obtaining the respective displacement integral expressions. After extending the integral variables to the complex domain and using contour integration to derive the same integral expressions as Lamb, Nakano took a different approach from Lamb. He employed Debye's method of steepest descent[13] to obtain the asymptotic expression of the integral.

Next, Nakano divided the analysis into two parts, "compressional origin" and "distortional origin", to separately consider the displacement fields generated by the potential functions corresponding to P-wave and S-wave in the same manner. Unlike Lamb, Nakano focused primarily on the case of buried sources (i.e., force sources located within the half-space). Following the method described by Lamb (1904), he first considered the solution in an infinite space. By placing an image source symmetric about the free surface, he ensured that the sum of the sources satisfied the

[12] This essentially overcame the second limitation of Lamb's work. Additionally, Nakano (1925)'s analysis using the method of steepest descent was not limited to the far field, thus addressing the fifth limitation as well.

[13] This is a complex variable integration method, known as the saddle point method, used to approximate integrals of the form

$$I(z) = \int_L g(\zeta)\mathrm{e}^{-zf(\zeta)}\,\mathrm{d}\zeta$$

when the absolute value of z is large. The basic idea is that the integral depends only on the endpoints of L and not on the path itself. By selecting an integration path L' so that the function $f(\zeta)$ changes rapidly along L', and considering that $|z|$ is very large, the real part $\mathrm{Re}\,[zf(\zeta)]$ will vary significantly. Consequently, the contribution to the integral can be approximated by the result near the minimum of $\mathrm{Re}\,[zf(\zeta)]$.

free surface boundary conditions. To achieve this, he also had to add a displacement at the free surface. This approach allowed him to derive an integral expression for the displacement field at the free surface generated by a source buried within the half-space. By carefully considering the steepest descent paths in three different scenarios, Nakano obtained a displacement expression as the sum of contributions from three parts.[14] One of these parts, representing the contribution of Rayleigh waves, must satisfy certain conditions to be non-zero (i.e., the conditions for generating Rayleigh waves, which are related to the ratio of epicentral distance to source depth[15]). After deriving the solution for a periodic force source, Nakano proceeded to consider the case of non-periodic force sources. By replacing the time-varying part of the potential functions at the boundary with arbitrary functions and discussing it in a similar manner, he obtained the corresponding results for non-periodic force sources.[16]

An important conclusion from Nakano's paper is the condition for the generation of Rayleigh waves. As he stated, while the exact location where Rayleigh waves become significant cannot be pinpointed, the range where they do not appear can be specified. In the presence of a free surface, Rayleigh waves result from the interaction between P-wave and S-wave at the surface. For general non-harmonic forces, interactions between motions of different periods cause Rayleigh waves to gradually reach their peak; thus, it is difficult to determine a precise "starting point" in a strict sense. Additionally, Nakano discovered that when the epicentral distance exceeds a certain value, two other types of waves emerge, which he called "surface S-wave" and "surface P-wave". These propagate along the surface but are not strictly free surface waves. Nakano suggested that surface P-wave generated by a distortional source are possible, although their onset is not as sharp as that of direct P-wave and S-wave. However, the authenticity of surface S-wave generated by a compressional source is uncertain since the results were obtained only for periodic sources.

It can be said that Nakano (1925) exemplified an in-depth study of a specific problem within the solution framework established by Lamb (1904). Nakano focused on the conditions for the appearance of Rayleigh waves, thus he set aside the com-

[14] The contributions from P-wave and S-wave are represented as integrals, while the contribution from Rayleigh waves is expressed using elementary functions. This is similar to Lamb's conclusions. However, as Nakano noted in his paper, this way of dividing the contributions has a certain degree of "arbitrariness". Choosing different integration paths might lead to different physical interpretations.

[15] According to Nakano's conclusion, this condition is related to whether the initial disturbance is a "compressional origin" or a "distortional origin". The respective conditions are:

$$\frac{x}{f} > \frac{V_3}{\sqrt{V_1^3 - V_3^3}} \quad \text{and} \quad \frac{x}{f} > \frac{V_3}{\sqrt{V_2^3 - V_3^3}}$$

where x and f are the epicentral distance and source depth, and V_1, V_2, and V_3 are the propagation speeds of P, S, and Rayleigh waves, respectively. It is important to note that this conclusion depends on the analysis method. In the second volume of this book, we will use the Cagniard-de Hoop method to derive different conditions for two-dimensional problems.

[16] Similar to Lamb's approach, the general time-varying force source is obtained by replacing the time-varying part of the displacement potential function, rather than by strictly solving the partial differential equations.

plex discussion of three-dimensional problems and concentrated on two-dimensional problems. Unlike Lamb (1904), Nakano thoroughly investigated the surface displacement integral expression for a force source located within a half-space. Although Lamb (1904) had also derived this integral expression, he did not delve into its computation. Nakano applied the method of steepest descent to analyze this integral expression in detail, overcoming Lamb's limitation of providing only approximate solutions for the far-field. This allowed Nakano to determine the conditions for Rayleigh wave generation and gain deeper insights into their properties. Unfortunately, Nakano's paper did not include synthesized seismograms or their discussion.

5.2.2 Lapwood's (1949) Study on Displacement of Step Function Source

H. Jeffreys mentioned shortly before the publication of Lapwood (1949) that most observers identified as S-wave motion in seismic records with epicentral distances less than 20° was not actually S-wave, but its true nature was unknown. Further investigation into the potential characteristics of surface S-wave, in light of Nakano's (1925) work, might help answer this question. This is the motivation behind Lapwood's (1949) study. Unlike Lamb (1904) and Nakano (1925), Lapwood (1949) focused primarily on the displacement field near the surface generated by a line source with a step function time dependence[17] (see Fig. 5.3). Technically, Lapwood (1949) employed a contour integration technique on the Riemann surface, originally proposed by A. Sommerfeld and applied by Jeffreys to geophysical problems, which was more convenient than the analyses by Lamb and Nakano.

One significant difference in Lapwood's (1949) study compared to Lamb (1904) and Nakano (1925) is the consideration of the time function. Any temporal disturbance can be seen either as a combination of harmonics of different periods or as a combination of step functions occurring at different times. Lapwood chose the latter, believing that simple harmonic oscillation was not a suitable representation of the seismic source. Due to this strict consideration of the time function, Lapwood's starting point was the wave equation expressed in terms of the displacement potential function, unlike Lamb (1904) and Nakano (1925), who used the Helmholtz equation. The potential function can be represented as a frequency integral of harmonic oscillation solutions (0-order Hankel functions). Following steps similar to those of Lamb (1904) and Nakano (1925), by placing image sources and adding displacements on the free surface, he obtained an integral expression for the displacement potential function. Since this involves an integral over frequencies, it is a double integral. Lapwood focused on handling this integral. By extending the integration variable into the complex domain, he found that the integrand required a four-sheet Riemann surface for its description. By limiting the real part of the multivalued func-

[17] Although Lapwood (1949) focused on a two-dimensional problem, his work also overcame the limitations mentioned in Sect. 5.1.3, items (1), (2), (4), and (5).

THE DISTURBANCE DUE TO A LINE SOURCE IN A SEMI-INFINITE ELASTIC MEDIUM

BY E. R. LAPWOOD, *Department of Geodesy and Geophysics, University of Cambridge, and Yenching University, Peiping West, China*

(*Communicated by H. Jeffreys, F.R.S.—Received* 2 *February* 1948—*Revised* 16 *October* 1948)

When a cylindrical pulse is emitted from a line source buried in a semi-infinite homogeneous elastic medium, the subsequent disturbance at any point near the surface is much more complex than for an incident plane pulse. The curvature of the wave-fronts produces diffraction effects, of which the Rayleigh-pulse is the most important.

In this paper the exact formal solution is given in terms of double integrals. These are evaluated approximately for the case when the depths of source and point of reception are small compared with their distance apart, allowing a description of the sequence of pulses which arrive at the point of reception. When that point is at the surface and distant from the epicentre, the disturbance there can be regarded as made up of the following pulses, in order of arrival: (*a*) for initial *P*-pulse at source: *P*-pulse, surface *S*-pulse and Rayleigh-pulse; (*b*) for initial *S*-pulse: surface *P*-pulse, *S*-pulse and Rayleigh-pulse. If the initial pulse has the form of a jerk in displacement, the *P*- and *S*-pulses arrive as similar jerks, whereas the Rayleigh-pulse is blunted, having no definite beginning or end. The surface *P*-pulse takes a minimum-time path and arrives with a jerk in velocity. The surface *S*-pulse, on the other hand, is confined to the neighbourhood of the surface and arrives as a blunted pulse. Moreover, part of the *S*-pulse arrives before the time at which it would be expected on geometrical theory.

Although derived on very restricting hypotheses, these results may throw some light on seismological problems. In particular, it is shown that when the sharp *S*-pulse of ray theory is converted by the presence of the surface *S*-pulse and the spreading of *S* into a blunted pulse, the duration of this composite pulse is of the same order of magnitude as the observed scatter of readings of *Sg* and other distortional pulses from near earthquakes.

Fig. 5.3 Lapwood (1949) published in *Philosophical Transactions of the Royal Society of London*

tion, he restricted the study area to one of these sheets.[18] According to the residue theorem, the integration path can be altered to paths along two branch cuts (Γ_α and Γ_β) and around the poles (Γ_γ). The path integral around the poles yields Rayleigh waves, similar to the conclusions of Lamb (1904) and Nakano (1925). However, by analyzing the integrals along the branch cuts, Lapwood identified eight types of waves (see Fig. 5.4), including reflected P-wave (PP), converted P-wave (PS), reflected S-wave (SS), and converted S-wave (SP), which are all present in the case of plane waves. Additionally, there are several other waves associated with the free surface: sPs-wave (initiated and terminated as S-wave but traveling mostly at P-wave speed), pSp-wave (initiated and terminated as P-waves but traveling mostly at S-wave speed), sP-wave (initiated as S-waves, terminated as P-wave, and traveling mostly at P-wave speed), and pS-wave (initiated as P-wave, terminated as S-wave,

[18] Different ways of limiting the values of the multivalued function will lead to different Riemann sheets. The various Riemann sheets are connected through branch cuts. When the branch cuts are drawn such that the real part of the root causing the multivaluedness is zero, the cuts form part of a hyperbola in the complex plane. Due to Lapwood's double integral involving frequency, which is also extended into the complex domain, the singularities of the integrand are not on the integration path. This differs from the problem considered by Lamb (1904), where the singularities are exactly on the integration path, necessitating the use of the principal value of the integral.

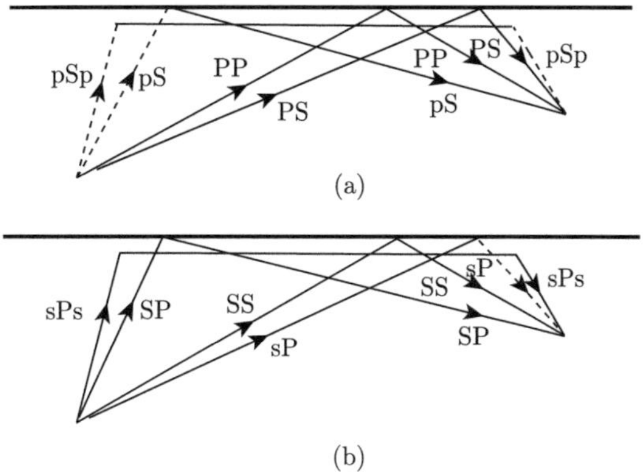

Fig. 5.4 Eight types of seismic phases near the free surface obtained by Lapwood (1949). **a** Waves generated by the P-source; **b** Waves generated by the SV-source

and traveling mostly at S-wave speed). Lapwood explained that these last four phases are related to diffraction effects when curved wavefronts encounter the free surface. Although this path-integral-based analysis allows for convenient physical interpretation, it is not suitable for directly obtaining the integral values. Lapwood used the method of stationary phase[19] to estimate the approximate values of these integrals in the far-field case.

One of the major highlights of Lapwood's (1949) work is his first rigorous study of the displacement field generated by non-harmonic sources, which is of significant importance. Since Rayleigh (1885), seismological studies have often been limited to cases where the source is a simple harmonic oscillator, primarily for simplification purposes. Although Lamb (1904) and Nakano (1925) both recognized this limitation and discussed non-periodic sources, their discussions were not rigorous. As Cagniard (1962) argued, while harmonic steady-state is evidently just an extreme case of seismic wave propagation, offering only incomplete and uncertain information about general wave propagation, the approach of superimposing sinusoidal solutions using Fourier transform, even if feasible, comes with limitations. This approach makes it difficult to study the properties of a function represented by Fourier integrals. Furthermore, Jeffreys pointed out in the early 1930s that simple step functions and

[19] The method of stationary phase is primarily used for evaluating integrals of the form

$$I = \int \varphi(x) e^{if(x)} \, dx$$

when $\varphi(x)$ changes slowly, and the periodic function $e^{if(x)}$ goes through many cycles. Therefore, except for the values of $f(x)$ that remain stationary, the contributions from various parts of the integral cancel each other out, resulting in a net integral of zero. Generally, stationary values do not cancel each other out, so at the stationary points of x, the integrand can be approximated as the constant $\varphi(x)$ multiplied by the function $e^{if(x)}$.

their resulting responses apply to a wide class of earthquakes.[20] Therefore, studying time function sources such as step functions holds significant practical importance for seismological problems.

5.2.3 Comments on Such Methods and Recent Studies

Using methods based on synthesizing different wave components, whether harmonic waves or step-function waves, the displacement solution is expressed as a double integral. Solving this double integral is the primary challenge of these methods. As shown in the discussion of the aforementioned classic works, obtaining quantitative results without computer assistance necessitated various simplifying assumptions and the use of asymptotic methods such as the method of steepest descent and the stationary phase method. Overcoming each simplifying assumption posed significant mathematical challenges. As a result, this type of research was relatively scarce and primarily concentrated in the first half of the 20th century. There are two main reasons for this:

(1) The introduction of the Cagniard-de Hoop method brought a breakthrough in solving the Lamb's problem. This ingenious approach allows for exact solutions in two-dimensional problems and expresses time-domain solutions for three-dimensional problems with only a single integral. A detailed analysis of this integral can even yield generalized closed-form analytical solutions. Therefore, this method has a significant advantage over current methods.[21] Consequently, since the 1950s, most research on the Lamb's problem has employed the Cagniard-de Hoop method or its variations.
(2) The advent of computers has made it possible to calculate integrals through programming. As mentioned earlier, solving double integrals was the greatest challenge. However, since the widespread adoption of computers in the 1950s, integrals can be solved using computers without relying on complex asymptotic analysis. Especially after the introduction of the Fast Fourier Transform (FFT) algorithm by J. W. Cooley and T. W. Tukey around 1965, frequency-domain seismic synthesis techniques rapidly emerged. On one hand, FFT can be used for frequency integration; on the other hand, the remaining wave number or slowness integral can be computed numerically. Since these techniques can be easily extended to layered media, a series of methods have been developed since the 1950s, such as the Thomson-Haskell method, the propagation matrix method,

[20] From today's perspective, this is indeed the case. Simulations of earthquake source dynamics show that the amount of slip on the fault plane evolves approximately according to a ramp function, and a step function can be considered a simplified version of the ramp function.

[21] Aki and Richards (2002) straightforwardly stated that the Cagniard-de Hoop method is the best solution for the Lamb's problem. In the second volume of this book, we will thoroughly explore the application of this method to the Lamb's problem.

the reflectivity method, and the generalized reflection-transmission coefficient method. Therefore, their application range is much broader compared to the Lamb's problem. The method introduced in Chap. 6 belongs to this category.

5.3 Time-Domain Solution Based on the Cagniard Method

Since Lamb's study in 1904, researchers have consistently aimed to obtain expressions for transient solutions in the time domain. The studies discussed earlier use synthetic approaches, either employing harmonic waves or step function-type waves, resulting in expressions that take the form of double integrals. These double integrals must be calculated directly through various methods to obtain solutions to the problems. In the 1930s, the French mathematician L. Cagniard introduced a groundbreaking new method, opening a new path for solving the Lamb's problem. This innovative approach led to numerous subsequent works that continuously improved the solutions to the Lamb's problem.

5.3.1 Proposal and Improvement of the Cagniard Method

Section 5.1.3 lists several limitations of Lamb (1904). Subsequent research often aimed to overcome one or more of these limitations, with the handling of time functions being a crucial aspect. Lamb's approach was to first study the steady-state[22] harmonic function solution and then superimpose these to form solutions corresponding to other time functions. However, analyzing the properties of a function represented by a Fourier integral is very challenging. Obtaining a *complete, rigorous, and precise solution* to the Lamb's problem has been a long-cherished goal of seismologists.

In 1939, the French mathematician L. Cagniard published a book in French titled *Reflection and Refraction of Progressive Seismic Waves*, summarizing the methods he had developed over the previous years. In this book, Cagniard addressed the problem of two isotropic, homogeneous, semi-infinite elastic media connected by a flat interface,[23] with an initial spherical wave present in one of the media. He aimed to determine the time-varying displacement at any point within the two media, with particular focus on the impact of the flat interface on the incident spherical wave.

---

[22] A steady state refers to a condition where variables remain stable over time, such as in simple harmonic motion, which occurs at a single frequency. The advantage of steady-state solutions is clear: the second-order partial derivatives with respect to t in the equation retain the same functional form, differing only by a coefficient. For homogeneous equations, it is even possible to eliminate the time-dependent terms at both ends, transforming the equation into a time-independent form, making it easier to solve.

[23] The Lamb's problem, where one side of the interface is an elastic medium and the other side is a vacuum, can be considered a simplified version of the problem studied by Cagniard.

Cagniard used the Laplace transform, turning the initial-boundary value problem into a boundary value problem in the Laplace domain. Whether using Lamb's analytical method or integral transforms, the solution to the boundary value problem still needs to be expressed in integral form. To obtain the time-domain solution, the inverse Laplace transform is necessary. The ingenuity of Cagniard's method lies in his handling of these integral transforms. Instead of directly dealing with the integrals through various asymptotic methods, he used variable substitution and altered the integration path in the complex plane, transforming the integral solution of the boundary value problem into a standard Laplace transform form. Thus, the time-domain solution could be obtained by inspection. In simple terms, Cagniard's method solves seismic wave problems through sophisticated complex variable analysis techniques. However, due to its complexity, even professional mathematicians found it difficult to master, which is why it did not gain widespread attention and application immediately after its publication. Additionally, Cagniard's work considered spherical wave sources rather than directional force sources, which may be another reason it was not quickly applied.

It should be noted that almost simultaneously with the development of Cagniard's method, Soviet scholars V. Smirnov and S. Sobolev independently discovered equivalent methods. Shortly thereafter, around 1940, American mathematician and physicist C. L. Pekeris used another method equivalent to Cagniard's to solve the surface pulse problem. Unlike Cagniard, Pekeris did not use the Laplace transform but instead employed a technique known as operational calculus. Therefore, this method is sometimes referred to as the Cagniard-Pekeris method.

Twenty years after the introduction of the Cagniard method, a significant improvement was made. In 1960, Dutch physicist A. T. de Hoop published a brief seven-page paper where he introduced the de Hoop transform, successfully applying the Cagniard method to solve seismic pulse problems. The Cagniard-Pekeris method utilized cylindrical coordinates, making the Hankel transform convenient but also adding mathematical complexity. In contrast, de Hoop used Cartesian coordinates, involving Fourier transforms of spatial coordinates in the solution process. The introduction of the de Hoop transform facilitated easier mathematical handling. Consequently, de Hoop's improvement retained the essence of the Cagniard method while making it simpler and more accessible to scientists across various fields of scientific and engineering research. In recognition of de Hoop's contributions, this method is more commonly referred to as the *Cagniard-de Hoop method*.

Putting aside the difficulty of the mathematical processing involved, considering that seismic waves are transient rather than steady-state, Fourier synthesis is feasible but not optimal. From this perspective, the Cagniard-de Hoop method is advantageous as it uses the Laplace transform rather than the Fourier transform, aligning better with the transient nature of seismic waves. However, the intrinsic characteristics of the problem dictate that regardless of the integral transform used, the final solution will be expressed as multiple integrals. Conventionally, one would tackle each integral head-on, facing significant mathematical challenges. The Cagniard-de Hoop method creatively employs an indirect approach, leveraging the freedom of complex path integrals in the complex plane. This successfully reformulates the multiple integrals into a standard Laplace transform form, thereby directly obtaining the transient solution in the time domain.

Following the introduction of the Cagniard method, numerous studies on the Lamb's problem based on this method were conducted, with notable contributions from C. L. Pekeris, W. W. Garvin, and C. C. Chao. Particularly after de Hoop's improvements, L. R. Johnson systematically summarized the solutions to three types of Lamb's problems in 1974. With the advancement of computers, seismologists have focused on calculating seismograms in models that more closely resemble Earth's media, such as multilayered media and laterally heterogeneous media. Additionally, the study of generalized closed-form solutions to the Lamb's problem has attracted significant interest from researchers.

5.3.2 *Researches Based on the Cagniard Method: From 1950s to Mid-1970s*

Cagniard's pioneering work paved the way for subsequent research in this field. However, as mentioned earlier, Cagniard did not directly address force source problems with practical seismological significance. It was not until several years later that the Israeli-American mathematician and physicist Pekeris published two important papers, which studied the surface responses of surface sources and buried sources. The force sources Pekeris considered had time functions that were step functions and were vertical loads. As previously noted, Pekeris used a method equivalent to the Cagniard method to tackle these problems.

In his first paper on surface sources (Pekeris 1955a), he investigated the transient solution for a vertical load acting on the surface (the $\mathcal{LP}_1$) and provided numerical results, as shown in Fig. 5.5.

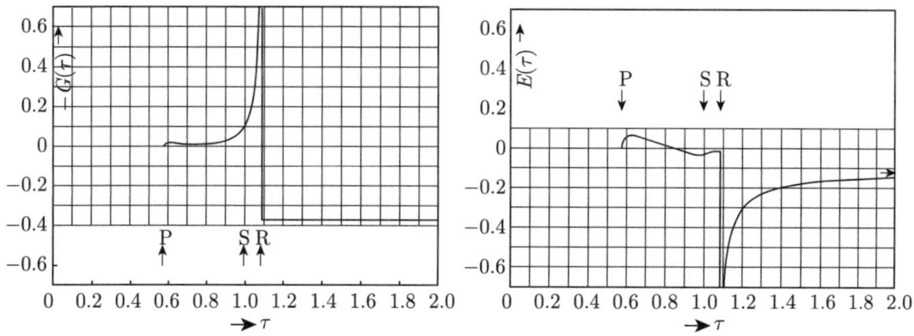

Fig. 5.5 Numerical results of Pekeris (1955a) on the $\mathcal{LP}_1$. The figure shows the vertical (left) and horizontal (right) displacements caused by a step function force acting vertically on the surface

This is the first time in the history of seismology that displacement images have been obtained for a single force source with a step function time dependence, $H(t)$.[24] The figure shows that after the arrival of the S-wave, an infinitely large amplitude Rayleigh wave immediately follows. This is a notable characteristic of the solution to the $\mathcal{LP}_1$.

In the second paper on buried sources (Pekeris 1955b), he further addressed the issue of a vertical load applied at a finite depth in a half-space, which is known as the $\mathcal{LP}_2$. For the $\mathcal{LP}_1$, Pekeris successfully expressed the results as a combination of elementary functions and elliptic integrals. However, for the $\mathcal{LP}_2$ with a buried source, he only provided an integral solution and did not achieve a generalized closed-form expression. Two years later, he and a colleague presented numerical results for the displacement field caused by an internal source at the surface (Pekeris and Lifson 1957), as shown in Fig. 5.6. The figure clearly illustrates how the vertical and horizontal displacements vary with different ratios of epicentral distance r to source depth H. When the source acts within the half-space, the amplitude of the Rayleigh wave is finite, and it increases with the ratio r/H, reaching the limiting case where the force acts on the free surface (the waveform at the bottom row of the figure, Fig. 5.5). These results are undoubtedly significant for understanding the characteristics of Lamb's problem solutions. It should be noted that Pekeris's series of works represent the first major advance in the study of Lamb's problems since the advent of the Cagniard method. It is important to point out that Pekeris's work was limited to Poisson solids with a Poisson's ratio $\nu = 0.25$, and only considered vertical force sources, which rendered the problem axisymmetric and facilitated mathematical treatment.

Around the same time, Garvin (1956) used the Cagniard method to study surface displacements caused by axisymmetric disturbances from a buried line source in a two-dimensional context (the $\mathcal{LP}_2$). He obtained expressions valid for any Poisson's ratio. Unlike the three-dimensional case, the solution for the two-dimensional problem can be fully expressed in closed form using elementary functions.

Shortly after Pekeris's paper was published, Chao expanded the study to include horizontal forces at the surface in 1960. The application of a horizontal force breaks the symmetry of the problem, making it more complex to handle compared to the axisymmetric vertical force. Chao (1960) noted that dealing with a horizontal force within a half-space is particularly challenging. Using Pekeris's approach, he derived the displacement solutions at the surface, as shown in Fig. 5.7. The figure presents the radial, tangential, and vertical displacement components over time. Notably, in the radial and vertical directions, where P-SV coupling occurs, there are prominent Rayleigh waves, similar to the displacement field caused by a vertical force shown in Fig. 5.5. The tangential component, which is mainly an SH component, does not exhibit Rayleigh wave characteristics. Like Pekeris's work, Chao's results are also limited to Poisson solids ($\nu = 0.25$).

[24] Although Lapwood (1949) studied the corresponding problem with a step function time dependence, he did not present the displacement-time variation curves.

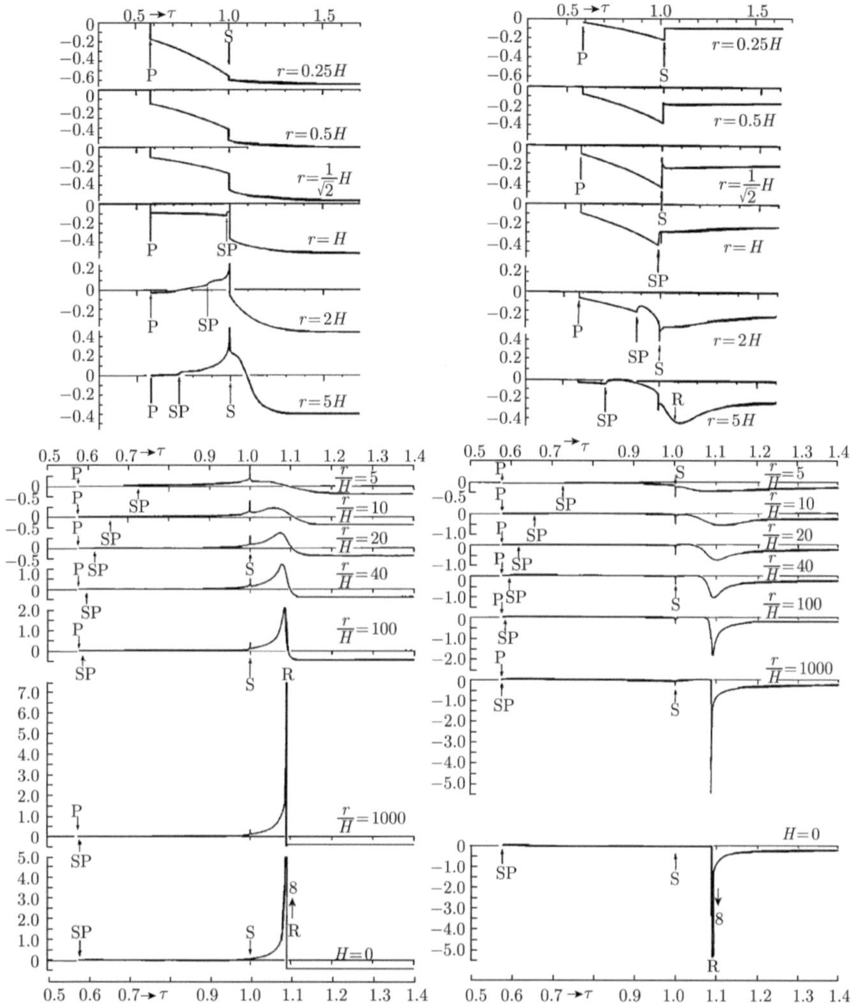

Fig. 5.6 Numerical results of Pekeris and Lifson (1957) on the $\mathcal{LP}_2$. The figure shows the vertical (left) and horizontal (right) displacements caused by a step function force acting vertically within a half-space for different ratios of epicentral distance r to source depth H. "P", "SP", "S", and "R" represent P-wave, SP-wave, S-wave, and Rayleigh wave, respectively

Most studies on the Lamb's problem have focused on the displacement field at the surface, which is understandable since most seismic instruments are placed there. However, what about the displacement field within the Earth's interior? This question has also intrigued seismologists. In 1966, W. M. Eason attempted to solve the internal displacement field caused by a vertical force applied on the surface of a half-space. He employed a different approach from the Cagniard-Pekeris method, involving contour integration along paths surrounding singularities and branch cuts

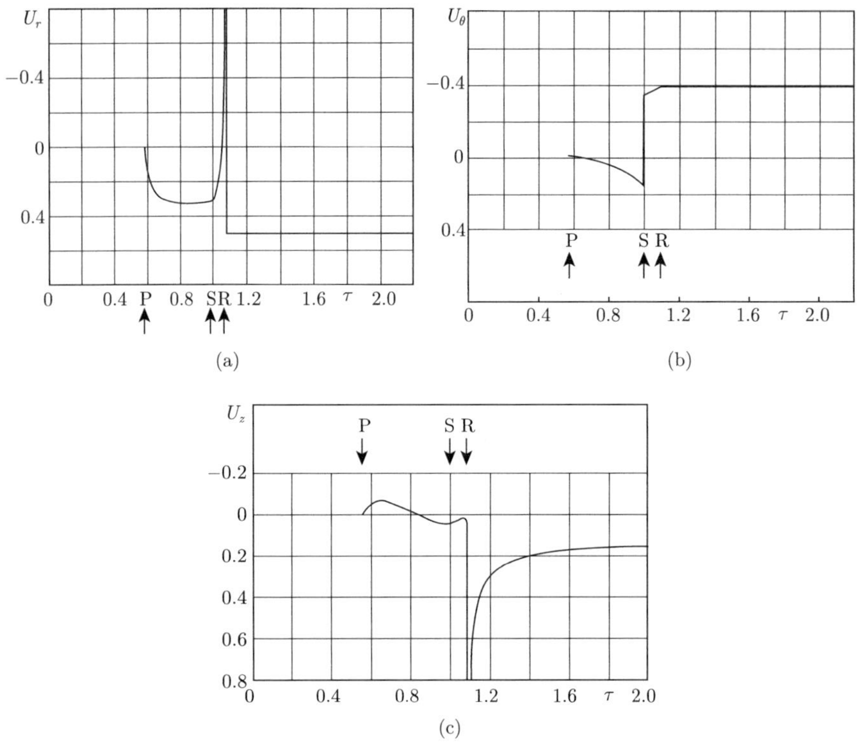

Fig. 5.7 Numerical results of Chao (1960) on the $\mathcal{LP}_1$. **a** Radial displacement U_r, **b** tangential displacement U_θ, and **c** vertical displacement U_z caused by a step-function force applied horizontally on the surface

in the complex plane. Eason's method did not rely on the assumption of a Poisson solid, making it applicable to materials with different Poisson's ratios. Compared to the displacement at the free surface, the internal displacement field is evidently more complex, and thus, Eason's final solution took the form of rather intricate finite integrals.

To address the limitation of previous studies that were restricted to Poisson solids, Mooney (1974) started from the integral form of Pekeris's solution (Pekeris 1955a). By modifying the integration path in the Cagniard-Pekeris method, Mooney extended Pekeris's work on the $\mathcal{LP}_1$ to general materials with arbitrary Poisson's ratios ν. Figure 5.8 shows the vertical displacements obtained by Mooney for different Poisson's ratios ($\nu = 0.15,\ 0.25,\ 0.33$ and 0.4).

By the early 1970s, the study of the Lamb's problem had become quite comprehensive. Researchers had addressed various limitations of Lamb's 1904 work from different perspectives, allowing a broader understanding of the Lamb's problem in general cases. For instance, seismologists had determined the displacement fields at the surface caused by horizontal and vertical forces, modeled as step functions,

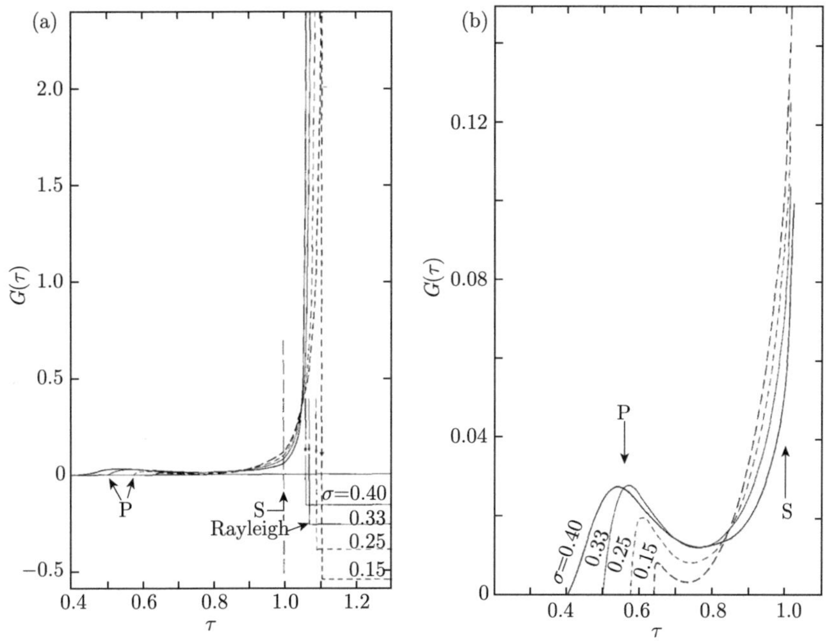

Fig. 5.8 Numerical results of Mooney (1974) on the $\mathcal{LP}_1$. The left figure shows the vertical displacement, while the right figure is a magnified view of it. In the figure, σ represents the Poisson's ratio, with its value 0.15, 0.25, 0.33, and 0.40

both on and below the surface without relying on far-field approximations. Although the $\mathcal{LP}_3$ remained unsolved, the overall understanding was fairly complete. As often happens in the history of science, once a critical mass of individual studies is reached from different angles, a comprehensive study emerges. The research on the Lamb's problem followed this pattern.

5.3.3 Johnson (1974): Complete Integral Solutions to Lamb's Problem

In 1974, American geophysicist L. R. Johnson published a paper titled "Green's Function for the Lamb's Problem", as shown in Fig. 5.9. In this seminal and comprehensive paper, Johnson systematically applied the Cagniard-de Hoop method to obtain precise integral expressions for the Green's functions of the three types of Lamb's problems. Furthermore, Johnson explicitly provided the integral representations for the spatial derivatives of the Green's functions involved in the source representation theorem. In essence, except for the results being in integral rather than closed form, Johnson's work perfectly met all expectations for solving the gen-

Geophys. J. R. astr. Soc. (1974) **37,** 99–131.

Green's Function for Lamb's Problem

Lane R. Johnson

(Received 1973 October 15)*

Summary

The complete solution to the three-dimensional Lamb's problem, the problem of determining the elastic disturbance resulting from a point force in a half space, is derived using the Cagniard–de Hoop method. In addition, spatial derivatives of this solution with respect to both the source co-ordinates and the receiver co-ordinates are derived. The solutions are quite amenable to numerical calculations and a few results of such calculations are given.

Fig. 5.9 Johnson (1974) published in *Geophysical Journal of the Royal Astronomical Society*

eral Lamb's problem: the results encompassed all source-field spatial combinations of the three types of Lamb's problems, were not limited to far-field, allowed forces to be applied in any manner, and accommodated any time function.

In the Cagniard-Pekeris method, the problem is solved in cylindrical coordinates. This presents two challenges: first, the representation of horizontal forces is more difficult compared to the axisymmetric vertical forces; second, the solution process inevitably involves the Hankel transform, which includes Bessel functions. In contrast, the Cagniard-de Hoop method solves the problem in Cartesian coordinates, naturally avoiding these two difficulties. Johnson's solution follows the standard Cagniard-de Hoop procedure, applying unilateral and bilateral Laplace transforms to the time and spatial variables, respectively. By introducing a variable substitution from de Hoop, he altered the integration path in the complex plane, transforming the double inverse Fourier transform of the spatial variables into the standard form of a Laplace transform. Johnson's integral results are presented as finite integrals. Although these integrals appear singular, simple substitutions can remove the singularities. Numerical integration facilitates the solution. Figures 5.10 and 5.11 show the numerical results of the Green's functions for the second and third types of Lamb's problems, respectively, under different source-field spatial combinations obtained by Johnson. Similar to Figs. 5.6, 5.10 clearly illustrates how surface displacement varies with different epicentral distances and source depth ratios. Additionally, Fig. 5.11, for the first time, clearly shows the waveform of the solution for the $\mathcal{LP}_3$, with both source and field located underground. Compared to the $\mathcal{LP}_2$, it exhibits more complex reflected and converted seismic phases.

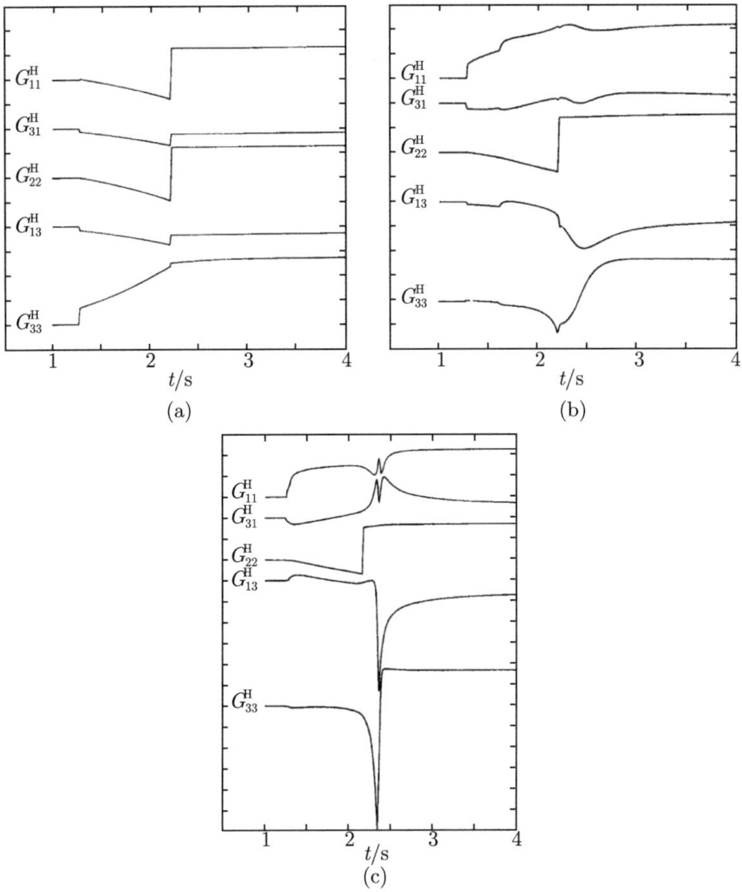

Fig. 5.10 Results of Green's functions in the $\mathcal{LP}_2$ in Johnson (1974d). **a** $\boldsymbol{\xi} = (2, 0, 0)$, $\boldsymbol{x} = (0, 0, 10)$; **b** $\boldsymbol{\xi} = (10, 0, 0)$, $\boldsymbol{x} = (0, 0, 2)$; **c** $\boldsymbol{\xi} = (10, 0, 0)$, $\boldsymbol{x} = (0, 0, 0.2)$. (Unit: km) $\boldsymbol{\xi}$ and $\boldsymbol{x}$ are position vectors of source and receiver, respectively

Unfortunately, Johnson's work did not receive the attention it deserved. A significant reason for this might be that Johnson provided only a very brief overview of the solution process without detailed steps. Despite this, Johnson's paper remains a classic and represents a comprehensive summary of the integral solutions to the Lamb's problem.

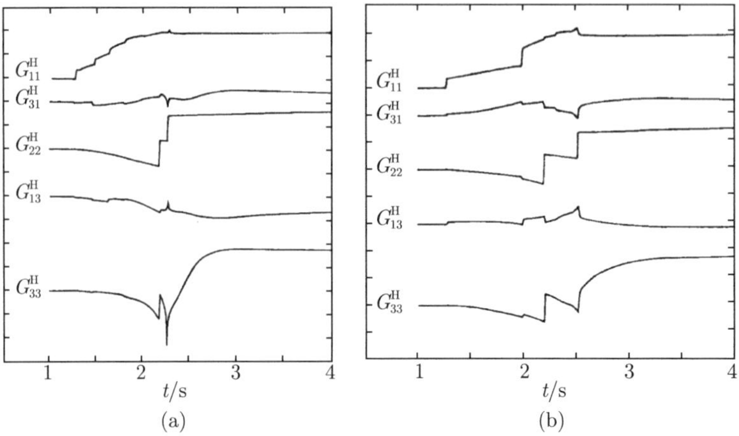

Fig. 5.11 Results of Green's functions in the $\mathcal{LP}_3$ in Johnson (1974). **a** $\boldsymbol{\xi} = (10, 0, 1)$, $\boldsymbol{x} = (0, 0, 2)$; **b** $\boldsymbol{\xi} = (10, 0, 4)$, $\boldsymbol{x} = (0, 0, 2)$. (Unit: km)

5.3.4 Researches on Closed-Form Solution After Johnson (1974)

Johnson's solution from 1974 is comprehensive, allowing for the general Lamb's problem to be tackled using computer algorithms. However, from a theoretical perspective, there's room for improvement by simplifying the integrals to achieve a generalized closed-form analytical solution. This represents the ultimate goal in the mathematical resolution of the Lamb's problem.

In the late 1970s, the field of source dynamics was experiencing significant growth. To provide Green's function solutions for the boundary integral equation method in displacement representation, P. Richards published a paper in 1979 titled "Elementary solutions to Lamb's Problem for a point source and their relevance to three-dimensional studies of spontaneous crack propagation", as shown in Fig. 5.12. The title of the paper clearly indicates Richards' research motivation. In this paper, Richards approached the $\mathcal{LP}_1$ by starting from Johnson's (1974) integral solution. Through variable substitution, he decomposed the integrals into several fundamental components and conducted detailed analyses for each, achieving closed-form solutions for some of these integrals. As shown by Pekeris (1955a), the solution to the $\mathcal{LP}_1$ includes elliptic integrals. Richards (1979) did not express the integrals, which could not be converted into closed-form solutions using elementary functions, in the standard form of elliptic integrals; instead, these integrals were solved numerically. Since Pekeris (1955a) only considered the special case of vertical force and Poisson solid, his solution was incomplete. Richards, on the other hand, started from Johnson's comprehensive integral expression, and thus, naturally, his solution was the most comprehensive generalized closed-form expression for the $\mathcal{LP}_1$ at the time.

Bulletin of the
Seismological Society of America

Vol. 69	August 1979	No. 4

ELEMENTARY SOLUTIONS TO LAMB'S PROBLEM FOR A POINT SOURCE AND THEIR RELEVANCE TO THREE-DIMENSIONAL STUDIES OF SPONTANEOUS CRACK PROPAGATION

BY PAUL G. RICHARDS

ABSTRACT

Certain exact solutions to Lamb's problem (the transient response of an elastic half-space to a force applied at a point) involve the computation merely of three square roots, and about ten arithmetic operations (+, −, ×, ÷). They arise when both source and receiver lie on the free surface. It is just these solutions which are needed in a method due to Hamano for obtaining the slip function (displacement discontinuity), as a function of space and time, for planar tension cracks and shear cracks which grow spontaneously with arbitrary shape. The solutions are described here in detail, for an elastic medium with general Poisson's ratio. They include perhaps the simplest-possible example of the $\bar{P}$-wave.

Fig. 5.12 Richards (1979) published in *Bulletin of the Seismological Society of America*

Unfortunately, Richards (1979) is not well known. As Kausel (2012) notes extensively in the introduction, there are a few reasons for this. Firstly, similar to Johnson (1974), Richards' paper lacks detailed derivations and only presents the main results. Secondly, and perhaps more importantly, since Richards was motivated by providing fundamental solutions for the boundary integral equation method, he might not have realized the significance of his work in the study of the Lamb's problem.

More than thirty years later, Kausel (2012) revisited this problem (Fig. 5.13). He further analyzed the parts of Richards (1979) that were not fully converted into standard elliptic integrals and obtained the simplest expression for the $\mathcal{LP}_1$.

Interestingly, much like Richards, we also encountered the need for second-order spatial derivatives of the half-space Green's function during our study, driven by the requirements of the boundary integral equation method in source dynamics research. Initially, we handled this by numerically computing wavenumber integrals in the frequency domain (Zhang 2004; Zhang and Chen 2006). Due to some technical limitations, we sought an alternative approach to derive the simplest expression directly in the time domain, which aligns with the ultimate goal of Lamb's problem research. In our source dynamics research, we need the Green's function solution for the $\mathcal{LP}_3$, making this a highly challenging task. Feng and Zhang (2018; 2021) obtained the generalized closed-form solutions for the $\mathcal{LP}_2$ and $\mathcal{LP}_3$, respectively, see Figs. 5.14 and 5.15. The basic idea is straightforward: to transform the integration through variable substitution. Similar to Richards' (1979) approach, we employed a strategy of

PROCEEDINGS
OF
THE ROYAL SOCIETY A

rspa.royalsocietypublishing.org

Research

CrossMark
click for updates

Cite this article: Kausel E. 2012 Lamb's problem at its simplest. Proc R Soc A 20120462. http://dx.doi.org/10.1098/rspa.2012.0462

Received: 2 August 2012
Accepted: 19 September 2012

Lamb's problem at its simplest

Eduardo Kausel

Civil and Environmental Engineering, Massachusetts Institute of Technology, Cambridge, MA 02139, USA

This article revisits the classical problem of horizontal and vertical point loads suddenly applied onto the surface of a homogeneous, elastic half-space, and provides a complete set of exact, explicit formulae which are cast in the most compact format and with the simplest possible structure. The formulae given are valid for the full range of Poisson's ratios from 0 to 0.5, and they treat real and complex poles alike, as a result of which a single set of formulae suffices and also exact formulae for dipoles can be given.

Fig. 5.13 Kausel (2012) published in *Proceedings of the Royal Society (A)*

Geophysical Journal International

Geophys. J. Int. (2018) **214**, 444–459 doi: 10.1093/gji/ggy131
Advance Access publication 2018 April 14
GJI Seismology

Exact closed-form solutions for Lamb's problem

Xi Feng and Haiming Zhang
Department of Geophysics, School of Earth and Space Sciences, Peking University, Beijing 100871, P. R. China. E-mail: zhanghm@pku.edu.cn

Accepted 2018 March 28. Received 2018 March 17; in original form 2017 May 18

SUMMARY

In this paper, we report on an exact closed-form solution for the displacement at the surface of an elastic half-space elicited by a buried point source that acts at some point underneath that surface. This is commonly referred to as the 3-D Lamb's problem for which previous solutions were restricted to sources and receivers placed at the free surface. By means of the reciprocity theorem, our solution should also be valid as a means to obtain the displacements at interior points when the source is placed at the free surface. We manage to obtain explicit results by expressing the solution in terms of elementary algebraic expression as well as elliptic integrals. We anchor our developments on Poisson's ratio 0.25 starting from Johnson's integral solutions which must be computed numerically. In the end, our closed-form results agree perfectly with the numerical results of Johnson, which strongly confirms the correctness of our explicit formulae. It is hoped that in due time, these formulae may constitute a valuable canonical solution that will serve as a yardstick against which other numerical solutions can be compared and measured.

Key words: Theoretical seismology; Wave propagation; Body waves.

Fig. 5.14 Feng and Zhang (2018) published in *Geophysical Journal International*

Geophysical Journal International

Geophys. J. Int. (2021) **224**, 517–532 doi: 10.1093/gji/ggaa485
Advance Access publication 2020 October 16
GJI Seismology

Exact closed-form solutions for Lamb's problem—III: the case for buried source and receiver

Xi Feng and Haiming Zhang

Department of Geophysics, School of Earth and Space Sciences, Peking University, Beijing 100871, P. R. China. E-mail: zhanghm@pku.edu.cn

Accepted 2020 October 9. Received 2020 September 9; in original form 2020 June 25

SUMMARY

In this paper, we derive the exact closed-form solution for the displacement in the interior of an elastic half-space due to a buried point force with Heaviside step function time history. It is referred to as the tensor Green's function for the elastic wave equation in a uniform half-space, also a natural generalization of the classical 3-D Lamb's problem, for which previous solutions have been restricted to the cases of either the source or the receiver or both are located on the free surface. Starting from the complex integral solutions of Johnson, we follow the similar procedures presented by Feng and Zhang to obtain the closed-form expressions in terms of elementary functions as well as elliptic integrals. Numerical results obtained from our closed-form expressions agree perfectly with those of Johnson, which validates our explicit formulae conclusively.

Key words: Theoretical seismology; Wave propagation.

Fig. 5.15 Feng and Zhang (2021) published in *Geophysical Journal International*

breaking down the transformed integrals into fundamental components, which fall into two categories: those involving square roots of quadratic polynomials and those involving square roots of quartic polynomials. The former can be expressed algebraically through analysis, while the latter, after further variable substitution, can be converted into combinations of three types of standard elliptic integrals, which can be efficiently computed using built-in functions of common mathematical software, such as Matlab and Maple. This topic will be discussed in detail in the second volume of this book.

Expressing the integral solution of the Lamb's problem in a generalized closed form is a complex and intricate task. Theoretically, its significance lies in two main aspects: firstly, the generalized closed-form solution are highly efficient in a way that numerical calculations cannot match, which is especially important for situations requiring extensive computation, such as computing the displacement due to slip on a finite fault by superimposing the contributions of point sources; secondly, these solutions facilitate theoretical analysis. For instance, we can completely isolate the components containing Rayleigh waves, which allows us to theoretically analyze the behaviors of Rayleigh wave.

The relevant studies based on the Cagniard method are summarized in Table 5.1.

Table 5.1 Representative studies on Lamb's problem based on the Cagniard method

Research work	Force direction	Source	Receiver	Solution form	Poisson's ratio
Pekeris (1955a)	Vertical	Surface	Surface	Generalized closed	0.25
Pekeris (1955b)	Vertical	Underground	Surface	Integral	0.25
Chao (1960)	Horizontal	Underground	Surface	Generalized closed	0.25
Eason (1966)	Vertical	Surface	Underground	Integral	(0, 0.5)
Mooney (1974)	Vertical	Surface	Surface	Integral	(0, 0.5)
Johnson (1974)	Arbitrary	Arbitrary	Arbitrary	Integral	(0, 0.5)
Richards (1979)	Arbitrary	Surface	Surface	Generalized closed	(0, 0.5)
Kausel (2012)	Arbitrary	Surface	Surface	Generalized closed	(0, 0.5)
Feng and Zhang (2018)	Arbitrary	Underground	Surface	Generalized closed	(0, 0.2631)
Feng and Zhang (2021)	Arbitrary	Underground	Underground	Generalized closed	(0, 0.2631)

5.4 Summary

Compared to the infinite space problem discussed in Chap. 4, the Lamb's problem is much more complex. This complexity arises from the intricate interactions of elastic waves at interfaces, with the generation of Rayleigh waves being a prime example, which is reflected in the mathematical treatment. The corresponding problem in infinite space, due to the absence of symmetry, is more conveniently solved using cylindrical or Cartesian coordinates. The former almost inevitably involves Bessel functions, which significantly complicates the solution process. Integral transforms are the indispensable tool for tackling the Lamb's problem. Depending on the integral transform used, research on the Lamb's problem generally falls into two categories: synthesis methods based on Fourier (and Hankel) transforms, and the Cagniard method based on Laplace transforms. As introduced in this chapter, the first method is often limited to special cases due to the mathematical difficulties in handling double integrals, such as the far-field situation in Lamb's study and the two-dimensional case in Nakano's study. On the other hand, the second method, by cleverly changing the integration path in complex variable integrals, has adeptly managed the handling of multiple integrals and has become the dominant approach in Lamb's problem research since the 1950s.

In the 1950s, the introduction of computers made numerical integration feasible. Especially with the emergence of FFT (Fast Fourier Transform) technology in the 1960s, the previously dormant first method gained new vitality. Leveraging the powerful computational capabilities of computers, researchers no longer needed to rely solely on complex asymptotic analysis techniques to obtain asymptotic solutions to problems; they could also explore issues in more complex models. For example, using the first method to synthesize theoretical seismograms for layered media models has seen new computational versions continuously developed since the 1950s. Because this method is crucial for synthesizing theoretical seismograms for earth models that are more realistic than the half-space model, we will lay the groundwork for further complex problem research by studying how to obtain solutions to the Lamb's problem in the frequency domain using Fourier transforms in the next two chapters. The integral form of the solution is no longer an obstacle for computing the complete seismic wavefield, as numerical integration and FFT can be utilized.[25]

[25] However, for the purpose of facilitating theoretical analysis, the Cagniard-de Hoop method remains the preferred choice, which will be the focus of the second volume of this book.

Chapter 6
Frequency-Domain Solution of Lamb's Problem (I): Theoretical Formulas

In the previous chapter's review of studies on the Lamb's problem, two methods were mentioned. One of these, originally used in Lamb (1904), involves first analyzing the steady-state case, then using Fourier transforms to synthesize the solution for any time function, resulting in a double integral for the final solution. In the first half of the 20th century, handling this double integral was the main challenge, forcing researchers to use various asymptotic analysis techniques to obtain approximate solutions. However, this situation naturally improved with the introduction of computers into the field of seismology. If our goal is simply to compute this double integral, rather than analyze the wave characteristics based on it, numerical calculations and Fast Fourier Transforms (FFT) can achieve this objective.

It should be noted that for the half-space model of the Lamb's problem, this solution method offers no advantage over the second type based on the Cagniard method. Its framework is relatively large and complex. However, since many important methods for synthesizing seismograms, such as the propagation matrix method, the reflectivity method, and reflection/transmission coefficients, have been based on this approach since the 1950s, it is necessary to demonstrate its application to the simple and specific case of the half-space to lay a solid foundation for further research on these methods. This chapter will first introduce the relevant theory.[1]

In this chapter, the analysis is conducted in the frequency domain. Unlike the method that directly applies integral transforms to spatial variables and analyzes in the transformed domain, here the function of interest is expanded using a set of complete orthogonal basis functions, leading to a system of ordinary differential equations for the expansion coefficients. By decomposing the coefficient matrix and incorporating boundary conditions at the free surface and infinity, the solution to the problem is obtained. We begin by detailing the frequency domain solution for the

[1] The content presented in this chapter is primarily based on the following works: Chen (1999), Zhang (2004, Chap. 4), Zhang and Chen (2006). Specifically, Sect. 6.5.2 is based on Zhang (2006, Chap. 2) and Zhang and Chen (2009), Sect. 6.5.3 is based on Zhou and Zhang (2020).

H. Zhang, *Lamb's Problem in Seismology (Volume 1)*,
https://doi.org/10.1007/978-981-95-3170-7_6

Green's function in the Lamb's problem, focusing on the construction of the crucial complete orthogonal basis. Based on this, the Green's function in the frequency domain is derived. Using this, we construct the displacement field for a dislocation point source. For the contribution at the free surface, the excitation formula for the Rayleigh wave in the frequency domain is derived through complex integral analysis of the wavenumber k. Finally, the special case of the static problem is discussed.

6.1 Problem Description and Resolving Thoughts

6.1.1 Description of Initial-Boundary Value Problem

The model of the problem is shown in Fig. 6.1. A coordinate system is chosen at the ground surface, with the vertical z axis passing through the source point, which is located at a depth z_s underground. At the source point, a body force $\boldsymbol{f}$ acts in any direction. The goal is to determine the displacement field at any point within the space due to this force.

We aim to solve the elastodynamic equations (see Eq. (2.2.21))

$$\rho\ddot{\boldsymbol{u}}(\boldsymbol{x},t) = (\lambda+2\mu)\nabla(\nabla\cdot\boldsymbol{u}(\boldsymbol{x},t)) - \mu\nabla\times(\nabla\times\boldsymbol{u}(\boldsymbol{x},t)) + \boldsymbol{f}(\boldsymbol{x},t)$$

under the following free surface and infinity boundary conditions

$$\hat{\boldsymbol{e}}_3\cdot\boldsymbol{\sigma}(\boldsymbol{x},t)\big|_{z=0} = \boldsymbol{0}, \quad \boldsymbol{u}(\boldsymbol{x},t)\big|_{z\to\infty} = \boldsymbol{0}$$

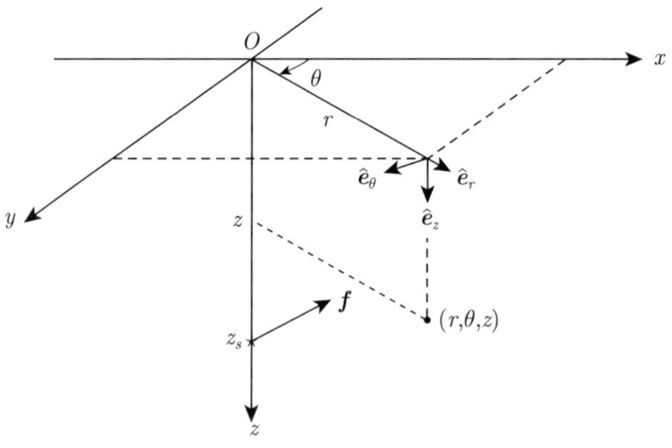

Fig. 6.1 Diagram illustrating the model for problem considered. The x axis and y axis lie within the free surface, while the z axis is directed downward perpendicular to the ground. At the source point $(0, 0, z_s)$ represented by $\star$, a body force $\boldsymbol{f}$ acts in any direction. We aim to determine the displacement field at any point (r, θ, z) within the half-space. $\hat{\boldsymbol{e}}_r$, $\hat{\boldsymbol{e}}_\theta$, and $\hat{\boldsymbol{e}}_z$ are the three basis vectors in the cylindrical coordinate system

Since the solution method developed in this chapter is carried out in the frequency domain, we first apply the Fourier transform to the above equations and boundary conditions to obtain the corresponding forms in the frequency domain

$$-\rho\omega^2\tilde{\boldsymbol{u}} =(\lambda+2\mu)\nabla\left(\nabla\cdot\tilde{\boldsymbol{u}}\right)-\mu\nabla\times\left(\nabla\times\tilde{\boldsymbol{u}}\right)+\tilde{\boldsymbol{f}} \tag{6.1.1a}$$

$$\hat{\boldsymbol{e}}_3\cdot\tilde{\boldsymbol{\sigma}}\Big|_{z=0}=\boldsymbol{0},\quad \tilde{\boldsymbol{u}}\Big|_{z\to\infty}=\boldsymbol{0} \tag{6.1.1b}$$

where $\tilde{\boldsymbol{u}}=\boldsymbol{u}(\boldsymbol{x},\omega)=\mathscr{F}\{\boldsymbol{u}(\boldsymbol{x},t)\}$, $\mathscr{F}\{\cdot\}$ denotes the Fourier transform, and the symbols $\tilde{\boldsymbol{f}}$ and $\tilde{\boldsymbol{\sigma}}$ follow similar notation. This convention will be used consistently throughout the rest of the chapter. The equations and boundary conditions in Eq. (6.1.1) define the current boundary value problem.

6.1.2 Resolving Thoughts

The first challenge is to select the appropriate coordinate system for solving the problem. Based on the geometric characteristics of the problem (see Fig. 6.1), two options are available: Cartesian coordinates and cylindrical coordinates. In this chapter, we choose the latter.[2] The advantage of using cylindrical coordinates is that it allows for an easier representation of different polarization directions—P, SV, and SH waves—providing a better physical context. However, using cylindrical coordinates inevitably introduces Bessel functions, which require special numerical treatment in certain specific cases (discussed in Chap. 7).

For solving the frequency-domain Lamb's problem, a basis function expansion method is adopted.[3] We start with a set of orthogonal and complete basis functions based on the three mutually perpendicular polarization directions: P, SV, and SH. The displacement $\boldsymbol{u}(\boldsymbol{x},\omega)$ and the body force $\boldsymbol{f}(\boldsymbol{x},\omega)$ are expanded using these basis functions. Since the form of the basis functions is predetermined, solving for the displacement $\boldsymbol{u}(\boldsymbol{x},\omega)$ is transformed into solving for the expansion coefficients. These coefficients satisfy a set of ordinary differential equations. By solving this set of equations and applying the boundary conditions at the free surface and at

---

[2] Of course, choosing Cartesian coordinates is also an option. In fact, in the second volume, we use Cartesian coordinates in the solution method based on the Cagniard-de Hoop technique, following Johnson (1974). However, for the solution method based on basis function expansion used in this chapter, Cartesian coordinates have the drawback of not aligning well with the physical characteristics of the problem. For example, the distinction between P-waves and S-waves in elastic wave propagation is not easily represented in Cartesian coordinates.

[3] In a function space, basis functions are similar to coordinate bases in a geometric space. Take three-dimensional geometric space as an example: the basis vectors $\hat{\boldsymbol{e}}_i$ of the Cartesian coordinate system form an orthogonal and complete set of bases. Orthogonality means $\hat{\boldsymbol{e}}_i\cdot\hat{\boldsymbol{e}}_j=\delta_{ij}$, and completeness means that any vector $\boldsymbol{a}$ in three-dimensional space can be represented using this set of bases: $\boldsymbol{a}=a_i\hat{\boldsymbol{e}}_i$, where $a_i=\boldsymbol{a}\cdot\hat{\boldsymbol{e}}_i$. The same concept applies to function spaces. If a set of functions satisfies the requirements of orthogonality and completeness in the function space, they form an orthogonal and complete basis set. This will be explained in more detail later.

infinity, we determine the unknown coefficients. Once the expansion coefficients are obtained, the displacement field $\boldsymbol{u}(\boldsymbol{x}, \omega)$ can be expressed as an integral over the wavenumber k. By numerically solving this integral and applying the inverse FFT, we obtain the time-domain displacement field $\boldsymbol{u}(\boldsymbol{x}, t)$.

6.2 Introduction and Properties of Basis Functions

In principle, any set of orthogonal and complete functions can be used as basis functions to expand displacement and body force vectors. However, different basis functions lead to varying levels of complexity in problem representation,[4] making the choice of appropriate basis functions crucial. Our current research focuses on seismic wave problems in elastic space, and selecting basis functions that align with the characteristics of seismic waves will facilitate easier problem-solving.

6.2.1 Decomposition of Elastic Waves: P-Wave, SV-Wave, and SH-Wave

According to the Lamé theorem in Sect. 4.2, if the displacement vector $\tilde{\boldsymbol{u}}$ in the frequency domain is decomposed as follows

$$\tilde{\boldsymbol{u}} = \nabla\phi + \nabla \times \boldsymbol{\Psi} \qquad (\nabla \cdot \boldsymbol{\Psi} = 0) \tag{6.2.1}$$

that is, the P-wave is represented by the gradient of the scalar potential function ϕ, while the S-wave is represented by the curl of the vector potential function $\boldsymbol{\Psi}$, the two potential functions, ϕ and $\boldsymbol{\Psi}$, each satisfy a Helmholtz equation. Consequently, Eq. (6.2.1) naturally provides a decomposition of the displacement vector into P-wave and S-wave, which exhibit irrotational and solenoidal properties, respectively,

[4] Let's take a familiar example from geometric space. In principle, the basis vectors of any orthogonal curvilinear coordinate system can be used as bases. However, the choice of basis vectors affects the ease of problem representation. For instance, in solving the wave equation in unbounded space discussed in Sect. 4.3, spherical coordinates are the most appropriate choice due to spherical symmetry, which simplifies the problem by making the terms involving θ and ϕ automatically zero. Using Cartesian or cylindrical coordinates would make it harder to exploit this symmetry. The situation is similar in function space.

meaning their curl and divergence are zero.[5] Clearly, the vector potential function, with its three components, is more complex than the scalar potential function. However, a closer examination reveals that for describing S-wave, only two components of the potential function are actually necessary.

If we further decompose the vector potential $\boldsymbol{\Psi} = (\Psi_x, \Psi_y, \Psi_z)$ for the S-wave as

$$\boldsymbol{\Psi} = (\Psi_x, \Psi_y, \Psi_z) = \boldsymbol{\Psi}^{\mathrm{SV}} + \boldsymbol{\Psi}^{\mathrm{SH}}$$

where

$$\boldsymbol{\Psi}^{\mathrm{SV}} = (\Psi_x, \Psi_y, 0), \quad \text{and} \quad \boldsymbol{\Psi}^{\mathrm{SH}} = (0, 0, \chi) \quad (\chi = \Psi_z)$$

the superscript "SV" and "SH" represent SV-wave and SH-wave, respectively, which are two different components of S-wave, it is easy to find that there exists a scalar potential function ψ, such that

$$\boldsymbol{\Psi}^{\mathrm{SV}} = \nabla \times (0, 0, \psi), \quad \text{i.e.} \quad \Psi_x = \frac{\partial \psi}{\partial y}, \quad \Psi_y = -\frac{\partial \psi}{\partial x}$$

this means that the SV-wave and SH-wave can each be represented by a scalar potential functions, ψ and χ, respectively. Thus, we can decompose the displacement field into the sum of three parts:

$$\tilde{\boldsymbol{u}} = \tilde{\boldsymbol{u}}^{\mathrm{P}} + \tilde{\boldsymbol{u}}^{\mathrm{SV}} + \tilde{\boldsymbol{u}}^{\mathrm{SH}}$$

in which,

$$\tilde{\boldsymbol{u}}^{\mathrm{P}} = \nabla \phi, \quad \tilde{\boldsymbol{u}}^{\mathrm{SH}} = \nabla \times (0, 0, \chi), \quad \tilde{\boldsymbol{u}}^{\mathrm{SV}} = \nabla \times \nabla \times (0, 0, \psi) \tag{6.2.2}$$

it is obvious that $\nabla \times \tilde{\boldsymbol{u}}^{\mathrm{P}} = 0$, $\nabla \cdot \tilde{\boldsymbol{u}}^{\mathrm{SH}} = \nabla \cdot \tilde{\boldsymbol{u}}^{\mathrm{SV}} = 0$.

[5] The decomposition of waves propagating in an elastic medium is not unique. Here is one method of decomposition. Another method involves decomposing based on the polarization directions of displacement vector, resulting in longitudinal and transverse waves. The former represents particle vibration in the same direction as wave propagation, while the latter represents vibration perpendicular to it. It is important to note that these two decomposition methods do not yield identical results, except in certain special cases. For example, in Sect. 4.5.1.2, we showed that the polarization directions of far-field P-wave and S-wave align with and are perpendicular to the direction of wave propagation, respectively, indicating they are longitudinal and transverse waves. However, this conclusion does not hold for near-field terms. Readers familiar with basic seismology might think that P-wave and S-wave are always longitudinal and transverse waves, respectively. Strictly speaking, such a statement is incorrect because the conclusion is contingent upon certain assumptions. This assumption is based on plane wave analysis, which corresponds to far-field conditions, assuming the source is at an infinite distance.

6.2.2 *Vector Helmholtz Equation and Construction of Basis Functions*

By substituting Eq. (6.2.2) into the corresponding homogeneous equation (6.1.1a),[6] it can be easily verified that $\tilde{\boldsymbol{u}}^{\mathrm{P}}$, $\tilde{\boldsymbol{u}}^{\mathrm{SV}}$, and $\tilde{\boldsymbol{u}}^{\mathrm{SH}}$ all satisfy the vector Helmholtz equation in the following form,

$$\nabla^2 \boldsymbol{A} + k_c^2 \boldsymbol{A} = 0, \quad k_c = \begin{cases} k_\alpha = \dfrac{\omega}{\alpha}, & \text{for } \tilde{\boldsymbol{u}}^{\mathrm{P}} \\ k_\beta = \dfrac{\omega}{\beta}, & \text{for } \tilde{\boldsymbol{u}}^{\mathrm{SV}} \text{ and } \tilde{\boldsymbol{u}}^{\mathrm{SH}} \end{cases} \tag{6.2.3}$$

and their corresponding potential functions ϕ, ψ, and χ each satisfy the scalar Helmholtz equation

$$\nabla^2 w + k_c^2 w = 0 \tag{6.2.4}$$

This means that for the non-homogeneous equation (6.1.1a), solutions of the vector Helmholtz equation (6.2.3) can be chosen as basis functions.

However, since we are currently choosing cylindrical coordinates as the solution coordinate system, and as a type of orthogonal curvilinear coordinate system, the adoption of cylindrical coordinates will face a difficulty. Note that the expression of the operator $\boldsymbol{\nabla}$ in cylindrical coordinates is

$$\boldsymbol{\nabla} = \hat{\boldsymbol{e}}_r \frac{\partial}{\partial r} + \hat{\boldsymbol{e}}_\theta \frac{\partial}{r\partial \theta} + \hat{\boldsymbol{e}}_z \frac{\partial}{\partial z} \tag{6.2.5}$$

Unlike the basis vectors in a Cartesian coordinate system, the basis vectors $\hat{\boldsymbol{e}}_r$ and $\hat{\boldsymbol{e}}_\theta$ in a cylindrical coordinate system vary with position, as shown in Fig. 6.1. This means that the spatial derivatives of the basis vectors are not necessarily zero. Consequently, in Eq. (6.2.3)

$$\nabla^2 \boldsymbol{A} = \left[\nabla^2 A_r - \frac{1}{r^2} A_r - \frac{2}{r^2}\frac{\partial A_\theta}{\partial \theta}\right]\hat{\boldsymbol{e}}_r + \left[\nabla^2 A_\theta - \frac{1}{r^2} A_\theta + \frac{2}{r^2}\frac{\partial A_r}{\partial \theta}\right]\hat{\boldsymbol{e}}_\theta + \nabla^2 A_z \hat{\boldsymbol{e}}_z$$

[6] Why do we only consider the homogeneous equation here? This is because basis functions reflect the intrinsic properties of the medium and are not influenced by external forces. Therefore, when selecting basis functions, we only need to study the corresponding homogeneous equation. This concept is actually quite familiar. Recall from studying mathematical physics equations that when introducing the classical method of separation of variables, two classic problems are often used as examples: the free vibration of a string fixed at both ends and the forced vibration of a string fixed at both ends. The vibration of the string can be described by a one-dimensional wave equation, but the non-homogeneous equation with a forcing term cannot be solved using the separation of variables method. A common approach is to first solve the homogeneous equation to obtain the eigenfunctions; then, for the non-homogeneous equation, expand the solution and the non-homogeneous term using the eigenfunctions of the corresponding homogeneous equation (Wu 2003, Sects. 14.1 and 14.5).

Based on the orthogonality of $\hat{\boldsymbol{e}}_r$, $\hat{\boldsymbol{e}}_\theta$, and $\hat{\boldsymbol{e}}_z$, the vector Helmholtz equation (6.2.3) can be transformed into the following system of equations

$$\nabla^2 A_r - \frac{1}{r^2} A_r - \frac{2}{r^2} \frac{\partial A_\theta}{\partial \theta} + k_c^2 A_r = 0$$
$$\nabla^2 A_\theta - \frac{1}{r^2} A_\theta + \frac{2}{r^2} \frac{\partial A_r}{\partial \theta} + k_c^2 A_\theta = 0$$
$$\nabla^2 A_z + k_c^2 A_z = 0$$

This is a coupled system of equations, which is difficult to solve.

Fortunately, there is a highly effective method for solving the vector Helmholtz equation, introduced by American physicist W. W. Hansen in the 1930s.[7] This method demonstrates that the solution w of the scalar Helmholtz equation, as given in Eq. (6.2.4), can be used to construct the solution to the vector Helmholtz equation (6.2.3) in the following manner:

$$\boldsymbol{S} = \nabla w, \quad \boldsymbol{T} = \nabla \times (\hat{\boldsymbol{a}} w), \quad \boldsymbol{R} = \frac{1}{k_c} \nabla \times \boldsymbol{T} \tag{6.2.6}$$

where $\hat{\boldsymbol{a}}$ is any arbitrary unit vector, and $\boldsymbol{S}$, $\boldsymbol{T}$, and $\boldsymbol{R}$ are linearly independent. A key advantage of this method is that these vectors have clear physical characteristics. Note that

$$\nabla \times \boldsymbol{S} = 0, \quad \nabla \cdot \boldsymbol{T} = \nabla \cdot \boldsymbol{R} = 0$$

which indicates that $\boldsymbol{S}$ is a curl-free field, while $\boldsymbol{T}$ and $\boldsymbol{R}$ are divergence-free fields. This perfectly matches the properties of P, SH, and SV waves as reflected in Eq. (6.2.2), making them ideal for constructing the basis functions we need.

Equation (6.2.6) involves two quantities: $\hat{\boldsymbol{a}}$ and w. Since $\hat{\boldsymbol{a}}$ can be any arbitrary unit vector, and considering that $\hat{\boldsymbol{e}}_z$ is the only one of the three basis vectors in cylindrical coordinates that does not change with position (see Fig. 6.1), it is appropriate to choose $\hat{\boldsymbol{a}} = \hat{\boldsymbol{e}}_z$, which simplifies the curl operation in the second equation of (6.2.6).

Consider the scalar Helmholtz equation (6.2.4), which is a homogeneous equation that can be solved using the standard method of separation of variables. Let $w(r, \theta, z) = R(r)\Theta(\theta)Z(z)$, where $R(r)$ and $\Theta(\theta)$ are functions in the horizontal plane, and $Z(z)$ is a function in the vertical plane. They satisfy the following equations respectively,

[7] The vector Helmholtz equation is frequently encountered in electromagnetics. Hansen developed this solution method while studying antenna radiation, publishing papers in 1935, 1936, and 1937. In honor of his contribution, solutions constructed using this method are referred to as the *Hansen vectors*.

$$\frac{1}{r}\frac{\mathrm{d}}{\mathrm{d}r}\left(r\frac{\mathrm{d}R}{\mathrm{d}r}\right)+\left(k^2-\frac{m^2}{r^2}\right)R=0 \tag{6.2.7a}$$

$$\frac{\mathrm{d}^2\Theta}{\mathrm{d}\theta^2}+m^2\Theta=0 \tag{6.2.7b}$$

$$\frac{\mathrm{d}^2 Z}{\mathrm{d}z^2}-(k^2-k_c^2)Z=0 \tag{6.2.7c}$$

where k and m are real constants that arise during the separation of variables. The solutions to the above three equations are, in order,

$$R(r)=J_m(kr),\quad \Theta(\theta)=\mathrm{e}^{\mathrm{i}m\theta},\quad Z(z)=\mathrm{e}^{\pm\nu z} \tag{6.2.8}$$

$J_m(kr)$ is the Bessel function of order m, and $\nu=\sqrt{k^2-k_c^2}$ (with $\mathrm{Re}(\nu)>0$). When $k>k_c$, to ensure the boundary condition at infinity is satisfied, we take $Z(z)=\mathrm{e}^{-\nu z}$. Since the displacement must be single-valued for the current problem, $\Theta(\theta)$ must satisfy $\Theta(\theta+2\pi)=\Theta(\theta)$, meaning m must be an integer: $m=0,\pm1,\pm2,\ldots$. Additionally, to fit the physical context of the problem, k is chosen as a positive real number: $k\in[0,\infty)$.[8]

For the purpose of constructing basis functions, it is convenient to make them independent of z. On one hand, the corresponding scalar function $w'(r,\theta)$ still satisfies the scalar Helmholtz equation, and the basis functions themselves still satisfy the vector Helmholtz equation. On the other hand, choosing the basis functions in this way means they depend only on the transverse coordinates r and θ. When expanded in terms of these basis functions, solving for the displacement field is reduced to solving for the displacement expansion coefficients, leaving the z-dependence to the coefficients of the basis functions. This transforms the equations into a system of ordinary differential equations in z. Note that in Eqs. (6.2.7a) and (6.2.7b), $w'(r,\theta)=R(r)\Theta(\theta)$ satisfies the following equation,

$$\nabla'^2 w'(r,\theta)+k^2 w'(r,\theta)=0\quad\left(\boldsymbol{\nabla}'=\hat{\boldsymbol{e}}_r\frac{\partial}{\partial r}+\hat{\boldsymbol{e}}_\theta\frac{\partial}{r\partial\theta}\right) \tag{6.2.9}$$

where $k^2=k_c^2+\nu^2$, and $\boldsymbol{\nabla}'$ denotes the part of $\boldsymbol{\nabla}$ that only involves r and θ (see Eq. (6.2.5)). The function $w'(r,\theta)=Y_m(kr,\theta)=J_m(kr)\mathrm{e}^{\mathrm{i}m\theta}$ is called a *cylindrical harmonic function*. Substituting $w'(r,\theta)$ into Eq. (6.2.6),[9] we get

---

[8] Since kr appears as the argument in the Bessel function and is dimensionless, and r is the radial distance in the horizontal plane with the dimension of length, k must have the dimension of wavenumber. For this reason, k is referred to as the *horizontal wavenumber*, representing the number of wave cycles per unit length, and is therefore a positive real number.

[9] Note that in this context, k_c in Eq. (6.2.6) should be replaced by k because the scalar potential function forming this set of basis functions is now $w'(r,\theta)$, which satisfies Eq. (6.2.9), rather than $w(r,\theta,z)$ from Eq. (6.2.4).

$$\boldsymbol{S}_m(kr,\theta) = a_S \nabla' Y_m(kr,\theta) = a_S \left[kJ'_m(kr)\hat{\boldsymbol{e}}_r + \frac{\mathrm{i}m}{r} J_m(kr)\hat{\boldsymbol{e}}_\theta \right] \mathrm{e}^{\mathrm{i}m\theta} \tag{6.2.10a}$$

$$\boldsymbol{T}_m(kr,\theta) = a_T \nabla' \times \left[\hat{\boldsymbol{e}}_z Y_m(kr,\theta)\right] = a_T \left[\frac{\mathrm{i}m}{r} J_m(kr)\hat{\boldsymbol{e}}_r - kJ'_m(kr)\hat{\boldsymbol{e}}_\theta \right] \mathrm{e}^{\mathrm{i}m\theta} \tag{6.2.10b}$$

$$\boldsymbol{R}_m(kr,\theta) = \frac{a_R}{k} \nabla' \times \left\{\nabla' \times \left[\hat{\boldsymbol{e}}_z Y_m(kr,\theta)\right]\right\} = a_R k \hat{\boldsymbol{e}}_z J_m(kr) \mathrm{e}^{\mathrm{i}m\theta} \tag{6.2.10c}$$

where a_T, a_S, and a_R are the normalization coefficients for $\boldsymbol{T}_m(kr,\theta)$, $\boldsymbol{S}_m(kr,\theta)$, and $\boldsymbol{R}_m(kr,\theta)$, respectively. These coefficients need to be determined based on orthogonality and completeness. Hansen's research played a crucial role in the methods described above, and hence, this set of basis functions is called *Hansen vectors*.

6.2.3 Properties of Basis Functions

6.2.3.1 Corresponding Relationship with Three Types of Polarized Waves

By comparing Eqs. (6.2.10) and (6.2.2), it is clear that the three basis vectors $\boldsymbol{S}_m(kr,\theta)$, $\boldsymbol{T}_m(kr,\theta)$, and $\boldsymbol{R}_m(kr,\theta)$ correspond to P wave, SH wave, and SV wave, respectively:

$$\boldsymbol{S}_m(kr,\theta) \leftrightsquigarrow \text{P wave}, \quad \boldsymbol{T}_m(kr,\theta) \leftrightsquigarrow \text{SH wave}, \quad \boldsymbol{R}_m(kr,\theta) \leftrightsquigarrow \text{SV wave}$$

It is worth noting that the "correspond to" mentioned here does not imply that they are exactly the same. The directions of the actual three types of polarized waves are clearly related to the position of source, whereas the three basis vectors depend only on r and θ and are independent of z. The "correspond to" here has two specific implications:

(1) Based on the expression of the basis function in Eq. (6.2.10), it is similar to the displacement forms of the three types of polarized waves shown in Eq. (6.2.2). Note that they are not exactly the same, because in the expression for the displacement fields of the actual P, SV, and SH waves in Eq. (6.2.2), the operator is ∇, whereas in Eq. (6.2.10), the operator is only the horizontal coordinate-related part of ∇'.
(2) As will be seen in Sect. 6.3.1, the displacement expansion based on this set of basis functions results in a system of ordinary differential equations for the expansion coefficients. The expansion coefficient corresponding to $\boldsymbol{T}_m(kr,\theta)$ independently satisfies a second-order ordinary differential equation, while the expansion coefficients for $\boldsymbol{S}_m(kr,\theta)$ and $\boldsymbol{R}_m(kr,\theta)$ satisfy two coupled ordinary differential equations. This characteristic aligns with the property that P-wave and SV-wave are coupled, while SH-wave is not coupled with them.

6.2.3.2 Orthogonality

A set of single-variable functions without parameters is called an *orthogonal system* if it satisfies the condition

$$\int_a^b f_i^*(x) f_j(x)\, \mathrm{d}x = \delta_{ij}$$

For the parameterized multivariable functions constructed using Hansen's method in Eq. (6.2.10), the definition of orthogonality is similar but more complex

$$\int_0^{2\pi}\int_0^{+\infty} \boldsymbol{T}_m(kr,\theta)\cdot \boldsymbol{T}_{m'}^*(k'r,\theta) r\, \mathrm{d}r\, \mathrm{d}\theta = \frac{2\pi}{\sqrt{k'k}}\delta_{m'm}\delta(k-k') \tag{6.2.11a}$$

$$\int_0^{2\pi}\int_0^{+\infty} \boldsymbol{S}_m(kr,\theta)\cdot \boldsymbol{S}_{m'}^*(k'r,\theta) r\, \mathrm{d}r\, \mathrm{d}\theta = \frac{2\pi}{\sqrt{k'k}}\delta_{m'm}\delta(k-k') \tag{6.2.11b}$$

$$\int_0^{2\pi}\int_0^{+\infty} \boldsymbol{R}_m(kr,\theta)\cdot \boldsymbol{R}_{m'}^*(k'r,\theta) r\, \mathrm{d}r\, \mathrm{d}\theta = \frac{2\pi}{\sqrt{k'k}}\delta_{m'm}\delta(k-k') \tag{6.2.11c}$$

$$\int_0^{2\pi}\int_0^{+\infty} \boldsymbol{T}_m(kr,\theta)\cdot \boldsymbol{S}_{m'}^*(k'r,\theta) r\, \mathrm{d}r\, \mathrm{d}\theta = 0 \tag{6.2.11d}$$

$$\int_0^{2\pi}\int_0^{+\infty} \boldsymbol{S}_m(kr,\theta)\cdot \boldsymbol{R}_{m'}^*(k'r,\theta) r\, \mathrm{d}r\, \mathrm{d}\theta = 0 \tag{6.2.11e}$$

$$\int_0^{2\pi}\int_0^{+\infty} \boldsymbol{T}_m(kr,\theta)\cdot \boldsymbol{R}_{m'}^*(k'r,\theta) r\, \mathrm{d}r\, \mathrm{d}\theta = 0 \tag{6.2.11f}$$

Proof Note the following equation

$$\int_0^{+\infty} J_m(kr) J_m(k'r) r\, \mathrm{d}r = \frac{1}{\sqrt{k'k}}\delta(k-k') \tag{6.2.12}$$

and recurrence relations satisfied by Bessel functions

$$J_m'(x) = \frac{J_{m-1}(x) - J_{m+1}(x)}{2} \tag{6.2.13a}$$

$$\frac{m J_m(x)}{x} = \frac{J_{m-1}(x) + J_{m+1}(x)}{2} \tag{6.2.13b}$$

therefore

$$\int_0^{2\pi}\int_0^{+\infty} \boldsymbol{T}_m(kr,\theta)\cdot \boldsymbol{T}^*_{m'}(k'r,\theta) r\,\mathrm{d}r\,\mathrm{d}\theta$$
$$= a_T^2 \int_0^{2\pi} \mathrm{e}^{\mathrm{i}(m-m')}\,\mathrm{d}\theta \int_0^{+\infty}\left[\frac{m'm}{r^2}J_m(kr)J_{m'}(k'r) + k'kJ'_m(kr)J'_{m'}(k'r)\right] r\,\mathrm{d}r$$
$$\xlongequal{(6.2.13)} \frac{\pi a_T^2 \delta_{m'm}k'k}{2}\int_0^{+\infty}\{[J_{m-1}(kr)+J_{m+1}(kr)][J_{m-1}(k'r)+J_{m+1}(k'r)]$$
$$+[J_{m-1}(kr)-J_{m+1}(kr)][J_{m-1}(k'r)-J_{m+1}(k'r)]\}r\,\mathrm{d}r$$
$$= \pi a_T^2\delta_{m'm}k'k\int_0^{+\infty}[J_{m-1}(kr)J_{m-1}(k'r)+J_{m+1}(kr)J_{m+1}(k'r)]r\,\mathrm{d}r$$
$$\xlongequal{(6.2.12)} a_T^2k^2\frac{2\pi}{\sqrt{k'k}}\delta_{m'm}\delta(k-k') = \frac{2\pi}{\sqrt{k'k}}\delta_{m'm}\delta(k-k')$$

$$\int_0^{2\pi}\int_0^{+\infty} \boldsymbol{S}_m(kr,\theta)\cdot \boldsymbol{S}^*_{m'}(k'r,\theta) r\,\mathrm{d}r\,\mathrm{d}\theta$$
$$= a_S^2 \int_0^{2\pi} \mathrm{e}^{\mathrm{i}(m-m')}\,\mathrm{d}\theta \int_0^{+\infty}\left[\frac{m'm}{r^2}J_m(kr)J_{m'}(k'r) + k'kJ'_m(kr)J'_{m'}(k'r)\right] r\,\mathrm{d}r$$
$$\xlongequal{\text{similarly}} a_S^2k^2\frac{2\pi}{\sqrt{k'k}}\delta_{m'm}\delta(k-k') = \frac{2\pi}{\sqrt{k'k}}\delta_{m'm}\delta(k-k')$$

$$\int_0^{2\pi}\int_0^{+\infty} \boldsymbol{R}_m(kr,\theta)\cdot \boldsymbol{R}^*_{m'}(k'r,\theta) r\,\mathrm{d}r\,\mathrm{d}\theta$$
$$= a_R^2k^2 \int_0^{2\pi} \mathrm{e}^{\mathrm{i}(m-m')}\,\mathrm{d}\theta \int_0^{+\infty} J_m(kr)J_{m'}(k'r)r\,\mathrm{d}r$$
$$= a_R^2k^2\frac{2\pi}{\sqrt{k'k}}\delta_{m'm}\delta(k-k') = \frac{2\pi}{\sqrt{k'k}}\delta_{m'm}\delta(k-k')$$

and $a_T = a_S = 1/k$, $a_R = -1/k$.[10] Thus, Eq. (6.2.10) is finally determined as

$$\begin{cases} \boldsymbol{S}_m(kr,\theta) = \dfrac{1}{k}\nabla' Y_m(kr,\theta) = \left[J'_m(kr)\hat{\boldsymbol{e}}_r + \dfrac{\mathrm{i}m}{kr}J_m(kr)\hat{\boldsymbol{e}}_\theta\right]\mathrm{e}^{\mathrm{i}m\theta} \\ \boldsymbol{T}_m(kr,\theta) = \dfrac{1}{k}\nabla'\times[\hat{\boldsymbol{e}}_z Y_m(kr,\theta)] = \left[\dfrac{\mathrm{i}m}{kr}J_m(kr)\hat{\boldsymbol{e}}_r - J'_m(kr)\hat{\boldsymbol{e}}_\theta\right]\mathrm{e}^{\mathrm{i}m\theta} \\ \boldsymbol{R}_m(kr,\theta) = -\hat{\boldsymbol{e}}_z Y_m(kr,\theta) = -\hat{\boldsymbol{e}}_z J_m(kr)\mathrm{e}^{\mathrm{i}m\theta} \end{cases} \tag{6.2.14}$$

Since $\hat{\boldsymbol{e}}_z\cdot\hat{\boldsymbol{e}}_r = \hat{\boldsymbol{e}}_z\cdot\hat{\boldsymbol{e}}_\theta = 0$, Eqs. (6.2.11e) and (6.2.11f) are evidently valid. Moreover,

[10] It is also possible to use the positive sign, as both options satisfy the normalization and orthogonality conditions. The difference lies in the sign of the coefficients when these functions are used as basis functions for expansion, which slightly affects the resulting ordinary differential equations, but the essence remains the same. In fact, Aki and Richards (2002, p.300) use the negative sign, while Ben-Menahem and Singh (1981, p.60) use the positive sign. Here, we follow the approach of the former.

$$\begin{aligned}&\int_0^{2\pi}\int_0^{+\infty} \boldsymbol{T}_m(kr,\theta)\cdot \boldsymbol{S}^*_{m'}(k'r,\theta) r\,\mathrm{d}r\,\mathrm{d}\theta\\&=\int_0^{2\pi} \mathrm{e}^{\mathrm{i}(m-m')}\,\mathrm{d}\theta \int_0^{+\infty}\left[\frac{\mathrm{i}m}{kr}J_m(kr)J'_{m'}(k'r)+\frac{\mathrm{i}m'}{k'r}J'_m(kr)J_{m'}(k'r)\right] r\,\mathrm{d}r\\&=\frac{2\pi\mathrm{i}m\delta_{m'm}}{k'k}\left[J_m(kr)J_m(k'r)\right]\Big|_{r=0}^{r=\infty}=0\end{aligned}$$

6.2.3.3 Completeness

Completeness for $\boldsymbol{S}_m(kr,\theta)$, $\boldsymbol{T}_m(kr,\theta)$, and $\boldsymbol{R}_m(kr,\theta)$ is defined as

$$\begin{aligned}&\frac{1}{2\pi}\sum_{m=-\infty}^{+\infty}\int_0^{+\infty}\left[\boldsymbol{T}_m(kr,\theta)\boldsymbol{T}^*_m(kr',\theta')+\boldsymbol{S}_m(kr,\theta)\boldsymbol{S}^*_m(kr',\theta')\right.\\&\qquad\left.+\boldsymbol{R}_m(kr,\theta)\boldsymbol{R}^*_m(kr',\theta')\right]k\,\mathrm{d}k=\left(\hat{\boldsymbol{e}}_r\hat{\boldsymbol{e}}_r+\hat{\boldsymbol{e}}_\theta\hat{\boldsymbol{e}}_\theta+\hat{\boldsymbol{e}}_z\hat{\boldsymbol{e}}_z\right)\delta(\boldsymbol{r}-\boldsymbol{r}')\end{aligned}$$

where the superscript "$*$" represents the complex conjugate.

Proof Using the following formula,

$$\sum_{m=-\infty}^{+\infty}\mathrm{e}^{\mathrm{i}m(\theta-\theta')}=2\pi\delta(\theta-\theta'),\quad \theta\in[0,2\pi) \tag{6.2.15a}$$

$$\delta(\boldsymbol{r}-\boldsymbol{r}')=\frac{1}{\sqrt{r'r}}\delta(r-r')\delta(\theta-\theta') \tag{6.2.15b}$$

we can obtain

$$\begin{aligned}&\frac{1}{2\pi}\sum_{m=-\infty}^{+\infty}\int_0^{+\infty}\left[\boldsymbol{T}_m(kr,\theta)\boldsymbol{T}^*_m(kr',\theta')+\boldsymbol{S}_m(kr,\theta)\boldsymbol{S}^*_m(kr',\theta')\right.\\&\left.+\boldsymbol{R}_m(kr,\theta)\boldsymbol{R}^*_m(kr',\theta')\right]k\,\mathrm{d}k\\&=\sum_{m=-\infty}^{+\infty}\frac{\mathrm{e}^{\mathrm{i}m(\theta-\theta')}}{2\pi}\int_0^{+\infty}\left\{\left[\frac{m^2J_m(kr)J_m(kr')}{k^2r'r}+J'_m(kr)J'_m(kr')\right](\hat{\boldsymbol{e}}_r\hat{\boldsymbol{e}}_{r'}+\hat{\boldsymbol{e}}_\theta\hat{\boldsymbol{e}}_{\theta'})\right.\\&-\mathrm{i}m\left[\frac{J_m(kr)J'_m(kr')}{kr}+\frac{J_m(kr')J'_m(kr)}{kr'}\right](\hat{\boldsymbol{e}}_r\hat{\boldsymbol{e}}_{\theta'}-\hat{\boldsymbol{e}}_\theta\hat{\boldsymbol{e}}_{r'})\\&\left.+J_m(kr)J_m(kr')\hat{\boldsymbol{e}}_z\hat{\boldsymbol{e}}_z\right\}k\,\mathrm{d}k\\&\overset{(6.2.13)}{=\!=}\delta(\theta-\theta')\int_0^{+\infty}\left\{\frac{1}{2}\left[J_{m-1}(kr)J_{m-1}(kr')+J_{m+1}(kr)J_{m+1}(kr')\right]\right.\\&\qquad\left.\cdot(\hat{\boldsymbol{e}}_r\hat{\boldsymbol{e}}_{r'}+\hat{\boldsymbol{e}}_\theta\hat{\boldsymbol{e}}_\theta)+J_m(kr)J_m(kr')\hat{\boldsymbol{e}}_z\hat{\boldsymbol{e}}_z\right\}k\,\mathrm{d}k\\&\overset{(6.2.12),(6.2.15b)}{=\!=}\left(\hat{\boldsymbol{e}}_r\hat{\boldsymbol{e}}_r+\hat{\boldsymbol{e}}_\theta\hat{\boldsymbol{e}}_\theta+\hat{\boldsymbol{e}}_z\hat{\boldsymbol{e}}_z\right)\delta(\boldsymbol{r}-\boldsymbol{r}')\end{aligned}$$

Completeness means that any vector function $\boldsymbol{A}(\boldsymbol{x}, \omega)$ can be represented using this set of basis functions,

$$\boldsymbol{A}(\boldsymbol{x}, \omega) = \frac{1}{2\pi} \sum_{m=-\infty}^{+\infty} \int_0^{+\infty} \Big[A_m^T(k, z, \omega) \boldsymbol{T}_m(kr, \theta) + A_m^S(k, z, \omega) \boldsymbol{S}_m(kr, \theta) + A_m^R(k, z, \omega) \boldsymbol{R}_m(kr, \theta) \Big] k \, \mathrm{d}k \tag{6.2.16}$$

where the expressions for the coefficients $A_m^T(k, z, \omega)$, $A_m^S(k, z, \omega)$, and $A_m^R(k, z, \omega)$ are given by

$$A_m^T(k, z, \omega) = \int_0^{2\pi} \int_0^{+\infty} \boldsymbol{A}(\boldsymbol{x}, \omega) \cdot \boldsymbol{T}_m^*(kr, \theta) r \, \mathrm{d}r \, \mathrm{d}\theta \tag{6.2.17a}$$

$$A_m^S(k, z, \omega) = \int_0^{2\pi} \int_0^{+\infty} \boldsymbol{A}(\boldsymbol{x}, \omega) \cdot \boldsymbol{S}_m^*(kr, \theta) r \, \mathrm{d}r \, \mathrm{d}\theta \tag{6.2.17b}$$

$$A_m^R(k, z, \omega) = \int_0^{2\pi} \int_0^{+\infty} \boldsymbol{A}(\boldsymbol{x}, \omega) \cdot \boldsymbol{R}_m^*(kr, \theta) r \, \mathrm{d}r \, \mathrm{d}\theta \tag{6.2.17c}$$

For a known function, such as the body force $\boldsymbol{f}(\boldsymbol{x}, \omega)$, the expansion coefficients can be determined using Eq. (6.2.17). However, for an unknown function, like the displacement vector $\boldsymbol{u}(\boldsymbol{x}, \omega)$, since the coefficients cannot be explicitly determined using Eq. (6.2.17), we must use the formal expression (6.2.16) to determine the coefficients. Notably, because the chosen basis functions retain only the parts related to r and θ, projecting $\boldsymbol{A}(\boldsymbol{x}, \omega)$ onto this basis and integrating over r and θ yields coefficients that depend solely on z (see Eq. (6.2.17)). This implies that substituting the expression for $\boldsymbol{A}(\boldsymbol{x}, \omega)$ into the governing equations results in a set of ordinary differential equations that are dependent only on z.

6.3 Ordinary Differential Equations and Their Solution

By using the basis function expansions for displacement and body force, and leveraging the orthogonality of the basis functions, the equations can be transformed into a series of ordinary differential equations involving the expansion coefficients. The solution to this system of ordinary differential equations can be obtained by exploiting the decomposition properties of the coefficient matrix.

6.3.1 Ordinary Differential Equations

Expanding the displacement function $\boldsymbol{u}(\boldsymbol{x},\omega)$ and the body force $\boldsymbol{f}(\boldsymbol{x},\omega)$ in the frequency domain using the basis functions in Eq. (6.2.14), we have

$$\begin{aligned}\boldsymbol{u}(\boldsymbol{x},\omega) =&\frac{1}{2\pi}\sum_{m=-\infty}^{+\infty}\int_0^{+\infty}\Big[u_m^T(k,z,\omega)\boldsymbol{T}_m(kr,\theta)+u_m^S(k,z,\omega)\boldsymbol{S}_m(kr,\theta)\\&+u_m^R(k,z,\omega)\boldsymbol{R}_m(kr,\theta)\Big]k\,\mathrm{d}k\end{aligned} \tag{6.3.1a}$$

$$\begin{aligned}\boldsymbol{f}(\boldsymbol{x},\omega) =&\frac{1}{2\pi}\sum_{m=-\infty}^{+\infty}\int_0^{+\infty}\Big[f_m^T(k,z,\omega)\boldsymbol{T}_m(kr,\theta)+f_m^S(k,z,\omega)\boldsymbol{S}_m(kr,\theta)\\&+f_m^R(k,z,\omega)\boldsymbol{R}_m(kr,\theta)\Big]k\,\mathrm{d}k\end{aligned} \tag{6.3.1b}$$

Substituting them into Eq. (6.1.1a), we obtain

$$\begin{aligned}&\frac{1}{2\pi}\sum_{m=-\infty}^{+\infty}\int_0^{+\infty}\Big\{\rho\omega^2\left(u_m^T\boldsymbol{T}_m+u_m^S\boldsymbol{S}_m+u_m^R\boldsymbol{R}_m\right)+(\lambda+2\mu)\nabla\nabla\cdot\left(u_m^T\boldsymbol{T}_m+u_m^S\boldsymbol{S}_m+u_m^R\boldsymbol{R}_m\right)\\&-\mu\nabla\times\nabla\times\left(u_m^T\boldsymbol{T}_m+u_m^S\boldsymbol{S}_m+u_m^R\boldsymbol{R}_m\right)+\left(f_m^T\boldsymbol{T}_m+f_m^S\boldsymbol{S}_m+f_m^R\boldsymbol{R}_m\right)\Big\}k\,\mathrm{d}k=0\end{aligned} \tag{6.3.2}$$

Since $Y_m(kr,\theta)$ satisfies the scalar Helmholtz equation (6.2.9),

$$\nabla'^2 Y_m(kr,\theta) = -k^2 Y_m(kr,\theta) \tag{6.3.3}$$

according to the definition in Eq. (6.2.14), we have

$$\begin{aligned}\nabla\cdot\boldsymbol{S}_m &= \frac{1}{k}\nabla\cdot\nabla' Y_m(kr,\theta)=\frac{1}{k}\nabla'^2Y_m(kr,\theta)\overset{(6.3.3)}{=\!=\!=}-kY_m(kr,\theta)\\ \nabla\cdot\boldsymbol{T}_m &= \frac{1}{k}\nabla\cdot\nabla'\times\left[\hat{\boldsymbol{e}}_zY_m(kr,\theta)\right]=\frac{1}{k}\nabla'\cdot\nabla'\times\left[\hat{\boldsymbol{e}}_zY_m(kr,\theta)\right]=0\\ \nabla\cdot\boldsymbol{R}_m &= \nabla\cdot\left[-\hat{\boldsymbol{e}}_zY_m(kr,\theta)\right]=-\frac{\partial}{\partial z}Y_m(kr,\theta)=0\\ \nabla\times\boldsymbol{S}_m &= \frac{1}{k}\nabla\times\nabla' Y_m(kr,\theta)=0\\ \nabla\times\boldsymbol{T}_m &= \frac{1}{k}\nabla\times\nabla'\times\left[\hat{\boldsymbol{e}}_zY_m(kr,\theta)\right]=-\frac{1}{k}\nabla'^2\left[\hat{\boldsymbol{e}}_zY_m(kr,\theta)\right]=-k\boldsymbol{R}_m\\ \nabla\times\boldsymbol{R}_m &= \nabla\times\left[-\hat{\boldsymbol{e}}_zY_m(kr,\theta)\right]=-k\boldsymbol{T}_m\end{aligned}$$

Noting that for the scalar functions u and v, and the vector function $\boldsymbol{f}$, the following relationships hold:

$$\begin{aligned}\nabla\cdot(u\boldsymbol{f}) &= u\nabla\cdot\boldsymbol{f}+(\nabla u)\cdot\boldsymbol{f}\\ \nabla\times(u\boldsymbol{f}) &= u\nabla\times\boldsymbol{f}+(\nabla u)\times\boldsymbol{f}\\ \nabla(uv) &= (\nabla u)v+(\nabla v)u\end{aligned}$$

thus,

$$\begin{aligned}\nabla\cdot\left(u_m^T\boldsymbol{T}_m\right) &= u_m^T\nabla\cdot\boldsymbol{T}_m+\left(\nabla u_m^T\right)\cdot\boldsymbol{T}_m=0\\ \nabla\cdot\left(u_m^S\boldsymbol{S}_m\right) &= u_m^S\nabla\cdot\boldsymbol{S}_m+\left(\nabla u_m^S\right)\cdot\boldsymbol{S}_m=-ku_m^SY_m\\ \nabla\cdot\left(u_m^R\boldsymbol{R}_m\right) &= u_m^R\nabla\cdot\boldsymbol{R}_m+\left(\nabla u_m^R\right)\cdot\boldsymbol{R}_m=-\frac{\mathrm{d}u_m^R}{\mathrm{d}z}Y_m\end{aligned}$$

Based on this, it can be further obtained

$$\nabla\nabla\cdot\left(u_m^T\boldsymbol{T}_m\right)=0 \tag{6.3.4a}$$

$$\nabla\nabla\cdot\left(u_m^S\boldsymbol{S}_m\right)=-k\nabla\left(u_m^SY_m\right)=k\frac{\mathrm{d}u_m^S}{\mathrm{d}z}\boldsymbol{R}_m-k^2u_m^S\boldsymbol{S}_m \tag{6.3.4b}$$

$$\nabla\nabla\cdot\left(u_m^R\boldsymbol{R}_m\right)=-\nabla\left(\frac{\mathrm{d}u_m^R}{\mathrm{d}z}Y_m\right)=\frac{\mathrm{d}^2u_m^R}{\mathrm{d}z^2}\boldsymbol{R}_m-k\frac{\mathrm{d}u_m^R}{\mathrm{d}z}\boldsymbol{S}_m \tag{6.3.4c}$$

Additionally, since

$$\begin{aligned}\hat{\boldsymbol{e}}_z\times\boldsymbol{T}_m &= \hat{\boldsymbol{e}}_z\times\left[\frac{\mathrm{i}m}{kr}J_m(kr)\hat{\boldsymbol{e}}_r-J_m'(kr)\hat{\boldsymbol{e}}_\theta\right]\mathrm{e}^{\mathrm{i}m\theta}=\boldsymbol{S}_m\\ \hat{\boldsymbol{e}}_z\times\boldsymbol{S}_m &= \hat{\boldsymbol{e}}_z\times\left[J_m'(kr)\hat{\boldsymbol{e}}_r+\frac{\mathrm{i}m}{kr}J_m(kr)\hat{\boldsymbol{e}}_\theta\right]\mathrm{e}^{\mathrm{i}m\theta}=-\boldsymbol{T}_m\\ \hat{\boldsymbol{e}}_z\times\boldsymbol{R}_m &= \hat{\boldsymbol{e}}_z\times\hat{\boldsymbol{e}}_zY_m=0\end{aligned}$$

we have

$$\begin{aligned}\nabla\times\left(u_m^T\boldsymbol{T}_m\right) &= u_m^T\nabla\times\boldsymbol{T}_m+\left(\nabla u_m^T\right)\times\boldsymbol{T}_m=-ku_m^T\boldsymbol{R}_m+\frac{\mathrm{d}u_m^T}{\mathrm{d}z}\boldsymbol{S}_m\\ \nabla\times\left(u_m^S\boldsymbol{S}_m\right) &= u_m^S\nabla\times\boldsymbol{S}_m+\left(\nabla u_m^S\right)\times\boldsymbol{S}_m=\frac{\mathrm{d}u_m^S}{\mathrm{d}z}\hat{\boldsymbol{e}}_z\times\boldsymbol{S}_m=-\frac{\mathrm{d}u_m^S}{\mathrm{d}z}\boldsymbol{T}_m\\ \nabla\times\left(u_m^R\boldsymbol{R}_m\right) &= u_m^R\nabla\times\boldsymbol{R}_m+\left(\nabla u_m^R\right)\times\boldsymbol{R}_m=-ku_m^R\boldsymbol{T}_m\end{aligned}$$

and thereby obtain

$$\begin{aligned}\nabla \times \nabla \times \left(u_m^T \boldsymbol{T}_m\right) &= -k\nabla \times \left(u_m^T \boldsymbol{R}_m\right) + \nabla \times \left(\frac{\mathrm{d}u_m^T}{\mathrm{d}z}\boldsymbol{S}_m\right) \\ &= k^2 u_m^T \boldsymbol{T}_m - \frac{\mathrm{d}^2 u_m^T}{\mathrm{d}z^2}\boldsymbol{T}_m \end{aligned} \tag{6.3.5a}$$

$$\nabla \times \nabla \times \left(u_m^S \boldsymbol{S}_m\right) = -\nabla \times \frac{\mathrm{d}u_m^S}{\mathrm{d}z}\boldsymbol{T}_m = k\frac{\mathrm{d}u_m^S}{\mathrm{d}z}\boldsymbol{R}_m - \frac{\mathrm{d}^2 u_m^S}{\mathrm{d}z^2}\boldsymbol{S}_m \tag{6.3.5b}$$

$$\nabla \times \nabla \times \left(u_m^R \boldsymbol{R}_m\right) = -k\nabla \times u_m^R \boldsymbol{R}_m = k^2 u_m^R \boldsymbol{R}_m - k\frac{\mathrm{d}u_m^R}{\mathrm{d}z}\boldsymbol{S}_m \tag{6.3.5c}$$

Substitute Eqs. (6.3.4) and (6.3.5) into Eq. (6.3.2). Due to the orthogonality of the basis vectors, their corresponding coefficients are zero, resulting in the following

$$\rho\omega^2 u_m^T + \mu\frac{\mathrm{d}^2 u_m^T}{\mathrm{d}z^2} - k^2\mu u_m^T + f_m^T = 0 \tag{6.3.6a}$$

$$\rho\omega^2 u_m^R + (\lambda + 2\mu)\frac{\mathrm{d}^2 u_m^R}{\mathrm{d}z^2} - k^2\mu u_m^R + (\lambda+\mu)k\frac{\mathrm{d}u_m^S}{\mathrm{d}z} + f_m^R = 0 \tag{6.3.6b}$$

$$\rho\omega^2 u_m^S + \mu\frac{\mathrm{d}^2 u_m^S}{\mathrm{d}z^2} - k^2(\lambda + 2\mu)u_m^S - (\lambda+\mu)k\frac{\mathrm{d}u_m^R}{\mathrm{d}z} + f_m^S = 0 \tag{6.3.6c}$$

The above equations represent a set of second-order ordinary differential equations for the three components of the displacement vector u_m^T, u_m^S, and u_m^R. A notable characteristic is that Eq. (6.3.6a) involves only u_m^T, whereas Eqs. (6.3.6b) and (6.3.6c) both include u_m^S and u_m^R. This indicates that the latter two form a coupled system, while the first is independent. Referring to Sect. 6.2.3.1, this can be interpreted as reflecting the characteristic where P-wave and SV-wave interact, while SH-wave is independent of them. Therefore, we refer to solutions involving u_m^T as the "SH case," and those involving u_m^S and u_m^R as the "P-SV case". This is merely a formal correspondence and does not imply actual SH, P, or SV waves.

Directly solving the coupled second-order equations is quite challenging. There are two possible approaches:

(1) By introducing intermediate functions, the aim is to reduce the number of functions to be solved. Refer to Sect. 2.2.4 for the stress solution method. This approach transforms the original problem of solving several equations for stress components into solving a biharmonic equation for a stress function. However, the drawback is that the order of the differential equations increases from second to fourth order. Finding the general solution for such high-order equations is difficult.

(2) In contrast to the previous approach, this method introduces more functions to be solved, thereby reducing the order of the differential equations by one. Although this increases the complexity of the differential equations, it allows us to solve the system using the properties of the coefficient matrix decomposition.

We adopt the second approach. Introducing the traction vector as the "additional functions to be solved" is appropriate for two reasons: first, there is a linear relationship between the traction and the spatial derivatives of the displacement, which facilitates the reduction of the order of the system of Eq. (6.3.6); second, the boundary condition at the free interface (6.1.1b) is naturally expressed in terms of the traction, making it easier to use the boundary condition to determine the unknown coefficients. The traction vector $\boldsymbol{\tau}(\boldsymbol{x},\omega)$ can be expanded using the same basis vectors as

$$\begin{aligned}\boldsymbol{\tau}(\boldsymbol{x},\omega)=&\frac{1}{2\pi}\sum_{m=-\infty}^{+\infty}\int_0^{+\infty}\Big[\tau_m^T(k,z,\omega)\boldsymbol{T}_m(kr,\theta)+\tau_m^S(k,z,\omega)\boldsymbol{S}_m(kr,\theta)\\&+\tau_m^R(k,z,\omega)\boldsymbol{R}_m(kr,\theta)\Big]k\,\mathrm{d}k\end{aligned}\tag{6.3.7}$$

According to the generalized Hooke's law (2.2.18),

$$\tilde{\boldsymbol{\tau}}=\boldsymbol{\sigma}\cdot\hat{\boldsymbol{e}}_z=\lambda(\nabla\cdot\tilde{\boldsymbol{u}})\hat{\boldsymbol{e}}_z+\mu\left[\nabla\left(\tilde{\boldsymbol{u}}\cdot\hat{\boldsymbol{e}}_z\right)+\left(\hat{\boldsymbol{e}}_z\cdot\nabla\right)\tilde{\boldsymbol{u}}\right]\tag{6.3.8}$$

and

$$\begin{aligned}\nabla\cdot\tilde{\boldsymbol{u}}&=-\frac{1}{2\pi}\sum_{m=-\infty}^{+\infty}\int_0^{+\infty}\left(ku_m^S+\frac{\mathrm{d}u_m^R}{\mathrm{d}z}\right)Y_m(kr,\theta)k\,\mathrm{d}k\\\nabla\left(\tilde{\boldsymbol{u}}\cdot\hat{\boldsymbol{e}}_z\right)&=\frac{1}{2\pi}\sum_{m=-\infty}^{+\infty}\int_0^{+\infty}\left(\frac{\mathrm{d}u_m^R}{\mathrm{d}z}\boldsymbol{R}_m-ku_m^R\boldsymbol{S}_m\right)k\,\mathrm{d}k\\\left(\hat{\boldsymbol{e}}_z\cdot\nabla\right)\tilde{\boldsymbol{u}}&=\frac{1}{2\pi}\sum_{m=-\infty}^{+\infty}\int_0^{+\infty}\left(\frac{\mathrm{d}u_m^S}{\mathrm{d}z}\boldsymbol{S}_m+\frac{\mathrm{d}u_m^R}{\mathrm{d}z}\boldsymbol{R}_m+\frac{\mathrm{d}u_m^T}{\mathrm{d}z}\boldsymbol{T}_m\right)k\,\mathrm{d}k\end{aligned}$$

substituting them into Eq. (6.3.8) and comparing with Eq. (6.3.7), we get

$$\tau_m^T=\mu\frac{\mathrm{d}u_m^T}{\mathrm{d}z},\quad\tau_m^S=\mu\left(\frac{\mathrm{d}u_m^T}{\mathrm{d}z}-ku_m^R\right),\quad\tau_m^R=(\lambda+2\mu)\frac{\mathrm{d}u_m^T}{\mathrm{d}z}+\lambda ku_m^S$$

By combining the above equation with Eq. (6.3.6), we can form two sets of ordinary differential equations. One set pertains to the SH case:

$$\begin{aligned}\frac{\mathrm{d}u_m^T}{\mathrm{d}z}&=\frac{1}{\mu}\tau_m^T\\\frac{\mathrm{d}\tau_m^T}{\mathrm{d}z}&=\left(\mu k^2-\rho\omega^2\right)u_m^T-f_m^T\end{aligned}$$

the corresponding matrix form is

$$\frac{\mathrm{d}}{\mathrm{d}z}\begin{bmatrix} u_m^T \\ \tau_m^T \end{bmatrix} = \begin{bmatrix} 0 & \frac{1}{\mu} \\ \mu k^2 - \rho\omega^2 & 0 \end{bmatrix}\begin{bmatrix} u_m^T \\ \tau_m^T \end{bmatrix} + \begin{bmatrix} 0 \\ -f_m^T \end{bmatrix} \tag{6.3.9}$$

The other set pertains to the P-SV case:

$$\begin{aligned}
\frac{\mathrm{d}u_m^S}{\mathrm{d}z} &= ku_m^R + \frac{1}{\mu}\tau_m^S \\
\frac{\mathrm{d}u_m^R}{\mathrm{d}z} &= -\frac{\lambda k}{\lambda + 2\mu}u_m^S + \frac{1}{\lambda + 2\mu}\tau_m^R \\
\frac{\mathrm{d}\tau_m^S}{\mathrm{d}z} &= \left[\frac{4\mu(\lambda + \mu)}{\lambda + 2\mu}k^2 - \rho\omega^2\right]u_m^S + \frac{\lambda k}{\lambda + 2\mu}\tau_m^R - f_m^S \\
\frac{\mathrm{d}\tau_m^R}{\mathrm{d}z} &= -\rho\omega^2 u_m^R - k\tau_m^S - f_m^R
\end{aligned}$$

the corresponding matrix form is

$$\frac{\mathrm{d}}{\mathrm{d}z}\begin{bmatrix} u_m^S \\ u_m^R \\ \tau_m^S \\ \tau_m^R \end{bmatrix} = \begin{bmatrix} 0 & k & \frac{1}{\mu} & 0 \\ -\frac{\lambda k}{\lambda + 2\mu} & 0 & 0 & \frac{1}{\lambda + 2\mu} \\ \frac{4\mu(\lambda + \mu)}{\lambda + 2\mu}k^2 - \rho\omega^2 & 0 & 0 & \frac{\lambda k}{\lambda + 2\mu} \\ 0 & -\rho\omega^2 & -k & 0 \end{bmatrix}\begin{bmatrix} u_m^S \\ u_m^R \\ \tau_m^S \\ \tau_m^R \end{bmatrix} + \begin{bmatrix} 0 \\ 0 \\ -f_m^S \\ -f_m^R \end{bmatrix} \tag{6.3.10}$$

Equations (6.3.9) and (6.3.10) can be uniformly written as follows:

$$\frac{\mathrm{d}}{\mathrm{d}z}\boldsymbol{y}(z) = \mathbf{A}\boldsymbol{y}(z) + \boldsymbol{f}(z) \tag{6.3.11}$$

For the SH case, $\boldsymbol{y}(z)$ and $\boldsymbol{f}(z)$ are 2×1 vectors, and $\mathbf{A}$ is a 2×2 matrix. In contrast, for the P-SV case, they are 4×1 vectors and a 4×4 matrix, respectively. The specific expressions can be found by referring to Eqs. (6.3.9) and (6.3.10), so they are not listed here. It is noteworthy that $\boldsymbol{y}(z)$ is a vector composed of the expansion coefficients of displacement and traction, known as the *displacement-traction vector*.[11] Equation (6.3.11) is actually a system of ordinary differential equations. It is clear that the

[11] It is important to note that despite the nominal name, it is not a vector made up of displacement and traction components. Instead, it consists of the coefficients from their expansion using basis functions. Moreover, the SH and P-SV cases are distinct, each having its own displacement-traction vector.

coefficient matrix $\mathbf{A}$ is crucial to the form of the solution. If $\mathbf{A}$ can be appropriately decomposed, it might be possible to transform the equations into a form that is easier to solve.

6.3.2 General Solution of Ordinary Differential Equations

To find the displacement $\tilde{\boldsymbol{u}}$, we need to solve Eq. (6.3.11) separately for the SH and P-SV cases to obtain the displacement expansion coefficients in the displacement-traction vector. This is a non-homogeneous system of ordinary differential equations. According to the theory of ordinary differential equations, the solution of Eq. (6.3.11) can be expressed as the sum of the general solution of the homogeneous equation, $\boldsymbol{y}_{\mathrm{g}}(z)$, and the particular solution of the non-homogeneous equation, $\boldsymbol{y}_{\mathrm{p}}(z)$,

$$\boldsymbol{y}(z) = \boldsymbol{y}_{\mathrm{g}}(z) + \boldsymbol{y}_{\mathrm{p}}(z) \tag{6.3.12}$$

6.3.2.1 General Solution of Homogeneous Equations: $\boldsymbol{y}_{\mathrm{g}}(z)$

$\boldsymbol{y}_{\mathrm{g}}(z)$ satisfies the following equation

$$\frac{\mathrm{d}}{\mathrm{d}z}\boldsymbol{y}_{\mathrm{g}}(z) = \mathbf{A}\boldsymbol{y}_{\mathrm{g}}(z) \tag{6.3.13}$$

According to linear algebra theory,[12] $\mathbf{A}$ can be decomposed in the following way

$$\mathbf{A} = \mathbf{EJE}^{-1} \tag{6.3.14}$$

[12] Any n-order complex matrix $\mathbf{A}$ is similar to some Jordan form matrix. Apart from the order of the Jordan blocks, this Jordan form matrix is uniquely determined by $\mathbf{A}$ and is known as the *Jordan canonical form* of $\mathbf{A}$. The elements on its main diagonal are all the roots of the characteristic polynomial of $\mathbf{A}$. The form is

$$J(\lambda, t) = \begin{bmatrix} \lambda & 0 & \ldots & 0 & 0 & 0 \\ 1 & \lambda & \ldots & 0 & 0 & 0 \\ \vdots & \vdots & & \vdots & \vdots & \vdots \\ 0 & 0 & \cdots & 1 & \lambda & 0 \\ 0 & 0 & \cdots & 0 & 1 & \lambda \end{bmatrix}_{t\times t}$$

where λ is a complex number, and this matrix is called a Jordan block. A block diagonal matrix consisting of several Jordan blocks is referred to as a *Jordan matrix*.

$\mathbf{J}$ is a Jordan form matrix with diagonal elements that are the eigenvalues of the coefficient matrix $\mathbf{A}$. $\mathbf{E}$ is the matrix formed by the eigenvectors of $\mathbf{A}$ as its column vectors. Specifically, when $\mathbf{A}$ has distinct eigenvalues,[13] $\mathbf{J}$ becomes a diagonal matrix

$$\mathbf{J} = \operatorname{diag}\{\lambda_1, \lambda_2, \dots, \lambda_n\}$$

where diag $\{\cdots\}$ represents a diagonal matrix, and λ_i $(i = 1, 2, \dots, n)$ are distinct eigenvalues. Substituting Eq. (6.3.14) into Eq. (6.3.13) gives

$$\frac{\mathrm{d}}{\mathrm{d}z}\boldsymbol{y}_{\mathrm{g}}(z) = \mathbf{EJE}^{-1}\boldsymbol{y}_{\mathrm{g}}(z) \quad \Rightarrow \quad \frac{\mathrm{d}}{\mathrm{d}z}\left[\mathbf{E}^{-1}\boldsymbol{y}_{\mathrm{g}}(z)\right] = \mathbf{J}\left[\mathbf{E}^{-1}\boldsymbol{y}_{\mathrm{g}}(z)\right]$$

If we let $\boldsymbol{Y}(z) = \mathbf{E}^{-1}\boldsymbol{y}_{\mathrm{g}}(z)$, then the above equation can be written as

$$\frac{\mathrm{d}}{\mathrm{d}z}\boldsymbol{Y}(z) = \mathbf{J}\boldsymbol{Y}(z) \quad \Rightarrow \quad \frac{\mathrm{d}}{\mathrm{d}z}Y_i(z) = \lambda_i Y_i(z) \quad \text{(without summation)}$$

Thus, $Y_i(z) = C_i \mathrm{e}^{\lambda_i z}$ $(i = 1, 2, \dots, n)$. In matrix form, this can be written as $\boldsymbol{Y}(z) = \mathbf{Q}(z)\boldsymbol{C}$, where

$$\mathbf{Q}(z) = \operatorname{diag}\left\{\mathrm{e}^{\lambda_1 z}, \mathrm{e}^{\lambda_2 z}, \dots, \mathrm{e}^{\lambda_n z}\right\}, \quad \boldsymbol{C} = (C_1, C_2, \dots, C_n)^{\mathrm{T}} \tag{6.3.15}$$

$\boldsymbol{C}$ is a vector composed of the undetermined coefficients C_i $(i = 1, 2, \dots, n)$. From this, we obtain

$$\boldsymbol{y}_{\mathrm{g}}(z) = \mathbf{E}\boldsymbol{Y}(z) = \mathbf{EQ}(z)\boldsymbol{C} \tag{6.3.16}$$

Equation (6.3.16) shows that the general solution of the homogeneous equation $\boldsymbol{y}_{\mathrm{g}}(z)$ is closely related to the eigenvalues and eigenvectors of the coefficient matrix $\mathbf{A}$. In Eq. (6.3.16), $\mathbf{E} = [\boldsymbol{e}_1, \boldsymbol{e}_2, \dots, \boldsymbol{e}_n]$, where $\boldsymbol{e}_i$ is the eigenvector corresponding to the eigenvalue λ_i. The expressions for $\mathbf{Q}(z)$ and $\boldsymbol{C}$ can be found in Eq. (6.3.15).

6.3.2.2 Particular Solution of Non-homogeneous Equations: $\boldsymbol{y}_{\mathrm{p}}(z)$

A common method for finding the particular solution of a non-homogeneous equation is the method of variation of parameters. This involves changing the undetermined constants $\boldsymbol{C}$ in Eq. (6.3.16) to functions $\boldsymbol{w}(z)$, thus forming the particular solution of the non-homogeneous equation. Specifically, the particular solution is assumed to be in the form $\boldsymbol{y}_{\mathrm{p}}(z) = \mathbf{EQ}(z)\boldsymbol{w}(z)$, which satisfies the equation

[13] The condition of distinct eigenvalues will be discussed later. In practice, this corresponds to the angular frequency $\omega \neq 0$. For the case where $\omega = 0$, the eigenvalues are degenerate (equal), and $\mathbf{J}$ cannot be written in diagonal matrix form. This will be specifically addressed in Sect. 6.7.

$$\frac{\mathrm{d}}{\mathrm{d}z}\boldsymbol{y}_\mathrm{p}(z) = \mathbf{A}\boldsymbol{y}_\mathrm{p}(z) + \boldsymbol{f}(z) \tag{6.3.17}$$

Noting that from Eq. (6.3.14), we have $\mathbf{EJ} = \mathbf{AE}$, thus, based on the form of the particular solution, it is not difficult to obtain

$$\begin{aligned}
\frac{\mathrm{d}}{\mathrm{d}z}\boldsymbol{y}_\mathrm{p}(z) &= \mathbf{E}\frac{\mathrm{d}\mathbf{Q}(z)}{\mathrm{d}z}\boldsymbol{w}(z) + \mathbf{EQ}(z)\frac{\mathrm{d}\boldsymbol{w}(z)}{\mathrm{d}z} \\
&= \mathbf{EJQ}(z)\boldsymbol{w}(z) + \mathbf{EQ}(z)\frac{\mathrm{d}\boldsymbol{w}(z)}{\mathrm{d}z} \\
&= \mathbf{AEQ}(z)\boldsymbol{w}(z) + \mathbf{EQ}(z)\frac{\mathrm{d}\boldsymbol{w}(z)}{\mathrm{d}z} \\
&= \mathbf{A}\boldsymbol{y}_\mathrm{p}(z) + \mathbf{EQ}(z)\frac{\mathrm{d}\boldsymbol{w}(z)}{\mathrm{d}z}
\end{aligned}$$

Comparing with Eq. (6.3.17), we obtain

$$\frac{\mathrm{d}}{\mathrm{d}z}\boldsymbol{w}(z) = \mathbf{Q}^{-1}(z)\mathbf{E}^{-1}\boldsymbol{f}(z) \quad\Rightarrow\quad \boldsymbol{w}(z) = \int^z \mathbf{Q}^{-1}(\eta)\mathbf{E}^{-1}\boldsymbol{f}(\eta)\,\mathrm{d}\eta$$

Therefore, the particular solution $\boldsymbol{y}_\mathrm{p}(z)$ of the non-homogeneous equation is

$$\boldsymbol{y}_\mathrm{p}(z) = \mathbf{EQ}(z)\int^z \mathbf{Q}^{-1}(\eta)\mathbf{E}^{-1}\boldsymbol{f}(\eta)\,\mathrm{d}\eta \tag{6.3.18}$$

Substituting Eqs. (6.3.16) and (6.3.18) into Eq. (6.3.12) gives us the solution to the differential equation system (6.3.11)

$$\boldsymbol{y}(z) = \mathbf{EQ}(z)\left\{\boldsymbol{C} + \int^z \mathbf{Q}^{-1}(\eta)\mathbf{E}^{-1}\boldsymbol{f}(\eta)\,\mathrm{d}\eta\right\}. \tag{6.3.19}$$

6.4 Specific Form of General Solution to Ordinary Differential Equations

Equation (6.3.19) indicates that the displacement-traction vector, composed of expansion coefficients of displacement and traction, is formed by $\mathbf{E}$ and $\mathbf{Q}$ (which are related to the eigenvalues and eigenvectors of the coefficient matrix $\mathbf{A}$ of the system of equations) and their inverses, as well as the expansion coefficients of the body force term $\boldsymbol{f}$. The term $\boldsymbol{C}$ in braces represents undetermined constants, while the integral term is related to the source. To determine the specific expressions for the expansion coefficients of the displacement in the displacement-traction vector $\boldsymbol{y}$, and thereby obtain the displacement, the specific expressions for the terms in Eq. (6.3.19) are needed.

6.4.1 *Expressions of* E, Q, $\mathbf{E}^{-1}$ *and* $\mathbf{Q}^{-1}$

According to Eq. (6.3.15), **Q** is related to the eigenvalues of the coefficient matrix **A**, while **E** is a matrix formed directly by the eigenvectors of **A** as column vectors. Therefore, it is necessary to provide specific expressions for **E**, **Q**, $\mathbf{E}^{-1}$, and $\mathbf{Q}^{-1}$ for both the SH and P-SV cases.

6.4.1.1 SH Case

In SH case,

$$\mathbf{A} = \begin{bmatrix} 0 & \frac{1}{\mu} \\ \mu k^2 - \rho\omega^2 & 0 \end{bmatrix}$$

Let the eigenvalue be p. According to $|p\mathbf{I} - \mathbf{A}| = 0$, it is easy to obtain $p = \pm\nu$, where[14]

$$\boxed{\nu = \sqrt{k^2 - \frac{\omega^2}{\beta^2}}, \quad \mathrm{Re}(\nu) > 0} \tag{6.4.1}$$

Let $p_1 = -\nu$, with the corresponding eigenvector $[1 \quad -\mu\nu]^{\mathrm{T}}$, and $p_2 = \nu$, with the corresponding eigenvector $[1 \quad \mu\nu]^{\mathrm{T}}$. Therefore,[15]

$$\mathbf{E} = \begin{bmatrix} 1 & 1 \\ -\mu\nu & \mu\nu \end{bmatrix} \quad \Rightarrow \quad \mathbf{E}^{-1} = \frac{1}{2\mu\nu}\begin{bmatrix} \mu\nu & -1 \\ \mu\nu & 1 \end{bmatrix} \tag{6.4.2a}$$

$$\mathbf{Q} = \begin{bmatrix} Q_1 & 0 \\ 0 & Q_2 \end{bmatrix} = \begin{bmatrix} \mathrm{e}^{-\nu z} & 0 \\ 0 & \mathrm{e}^{\nu z} \end{bmatrix} \quad \Rightarrow \quad \mathbf{Q}^{-1} = \begin{bmatrix} \mathrm{e}^{\nu z} & 0 \\ 0 & \mathrm{e}^{-\nu z} \end{bmatrix} \tag{6.4.2b}$$

Substituting Eq. (6.4.2) into Eq. (6.3.19) and letting $\boldsymbol{C} = [C_1 \quad C_2]^{\mathrm{T}}$, we get

$$\begin{bmatrix} u_m^T(k,z) \\ \tau_m^T(k,z) \end{bmatrix} = \begin{bmatrix} E_{11} & E_{12} \\ E_{21} & E_{22} \end{bmatrix} \begin{bmatrix} Q_1(z) & 0 \\ 0 & Q_2(z) \end{bmatrix} \begin{bmatrix} C_1 + F_1(z) \\ C_2 + F_2(z) \end{bmatrix}$$

[14] Since the expression under the radical is not necessarily positive, and as mentioned later, complex frequencies are used in actual calculations, this involves the square root of complex numbers, leading to multivaluedness. To ensure analyticity, its values need to be specified. Following the common convention, the real part is generally restricted to be greater than zero. For purely imaginary numbers with a zero real part, such as $\nu = \sqrt{-a}$, where a is a positive real number, we take $\nu = \mathrm{i}\sqrt{a}$.

[15] The non-normalized form is used here for simplicity. Although a normalized form could also be used, it would be more complex in appearance and would not affect the final result.

that is,

$$u_m^T(k,z) = E_{11}Q_1(z)\big[C_1 + F_1(z)\big] + E_{12}Q_2(z)\big[C_2 + F_2(z)\big] \tag{6.4.3a}$$

$$\tau_m^T(k,z) = E_{21}Q_1(z)\big[C_1 + F_1(z)\big] + E_{22}Q_2(z)\big[C_2 + F_2(z)\big] \tag{6.4.3b}$$

where

$$F_1(z) = \int_{z_s^-}^{z} Q_2(\eta)\left[\mathbf{E}^{-1}\right]_{12} f_2(\eta)\,\mathrm{d}\eta, \quad F_2(z) = \int_{z_s^-}^{z} Q_1(\eta)\left[\mathbf{E}^{-1}\right]_{22} f_2(\eta)\,\mathrm{d}\eta \tag{6.4.4}$$

are terms related to the source, $f_2(z) = -f_m^T(z)$ (refer to Eq. (6.3.9)), $z_s^- = z_s - \varepsilon$ $(\varepsilon \to 0)$.

6.4.1.2 P-SV Case

In P-SV case,

$$\mathbf{A} = \begin{bmatrix} 0 & k & 1/\mu & 0 \\ -\lambda k\zeta & 0 & 0 & \zeta \\ 4\mu(\lambda+\mu)k^2\zeta - \rho\omega^2 & 0 & 0 & \lambda k\zeta \\ 0 & -\rho\omega^2 & -k & 0 \end{bmatrix} \quad \left(\zeta = \frac{1}{\lambda + 2\mu}\right)$$

Let the eigenvalue be p. Based on the equation $|p\mathbf{I} - \mathbf{A}| = 0$, we can rearrange it to obtain

$$\left(p^2 - \gamma^2\right)\left(p^2 - \nu^2\right) = 0^1$$

where[16]

$$\boxed{\gamma = \sqrt{k^2 - \frac{\omega^2}{\alpha^2}}, \quad \mathrm{Re}(\gamma) > 0} \tag{6.4.5}$$

---

[16] This is a quartic equation, which is generally difficult to solve. For conventional eigenvalue problems like this, we can use mathematical software capable of symbolic computation. For example, in Maple, you would input the following commands:

```
> with(LinearAlgebra);
> A:= Matrix([[0,k,1/mu,0], [-lambda*k/(lambda+2*mu),0,0,1/(lambda+2*mu)],
(continued) [4*mu*(lambda+mu)*k^2/(lambda+2*mu)-rho*omega^2,0,0,lambda*k/(lambda+2*mu)],
(continued) [0,-rho*omega^2,-k,0]]);
> Eigenvalues(A);
```

By slightly reorganizing the output, we obtain four eigenvalues.

Therefore, $p = \pm\gamma, \pm\nu$. Let $p_1 = -\gamma$, $p_2 = -\nu$, $p_3 = \gamma$, and $p_4 = \nu$. The corresponding eigenvectors can be found, and by taking them as column vectors, we can form the matrix $\mathbf{E}$

$$\mathbf{E} = \begin{bmatrix} \mathbf{E}_{11} & \mathbf{E}_{12} \\ \mathbf{E}_{21} & \mathbf{E}_{22} \end{bmatrix} \tag{6.4.6}$$

where

$$\mathbf{E}_{11} \triangleq \frac{1}{\omega}\begin{bmatrix} \alpha k & \beta\nu \\ \alpha\gamma & \beta k \end{bmatrix}, \qquad \mathbf{E}_{12} \triangleq \frac{1}{\omega}\begin{bmatrix} \alpha k & \beta\nu \\ -\alpha\gamma & -\beta k \end{bmatrix}$$

$$\mathbf{E}_{21} \triangleq \frac{1}{\omega}\begin{bmatrix} -2\alpha\mu k\gamma & -\beta\mu\chi \\ -\alpha\mu\chi & -2\beta\mu k\nu \end{bmatrix}, \qquad \mathbf{E}_{22} \triangleq \frac{1}{\omega}\begin{bmatrix} 2\alpha\mu k\gamma & \beta\mu\chi \\ -\alpha\mu\chi & -2\beta\mu k\nu \end{bmatrix} \qquad (\chi = k^2 + \nu^2)$$

Similar to the SH case, the eigenvectors here are not normalized.[17] Each column vector can be multiplied by any non-zero constant. The specific form of the $\mathbf{E}$ matrix will directly determine the form of $\mathbf{E}^{-1}$, but it won't affect the final result. The current form is chosen to keep $\mathbf{E}^{-1}$ as simple as possible. It is worth noting that in the current expression of the matrix $\mathbf{E}$, ω appears in the denominator of the coefficients, so $\omega \neq 0$ is required. According to the definitions of ν and γ in Eqs. (6.4.1) and (6.4.5), when $\omega = 0$, $\gamma = \nu$, leading to eigenvalue degeneracy. As mentioned in Sect. 6.3.2.1, for the general solution of the homogeneous equation, $\mathbf{J}$ is diagonal only if the eigenvalues of the coefficient matrix $\mathbf{A}$ are distinct. Therefore, the current solution does not apply to the case where $\omega = 0$. In other words, the current solution does not include static components, which will be specifically discussed in Sect. 6.7.

Based on the selection of p_i $(i = 1, 2, \ldots, 4)$, $\mathbf{Q}$ is

$$\mathbf{Q} = \begin{bmatrix} \mathbf{Q}_1 & \mathbf{0} \\ \mathbf{0} & \mathbf{Q}_2 \end{bmatrix}, \quad \mathbf{Q}_1 \triangleq \begin{bmatrix} e^{-\gamma z} & 0 \\ 0 & e^{-\nu z} \end{bmatrix}, \quad \mathbf{Q}_2 \triangleq \begin{bmatrix} e^{\gamma z} & 0 \\ 0 & e^{\nu z} \end{bmatrix} \tag{6.4.7}$$

Noting that $\mathbf{Q}_1$ and $\mathbf{Q}_2$ are inverse matrices of each other, therefore,

$$\mathbf{Q}^{-1} = \begin{bmatrix} \mathbf{Q}_1^{-1} & \mathbf{0} \\ \mathbf{0} & \mathbf{Q}_2^{-1} \end{bmatrix} = \begin{bmatrix} \mathbf{Q}_2 & \mathbf{0} \\ \mathbf{0} & \mathbf{Q}_1 \end{bmatrix} \tag{6.4.8}$$

[17] Finding eigenvectors for a fourth-order matrix is significantly more complex than for a second-order matrix in the SH case. We can use Maple to solve this problem. In Maple, the command to find the eigenvectors of matrix $\mathbf{A}$ is `Eigenvectors(A)`.

From the expression of $\mathbf{E}$ in Eq. (6.4.6), its inverse matrix can be determined as[18]

$$\mathbf{E}^{-1} = \frac{\beta}{2\alpha\mu\nu\gamma\omega}\begin{bmatrix} 2\beta k\mu\nu\gamma & -\beta\mu\nu\chi & -\beta k\nu & \beta\nu\gamma \\ -\alpha\mu\gamma\chi & 2\alpha k\mu\nu\chi & \alpha\gamma\nu & -\alpha k\gamma \\ 2\beta k\mu\nu\gamma & \beta\mu\nu\chi & \beta k\nu & \beta\nu\gamma \\ -\alpha\mu\gamma\chi & -2\alpha k\mu\nu\chi & -\alpha\gamma\nu & -\alpha k\gamma \end{bmatrix} = \begin{bmatrix} (\mathbf{E}^{-1})_{11} & (\mathbf{E}^{-1})_{12} \\ (\mathbf{E}^{-1})_{21} & (\mathbf{E}^{-1})_{22} \end{bmatrix} \tag{6.4.9}$$

in which,

$$(\mathbf{E}^{-1})_{11} = \frac{\beta}{2\alpha\mu\nu\gamma\omega}\begin{bmatrix} 2\beta k\mu\nu\gamma & -\beta\mu\nu\chi \\ -\alpha\mu\gamma\chi & 2\alpha k\mu\nu\chi \end{bmatrix},\quad (\mathbf{E}^{-1})_{12} = \frac{\beta}{2\alpha\mu\nu\gamma\omega}\begin{bmatrix} -\beta k\nu & \beta\nu\gamma \\ \alpha\gamma\nu & -\alpha k\gamma \end{bmatrix}$$

$$(\mathbf{E}^{-1})_{21} = \frac{\beta}{2\alpha\mu\nu\gamma\omega}\begin{bmatrix} 2\beta k\mu\nu\gamma & \beta\mu\nu\chi \\ -\alpha\mu\gamma\chi & -2\alpha k\mu\nu\chi \end{bmatrix},\quad (\mathbf{E}^{-1})_{22} = \frac{\beta}{2\alpha\mu\nu\gamma\omega}\begin{bmatrix} \beta k\nu & \beta\nu\gamma \\ -\alpha\gamma\nu & -\alpha k\gamma \end{bmatrix}$$

Substitute Eqs. (6.4.6) to (6.4.9) into Eq. (6.3.19), and let

$$\boldsymbol{y}_1 \triangleq \begin{bmatrix} u_m^S \\ u_m^R \end{bmatrix},\quad \boldsymbol{y}_2 \triangleq \begin{bmatrix} \tau_m^S \\ \tau_m^R \end{bmatrix},\quad \boldsymbol{f}_2 \triangleq -\begin{bmatrix} f_m^S \\ f_m^R \end{bmatrix},\quad \boldsymbol{C} \triangleq \begin{bmatrix} \boldsymbol{C}_1 \\ \boldsymbol{C}_2 \end{bmatrix},\quad \boldsymbol{C}_1 \triangleq \begin{bmatrix} C_1 \\ C_2 \end{bmatrix},\quad \boldsymbol{C}_2 \triangleq \begin{bmatrix} C_3 \\ C_4 \end{bmatrix}$$

then for the P-SV case, we have

$$\begin{bmatrix} \boldsymbol{y}_1(k,z) \\ \boldsymbol{y}_2(k,z) \end{bmatrix} = \begin{bmatrix} \mathbf{E}_{11} & \mathbf{E}_{12} \\ \mathbf{E}_{21} & \mathbf{E}_{22} \end{bmatrix}\begin{bmatrix} \mathbf{Q}_1(z) & 0 \\ 0 & \mathbf{Q}_2(z) \end{bmatrix}\begin{bmatrix} \boldsymbol{C}_1 + \boldsymbol{F}_1(z) \\ \boldsymbol{C}_2 + \boldsymbol{F}_2(z) \end{bmatrix}$$

that is,

$$\boldsymbol{y}_1(k,z) = \mathbf{E}_{11}\mathbf{Q}_1(z)\left[\boldsymbol{C}_1 + \boldsymbol{F}_1(z)\right] + \mathbf{E}_{12}\mathbf{Q}_2(z)\left[\boldsymbol{C}_2 + \boldsymbol{F}_2(z)\right] \tag{6.4.10a}$$

$$\boldsymbol{y}_2(k,z) = \mathbf{E}_{21}\mathbf{Q}_1(z)\left[\boldsymbol{C}_1 + \boldsymbol{F}_1(z)\right] + \mathbf{E}_{22}\mathbf{Q}_2(z)\left[\boldsymbol{C}_2 + \boldsymbol{F}_2(z)\right] \tag{6.4.10b}$$

where

$$\boldsymbol{F}_1(z) = \int_{z_s^-}^{z} \mathbf{Q}_2(\eta)\left[\mathbf{E}^{-1}\right]_{12}\boldsymbol{f}_2(\eta)\,\mathrm{d}\eta,\quad \boldsymbol{F}_2(z) = \int_{z_s^-}^{z} \mathbf{Q}_1(\eta)\left[\mathbf{E}^{-1}\right]_{22}\boldsymbol{f}_2(\eta)\,\mathrm{d}\eta \tag{6.4.11}$$

---

[18] Using Maple to solve, input the following commands in sequence:

```
> with(LinearAlgebra);
> E:=Matrix([[alpha*k,beta*nu,alpha*k,beta*nu], [alpha*lambda,beta*k,-alpha*lambda,-beta*k],
(continued) [-2*alpha*mu*k*gamma,-beta*mu*chi,2*alpha*mu*k*gamma,beta*mu*chi],
(continued) [-alpha*mu*chi,-2*beta*mu*k*nu,-alpha*mu*chi,-2*beta*mu*k*nu]])/omega;
> MatrixInverse(E);
```

Slightly rearrangement of the expression of output yields the result of Eq. (6.4.9).

6.4.2 Expressions of $F_\zeta(z)$ and $\boldsymbol{F}_\zeta(z)$

In Eqs. (6.4.3) for the SH case and (6.4.10) for the P-SV case, the source terms $F_\zeta(z)$ and $\boldsymbol{F}_\zeta(z)$ are included, respectively. Before further solving for the coefficients C_1, C_2, $\boldsymbol{C}_1$, and $\boldsymbol{C}_2$, their specific expressions must be obtained.

Consider a concentrated impulsive force $\boldsymbol{f}(\boldsymbol{x}, t) = \delta_{ij}\delta(\boldsymbol{x} - \boldsymbol{\xi})\delta(t)\hat{\boldsymbol{e}}_i$. Noting that the spectrum of the δ function is a constant 1, the frequency domain representation of the body force is given by $\boldsymbol{f}(\boldsymbol{x}, \omega) = \delta_{ij}\delta(\boldsymbol{x} - \boldsymbol{\xi})\hat{\boldsymbol{e}}_i$. Since $\delta(\boldsymbol{x} - \boldsymbol{\xi}) = \dfrac{1}{2\pi r}\delta(r)\delta(z - z_s)$,[19] therefore

$$\boldsymbol{f}(\boldsymbol{x}, \omega) = \frac{1}{2\pi r}\delta(r)\delta(z - z_s)\delta_{ij}\hat{\boldsymbol{e}}_i \tag{6.4.12}$$

Given the specific body force expression (6.4.12), we can determine its coefficients in terms of the basis functions using Eq. (6.2.17). Consequently, we can use Eqs. (6.4.4) and (6.4.11) to obtain $F_\zeta(z)$ and $\boldsymbol{F}_\zeta(z)$.

Noting the relationship between cylindrical coordinate basis vectors and Cartesian coordinate basis vectors:

$$\hat{\boldsymbol{e}}_r = \hat{\boldsymbol{e}}_x \cos\theta + \hat{\boldsymbol{e}}_y \sin\theta, \quad \hat{\boldsymbol{e}}'_\theta = -\hat{\boldsymbol{e}}_x \sin\theta + \hat{\boldsymbol{e}}_y \cos\theta \tag{6.4.13}$$

and the properties of Bessel functions:

$$J'_1(x) + \frac{1}{x}J_1(x) = J_0(x) \tag{6.4.14a}$$

$$J'_{-1}(x) + \frac{1}{x}J_{-1}(x) = -J_0(x) \tag{6.4.14b}$$

$$\int_0^\infty \delta(r)J_0(kr)\,\mathrm{d}r = J_0(0) = 1 \tag{6.4.14c}$$

[19] $\delta(\boldsymbol{x} - \boldsymbol{\xi}) = \delta(\boldsymbol{r} - \boldsymbol{r}_s)\delta(z - z_s)$, where $\boldsymbol{r}$ and $\boldsymbol{r}_s$ represent coordinates in the horizontal plane. Noting that the source is located on the z axis, it exhibits symmetry in the θ direction. Therefore,

$$\iiint_V \delta(\boldsymbol{x} - \boldsymbol{\xi})\,\mathrm{d}V(\boldsymbol{\xi}) = \int_0^\infty\int_0^{2\pi}\int_0^\infty \delta(\boldsymbol{r} - \boldsymbol{r}_s)\delta(z - z_s)r\,\mathrm{d}r\,\mathrm{d}\theta\,\mathrm{d}z = 2\pi\int_0^\infty \frac{\delta(r)}{2\pi r}r\,\mathrm{d}r = 1$$

Thus, $\delta(\boldsymbol{x} - \boldsymbol{\xi}) = \dfrac{1}{2\pi r}\delta(r)\delta(z - z_s)$.

Substituting Eq. (6.4.12) into Eq. (6.2.17), we have:

$$\begin{aligned}
f_m^T &\xlongequal{a=\delta(z-z_s)} -\int_0^{2\pi}\int_0^{\infty}\frac{a\delta(r)}{2\pi r}\left(\delta_{1j}\hat{\boldsymbol{e}}_x+\delta_{2j}\hat{\boldsymbol{e}}_y+\delta_{3j}\hat{\boldsymbol{e}}_z\right) \\
&\quad\cdot\left[\frac{\mathrm{i}m}{kr}J_m(kr)\hat{\boldsymbol{e}}_r+J_m'(kr)\hat{\boldsymbol{e}}_\theta\right]\mathrm{e}^{-\mathrm{i}m\theta}r\,\mathrm{d}r\,\mathrm{d}\theta \\
&\xlongequal{(6.4.13)} -\frac{a}{2\pi}\int_0^{2\pi}\int_0^{\infty}\delta(r)\left\{\delta_{1j}\left[\frac{\mathrm{i}m}{kr}J_m(kr)\frac{\mathrm{e}^{\mathrm{i}\theta}+\mathrm{e}^{-\mathrm{i}\theta}}{2}-J_m'(kr)\frac{\mathrm{e}^{\mathrm{i}\theta}-\mathrm{e}^{-\mathrm{i}\theta}}{2\mathrm{i}}\right]\right. \\
&\quad\left.+\delta_{2j}\left[\frac{\mathrm{i}m}{kr}J_m(kr)\frac{\mathrm{e}^{\mathrm{i}\theta}-\mathrm{e}^{-\mathrm{i}\theta}}{2\mathrm{i}}+J_m'(kr)\frac{\mathrm{e}^{\mathrm{i}\theta}+\mathrm{e}^{-\mathrm{i}\theta}}{2}\right]\right\}\mathrm{e}^{-\mathrm{i}m\theta}\,\mathrm{d}r\,\mathrm{d}\theta \\
&= -\frac{a}{4\pi}\int_0^{2\pi}\int_0^{\infty}\delta(r)\left\{\left[(\delta_{1j}-\mathrm{i}\delta_{2j})\frac{\mathrm{i}m}{kr}J_m(kr)+(\delta_{2j}+\mathrm{i}\delta_{1j})J_m'(kr)\right]\right. \\
&\quad\left.\cdot\mathrm{e}^{-\mathrm{i}(m-1)\theta}+\left[(\delta_{1j}+\mathrm{i}\delta_{2j})\frac{\mathrm{i}m}{kr}J_m(kr)+(\delta_{2j}-\mathrm{i}\delta_{1j})J_m'(kr)\right]\mathrm{e}^{-\mathrm{i}(m+1)\theta}\right\}\mathrm{d}r\,\mathrm{d}\theta \\
&= -\frac{a}{2}\left\{\delta_{m,1}\int_0^{\infty}\delta(r)\left[(\delta_{1j}-\mathrm{i}\delta_{2j})\frac{\mathrm{i}}{kr}J_1(kr)+(\delta_{2j}+\mathrm{i}\delta_{1j})J_1'(kr)\right]\mathrm{d}r\right. \\
&\quad\left.+\delta_{m,-1}\int_0^{\infty}\delta(r)\left[-(\delta_{1j}+\mathrm{i}\delta_{2j})\frac{\mathrm{i}}{kr}J_{-1}(kr)+(\delta_{2j}-\mathrm{i}\delta_{1j})J_{-1}'(kr)\right]\mathrm{d}r\right\} \\
&\xlongequal{(6.4.14\mathrm{a}),(6.4.14\mathrm{b})} -\frac{a}{2}\left[(\mathrm{i}\delta_{1j}+\delta_{2j})\delta_{m,1}+(\mathrm{i}\delta_{1j}-\delta_{2j})\delta_{m,-1}\right]\int_0^{\infty}\delta(r)J_0(kr)\,\mathrm{d}r \\
&\xlongequal{(6.4.14\mathrm{c})} \mathscr{F}_m^T\delta(z-z_s)
\end{aligned}$$

where

$$\mathscr{F}_m^T \triangleq -\frac{1}{2}\left[(\mathrm{i}\delta_{1j}+\delta_{2j})\delta_{m,1}+(\mathrm{i}\delta_{1j}-\delta_{2j})\delta_{m,-1}\right] \tag{6.4.15}$$

Similarly, we can obtain

$$f_m^S=\mathscr{F}_m^S\delta(z-z_s),\quad f_m^R=\mathscr{F}_m^R\delta(z-z_s)$$

where

$$\mathscr{F}_m^S \triangleq \frac{1}{2}\left[(\delta_{1j}-\mathrm{i}\delta_{2j})\delta_{m,1}-(\delta_{1j}+\mathrm{i}\delta_{2j})\delta_{m,-1}\right],\quad \mathscr{F}_m^R\triangleq-\delta_{3j}\delta_{m,0} \tag{6.4.16}$$

Substituting Eq. (6.4.15) along with Eq. (6.4.2) into Eq. (6.4.4), we get

$$F_1(z)=\int_{z_s^-}^{z}\mathrm{e}^{\nu\eta}\frac{1}{2\mu\nu}\mathscr{F}_m^T\delta(\eta-z_s)\,\mathrm{d}\eta=H(z-z_s)s_1 \tag{6.4.17a}$$

$$F_2(z)=-\int_{z_s^-}^{z}\mathrm{e}^{-\nu\eta}\frac{1}{2\mu\nu}\mathscr{F}_m^T\delta(\eta-z_s)\,\mathrm{d}\eta=H(z-z_s)s_2 \tag{6.4.17b}$$

where $H(\cdot)$ is step function, and

$$s_1 \triangleq \frac{e^{\nu z_s}}{2\mu\nu}\mathscr{F}_m^T, \quad s_2 \triangleq -\frac{e^{-\nu z_s}}{2\mu\nu}\mathscr{F}_m^T$$

Similarly, substituting Eq. (6.4.16) along with Eqs. (6.4.7) through (6.4.9) into Eq. (6.4.11), we obtain

$$\begin{aligned} \boldsymbol{F}_1(z) &= -\frac{\beta}{2\alpha\mu\nu\gamma\omega}\int_{z_s^-}^{z}\begin{bmatrix} e^{\gamma\eta} & 0 \\ 0 & e^{\nu\eta}\end{bmatrix}\begin{bmatrix} -\beta k\nu & \beta\nu\gamma \\ \alpha\nu\gamma & -\alpha k\gamma\end{bmatrix}\begin{bmatrix}\mathscr{F}_m^S \\ \mathscr{F}_m^R\end{bmatrix}\delta(\eta - z_s)\,d\eta \\ &= H(z - z_s)\boldsymbol{s}_1 \end{aligned} \tag{6.4.18a}$$

$$\begin{aligned} \boldsymbol{F}_2(z) &= -\frac{\beta}{2\alpha\mu\nu\gamma\omega}\int_{z_s^-}^{z}\begin{bmatrix} e^{-\gamma\eta} & 0 \\ 0 & e^{-\nu\eta}\end{bmatrix}\begin{bmatrix} \beta k\nu & \beta\nu\gamma \\ -\alpha\nu\gamma & -\alpha k\gamma\end{bmatrix}\begin{bmatrix}\mathscr{F}_m^S \\ \mathscr{F}_m^R\end{bmatrix}\delta(\eta - z_s)\,d\eta \\ &= H(z - z_s)\boldsymbol{s}_2 \end{aligned} \tag{6.4.18b}$$

where

$$\boldsymbol{s}_1 = \frac{\beta}{2\alpha\mu\nu\gamma\omega}\begin{bmatrix} \beta k\nu e^{\gamma z_s}\mathscr{F}_m^S - \beta\nu\gamma e^{\gamma z_s}\mathscr{F}_m^R \\ -\alpha\nu\gamma e^{\nu z_s}\mathscr{F}_m^S + \alpha k\gamma e^{\nu z_s}\mathscr{F}_m^R \end{bmatrix}$$

$$\boldsymbol{s}_2 = \frac{\beta}{2\alpha\mu\nu\gamma\omega}\begin{bmatrix} -\beta k\nu e^{-\gamma z_s}\mathscr{F}_m^S - \beta\nu\gamma e^{-\gamma z_s}\mathscr{F}_m^R \\ \alpha\nu\gamma e^{-\nu z_s}\mathscr{F}_m^S + \alpha k\gamma e^{-\nu z_s}\mathscr{F}_m^R \end{bmatrix}$$

6.4.3 Expressions of C_ζ and $\boldsymbol{C}_\zeta$

Once the specific expressions for $\mathbf{E}$, $\mathbf{Q}$, $F_\zeta(z)$, and $\boldsymbol{F}_\zeta(z)$ are obtained, the only remaining undetermined coefficients in Eq. (6.3.19) are C_ζ and $\boldsymbol{C}_\zeta$, which need to be determined based on the boundary conditions.

Substitute the displacements and tractions into the boundary conditions (6.1.1) using the expansions given in Eqs. (6.3.1) and (6.3.7). By leveraging the orthogonality of the basis functions, it is straightforward to derive the boundary conditions that each expansion coefficient must satisfy:

$$u_m^T(k, z)\big|_{z\to\infty} = 0, \quad \boldsymbol{y}_1(k, z)\big|_{z\to\infty} = \boldsymbol{0} \tag{6.4.19a}$$

$$\tau_m^T(k, z)\big|_{z=0} = 0, \quad \boldsymbol{y}_2(k, z)\big|_{z=0} = \boldsymbol{0} \tag{6.4.19b}$$

Substitute Eqs. (6.4.2) and (6.4.17) into Eq. (6.4.3) for the SH case, and Eqs. (6.4.6), (6.4.7), and (6.4.18) into Eq. (6.4.10) for the P-SV case. Based on Eq. (6.4.19a), we can obtain

$$u_m^T(k, z)\big|_{z\to\infty} = 0\cdot(C_1 + s_1) + \infty\cdot(C_2 + s_2) = 0 \Rightarrow C_2 = -s_2 \tag{6.4.20a}$$

$$\boldsymbol{y}_1(k, z)\big|_{z\to\infty} = \boldsymbol{0}\cdot(\boldsymbol{C}_1 + \boldsymbol{s}_1) + \infty\cdot(\boldsymbol{C}_2 + \boldsymbol{s}_2) = \boldsymbol{0} \Rightarrow \boldsymbol{C}_2 = -\boldsymbol{s}_2 \tag{6.4.20b}$$

According to Eq. (6.4.19b), we obtain[20]

$$\tau_m^T(k,z)\big|_{z=0} = -\mu\nu C_1 + \mu\nu C_2 = 0 \Rightarrow C_1 = -s_2 \tag{6.4.21a}$$

$$\mathbf{y}_2(k,z)\big|_{z=0} = \mathbf{E}_{21}\mathbf{C}_1 + \mathbf{E}_{22}\mathbf{C}_2 = \mathbf{0} \Rightarrow \mathbf{C}_1 = \mathbf{E}_{21}^{-1}\mathbf{E}_{22}\mathbf{s}_2 \tag{6.4.21b}$$

Once we determine the undetermined coefficient $\boldsymbol{C}$, all the terms on the right-hand side of Eq. (6.3.19) are obtained. After a bit of simplification, the expansion coefficients of the displacement can be expressed as

$$u_m^T(k,z) = \frac{1}{2\mu\nu} A^T(k,z)\mathscr{F}_m^T \tag{6.4.22a}$$

$$u_m^S(k,z) = \frac{\beta^2}{2\mu\gamma\omega^2}\left[A^{SS}(k,z)\mathscr{F}_m^S - A^{SR}(k,z)\mathscr{F}_m^R\right] \tag{6.4.22b}$$

$$u_m^R(k,z) = \frac{\beta^2}{2\mu\nu\omega^2}\left[A^{RS}(k,z)\mathscr{F}_m^S - A^{RR}(k,z)\mathscr{F}_m^R\right] \tag{6.4.22c}$$

where

$$A^{\mathrm{T}}(k,z) = \mathrm{e}^{\mathrm{S}} + \mathrm{e}^{\mathrm{SS}} \tag{6.4.23a}$$

$$A^{\mathrm{SS}}(k,z) = \left(k^2\mathrm{e}^{\mathrm{P}} - \nu\gamma\mathrm{e}^{\mathrm{S}}\right) - \frac{R^*(k)}{R(k)}\left(k^2\mathrm{e}^{\mathrm{PP}} + \nu\gamma\mathrm{e}^{\mathrm{SS}}\right) + \frac{4k^2\nu\gamma\chi}{R(k)}\left(\mathrm{e}^{\mathrm{SP}} + \mathrm{e}^{\mathrm{PS}}\right) \tag{6.4.23b}$$

$$A^{\mathrm{SR}}(k,z) = k\gamma\left[\zeta(z)\left(\mathrm{e}^{\mathrm{P}} - \mathrm{e}^{\mathrm{S}}\right) + \frac{R^*(k)}{R(k)}\left(\mathrm{e}^{\mathrm{PP}} + \mathrm{e}^{\mathrm{SS}}\right) - \frac{4\chi}{R(k)}\left(k^2\mathrm{e}^{\mathrm{SP}} + \nu\gamma\mathrm{e}^{\mathrm{PS}}\right)\right] \tag{6.4.23c}$$

$$A^{\mathrm{RS}}(k,z) = k\nu\left[\zeta(z)\left(\mathrm{e}^{\mathrm{P}} - \mathrm{e}^{\mathrm{S}}\right) - \frac{R^*(k)}{R(k)}\left(\mathrm{e}^{\mathrm{PP}} + \mathrm{e}^{\mathrm{SS}}\right) + \frac{4\chi}{R(k)}\left(\nu\gamma\mathrm{e}^{\mathrm{SP}} + k^2\mathrm{e}^{\mathrm{PS}}\right)\right] \tag{6.4.23d}$$

$$A^{\mathrm{RR}}(k,z) = \left(\nu\gamma\mathrm{e}^{\mathrm{P}} - k^2\mathrm{e}^{\mathrm{S}}\right) + \frac{R^*(k)}{R(k)}\left(\nu\gamma\mathrm{e}^{\mathrm{PP}} + k^2\mathrm{e}^{\mathrm{SS}}\right) - \frac{4k^2\nu\gamma\chi}{R(k)}\left(\mathrm{e}^{\mathrm{SP}} + \mathrm{e}^{\mathrm{PS}}\right) \tag{6.4.23e}$$

$$\mathrm{e}^{\mathrm{P}} = \mathrm{e}^{-\gamma|z-z_s|},\quad \mathrm{e}^{\mathrm{S}} = \mathrm{e}^{-\nu|z-z_s|},\quad \mathrm{e}^{\mathrm{PP}} = \mathrm{e}^{-\gamma(z+z_s)}$$
$$\mathrm{e}^{\mathrm{SS}} = \mathrm{e}^{-\nu(z+z_s)},\quad \mathrm{e}^{\mathrm{PS}} = \mathrm{e}^{-(\gamma z_s+\nu z)},\quad \mathrm{e}^{\mathrm{SP}} = \mathrm{e}^{-(\gamma z+\nu z_s)}$$
$$R(k) = \chi^2 - 4k^2\nu\gamma,\ R^*(k) = \chi^2 + 4k^2\nu\gamma,\ \zeta(z) = \mathrm{sgn}(z - z_s)$$

[20] Here, we use the conclusion that when $z = z_s = 0$, $H(z - z_s) = 0$. Mathematically, $H(0) = 1/2$, but we need to consider the physical context of the problem. The reasoning is mainly based on two points:

(1) In Eqs. (6.4.17) and (6.4.18), the lower limit of the integral z_s^- is defined for the general case of $z_s > 0$. For the special case of $z_s = 0$, since $z_s^- < 0$ is meaningless, the lower limit of the integral must be taken as 0. With both the upper and lower limits of the integral being 0, the result of the integral is 0.
(2) The value of $H(0)$ also needs to account for the continuity of displacement. This means that the result for $z = z_s = 0$ should match the limit case when z approaches 0 and z_s tends to 0. Not taking $H(0) = 1/2$ for $z = z_s = 0$ would lead to a failure in meeting this condition.

The sgn(·) denotes the sign function. Note that A^{T}, A^{SS}, A^{SR}, A^{RS}, and A^{RR} can all be expressed as sums of six terms, each characterized by exponential factors. The terms containing e^{P}, e^{S}, e^{PP}, e^{SS}, e^{PS}, and e^{SP} respectively represent direct P-wave (P), direct S-wave (S), reflected P-wave (PP), reflected S-wave (SS), converted P-wave (PS), and converted S-wave (SP). The first two terms relate to the wave components associated with the properties of the medium itself, which corresponds to the infinite medium results discussed in Chap. 4. The latter four terms are related to the wave components associated with the free surface. Notably, since $R(k)$ appears in the denominator, it can approach zero under certain conditions, leading to a wave type closely related to the free surface: Rayleigh wave. In Sect. 6.5.3, we will examine issues related to Rayleigh wave in the frequency domain.

6.5 Frequency-Domain Solution of Lamb's Problem

With the specific expressions of each term in the general solution of the differential equations system (6.3.10), we can derive the final explicit form of the frequency domain Green's function for the Lamb's problem. In the final solution, it is clear that the components of the direct waves and those generated by the free surface are separated. We will rigorously prove that the former is equivalent to the Green's function for an infinite medium obtained in Chap. 4. This means that the Green's function for the Lamb's problem can be explicitly expressed as the sum of the Green's function for an infinite space and the contribution from the free surface, which is significant for analyzing certain seismological problems. Additionally, in the integral expression of the solution, there are terms from Eq. (6.4.23), where the denominator $R(k)$ may be zero in some cases. We will conduct a detailed analysis of this issue to derive the excitation formula for Rayleigh wave.

6.5.1 *Expressions for Frequency-Domain Green's Function*

Equation (6.4.22) provides the explicit expression for the expansion coefficients of the displacement components. Substituting these into Eq. (6.3.1), and given that the basis functions are represented in cylindrical coordinates (see Eq. (6.2.14)), we directly obtain the displacement components $u_r(r, \theta, z, \omega)$, $u_\theta(r, \theta, z, \omega)$, and $u_z(r, \theta, z, \omega)$ in cylindrical coordinates:

$$\begin{aligned}\tilde{u}_r =&\frac{1}{2\pi}\sum_{m=-\infty}^{+\infty}\mathrm{e}^{\mathrm{i}m\theta}\int_0^\infty\left[u_m^S(k,z)J_m'(kr)+\frac{\mathrm{i}m}{kr}u_m^T(k,z)J_m(kr)\right]k\,\mathrm{d}k\\ =&\frac{1}{4\pi\mu}\int_0^\infty\left\{\frac{\beta^2k}{\gamma\omega^2}\left[A^{\mathrm{SS}}P_1(k,\theta)-A^{\mathrm{SR}}P_2(k,\theta)\right]+\frac{A^{\mathrm{T}}}{\nu r}P_3(k,\theta)\right\}\mathrm{d}k\end{aligned}$$

$$
\begin{aligned}
\tilde{u}_\theta =& \frac{1}{2\pi}\sum_{m=-\infty}^{+\infty} \mathrm{e}^{im\theta}\int_0^\infty \left[\frac{im}{kr}u_m^S(k,z)J_m(kr) - u_m^T(k,z)J_m'(kr)\right]k\,\mathrm{d}k \\
=& \frac{1}{4\pi\mu}\int_0^\infty \left\{\frac{\beta^2}{\gamma\omega^2 r}\left[A^{\mathrm{SS}}P_4(k,\theta) - A^{\mathrm{SR}}P_5(k,\theta)\right] - \frac{kA^{\mathrm{T}}}{\nu}P_6(k,\theta)\right\}\mathrm{d}k \\
\tilde{u}_z =& -\frac{1}{2\pi}\sum_{m=-\infty}^{+\infty} \mathrm{e}^{im\theta}\int_0^\infty u_m^R(k,z)J_m(kr)k\,\mathrm{d}k \\
=& -\frac{1}{4\pi\mu}\int_0^\infty \frac{\beta^2 k}{\nu\omega^2}\left[A^{\mathrm{RS}}P_7(k,\theta) - A^{\mathrm{RR}}P_8(k,\theta)\right]\mathrm{d}k
\end{aligned}
$$

where[21]

$$
\begin{aligned}
P_1(k,\theta) &= \sum_{m=-\infty}^{+\infty} \mathrm{e}^{im\theta}\mathscr{F}_m^S J_m'(kr) \xlongequal{(6.4.16),\ J_{-1}=-J_1} (\delta_{1j}\cos\theta + \delta_{2j}\sin\theta)J_1'(kr) \\
P_2(k,\theta) &= \sum_{m=-\infty}^{+\infty} \mathrm{e}^{im\theta}\mathscr{F}_m^R J_m'(kr) \xlongequal{(6.4.16)} -\delta_{3j}J_0'(kr) \\
P_3(k,\theta) &= \sum_{m=-\infty}^{+\infty} im\mathrm{e}^{im\theta}\mathscr{F}_m^T J_m(kr) \xlongequal{(6.4.15),\ J_{-1}=-J_1} (\delta_{1j}\cos\theta + \delta_{2j}\sin\theta)J_1(kr) \\
P_4(k,\theta) &= \sum_{m=-\infty}^{+\infty} im\mathrm{e}^{im\theta}\mathscr{F}_m^S J_m(kr) = (\delta_{2j}\cos\theta - \delta_{1j}\sin\theta)J_1(kr) \\
P_5(k,\theta) &= \sum_{m=-\infty}^{+\infty} im\mathrm{e}^{im\theta}\mathscr{F}_m^R J_m(kr) = 0 \\
P_6(k,\theta) &= \sum_{m=-\infty}^{+\infty} \mathrm{e}^{im\theta}\mathscr{F}_m^T J_m'(kr) = (\delta_{1j}\sin\theta - \delta_{2j}\cos\theta)J_1(kr) \\
P_7(k,\theta) &= \sum_{m=-\infty}^{+\infty} \mathrm{e}^{im\theta}\mathscr{F}_m^S J_m(kr) = (\delta_{1j}\cos\theta + \delta_{2j}\sin\theta)J_1(kr) \\
P_8(k,\theta) &= \sum_{m=-\infty}^{+\infty} \mathrm{e}^{im\theta}\mathscr{F}_m^R J_m(kr) = -\delta_{3j}J_0(kr)
\end{aligned}
$$

[21] We can see that the summation over m can be explicitly expressed. This is because, under the basis functions currently chosen, the expansion coefficients for the body forces are non-zero only when $m = 0, \pm 1$. This is also one of the important reasons why we have chosen the current basis functions.

Using the transformation relationships between the displacement components in cylindrical and Cartesian coordinate systems:

$$\tilde{u}_1 = \tilde{u}_r \cos\theta - \tilde{u}_\theta \sin\theta, \quad \tilde{u}_2 = \tilde{u}_r \sin\theta + \tilde{u}_\theta \cos\theta, \quad \tilde{u}_3 = \tilde{u}_z$$

we can obtain the three displacement components $\tilde{u}_i$ in the Cartesian coordinate system. Since the expression for body force (6.4.12) includes j, which is the direction of the body force, if we consider j as a variable free index, then the resulting displacement components in the Cartesian coordinate system are the Green's function:

$$\begin{aligned}
G_{11}(\boldsymbol{x}, \boldsymbol{x}', \omega) =& \frac{1}{4\pi\mu}\left\{\cos^2\theta\left[\int_0^\infty \frac{\beta^2 k}{\gamma\omega^2} A^{\mathrm{SS}} J_1'(kr)\,\mathrm{d}k + \int_0^\infty \frac{1}{\nu r} A^{\mathrm{T}} J_1(kr)\,\mathrm{d}k\right]\right.\\
&\left. + \sin^2\theta\left[\int_0^\infty \frac{\beta^2}{\gamma\omega^2 r} A^{\mathrm{SS}} J_1(kr)\,\mathrm{d}k + \int_0^\infty \frac{k}{\nu} A^{\mathrm{T}} J_1'(kr)\,\mathrm{d}k\right]\right\}\\
G_{12}(\boldsymbol{x}, \boldsymbol{x}', \omega) =& G_{21}(\boldsymbol{x}, \boldsymbol{x}', \omega) = \frac{1}{4\pi\mu} \sin\theta\cos\theta\left[\int_0^\infty \frac{\beta^2 k}{\gamma\omega^2} A^{\mathrm{SS}} J_1'(kr)\,\mathrm{d}k\right.\\
&\left. + \int_0^\infty \frac{1}{\nu r} A^{\mathrm{T}} J_1(kr)\,\mathrm{d}k - \int_0^\infty \frac{\beta^2}{\gamma\omega^2 r} A^{\mathrm{SS}} J_1(kr)\,\mathrm{d}k - \int_0^\infty \frac{k}{\nu} A^{\mathrm{T}} J_1'(kr)\,\mathrm{d}k\right]\\
G_{13}(\boldsymbol{x}, \boldsymbol{x}', \omega) =& -\frac{1}{4\pi\mu}\cos\theta \int_0^\infty \frac{\beta^2 k}{\gamma\omega^2} A^{\mathrm{SR}} J_1(kr)\,\mathrm{d}k\\
G_{22}(\boldsymbol{x}, \boldsymbol{x}', \omega) =& \frac{1}{4\pi\mu}\left\{\sin^2\theta\left[\int_0^\infty \frac{\beta^2 k}{\gamma\omega^2} A^{\mathrm{SS}} J_1'(kr)\,\mathrm{d}k + \int_0^\infty \frac{1}{\nu r} A^{\mathrm{T}} J_1(kr)\,\mathrm{d}k\right]\right.\\
&\left. + \cos^2\theta\left[\int_0^\infty \frac{\beta^2}{\gamma\omega^2 r} A^{\mathrm{SS}} J_1(kr)\,\mathrm{d}k + \int_0^\infty \frac{k}{\nu} A^{\mathrm{T}} J_1'(kr)\,\mathrm{d}k\right]\right\}\\
G_{23}(\boldsymbol{x}, \boldsymbol{x}', \omega) =& -\frac{1}{4\pi\mu}\sin\theta \int_0^\infty \frac{\beta^2 k}{\gamma\omega^2} A^{\mathrm{SR}} J_1(kr)\,\mathrm{d}k\\
G_{31}(\boldsymbol{x}, \boldsymbol{x}', \omega) =& -\frac{1}{4\pi\mu}\cos\theta \int_0^\infty \frac{\beta^2 k}{\nu\omega^2} A^{\mathrm{RS}} J_1(kr)\,\mathrm{d}k\\
G_{32}(\boldsymbol{x}, \boldsymbol{x}', \omega) =& -\frac{1}{4\pi\mu}\sin\theta \int_0^\infty \frac{\beta^2 k}{\nu\omega^2} A^{\mathrm{RS}} J_1(kr)\,\mathrm{d}k\\
G_{33}(\boldsymbol{x}, \boldsymbol{x}', \omega) =& -\frac{1}{4\pi\mu} \int_0^\infty \frac{\beta^2 k}{\nu\omega^2} A^{\mathrm{RR}} J_0(kr)\,\mathrm{d}k
\end{aligned}$$

Note that, up to this point, we have analyzed problems where the source point is located on the z axis (see Fig. 6.1). To generalize to the case where the source point is at $\boldsymbol{x}' = (x_1', x_2', x_3')$, we have

$$\cos\theta = \frac{x_1 - x_1'}{r}, \quad \sin\theta = \frac{x_2 - x_2'}{r}$$

where $r = \sqrt{(x_1 - x_1')^2 + (x_2 - x_2')^2}$. We then obtain the frequency-domain Green's function for the Lamb's problem as

$$\mathbf{G}(\boldsymbol{x}, \boldsymbol{x}', \omega) = \frac{1}{4\pi\mu} \begin{bmatrix} I_1 \cos^2\theta + I_2 \sin^2\theta & (I_1 - I_2)\cos\theta \sin\theta & I_3 \cos\theta \\ (I_1 - I_2)\cos\theta \sin\theta & I_1 \sin^2\theta + I_2 \cos^2\theta & I_3 \sin\theta \\ I_4 \cos\theta & I_4 \sin\theta & I_5 \end{bmatrix} \tag{6.5.1}$$

where,

$$I_1 = \int_0^\infty \left[\frac{\beta^2 k}{\gamma\omega^2} A^{\mathrm{SS}} J_1'(kr) + \frac{1}{\nu r} A^{\mathrm{T}} J_1(kr) \right] \mathrm{d}k \tag{6.5.2a}$$

$$I_2 = \int_0^\infty \left[\frac{\beta^2}{\gamma\omega^2 r} A^{\mathrm{SS}} J_1(kr) + \frac{k}{\nu} A^{\mathrm{T}} J_1'(kr) \right] \mathrm{d}k \tag{6.5.2b}$$

$$I_3 = -\int_0^\infty \frac{\beta^2 k}{\gamma\omega^2} A^{\mathrm{SR}} J_1(kr)\, \mathrm{d}k \tag{6.5.2c}$$

$$I_4 = -\int_0^\infty \frac{\beta^2 k}{\nu\omega^2} A^{\mathrm{RS}} J_1(kr)\, \mathrm{d}k \tag{6.5.2d}$$

$$I_5 = -\int_0^\infty \frac{\beta^2 k}{\nu\omega^2} A^{\mathrm{RR}} J_0(kr)\, \mathrm{d}k \tag{6.5.2e}$$

Equations (6.5.1) and (6.5.2), along with Eq. (6.4.23), form the frequency-domain Green's function.

The Green's function expressed in Eq. (6.5.1) is given as an integral over the wavenumber k. For a fixed frequency, this wavenumber integral can be solved numerically, and then the result can be transformed back to the time domain using an inverse Fourier transform. Because the time-domain displacement results require an inverse Fourier transform and because the relationship between time signals and their frequency components is not one-to-one, it is difficult to theoretically analyze the wave properties directly from Eq. (6.5.1). Unlike the radiation pattern of the Green's function in infinite space obtained in Chap. 5, we cannot obtain such patterns through theoretical analysis without numerical computation. Nevertheless, useful insights can be drawn by analyzing Eq. (6.4.23).

As mentioned in Sect. 6.4.3, since A^T, A^{SS}, A^{SR}, A^{RS}, and A^{RR} are explicitly expressed as sums of six terms, each with a phase factor (exponential term) that has clear physical meaning, the current solution to the Lamb's problem clearly separates the contributions of the free surface from those of the infinite medium. Rayleigh waves appear only in the terms related to the free surface. Next, we will study the contributions of the first two terms (infinite medium) and the last four terms (free surface) separately.

6.5.2 Equivalence of Direct Wave Components and Green's Function in Infinite Medium

This section mathematically proves that the contribution of the direct wave components (the terms containing e^{P} and e^{S}) is equivalent to the Green's function in an infinite medium given by Eq. (4.4.11) (Zhang and Chen 2009). According to Eq. (6.4.23), the form that only retains the direct wave components is as follows:

$$
\begin{aligned}
A^{\mathrm{T}} &= \mathrm{e}^{-\nu|z|}, \quad A^{\mathrm{SS}} = k^2\mathrm{e}^{-\gamma|z|} - \nu\gamma\mathrm{e}^{-\nu|z|}, \quad A^{\mathrm{RR}} = \nu\gamma\mathrm{e}^{-\gamma|z|} - k^2\mathrm{e}^{-\nu|z|} \\
A^{\mathrm{SR}} &= k\gamma\,\mathrm{sgn}(z)(\mathrm{e}^{-\gamma|z|} - \mathrm{e}^{-\nu|z|}), \quad A^{\mathrm{RS}} = k\nu\,\mathrm{sgn}(z)(\mathrm{e}^{-\gamma|z|} - \mathrm{e}^{-\nu|z|})
\end{aligned}
$$

where $z = x_3 - x_3'$.

Define the integral in the following form:

$$
\begin{aligned}
K_1(c) &\triangleq \int_0^{+\infty} \frac{k^3}{\eta(c)}\mathrm{e}^{-\eta(c)|z|} J_1'(kr)\,\mathrm{d}k, \quad K_2(c) \triangleq \int_0^{+\infty} k\eta(c)\mathrm{e}^{-\eta(c)|z|} J_1'(kr)\,\mathrm{d}k \\
K_3(c) &\triangleq \int_0^{+\infty} \frac{1}{\eta(c)}\mathrm{e}^{-\eta(c)|z|} J_1(kr)\,\mathrm{d}k, \quad K_4(c) \triangleq \int_0^{+\infty} \frac{k^2}{\eta(c)}\mathrm{e}^{-\eta(c)|z|} J_1(kr)\,\mathrm{d}k \\
K_5(c) &\triangleq \int_0^{+\infty} \eta(c)\mathrm{e}^{-\eta(c)|z|} J_1(kr)\,\mathrm{d}k, \quad K_6(c) \triangleq \int_0^{+\infty} \frac{k}{\eta(c)}\mathrm{e}^{-\eta(c)|z|} J_1'(kr)\,\mathrm{d}k \\
K_7(c) &\triangleq \int_0^{+\infty} k^2\mathrm{e}^{-\eta(c)|z|} J_1(kr)\,\mathrm{d}k, \quad K_8(c) \triangleq \int_0^{+\infty} k\eta(c)\mathrm{e}^{-\eta(c)|z|} J_0(kr)\,\mathrm{d}k \\
K_9(c) &\triangleq \int_0^{+\infty} \frac{k^3}{\eta(c)}\mathrm{e}^{-\eta(c)|z|} J_0(kr)\,\mathrm{d}k
\end{aligned} \tag{6.5.3}
$$

where

$$
\eta(c) = \sqrt{k^2 - \frac{\omega^2}{c^2}} \quad \Rightarrow \quad \gamma = \eta(\alpha), \ \nu = \eta(\beta)
$$

Therefore, the I_i $(i = 1, 2, \ldots, 5)$ containing only the direct wave components are

$$
I_1 = \frac{\beta^2}{\omega^2}\big[K_1(\alpha) - K_2(\beta)\big] + \frac{1}{r}K_3(\beta) \tag{6.5.4a}
$$

$$
I_2 = \frac{\beta^2}{r\omega^2}\big[K_4(\alpha) - K_5(\beta)\big] + K_6(\beta) \tag{6.5.4b}
$$

$$
I_3 = I_4 = \frac{\beta^2}{\omega^2}\mathrm{sgn}(z)\big[K_7(\alpha) - K_7(\beta)\big] \tag{6.5.4c}
$$

$$
I_5 = \frac{\beta^2}{\omega^2}\big[K_8(\alpha) - K_9(\beta)\big] \tag{6.5.4d}
$$

To find the direct wave components of the frequency-domain Green's function, it is simply required to calculate the K_i $(i = 1, 2, \ldots, 9)$ defined in Eq. (6.5.3), substitute them into Eq. (6.5.4), and then into Eq. (6.5.1). By performing an inverse Fourier transform, we can obtain the time-domain Green's function.

Note the two integrals containing Bessel functions (Prudnikov et al. 1986):

$$\int_0^{+\infty} \frac{k}{\sqrt{k^2-a^2}} \mathrm{e}^{-\sqrt{k^2-a^2}|z|} J_0(kr)\, \mathrm{d}k = \frac{1}{R} \mathrm{e}^{-\mathrm{i}aR} \tag{6.5.5a}$$

$$\int_0^{+\infty} \frac{1}{\sqrt{k^2-a^2}} \mathrm{e}^{-\sqrt{k^2-a^2}|z|} J_1(kr)\, \mathrm{d}k = \frac{1}{\mathrm{i}ar} \left(\mathrm{e}^{-\mathrm{i}|z|a} - \mathrm{e}^{-\mathrm{i}Ra}\right) \tag{6.5.5b}$$

where $R = \sqrt{r^2+z^2}$. The integral in Eq. (6.5.5a) is also known as the *Sommerfeld integral*. By differentiating Eq. (6.5.5) with respect to z or r, and using the recurrence relations of Bessel functions, the integrals $K_i(c)$ ($i = 1, 2, \ldots, 9$) can be obtained:

$$K_1(c) = \frac{1}{R^5} \left[R^2 - 3r^2 + \frac{\omega^2}{c^2} R^2 r^2 + \mathrm{i}\frac{\omega}{c} R \left(R^2 - 3r^2\right) \right] \mathrm{e}^{-\mathrm{i}\frac{\omega}{c}R} \tag{6.5.6a}$$

$$\begin{aligned} K_2(c) = &-\frac{1}{r^2 R^5} \left\{ \mathrm{i}\frac{\omega}{c} R^5 \mathrm{e}^{-\mathrm{i}\frac{\omega}{c}|z|} + \left[\frac{\omega^2}{c^2} z^2 r^2 R^2 + \mathrm{i}\frac{\omega}{c} R \left(r^4 - z^4 - 3r^2 z^2\right) \right.\right. \\ &\left.\left. + r^2 \left(2r^2 - z^2\right) \right] \mathrm{e}^{-\mathrm{i}\frac{\omega}{c}R} \right\} \end{aligned} \tag{6.5.6b}$$

$$K_3(c) = \frac{c}{\mathrm{i}r\omega} \left(\mathrm{e}^{-\mathrm{i}\frac{\omega}{c}|z|} - \mathrm{e}^{-\mathrm{i}\frac{\omega}{c}R}\right) \tag{6.5.6c}$$

$$K_4(c) = \frac{r}{R^3} \left(1 + \mathrm{i}\frac{\omega}{c} R\right) \mathrm{e}^{-\mathrm{i}\frac{\omega}{c}R} \tag{6.5.6d}$$

$$K_5(c) = \frac{1}{R^3 r} \left[\mathrm{i}\frac{\omega}{c} R^3 \mathrm{e}^{-\mathrm{i}\frac{\omega}{c}|z|} + \left(r^2 - \mathrm{i}\frac{\omega}{c} R z^2\right) \mathrm{e}^{-\mathrm{i}\frac{\omega}{c}R} \right] \tag{6.5.6e}$$

$$K_6(c) = \frac{c}{r^2 R\omega} \left[\mathrm{i}R\mathrm{e}^{-\mathrm{i}\frac{\omega}{c}|z|} + \left(r^2 \frac{\omega}{c} - \mathrm{i}R\right) \mathrm{e}^{-\mathrm{i}\frac{\omega}{c}R} \right] \tag{6.5.6f}$$

$$K_7(c) = \frac{|z|\,r}{R^5} \left(3 - \frac{\omega^2}{c^2} R^2 + 3\mathrm{i}\frac{\omega}{c} R\right) \mathrm{e}^{-\mathrm{i}\frac{\omega}{c}R} \tag{6.5.6g}$$

$$K_8(c) = \frac{1}{R^5} \left[3z^2 - R^2 - \frac{\omega^2}{c^2} R^2 z^2 + \mathrm{i}\frac{\omega}{c} R \left(3z^2 - R^2\right) \right] \mathrm{e}^{-\mathrm{i}\frac{\omega}{c}R} \tag{6.5.6h}$$

$$K_9(c) = \frac{1}{R^5} \left[3z^2 - R^2 + \frac{\omega^2}{c^2} R^2 r^2 + \mathrm{i}\frac{\omega}{c} R \left(3z^2 - R^2\right) \right] \mathrm{e}^{-\mathrm{i}\frac{\omega}{c}R} \tag{6.5.6i}$$

Substituting Eq. (6.5.6) into Eq. (6.5.4), we obtain

$$\begin{aligned} I_1 = &\left(\frac{\beta^2}{\omega^2} \frac{R^2 - 3r^2}{R^5} + \frac{\beta^2 r^2}{\alpha^2 R^3} - \frac{\beta^2}{\mathrm{i}\alpha\omega} \frac{R^2 - 3r^2}{R^4} \right) \Phi_\alpha \\ &+ \left(\frac{z^2}{R^3} + \frac{\beta^2}{\omega^2} \frac{2r^2 - z^2}{R^5} - \frac{\beta}{\mathrm{i}\omega} \frac{2r^2 - z^2}{R^4} \right) \Phi_\beta \end{aligned} \tag{6.5.7a}$$

$$I_2 = \left(\frac{\beta^2}{\omega^2} \frac{1}{R^3} - \frac{\beta^2}{\mathrm{i}\alpha\omega} \frac{1}{R^2} \right) \Phi_\alpha - \left(\frac{\beta^2}{\omega^2} \frac{1}{R^3} - \frac{1}{R} - \frac{\beta}{\mathrm{i}\omega} \frac{1}{R^2} \right) \Phi_\beta \tag{6.5.7b}$$

$$I_3 = I_4 = \frac{\beta^2}{\omega^2}\frac{zr}{R^5}\left\{\left(3-\frac{\omega^2}{\alpha^2}R^2+3\,\mathrm{i}\frac{\omega}{\alpha}R\right)\Phi_\alpha-\left(3-\frac{\omega^2}{\beta^2}R^2+3\,\mathrm{i}\frac{\omega}{\beta}R\right)\Phi_\beta\right\} \tag{6.5.7c}$$

$$\begin{aligned} I_5 =& \frac{\beta^2}{\omega^2}\frac{1}{R^5}\left\{\left[3z^2-R^2-\frac{\omega^2}{\alpha^2}R^2z^2+\mathrm{i}\frac{\omega}{\alpha}R\left(3z^2-R^2\right)\right]\Phi_\alpha\right. \\ &\left.-\left[3z^2-R^2+\frac{\omega^2}{\beta^2}R^2r^2+\mathrm{i}\frac{\omega}{\beta}R\left(3z^2-R^2\right)\right]\Phi_\beta\right\} \end{aligned} \tag{6.5.7d}$$

in which, $\Phi_\alpha = \mathrm{e}^{-\mathrm{i}\frac{\omega}{\alpha}R}$ and $\Phi_\beta = \mathrm{e}^{-\mathrm{i}\frac{\omega}{\beta}R}$. Substituting Eq. (6.5.7) into Eq. (6.5.1), and noting that

$$\frac{x_\alpha-\xi_\alpha}{r}=\frac{x_\alpha-\xi_\alpha}{R}\frac{R}{r}=\gamma_\alpha\frac{R}{r}$$

we can simplify and obtain the Green's function for the Lamb's problem containing only the direct wave, $\tilde{G}_{ij}^D$ (where the superscript "D" represents direct wave), as follows:

$$G_{ii}^D(\boldsymbol{x},\boldsymbol{x}',\omega)=\frac{1}{4\pi\mu}\left[\frac{3\gamma_i^2-1}{R^2}Q(\omega)+\frac{\beta^2\gamma_i^2}{\alpha^2R}\Phi_\alpha-\frac{\gamma_i^2-1}{R}\Phi_\beta\right]\quad\text{(without summation)} \tag{6.5.8a}$$

$$G_{ij}^D(\boldsymbol{x},\boldsymbol{x}',\omega)=G_{ji}^D(\boldsymbol{x},\boldsymbol{\xi},\omega)=\frac{\gamma_i\gamma_j}{4\pi\mu}\left[\frac{3}{R^2}Q(\omega)+\frac{\beta^2}{\alpha^2R}\Phi_\alpha-\frac{1}{R}\Phi_\beta\right]\quad(i\neq j) \tag{6.5.8b}$$

where

$$Q(\omega)=-\frac{\beta^2}{R\omega^2}\left(\Phi_\alpha-\Phi_\beta\right)+\frac{\beta}{\mathrm{i}\alpha\omega}\left(\beta\Phi_\alpha-\alpha\Phi_\beta\right)$$

Using the Fourier transform pairs in Table 6.1, the inverse Fourier transform of $Q(\omega)$ can be obtained as follows:

Table 6.1 Several commonly used Fourier transform pairs

Time-domain function $f(t)$	Frequency-domain $F(\omega)$
$\delta(t)$	1
1	$2\pi\delta(\omega)$
$H(t)$	$\frac{1}{\mathrm{i}\omega}+\pi\delta(\omega)$
$tH(t)$	$-\frac{1}{\omega^2}+\mathrm{i}\pi\delta'(\omega)$
$\frac{t}{2}$	$\mathrm{i}\pi\delta'(\omega)$
$f(t-t_0)$	$F(\omega)\mathrm{e}^{-i\omega t_0}$

$$
\begin{aligned}
Q(t) &= \mathscr{F}^{-1}\{Q(\omega)\} \\
&= \mathscr{F}^{-1}\left\{\frac{\beta^2}{R}\left[-\frac{1}{\omega^2}+\mathrm{i}\pi\delta'(\omega)\right](\Phi_\alpha-\Phi_\beta)-\frac{\beta^2}{R}\mathrm{i}\pi\delta'(\omega)(\Phi_\alpha-\Phi_\beta)\right. \\
&\quad \left.+\beta\left[\frac{1}{\mathrm{i}\omega}+\pi\delta(\omega)\right]\frac{\beta\Phi_\alpha-\alpha\Phi_\beta}{\alpha}-\beta\pi\delta(\omega)\frac{\beta\Phi_\alpha-\alpha\Phi_\beta}{\alpha}\right\} \\
&= \frac{\beta^2}{R}\left\{\left(t-\frac{R}{\alpha}\right)H\left(t-\frac{R}{\alpha}\right)-\left(t-\frac{R}{\beta}\right)H\left(t-\frac{R}{\beta}\right)\right\}-\beta\left(\frac{\beta}{2\alpha}-\frac{1}{2}\right) \\
&\quad -\frac{\beta^2}{R}\left\{\frac{1}{2}\left(t-\frac{R}{\alpha}\right)-\frac{1}{2}\left(t-\frac{R}{\beta}\right)\right\}+\beta\left\{\frac{\beta}{\alpha}H\left(t-\frac{R}{\alpha}\right)-H\left(t-\frac{R}{\beta}\right)\right\} \\
&= \frac{\beta^2}{R}t\left[H\left(t-\frac{R}{\alpha}\right)-H\left(t-\frac{R}{\beta}\right)\right]
\end{aligned}
\tag{6.5.9}
$$

where the operator $\mathscr{F}^{-1}$ denotes the inverse Fourier transform. Combining with Eq. (6.5.9), applying the inverse Fourier transform to Eq. (6.5.8) ultimately yields:

$$
\begin{aligned}
G_{ij}^D(\boldsymbol{x},\boldsymbol{x}',t) =& \frac{3\gamma_i\gamma_j-\delta_{ij}}{4\pi\rho R^3}t\left[H\left(t-\frac{R}{\alpha}\right)-H\left(t-\frac{R}{\beta}\right)\right] \\
&+\frac{\gamma_i\gamma_j}{4\pi\rho\alpha^2 R}\delta\left(t-\frac{R}{\alpha}\right)-\frac{\gamma_i\gamma_j-\delta_{ij}}{4\pi\rho\beta^2 R}\delta\left(t-\frac{R}{\beta}\right)
\end{aligned}
\tag{6.5.10}
$$

Comparing Eq. (6.5.10) with Eq. (4.4.11), we find that the two expressions are identical. This indicates that the direct wave component of the Green's function solution for the Lamb's problem in Eq. (6.5.1) is exactly the same as the solution for the Green's function in infinite space.

6.5.3 Anaysis of Rayleigh Wave Based on Green's Function

As mentioned in Sect. 6.4.3, the terms A^{T}, A^{SS}, A^{SR}, A^{RS}, and A^{RR} defined in Eq. (6.4.23) can each be expressed as a sum of six components. The first two components represent the contribution of the direct wave, which was proven in Sect. 6.5.2 to be identical to the solution for the Green's function in infinite space. The remaining four components, which have more complex forms, represent the contribution from the free surface. This complexity is mainly due to the coefficients, all of which are expressed as fractions with denominators that include $R(k)$. Under certain combinations of ω and k, $R(k)$ can approach zero, leading to the formation of a special type of wave known as the Rayleigh wave. For this reason, $R(k)$ is called the *Rayleigh function*. The generation of these waves based on the frequency-domain Green's function will be examined in this section.

6.5.3.1 Free-Surface Part in Frequency-Domain Green's Function

Since the generation of Rayleigh wave is closely related to the term $R(k)$ in the denominator, any terms that do not contain $R(k)$ can be disregarded when analyzing the generation of Rayleigh wave. By substituting the terms containing $R(k)$ from Eq. (6.4.23) into Eq. (6.5.2), and using the relationships

$$J_1'(x) = \frac{J_0(x) - J_2(x)}{2}, \quad \frac{J_1(x)}{x} = \frac{J_0(x) + J_2(x)}{2}$$

we obtain

$$I_i = \int_0^\infty \frac{F_1(k)}{R(k)} \left[J_0(kr) + (-1)^i J_2(kr)\right] \mathrm{d}k \quad (i = 1, 2) \tag{6.5.11a}$$

$$I_j = \int_0^\infty \frac{F_j(k)}{R(k)} J_1(kr)\, \mathrm{d}k \quad (j = 3, 4), \quad I_5 = \int_0^\infty \frac{F_5(k)}{R(k)} J_0(kr)\, \mathrm{d}k \tag{6.5.11b}$$

where,

$$F_1(k) = -\frac{k\beta^2}{2\omega^2}\left[\frac{R^*(k)}{\gamma}\left(k^2 \mathrm{e}^{\mathrm{PP}} + \nu\gamma \mathrm{e}^{\mathrm{SS}}\right) - 4k^2\nu\chi\left(\mathrm{e}^{\mathrm{SP}} + \mathrm{e}^{\mathrm{PS}}\right)\right] \tag{6.5.12a}$$

$$F_3(k) = -\frac{k^2\beta^2}{\omega^2}\left[R^*(k)\left(\mathrm{e}^{\mathrm{PP}} + \mathrm{e}^{\mathrm{SS}}\right) - 4\chi\left(k^2\mathrm{e}^{\mathrm{SP}} + \nu\gamma \mathrm{e}^{\mathrm{PS}}\right)\right] \tag{6.5.12b}$$

$$F_4(k) = +\frac{k^2\beta^2}{\omega^2}\left[R^*(k)\left(\mathrm{e}^{\mathrm{PP}} + \mathrm{e}^{\mathrm{SS}}\right) - 4\chi\left(\nu\gamma\mathrm{e}^{\mathrm{SP}} + k^2 \mathrm{e}^{\mathrm{PS}}\right)\right] \tag{6.5.12c}$$

$$F_5(k) = -\frac{k\beta^2}{\omega^2}\left[\frac{R^*(k)}{\nu}\left(\nu\gamma \mathrm{e}^{\mathrm{PP}} + k^2 \mathrm{e}^{\mathrm{SS}}\right) - 4k^2\gamma\chi\left(\mathrm{e}^{\mathrm{SP}} + \mathrm{e}^{\mathrm{PS}}\right)\right] \tag{6.5.12d}$$

Note that $F_1(k)$ and $F_5(k)$ are odd functions of k, while $F_3(k)$ and $F_4(k)$ are even functions of k. According to the properties of Bessel functions $J_n(-x) = (-1)^n J_n(x)$, $J_0(kr)$ and $J_2(kr)$ are even functions of k, whereas $J_1(kr)$ is an odd function of k. For easier analysis, the integration limits in Eq. (6.5.11) are extended to range from $-\infty$ to ∞. Using the relationship between Bessel and Hankel functions, along with the properties of Hankel functions,

$$J_n(x) = \frac{H_n^{(1)}(x) + H_n^{(2)}(x)}{2} \tag{6.5.13a}$$

$$H_n^{(2)}(-x) = (-1)^{n+1} H_n^{(1)}(x) \tag{6.5.13b}$$

for the case where $F_i(k)$ is an odd function and the Bessel function is an even function (e.g., $J_0(kr)$), we obtain

$$
\begin{aligned}
&\frac{1}{2}\int_{-\infty}^{\infty}\frac{F_i(k)}{R(k)}H_0^{(2)}(kr)\,\mathrm{d}k\\
&=\frac{1}{2}\int_{-\infty}^{0}\frac{F_i(k)}{R(k)}H_0^{(2)}(kr)\,\mathrm{d}k+\frac{1}{2}\int_{0}^{\infty}\frac{F_i(k)}{R(k)}H_0^{(2)}(kr)\,\mathrm{d}k\\
&\overset{k=-k'}{=\!=}\frac{1}{2}\int_{-\infty}^{0}\frac{F_i(-k')}{R(-k')}H_0^{(2)}(-k'r)\,\mathrm{d}(-k')+\frac{1}{2}\int_{0}^{\infty}\frac{F_i(k)}{R(k)}H_0^{(2)}(kr)\,\mathrm{d}k\\
&\overset{(6.5.13b)}{=\!=}\frac{1}{2}\int_{0}^{\infty}\frac{F_i(k')}{R(k')}H_0^{(1)}(k'r)\,\mathrm{d}k'+\frac{1}{2}\int_{0}^{\infty}\frac{F_i(k)}{R(k)}H_0^{(2)}(kr)\,\mathrm{d}k\\
&\overset{(6.5.13a)}{=\!=}\int_{0}^{\infty}\frac{F_i(k)}{R(k)}J_0(kr)\,\mathrm{d}k
\end{aligned}
$$

and for the case where $F_i(k)$ is an even function and the Bessel function is an odd function, we have

$$
\begin{aligned}
&\frac{1}{2}\int_{-\infty}^{\infty}\frac{F_i(k)}{R(k)}H_1^{(2)}(kr)\,\mathrm{d}k\\
&=\frac{1}{2}\int_{-\infty}^{0}\frac{F_i(k)}{R(k)}H_1^{(2)}(kr)\,\mathrm{d}k+\frac{1}{2}\int_{0}^{\infty}\frac{F_i(k)}{R(k)}H_1^{(2)}(kr)\,\mathrm{d}k\\
&\overset{k=-k'}{=\!=}\frac{1}{2}\int_{-\infty}^{0}\frac{F_i(-k')}{R(-k')}H_1^{(2)}(-k'r)\,\mathrm{d}(-k')+\frac{1}{2}\int_{0}^{\infty}\frac{F_i(k)}{R(k)}H_1^{(2)}(kr)\,\mathrm{d}k\\
&\overset{(6.5.13b)}{=\!=}\frac{1}{2}\int_{0}^{\infty}\frac{F_i(k')}{R(k')}H_1^{(1)}(k'r)\,\mathrm{d}k'+\frac{1}{2}\int_{0}^{\infty}\frac{F_i(k)}{R(k)}H_1^{(2)}(kr)\,\mathrm{d}k\\
&\overset{(6.5.13a)}{=\!=}\int_{0}^{\infty}\frac{F_i(k)}{R(k)}J_1(kr)\,\mathrm{d}k
\end{aligned}
$$

Therefore, Eq. (6.5.11) can be rewritten as

$$
I_i=\frac{1}{2}\int_{-\infty}^{\infty}\frac{F_1(k)}{R(k)}\left[H_0^{(2)}(kr)+(-1)^i H_2^{(2)}(kr)\right]\mathrm{d}k\quad (i=1,2) \tag{6.5.14a}
$$

$$
I_j=\frac{1}{2}\int_{-\infty}^{\infty}\frac{F_j(k)}{R(k)}H_1^{(2)}(kr)\,\mathrm{d}k\quad (j=3,4) \tag{6.5.14b}
$$

$$
I_5=\frac{1}{2}\int_{-\infty}^{\infty}\frac{F_5(k)}{R(k)}H_0^{(2)}(kr)\,\mathrm{d}k \tag{6.5.14c}
$$

By substituting Eq. (6.5.14) into Eq. (6.5.1), we obtain the contribution part of the Green's function for the free surface.

6.5.3.2 Integrals in the Complex Plane of k

The next task is to analyze the integral in Eq. (6.5.14) and separate out the Rayleigh-wave component. To achieve this, we consider these integrals within the complex domain. Extending a real-valued integral to the complex domain is a common method in integral analysis. Although it might seem to complicate the problem, it actually doesn't. In the complex domain, we can fully utilize the residue theorem[22] and derive useful conclusions through integrals along different paths.

To study the integral in Eq. (6.5.14) in the complex plane of k, we first need to note that the integrand contains $\gamma = \sqrt{k^2 - k_\alpha^2}$ and $\nu = \sqrt{k^2 - k_\beta^2}$, where $k_\alpha = \omega/\alpha$ and $k_\beta = \omega/\beta$. Clearly, there are four branch points: $k = \pm k_\alpha$ and $\pm k_\beta$. Both γ and ν are of the form[23]

$$f(z) = \sqrt{z^2 - z_0^2} \quad (z_0 = a - \mathrm{i}b,\ a, b \in \mathbb{R}^+)$$

Since we are considering complex frequencies, the branch points $\pm k_\alpha$ and $\pm k_\beta$ are no longer on the real axis but instead lie in the second and fourth quadrants of the complex plane, respectively. Given the four possible combinations of the signs of $\mathrm{Re}(\gamma)$ and $\mathrm{Re}(\nu)$), the problem involves a four-sheeted Riemann surface. We select the sheet where $\mathrm{Re}(\gamma) \geq 0$ and $\mathrm{Re}(\nu) \geq 0$ to ensure the analyticity of the function. Therefore, the branch cuts are determined by $\mathrm{Re}(\gamma) = 0$ and $\mathrm{Re}(\nu) = 0$. Appendix A provides a detailed explanation of how to draw the branch cuts for this type of complex radical. Figure 6.2 illustrates the distribution of branch points and branch cuts in the complex plane of k. The branch cuts are parts of hyperbolas, and on them, $\mathrm{Re}(\gamma) = 0$ or $\mathrm{Re}(\nu) = 0$. On different sides of the branch cuts, the arguments are

[22] Let C be a piecewise smooth simple closed curve that forms the boundary of region G. Suppose the function $f(z)$ is single-valued and analytic within G except for a finite number of isolated singularities b_k $(k = 1, 2, \ldots, n)$. Further, $f(z)$ is continuous on $\bar{G}$ and has no singularities on C. Then, we have

$$\oint_C f(z)\,\mathrm{d}z = 2\pi\mathrm{i} \sum_{k=1}^{n} \mathrm{res}\, f(b_k)$$

The term $\mathrm{res}\, f(b_k)$ is called the residue of $f(z)$ at b_k. It is equal to the coefficient of $(z - b_k)^{-1}$ in the Laurent series expansion of $f(z)$ around b_k:

$$f(z) = \sum_{l=-\infty}^{\infty} a_l^{(k)} (z - b_k)^l$$

The residue $\mathrm{res}\, f(b_k)$ is thus given by $a_{-1}^{(k)}$.

[23] In general, the angular frequency ω in the Fourier transform domain is a real number. However, in practical computations, a complex frequency $\omega^* = \omega - \mathrm{i}\omega_\mathrm{I}$ is used, where $\omega_\mathrm{I} > 0$. As a result, k_α and k_β are actually complex. Moreover, for non-ideal elastic media considering attenuation, the P-wave speed α and S-wave speed β also become complex after introducing the attenuation factor Q. The immediate effect of k_α and k_β being complex is that the branch points are no longer on the real axis of the k complex plane but are instead located in the second and fourth quadrants.

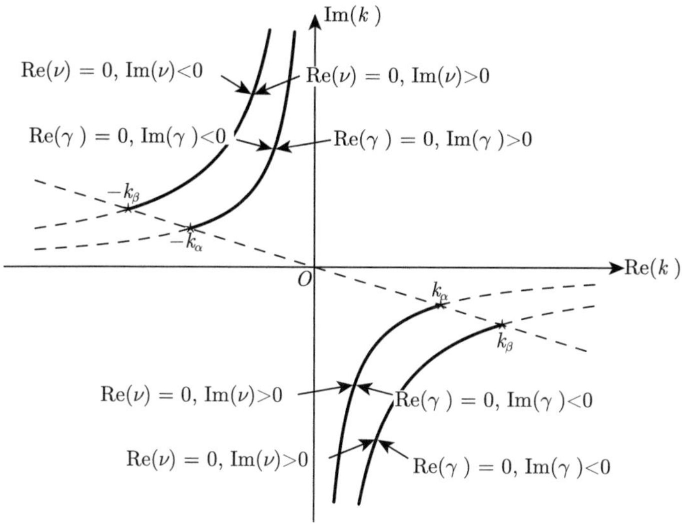

Fig. 6.2 Branch points and branch cuts on the complex plane of k. There are four branch points: $\pm k_\alpha$ and $\pm k_\beta$. The branch cuts are shown as parts of a hyperbola in bold black lines. The values of γ or ν on different sides of each branch cut are indicated in the figure

$\pm\pi/2$, indicating opposite signs of the imaginary parts of γ or ν. We choose the Riemann sheet where $\text{Re}(\gamma) \geq 0$ and $\text{Re}(\nu) \geq 0$ for further analysis.

Let $x = k^2\beta^2/\omega^2$, then

$$R(k) = \left(2k^2 - k_\beta^2\right)^2 - 4k^2\sqrt{k^2 - k_\alpha^2}\sqrt{k^2 - k_\beta^2}$$
$$= \frac{\omega^4}{\beta^4}\left[(1-2x)^2 + 4x\sqrt{m-x}\sqrt{1-x}\right] \quad \left(m = \frac{\mu}{\lambda + 2\mu} = \frac{1-2\nu}{2(1-\nu)}\right)$$

where, $0 < \nu < 0.5$ represents the Poisson's ratio. Appendix B provides a detailed discussion on the roots of $R(x) = (1-2x)^2 + 4x\sqrt{m-x}\sqrt{1-x}$ when $0 < m < 0.5$. The conclusion is that there is exactly one real root $x_{_R}$ greater than 1, and the corresponding $k = \kappa = \pm\omega\sqrt{x_{_R}}/\beta$. When ω is a real number, κ lies on the real axis; when ω is a complex number, κ is located in the second and fourth quadrants, on the same line as k_α and k_β.

To use the residue theorem to calculate the integral in Eq. (6.5.14), a closed contour needs to be constructed, as shown in Fig. 6.3. The integration path in Eq. (6.5.14) is represented by C in the figure. We introduce a semicircle C_R centered at the origin with radius $R \to \infty$ in the third and fourth quadrants. Due to the presence of branch cuts, paths $C_\beta^{(1)}$, $C_\beta^{(2)}$, $C_\alpha^{(1)}$, and $C_\alpha^{(2)}$ are formed along both sides of the branch cuts extending from k_β and k_α. Thus, $C + C_R + C_\beta^{(1)} + C_\beta^{(2)} + C_\alpha^{(1)} + C_\alpha^{(2)}$ together form a complete closed loop.

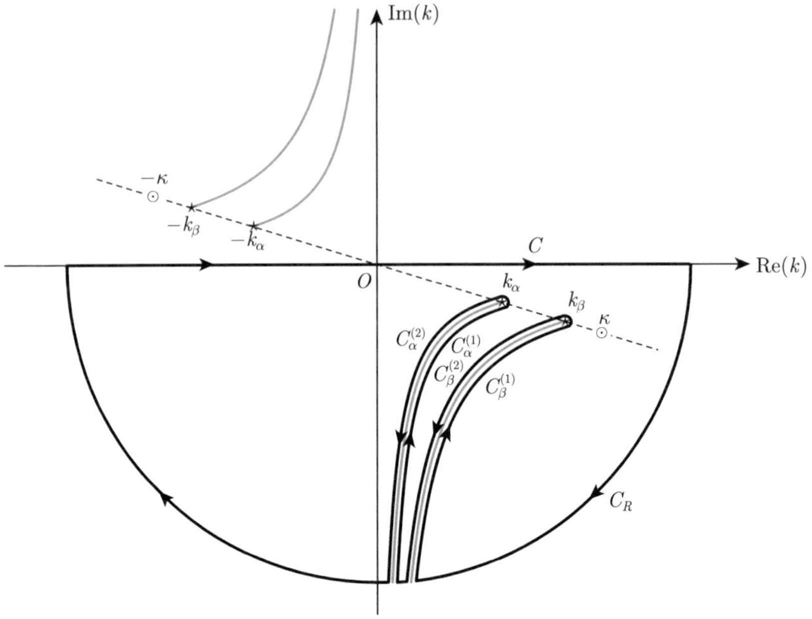

Fig. 6.3 Integration path in complex plane of k. The four branch points $\pm k_\alpha$, $\pm k_\beta$, and the zero of the Rayleigh function κ lie on the same line. The branch cuts are shown in gray, and the integration paths are shown in bold. C is the original integration path, C_R is the large arc of radius $R \to \infty$, $C_\alpha^{(1)}$ and $C_\alpha^{(2)}$ are the integration paths on either side of the branch cut at k_α, and $C_\beta^{(1)}$ and $C_\beta^{(2)}$ are the integration paths on either side of the branch cut at k_β

For the integral over the large arc, due to the presence of the phase factor, it can be verified by the large arc lemma[24] that $kf(k) \to 0$, where $f(k)$ is the integrand in Eq. (6.5.14). Therefore, the integral over the large arc C_R is zero. Within the region enclosed by the closed contour, there is only one simple pole at $k = \kappa$, so by the residue theorem, we get

$$\int_C f(k)\,\mathrm{d}k + \int_{C_\beta^{(1)}} f(k)\,\mathrm{d}k + \int_{C_\beta^{(2)}} f(k)\,\mathrm{d}k + \int_{C_\alpha^{(1)}} f(k)\,\mathrm{d}k + \int_{C_\alpha^{(2)}} f(k)\,\mathrm{d}k$$
$$= -2\pi\mathrm{i}\Big[\operatorname{res} f(\kappa)\Big] \tag{6.5.15}$$

[24] Let $f(z)$ be continuous in the neighborhood of $z = \infty$. If $zf(z)$ uniformly tends to K as $z \to \infty$ within $\theta_1 \le \arg z \le \theta_2$, then

$$\lim_{R\to\infty} \int_{C_R} f(z)\,\mathrm{d}z = \mathrm{i}K(\theta_2 - \theta_1)$$

where C_R is an arc centered at the origin with radius R and an angle $\theta_2 - \theta_1$, i.e., $|z| = R$ and $\theta_1 \le \arg z \le \theta_2$. See Wu (2003, p. 28).

The negative sign on the right side of the equation is due to the integration path direction in Fig. 6.3 being opposite to the conventionally positive direction. Since the arguments of γ or ν on either side of the branch cuts are not equal, the integrals along the opposite paths on the branch cuts do not cancel each other out. In Eq. (6.5.15), the integrals along $C_\beta^{(1)}$, $C_\beta^{(2)}$, $C_\alpha^{(1)}$, and $C_\alpha^{(2)}$ represent the contributions of body waves caused by the free interface, while the residue term related to the zero of the Rayleigh function on the right side of the equation corresponds to the contribution of Rayleigh surface wave. Note that

$$\operatorname{res} f(\kappa) = \lim_{k\to\kappa}(k-\kappa)f(k) = \lim_{k\to\kappa}(k-\kappa)\frac{P(k)}{Q(k)} = \frac{P(\kappa)}{Q'(\kappa)}$$

Therefore, applying Eqs. (6.5.15) to (6.5.14) and substituting into Eq. (6.5.1), we obtain the Rayleigh wave contribution to the Green's function, $G_{ij}^R(\boldsymbol{x}, \boldsymbol{x}', \omega)$,

$$G_{11}^R(\boldsymbol{x}, \boldsymbol{x}', \omega) = -\frac{\mathrm{i}F_1(\kappa)}{4\mu R'(\kappa)}\left[H_0^{(2)}(\kappa r) - H_2^{(2)}(\kappa r)\cos 2\theta\right] \tag{6.5.16a}$$

$$G_{12}^R(\boldsymbol{x}, \boldsymbol{x}', \omega) = G_{21}^R(\boldsymbol{x}, \boldsymbol{x}', \omega) = \frac{\mathrm{i}F_1(\kappa)\sin 2\theta}{4\mu R'(\kappa)}H_2^{(2)}(\kappa r) \tag{6.5.16b}$$

$$G_{22}^R(\boldsymbol{x}, \boldsymbol{x}', \omega) = -\frac{\mathrm{i}F_1(\kappa)}{4\mu R'(\kappa)}\left[H_0^{(2)}(\kappa r) + H_2^{(2)}(\kappa r)\cos 2\theta\right] \tag{6.5.16c}$$

$$G_{13}^R(\boldsymbol{x}, \boldsymbol{x}', \omega) = -\frac{\mathrm{i}F_3(\kappa)\cos\theta}{4\mu R'(\kappa)}H_1^{(2)}(\kappa r) \tag{6.5.16d}$$

$$G_{23}^R(\boldsymbol{x}, \boldsymbol{x}', \omega) = -\frac{\mathrm{i}F_3(\kappa)\sin\theta}{4\mu R'(\kappa)}H_1^{(2)}(\kappa r) \tag{6.5.16e}$$

$$G_{31}^R(\boldsymbol{x}, \boldsymbol{x}', \omega) = -\frac{\mathrm{i}F_4(\kappa)\cos\theta}{4\mu R'(\kappa)}H_1^{(2)}(\kappa r) \tag{6.5.16f}$$

$$G_{32}^R(\boldsymbol{x}, \boldsymbol{x}', \omega) = -\frac{\mathrm{i}F_4(\kappa)\sin\theta}{4\mu R'(\kappa)}H_1^{(2)}(\kappa r) \tag{6.5.16g}$$

$$G_{33}^R(\boldsymbol{x}, \boldsymbol{x}', \omega) = -\frac{\mathrm{i}F_5(\kappa)}{4\mu R'(\kappa)}H_0^{(2)}(\kappa r) \tag{6.5.16h}$$

Equation (6.5.16) is the closed-form solution for Rayleigh waves in the frequency domain.

6.5.3.3 Far-Field Approximation

Although Eq. (6.5.16) provides results in the frequency domain, making it less convenient for directly analyzing wave behavior in the time domain, it can still be used to determine the properties of Rayleigh wave at a single frequency. For instance, when the distance between the source and the observation point is large, the expected waveform should resemble that of a plane wave.

It is important to note that the asymptotic solution of the Hankel function as $x \to \infty$ is given by

$$H_m^{(2)}(x) \sim \sqrt{\frac{2}{\pi x}} \exp\left[-\mathrm{i}\left(x - \frac{m\pi}{2} - \frac{\pi}{4}\right)\right]$$

therefore,

$$H_0^{(2)}(\kappa r) \sim \sqrt{\frac{2}{\pi \kappa r}} \exp\left[-\mathrm{i}\left(\kappa r - \frac{\pi}{4}\right)\right] \tag{6.5.17a}$$

$$H_1^{(2)}(\kappa r) \sim \sqrt{\frac{2}{\pi \kappa r}} \exp\left[-\mathrm{i}\left(\kappa r - \frac{\pi}{2} - \frac{\pi}{4}\right)\right] = \mathrm{i} H_0^{(2)}(\kappa r) \tag{6.5.17b}$$

$$H_2^{(2)}(\kappa r) \sim \sqrt{\frac{2}{\pi \kappa r}} \exp\left[-\mathrm{i}\left(\kappa r - \pi - \frac{\pi}{4}\right)\right] = -H_0^{(2)}(\kappa r) \tag{6.5.17c}$$

Furthermore, since κ is a zero of the Rayleigh function, $R(\kappa) = \chi^2 - 4\kappa^2 \nu\gamma = 0$, that is,

$$\chi^2(\kappa) = 4\kappa^2 \nu(\kappa)\gamma(\kappa), \quad \text{and} \quad \chi(\kappa) = 2\kappa^2 - \frac{\omega^2}{\beta^2} \tag{6.5.18}$$

Substituting into Eq. (6.5.12), we obtain

$$F_1(\kappa) = -\frac{2\kappa^3 \nu \beta^2}{\omega^2}\left[2\left(\kappa^2 \mathrm{e}^{\mathrm{PP}} + \nu\gamma \mathrm{e}^{\mathrm{SS}}\right) - \chi\left(\mathrm{e}^{\mathrm{SP}} + \mathrm{e}^{\mathrm{PS}}\right)\right] \tag{6.5.19a}$$

$$F_3(\kappa) = -\frac{2\kappa^2 \beta^2 \chi}{\omega^2}\left[\chi\left(\mathrm{e}^{\mathrm{PP}} + \mathrm{e}^{\mathrm{SS}}\right) - 2\left(\kappa^2 \mathrm{e}^{\mathrm{SP}} + \nu\gamma \mathrm{e}^{\mathrm{PS}}\right)\right] \tag{6.5.19b}$$

$$F_4(\kappa) = +\frac{2\kappa^2 \beta^2 \chi}{\omega^2}\left[\chi\left(\mathrm{e}^{\mathrm{PP}} + \mathrm{e}^{\mathrm{SS}}\right) - 2\left(\nu\gamma \mathrm{e}^{\mathrm{SP}} + \kappa^2 \mathrm{e}^{\mathrm{PS}}\right)\right] \tag{6.5.19c}$$

$$F_5(\kappa) = -\frac{4\kappa^3 \gamma \beta^2}{\omega^2}\left[2\left(\nu\gamma \mathrm{e}^{\mathrm{PP}} + \kappa^2 \mathrm{e}^{\mathrm{SS}}\right) - \chi\left(\mathrm{e}^{\mathrm{SP}} + \mathrm{e}^{\mathrm{PS}}\right)\right] \tag{6.5.19d}$$

Under the conditions $x_3 = 0$ and $x_3' \geq 0$, as well as $x_3' = 0$ and $x_3 \geq 0$, the above equation can be further simplified into a more compact form. Define

$$\phi_{\mathrm{P}} \triangleq \mathrm{e}^{-\gamma x_3}, \quad \phi_{\mathrm{S}} \triangleq \mathrm{e}^{-\nu x_3}, \quad \phi_{\mathrm{P}}' \triangleq \mathrm{e}^{-\gamma x_3'}, \quad \phi_{\mathrm{S}}' \triangleq \mathrm{e}^{-\nu x_3'}$$

and note that

$$\frac{\beta^2}{\omega^2}\chi(\chi - 2\nu\gamma) \xlongequal{(6.5.18)} \frac{\beta^2}{\omega^2}\chi\left(\chi - \frac{\chi^2}{2\kappa^2}\right) = \frac{\chi^2}{2\kappa^2} = 2\nu\gamma$$

the specific expressions listed in Table 6.2 can be obtained.

Table 6.2 Expressions of $F_i(\kappa)$ $(i = 1, 3, 4, 5)$ in two different cases

	$x_3 = 0$ and $x_3' \geq 0$	$x_3' = 0$ and $x_3 \geq 0$
$F_1(\kappa)$	$-\kappa\nu\left(2\kappa^2\phi_{\mathrm{P}}' - \chi\phi_{\mathrm{S}}'\right)$	$-\kappa\nu\left(2\kappa^2\phi_{\mathrm{P}} - \chi\phi_{\mathrm{S}}\right)$
$F_3(\kappa)$	$-2\kappa^2\left(2\nu\gamma\phi_{\mathrm{P}}' - \chi\phi_{\mathrm{S}}'\right)$	$2\kappa^2\left(\chi\phi_{\mathrm{P}} - 2\nu\gamma\phi_{\mathrm{S}}\right)$
$F_4(\kappa)$	$-2\kappa^2\left(\chi\phi_{\mathrm{P}}' - 2\nu\gamma\phi_{\mathrm{S}}'\right)$	$2\kappa^2\left(2\nu\gamma\phi_{\mathrm{P}} - \chi\phi_{\mathrm{S}}\right)$
$F_5(\kappa)$	$2\kappa\gamma\left(\chi\phi_{\mathrm{P}}' - 2\kappa^2\phi_{\mathrm{S}}'\right)$	$2\kappa\gamma\left(\chi\phi_{\mathrm{P}} - 2\kappa^2\phi_{\mathrm{S}}\right)$

(1) For the far-field case where the source is on the surface and the observation point varies with depth ($x_3' = 0$ and $x_3 \geq 0$), substituting the results from Eq. (6.5.17) and the last column of Table 6.2 into Eq. (6.5.16) for the components $\tilde{G}_{11}^R$, $\tilde{G}_{31}^R$, $\tilde{G}_{13}^R$, and $\tilde{G}_{33}^R$, we get

$$G_{11}^R(\boldsymbol{x}, \boldsymbol{x}', \omega)\Big|_{x_3'=0} \sim \sqrt{\frac{\kappa}{2\pi r}}\frac{\nu\cos^2\theta}{\mu R'(\kappa)}\left(2\kappa^2\mathrm{e}^{-\gamma x_3} - \chi\mathrm{e}^{-\nu x_3}\right)\mathrm{e}^{-\mathrm{i}\left(\kappa r - \frac{3\pi}{4}\right)} \tag{6.5.20a}$$

$$G_{31}^R(\boldsymbol{x}, \boldsymbol{x}', \omega)\Big|_{x_3'=0} \sim \sqrt{\frac{\kappa}{2\pi r}}\frac{\kappa\cos\theta}{\mu R'(\kappa)}\left(2\nu\gamma\mathrm{e}^{-\gamma x_3} - \chi\mathrm{e}^{-\nu x_3}\right)\mathrm{e}^{-\mathrm{i}\left(\kappa r - \frac{\pi}{4}\right)} \tag{6.5.20b}$$

$$G_{13}^R(\boldsymbol{x}, \boldsymbol{x}', \omega)\Big|_{x_3'=0} \sim \sqrt{\frac{\kappa}{2\pi r}}\frac{\kappa\cos\theta}{\mu R'(\kappa)}\left(\chi\mathrm{e}^{-\gamma x_3} - 2\nu\gamma\mathrm{e}^{-\nu x_3}\right)\mathrm{e}^{-\mathrm{i}\left(\kappa r - \frac{\pi}{4}\right)} \tag{6.5.20c}$$

$$G_{33}^R(\boldsymbol{x}, \boldsymbol{x}', \omega)\Big|_{x_3'=0} \sim \sqrt{\frac{\kappa}{2\pi r}}\frac{\gamma}{\mu R'(\kappa)}\left(\chi\mathrm{e}^{-\gamma x_3} - 2\kappa^2\mathrm{e}^{-\nu x_3}\right)\mathrm{e}^{-\mathrm{i}\left(\kappa r + \frac{\pi}{4}\right)} \tag{6.5.20d}$$

where $\tilde{G}_{11}^R$ and $\tilde{G}_{31}^R$ represent the displacement components in the horizontal and vertical directions caused by a horizontal force, respectively. Similarly, $\tilde{G}_{13}^R$ and $\tilde{G}_{33}^R$ represent the displacement components in the horizontal and vertical directions caused by a vertical force, respectively. By multiplying these quantities by $\mathrm{e}^{\mathrm{i}\omega t}$[25] and taking the real part, we can observe their variations over time t. For the case of a harmonic wave propagating along the positive x_1 axis, $\theta = 0$, thus:

[25] $\tilde{G}_{ij}^R$ itself does not include the time variable. To account for its single-frequency oscillation characteristic, it needs to be multiplied by a time-dependent factor. In fact, according to the inverse Fourier transform, its corresponding time-domain result is given by

$$G_{ij}^R(\boldsymbol{x}, \boldsymbol{x}', t) = \frac{1}{2\pi}\int_{-\infty}^{\infty} G_{ij}^R(\boldsymbol{x}, \boldsymbol{x}', \omega)\mathrm{e}^{\mathrm{i}\omega t}\,\mathrm{d}\omega$$

This represents a summation over all frequency components, with $\mathrm{e}^{\mathrm{i}\omega t}$ in the integrand indicating the harmonic variation with time.

$$\mathrm{Re}\left[\tilde{G}_{11}^{R}\mathrm{e}^{\mathrm{i}\omega t}\right]\Big|_{x_3'=0} \sim \nu C\left(2\kappa^2\mathrm{e}^{-\gamma x_3} - \chi\mathrm{e}^{-\nu x_3}\right)\cos\left(\omega t - \kappa r + \frac{3\pi}{4}\right) \tag{6.5.21a}$$

$$\mathrm{Re}\left[\tilde{G}_{31}^{R}\mathrm{e}^{\mathrm{i}\omega t}\right]\Big|_{x_3'=0} \sim \kappa C\left(2\nu\gamma\mathrm{e}^{-\gamma x_3} - \chi\mathrm{e}^{-\nu x_3}\right)\cos\left(\omega t - \kappa r + \frac{\pi}{4}\right) \tag{6.5.21b}$$

$$\mathrm{Re}\left[\tilde{G}_{13}^{R}\mathrm{e}^{\mathrm{i}\omega t}\right]\Big|_{x_3'=0} \sim \kappa C\left(\chi\mathrm{e}^{-\gamma x_3} - 2\nu\gamma\mathrm{e}^{-\nu x_3}\right)\cos\left(\omega t - \kappa r + \frac{\pi}{4}\right) \tag{6.5.21c}$$

$$\mathrm{Re}\left[\tilde{G}_{33}^{R}\mathrm{e}^{\mathrm{i}\omega t}\right]\Big|_{x_3'=0} \sim \gamma C\left(\chi\mathrm{e}^{-\gamma x_3} - 2\kappa^2\mathrm{e}^{-\nu x_3}\right)\cos\left(\omega t - \kappa r - \frac{\pi}{4}\right) \tag{6.5.21d}$$

where $C = \sqrt{\kappa/(2\pi r)}/[\mu R'(\kappa)]$. By plotting the horizontal and vertical displacements caused by horizontal or vertical forces as the x axis and y axis, respectively, we can obtain the trajectory of particle motion.

(2) For the case where the observation point is on the surface and the source point varies with depth ($x_3 = 0$ and $x_3' \geq 0$), substituting the results from Eq. (6.5.17) and the second column of Table 6.2 into Eq. (6.5.16) similarly yields:

$$G_{11}^{R}(\boldsymbol{x},\boldsymbol{x}',\omega)\Big|_{x_3=0} \sim \sqrt{\frac{\kappa}{2\pi r}}\frac{\nu\cos^2\theta}{\mu R'(\kappa)}\left(2\kappa^2\mathrm{e}^{-\gamma x_3'} - \chi\mathrm{e}^{-\nu x_3'}\right)\mathrm{e}^{-\mathrm{i}\left(\kappa r - \frac{3\pi}{4}\right)}$$

$$G_{31}^{R}(\boldsymbol{x},\boldsymbol{x}',\omega)\Big|_{x_3=0} \sim \sqrt{\frac{\kappa}{2\pi r}}\frac{\kappa\cos\theta}{\mu R'(\kappa)}\left(\chi\mathrm{e}^{-\gamma x_3'} - 2\nu\gamma\mathrm{e}^{-\nu x_3'}\right)\mathrm{e}^{-\mathrm{i}\left(\kappa r - \frac{5\pi}{4}\right)}$$

$$G_{13}^{R}(\boldsymbol{x},\boldsymbol{x}',\omega)\Big|_{x_3=0} \sim \sqrt{\frac{\kappa}{2\pi r}}\frac{\kappa\cos\theta}{\mu R'(\kappa)}\left(2\nu\gamma\mathrm{e}^{-\gamma x_3'} - \chi\mathrm{e}^{-\nu x_3'}\right)\mathrm{e}^{-\mathrm{i}\left(\kappa r - \frac{5\pi}{4}\right)}$$

$$G_{33}^{R}(\boldsymbol{x},\boldsymbol{x}',\omega)\Big|_{x_3=0} \sim \sqrt{\frac{\kappa}{2\pi r}}\frac{\gamma}{\mu R'(\kappa)}\left(\chi\mathrm{e}^{-\gamma x_3'} - 2\kappa^2\mathrm{e}^{-\nu x_3'}\right)\mathrm{e}^{-\mathrm{i}\left(\kappa r + \frac{\pi}{4}\right)}$$

Thus, for a harmonic wave propagating in the positive x_1 axis direction ($\theta = 0$),

$$\mathrm{Re}\left[\tilde{G}_{11}^{R}\mathrm{e}^{\mathrm{i}\omega t}\right]\Big|_{x_3=0} \sim \nu C\left(2\kappa^2\mathrm{e}^{-\gamma x_3'} - \chi\mathrm{e}^{-\nu x_3'}\right)\cos\left(\omega t - \kappa r + \frac{3\pi}{4}\right) \tag{6.5.22a}$$

$$\mathrm{Re}\left[\tilde{G}_{31}^{R}\mathrm{e}^{\mathrm{i}\omega t}\right]\Big|_{x_3=0} \sim \kappa C\left(\chi\mathrm{e}^{-\gamma x_3'} - 2\nu\gamma\mathrm{e}^{-\nu x_3'}\right)\cos\left(\omega t - \kappa r + \frac{5\pi}{4}\right) \tag{6.5.22b}$$

$$\mathrm{Re}\left[\tilde{G}_{13}^{R}\mathrm{e}^{\mathrm{i}\omega t}\right]\Big|_{x_3=0} \sim \kappa C\left(\chi\mathrm{e}^{-\gamma x_3'} - 2\nu\gamma\mathrm{e}^{-\nu x_3'}\right)\cos\left(\omega t - \kappa r + \frac{5\pi}{4}\right) \tag{6.5.22c}$$

$$\mathrm{Re}\left[\tilde{G}_{33}^{R}\mathrm{e}^{\mathrm{i}\omega t}\right]\Big|_{x_3=0} \sim \gamma C\left(\chi\mathrm{e}^{-\gamma x_3'} - 2\kappa^2\mathrm{e}^{-\nu x_3'}\right)\cos\left(\omega t - \kappa r - \frac{\pi}{4}\right) \tag{6.5.22d}$$

(3) For the case where both the observation point and the source point are located underground, substituting Eqs. (6.5.17) and (6.5.19) into Eq. (6.5.16) yields:

$$\tilde{G}_{11}^{R} \sim \sqrt{\frac{\kappa}{\pi r}} \frac{2\kappa^2 \beta^2 \nu \cos^2\theta}{\mu\omega^2 R'(\kappa)} \left[2\left(\kappa^2 e^{PP} + \nu\gamma e^{SS}\right) - \chi\left(e^{SP} + e^{PS}\right)\right] e^{-i\left(\kappa r - \frac{3\pi}{4}\right)}$$

$$\tilde{G}_{31}^{R} \sim \sqrt{\frac{\kappa}{2\pi r}} \frac{\kappa\beta^2 \chi \cos\theta}{\mu\omega^2 R'(\kappa)} \left[\chi\left(e^{PP} + e^{SS}\right) - 2\left(\nu\gamma e^{SP} + \kappa^2 e^{PS}\right)\right] e^{-i\left(\kappa r - \frac{\pi}{4}\right)}$$

$$\tilde{G}_{13}^{R} \sim \sqrt{\frac{\kappa}{2\pi r}} \frac{\kappa\beta^2 \chi \cos\theta}{\mu\omega^2 R'(\kappa)} \left[\chi\left(e^{PP} + e^{SS}\right) - 2\left(\kappa^2 e^{SP} + \nu\gamma e^{PS}\right)\right] e^{-i\left(\kappa r - \frac{5\pi}{4}\right)}$$

$$\tilde{G}_{33}^{R} \sim \sqrt{\frac{2\kappa}{\pi r}} \frac{\kappa^2 \beta^2 \gamma}{\mu\omega R'(\kappa)} \left[2\left(\nu\gamma e^{PP} + \kappa^2 e^{SS}\right) - \chi\left(e^{SP} + e^{PS}\right)\right] e^{-i\left(\kappa r - \frac{3\pi}{4}\right)}$$

Thus, for a harmonic wave propagating in the positive x_1 axis direction ($\theta = 0$), we have

$$\mathrm{Re}\left[\tilde{G}_{11}^{R} e^{i\omega t}\right] \sim 2\kappa\nu C' \left[2\left(\kappa^2 e^{PP} + \nu\gamma e^{SS}\right) - \chi\left(e^{SP} + e^{PS}\right)\right] \cos\left(\omega t - \kappa r + \frac{3\pi}{4}\right) \tag{6.5.23a}$$

$$\mathrm{Re}\left[\tilde{G}_{31}^{R} e^{i\omega t}\right] \sim \chi C' \left[\chi\left(e^{PP} + e^{SS}\right) - 2\left(\nu\gamma e^{SP} + \kappa^2 e^{PS}\right)\right] \cos\left(\omega t - \kappa r + \frac{\pi}{4}\right) \tag{6.5.23b}$$

$$\mathrm{Re}\left[\tilde{G}_{13}^{R} e^{i\omega t}\right] \sim \chi C' \left[\chi\left(e^{PP} + e^{SS}\right) - 2\left(\kappa^2 e^{SP} + \nu\gamma e^{PS}\right)\right] \cos\left(\omega t - \kappa r + \frac{5\pi}{4}\right) \tag{6.5.23c}$$

$$\mathrm{Re}\left[\tilde{G}_{33}^{R} e^{i\omega t}\right] \sim 2\kappa\gamma C' \left[2\left(\nu\gamma e^{PP} + \kappa^2 e^{SS}\right) - \chi\left(e^{SP} + e^{PS}\right)\right] \cos\left(\omega t - \kappa r + \frac{3\pi}{4}\right) \tag{6.5.23d}$$

where $C' = \sqrt{\kappa/(2\pi r)}\kappa\beta^2/[\mu\omega^2 R'(\kappa)]$.

6.5.3.4 Near-Field Rayleigh Wave

The particle motion equations for far-field Rayleigh wave under different combinations of source and field depths are examined in detail in the previous sections. In Chap. 7, we will present the results calculated based on these equations. We will see that, although the particle motion amplitudes vary with source depth, the overall characteristics of the particle motion are consistent with the classical planar Rayleigh wave results. But what happens when the observation point is close to the source? This involves the characteristics of near-field Rayleigh wave.

In the near-field case, the asymptotic expression of the Hankel function in Eq. (6.5.17) is no longer valid, so the exact Hankel function representation must be used. Note that the second type of Hankel function is given by

$$H_m^{(2)}(x) = J_m(x) - \mathrm{i}Y_m(x)$$

where $J_m(x)$ and $Y_m(x)$ are the Bessel functions of the first and second types of order m, respectively. Therefore, according to Eq. (6.5.16), we have

$$\begin{aligned}\mathrm{Re}\left[\tilde{G}_{11}^{R}\mathrm{e}^{\mathrm{i}\omega t}\right] &= -\frac{1}{4\mu R'(\kappa)}\left\{\left[Y_0(\kappa r) - Y_2(\kappa r)\cos 2\theta\right]\cos\omega t\right.\\ &\quad \left. - \left[J_0(\kappa r) - J_2(\kappa r)\cos 2\theta\right]\sin\omega t\right\}F_1(\kappa r) \qquad (6.5.24\text{a})\end{aligned}$$

$$\mathrm{Re}\left[\tilde{G}_{31}^{R}\mathrm{e}^{\mathrm{i}\omega t}\right] = -\frac{\cos\theta}{4\mu R'(\kappa)}\left[Y_1(\kappa r)\cos\omega t - J_1(\kappa r)\sin\omega t\right]F_4(\kappa r) \qquad (6.5.24\text{b})$$

$$\mathrm{Re}\left[\tilde{G}_{13}^{R}\mathrm{e}^{\mathrm{i}\omega t}\right] = -\frac{\cos\theta}{4\mu R'(\kappa)}\left[Y_1(\kappa r)\cos\omega t - J_1(\kappa r)\sin\omega t\right]F_3(\kappa r) \qquad (6.5.24\text{c})$$

$$\mathrm{Re}\left[\tilde{G}_{33}^{R}\mathrm{e}^{\mathrm{i}\omega t}\right] = -\frac{1}{4\mu R'(\kappa)}\left[Y_0(\kappa r)\cos\omega t - J_0(\kappa r)\sin\omega t\right]F_5(\kappa r) \qquad (6.5.24\text{d})$$

According to Eq. (6.5.24), to calculate the particle motion trajectories of near-field Rayleigh wave, the values of the first and second types of Bessel functions must be computed precisely. In Sect. 7.5.2, we will present the relevant numerical results.

6.6 Solutions to Shear Dislocation Point Source

In Sect. 4.6, the problem of the seismic wave field caused by a shear dislocation point source in an infinite medium was considered. According to Eq. (4.6.1), in the frequency domain, we have

$$u_n(\boldsymbol{x}, \omega) = M_{pq}(\omega)G_{np,q'}(\boldsymbol{x}, \boldsymbol{x}', \omega) \qquad (6.6.1)$$

where the spectral representation of the seismic moment tensor $M_{pq}(\omega)$ is given by the spectrum corresponding to Eq. (3.4.8a), with $M_0(\omega) = M_0 S(\omega)$, where M_0 is the scalar seismic moment and $S(\omega)$ is the spectrum of the source time function. To distinguish between derivatives with respect to the field point coordinates and the source point coordinates, the derivative of the Green's function with respect to the spatial coordinates is written as $G_{np,q'}$ in Eq. (6.6.1), indicating it is the derivative with respect to the source point coordinate x_q'. For the problem of calculating the radiated displacement field, $M_{pq}(\omega)$ in Eq. (6.6.1) is a known quantity. Therefore, it is only necessary to compute the spatial derivatives of the Green's function components $G_{ij}(\boldsymbol{x}, \boldsymbol{x}', \omega)$ as previously derived.

6.6.1 Displacement Generated by Shear Dislocation Point Source

According to Eqs. (6.5.1), (6.5.2), and (6.4.23), it is clear that A^{T}, A^{SS}, A^{SR}, A^{RS}, and A^{RR} are only related to the vertical coordinates x_3 and x_3'. The terms related to the horizontal coordinates are contained within the azimuthal factor γ_i and the Bessel function-related terms.

Taking the derivative of Eq. (6.4.23) with respect to x_3', we get

$$A^{\mathrm{T}}_{,3'} = \nu\left[\zeta(z)\mathrm{e}^{\mathrm{S}} - \mathrm{e}^{\mathrm{SS}}\right] \tag{6.6.2a}$$

$$A^{\mathrm{SS}}_{,3'} = \gamma\zeta(z)\left(k^2\mathrm{e}^{\mathrm{P}} - \nu^2\mathrm{e}^{\mathrm{S}}\right) + \frac{\gamma R^*(k)}{R(k)}\left(k^2\mathrm{e}^{\mathrm{PP}} + \nu^2\mathrm{e}^{\mathrm{SS}}\right) - \frac{4k^2\nu\gamma\chi}{R(k)}\left(\nu\mathrm{e}^{\mathrm{SP}} + \gamma\mathrm{e}^{\mathrm{PS}}\right) \tag{6.6.2b}$$

$$A^{\mathrm{SR}}_{,3'} = k\gamma\left[\left(\gamma\mathrm{e}^{\mathrm{P}} - \nu\mathrm{e}^{\mathrm{S}}\right) - \frac{R^*(k)}{R(k)}\left(\gamma\mathrm{e}^{\mathrm{PP}} + \nu\mathrm{e}^{\mathrm{SS}}\right) + \frac{4\nu\chi}{R(k)}\left(k^2\mathrm{e}^{\mathrm{SP}} + \gamma^2\mathrm{e}^{\mathrm{PS}}\right)\right] \tag{6.6.2c}$$

$$A^{\mathrm{RS}}_{,3'} = k\nu\left[\left(\gamma\mathrm{e}^{\mathrm{P}} - \nu\mathrm{e}^{\mathrm{S}}\right) + \frac{R^*(k)}{R(k)}\left(\gamma\mathrm{e}^{\mathrm{PP}} + \nu\mathrm{e}^{\mathrm{SS}}\right) - \frac{4\gamma\chi}{R(k)}\left(\nu^2\mathrm{e}^{\mathrm{SP}} + k^2\mathrm{e}^{\mathrm{PS}}\right)\right] \tag{6.6.2d}$$

$$A^{\mathrm{RR}}_{,3'} = \nu\zeta(z)\left(\gamma^2\mathrm{e}^{\mathrm{P}} - k^2\mathrm{e}^{\mathrm{S}}\right) - \frac{\nu R^*(k)}{R(k)}\left(\gamma^2\mathrm{e}^{\mathrm{PP}} + k^2\mathrm{e}^{\mathrm{SS}}\right) + \frac{4k^2\nu\gamma\chi}{R(k)}\left(\nu\mathrm{e}^{\mathrm{SP}} + \gamma\mathrm{e}^{\mathrm{PS}}\right) \tag{6.6.2e}$$

Differentiating Eq. (6.5.2) with respect to x_i', and noting that $r_{,\alpha'} = -\gamma_\alpha$, we use the recurrence relations for the Bessel functions:

$$J_m'(x) = \frac{J_{m-1}(x) - J_{m+1}(x)}{2}, \quad \frac{J_m(x)}{x} = \frac{J_{m-1}(x) + J_{m+1}(x)}{2m}$$

and $J_{-m}(x) = (-1)^m J_m(x)$ to obtain

$$I_{1,\alpha'} = -\frac{\gamma_\alpha}{4}\int_0^\infty \left\{\frac{\beta^2 A^{\mathrm{SS}}}{\gamma\omega^2}\left[J_3(kr) - 3J_1(kr)\right] - \frac{A^{\mathrm{T}}}{\nu}\left[J_3(kr) + J_1(kr)\right]\right\} k^2\,\mathrm{d}k \tag{6.6.3a}$$

$$I_{2,\alpha'} = \frac{\gamma_\alpha}{4}\int_0^\infty \left\{\frac{\beta^2 A^{\mathrm{SS}}}{\gamma\omega^2}\left[J_3(kr) + J_1(kr)\right] - \frac{A^{\mathrm{T}}}{\nu}\left[J_3(kr) - 3J_1(kr)\right]\right\} k^2\,\mathrm{d}k \tag{6.6.3b}$$

$$I_{3,\alpha'} = \frac{\gamma_\alpha\beta^2}{2\omega^2}\int_0^\infty \frac{A^{\mathrm{SR}}}{\gamma}\left[J_0(kr) - J_2(kr)\right]k^2\,\mathrm{d}k \tag{6.6.3c}$$

$$I_{4,\alpha'} = \frac{\gamma_\alpha\beta^2}{2\omega^2}\int_0^\infty \frac{A^{\mathrm{RS}}}{\nu}\left[J_0(kr) - J_2(kr)\right]k^2\,\mathrm{d}k \tag{6.6.3d}$$

$$I_{5,\alpha'} = -\frac{\gamma_\alpha \beta^2}{\omega^2}\int_0^\infty \frac{A^{\mathrm{RR}}}{\nu} J_1(kr)k^2\,\mathrm{d}k \tag{6.6.3e}$$

$$I_{1,3'} = \frac{1}{2}\int_0^\infty \left\{\frac{\beta^2 A^{\mathrm{SS}}_{,3'}}{\gamma\omega^2}\big[J_0(kr) - J_2(kr)\big] + \frac{A^{\mathrm{T}}_{,3'}}{\nu}\big[J_0(kr) + J_2(kr)\big]\right\} k\,\mathrm{d}k \tag{6.6.3f}$$

$$I_{2,3'} = \frac{1}{2}\int_0^\infty \left\{\frac{\beta^2 A^{\mathrm{SS}}_{,3'}}{\gamma\omega^2}\big[J_0(kr) + J_2(kr)\big] + \frac{A^{\mathrm{T}}_{,3'}}{\nu}\big[J_0(kr) - J_2(kr)\big]\right\} k\,\mathrm{d}k \tag{6.6.3g}$$

$$I_{3,3'} = -\frac{\beta^2}{\gamma\omega^2}\int_0^\infty \frac{A^{\mathrm{SR}}_{,3'}}{\gamma} J_1(kr)k\,\mathrm{d}k \tag{6.6.3h}$$

$$I_{4,3'} = -\frac{\beta^2}{\nu\omega^2}\int_0^\infty \frac{A^{\mathrm{RS}}_{,3'}}{\nu} J_1(kr)k\,\mathrm{d}k \tag{6.6.3i}$$

$$I_{5,3'} = -\frac{\beta^2}{\nu\omega^2}\int_0^\infty \frac{A^{\mathrm{RR}}_{,3'}}{\nu} A^{\mathrm{RR}}_{,3'} J_0(kr)k\,\mathrm{d}k \tag{6.6.3j}$$

where $\alpha = 1, 2$.

Noting that $\gamma_{n,q'} = \gamma_n\gamma_q - \delta_{nq}/r$, differentiating Eq. (6.5.1) with respect to x_i', we get

$$\tilde{G}_{11,1'} = \varpi\left[(I_2 - I_1)\sin\theta\sin 2\theta + r\left(I_{1,1'}\cos^2\theta + I_{2,1'}\sin^2\theta\right)\right] \tag{6.6.4a}$$

$$\tilde{G}_{12,1'} = \tilde{G}_{21,1'} = \varpi\left[(I_1 - I_2)\cos 2\theta + \left(I_{1,1'} - I_{2,1'}\right) r\cos\theta\right]\sin\theta \tag{6.6.4b}$$

$$\tilde{G}_{13,1'} = -\varpi\left(I_3\sin^2\theta - I_{3,1'}r\cos\theta\right) \tag{6.6.4c}$$

$$\tilde{G}_{22,1'} = \varpi\left[(I_1 - I_2)\sin\theta\sin 2\theta + r\left(I_{1,1'}\sin^2\theta + I_{2,1'}\cos^2\theta\right)\right] \tag{6.6.4d}$$

$$\tilde{G}_{23,1'} = \varpi\left(I_3\cos\theta + I_{3,1'}r\right)\sin\theta \tag{6.6.4e}$$

$$\tilde{G}_{31,1'} = -\varpi\left(I_4\sin^2\theta - I_{4,1'}r\cos\theta\right) \tag{6.6.4f}$$

$$\tilde{G}_{32,1'} = \varpi\left(I_4\cos\theta + I_{4,1'}r\right)\sin\theta, \quad \tilde{G}_{33,1'} = \varpi r I_{5,1'} \tag{6.6.4g}$$

$$\tilde{G}_{11,2'} = \varpi\left[(I_1 - I_2)\cos\theta\sin 2\theta + r\left(I_{1,2'}\cos^2\theta + I_{2,2'}\sin^2\theta\right)\right] \tag{6.6.4h}$$

$$\tilde{G}_{12,2'} = \tilde{G}_{21,2'} = \varpi\left[(I_2 - I_1)\cos 2\theta + \left(I_{1,2'} - I_{2,2'}\right) r\sin\theta\right]\cos\theta \tag{6.6.4i}$$

$$\tilde{G}_{13,2'} = \varpi\left(I_3\sin\theta + I_{3,2'}r\right)\cos\theta \tag{6.6.4j}$$

$$\tilde{G}_{23,2'} = -\varpi\left(I_3\cos^2\theta - I_{3,2'}r\sin\theta\right) \tag{6.6.4k}$$

$$\tilde{G}_{22,2'} = \varpi\left[(I_2 - I_1)\cos\theta\sin 2\theta + r\left(I_{1,2'}\sin^2\theta + I_{2,2'}\cos^2\theta\right)\right] \tag{6.6.4l}$$

$$\tilde{G}_{31,2'} = \varpi\left(I_4\sin\theta + I_{4,2'}r\right)\cos\theta \tag{6.6.4m}$$

$$\tilde{G}_{32,2'} = -\varpi\left(I_4\cos^2\theta - I_{4,2'}r\sin\theta\right), \quad \tilde{G}_{33,2'} = \varpi r I_{5,2'} \tag{6.6.4n}$$

where $\varpi = 1/(4\pi\mu r)$, and

$$\tilde{\mathbf{G}}_{,3} = \frac{1}{4\pi\mu}\begin{bmatrix} I_{1,3'}\cos^2\theta + I_{2,3'}\sin^2\theta & (I_{1,3'} - I_{2,3'})\cos\theta\sin\theta & I_{3,3'}\cos\theta \\ (I_{1,3'} - I_{2,3'})\cos\theta\sin\theta & I_{1,3'}\sin^2\theta + I_{2,3'}\cos^2\theta & I_{3,3'}\sin\theta \\ I_{4,3'}\cos\theta & I_{4,3'}\sin\theta & I_{5,3'} \end{bmatrix} \tag{6.6.5}$$

Equations (6.6.4) and (6.6.5), along with Eqs. (6.6.3), (6.6.2), (6.5.2), and (6.4.23), form a comprehensive formula for calculating the spatial derivative $G_{np,q'}$ of the frequency-domain half-space Green's function. The specific substitution relationships are shown below:

$$\begin{array}{ccccc} A^{\mathrm{T}},\ A^{\mathrm{SS}},\ A^{\mathrm{SR}},\ A^{\mathrm{RS}},\ A^{\mathrm{RR}} & \longrightarrow & I_i & \xrightarrow{\ \frac{\partial}{\partial x'_\alpha}\ } & I_{i,\alpha'} \longrightarrow \tilde{G}_{ij,k'} \\ \Big\downarrow \frac{\partial}{\partial x'_3} & & & & \nearrow \\ A^{\mathrm{T}}_{,3'},\ A^{\mathrm{SS}}_{,3'},\ A^{\mathrm{SR}}_{,3'},\ A^{\mathrm{RS}}_{,3'},\ A^{\mathrm{RR}}_{,3'} & & \longrightarrow & & I_{i,3'} \end{array}$$

Substituting Eqs. (6.6.4) and (6.6.5) into Eq. (6.6.1) yields the displacement field spectrum for a shear dislocation point source. The time-domain displacement is then obtained through inverse Fourier transform

$$u_n(\boldsymbol{x}, t) = \mathscr{F}^{-1}\{u_n(\boldsymbol{x}, \omega)\} = \mathscr{F}^{-1}\left\{M_{pq}(\omega)G_{np,q'}(\boldsymbol{x}, \boldsymbol{x}', \omega)\right\} \tag{6.6.6}$$

Since the inverse Fourier transform requires computer computation, it is difficult to perform a theoretical analysis similar to that done for the infinite-space Green's function in Chap. 4, starting from Eq. (6.6.6). In the next chapter, we will explore the relevant properties through numerical examples.

6.6.2 Rayleigh Wave Generated by Shear Dislocation Point Source

Based on Eq. (6.5.16), differentiating with respect to the source coordinate x'_i and substituting into Eq. (6.6.1) yields the Rayleigh wave generated by a shear dislocation source. Differentiating Eq. (6.5.19) with respect to x'_3 gives

$$F_{1,3'}(\kappa) = \frac{2\kappa^3\nu\beta^2}{\omega^2}\left[2\gamma\left(\kappa^2 e^{\mathrm{PP}} + \nu^2 e^{\mathrm{SS}}\right) - \chi\left(\nu e^{\mathrm{SP}} + \gamma e^{\mathrm{PS}}\right)\right] \tag{6.6.7a}$$

$$F_{3,3'}(\kappa) = \frac{2\kappa^2\beta^2\chi}{\omega^2}\left[\chi\left(\gamma e^{\mathrm{PP}} + \nu e^{\mathrm{SS}}\right) - 2\nu\left(\kappa^2 e^{\mathrm{SP}} + \gamma^2 e^{\mathrm{PS}}\right)\right] \tag{6.6.7b}$$

$$F_{4,3'}(\kappa) = -\frac{2\kappa^2\beta^2\chi}{\omega^2}\left[\chi\left(\gamma e^{\mathrm{PP}} + \nu e^{\mathrm{SS}}\right) - 2\gamma\left(\nu^2 e^{\mathrm{SP}} + \kappa^2 e^{\mathrm{PS}}\right)\right] \tag{6.6.7c}$$

$$F_{5,3'}(\kappa) = \frac{4\kappa^3\gamma\beta^2}{\omega^2}\left[2\nu\left(\gamma^2 e^{\mathrm{PP}} + \kappa^2 e^{\mathrm{SS}}\right) - \chi\left(\nu e^{\mathrm{SP}} + \gamma e^{\mathrm{PS}}\right)\right] \tag{6.6.7d}$$

Notice that

$$r_{,q'} = -\gamma_q, \quad \gamma_{n,q'} = \frac{\gamma_n \gamma_q - \delta_{nq}}{r}$$

and

$$\frac{\mathrm{d}}{\mathrm{d}x} H_m^{(2)}(x) = \frac{H_{m-1}^{(2)}(x) - H_{m+1}^{(2)}(x)}{2}, \quad H_{-1}^{(2)}(x) = -H_1^{(2)}(x)$$

it can be obtained from Eq. (6.5.16) that

$$\begin{aligned}
\tilde{G}^R_{11,1'} &= \vartheta_1 F_1(\kappa)\left[2H_1^{(2)}(x)x + 8H_2^{(2)}(x)\sin^2\theta + M_{13}(x)x\cos 2\theta\right]\cos\theta \\
\tilde{G}^R_{12,1'} &= \tilde{G}^R_{21,1'} = -\vartheta_1 F_1(\kappa)\left[4H_2^{(2)}(x)\sin\theta\cos 2\theta + M_{31}(x)x\cos\theta\sin 2\theta\right] \\
\tilde{G}^R_{13,1'} &= -\vartheta_1 F_3(\kappa)\left[2H_1^{(2)}(x)\sin^2\theta + M_{02}(x)x\cos^2\theta\right] \\
\tilde{G}^R_{22,1'} &= \vartheta_1 F_1(\kappa)\left[2H_1^{(2)}(x)x - 8H_2^{(2)}(x)\sin^2\theta - M_{13}(x)x\cos 2\theta\right]\cos\theta \\
\tilde{G}^R_{23,1'} &= \tilde{G}^R_{13,2'} = \vartheta_1 F_3(\kappa)\left[2H_1^{(2)}(x) - M_{02}(x)x\right]\cos\theta\sin\theta \\
\tilde{G}^R_{31,1'} &= -\vartheta_1 F_4(\kappa)\left[2H_1^{(2)}(x)\sin^2\theta + M_{02}(x)x\cos^2\theta\right] \\
\tilde{G}^R_{32,1'} &= \tilde{G}^R_{31,2'} = \vartheta_1 F_4(\kappa)\left[2H_1^{(2)}(x) - M_{02}(x)x\right]\cos\theta\sin\theta \\
\tilde{G}^R_{33,1'} &= \vartheta_1 F_5(\kappa) 2H_1^{(2)}(x)x\cos\theta \\
\tilde{G}^R_{11,2'} &= \vartheta_1 F_1(\kappa)\left[2H_1^{(2)}(x)x - 8H_2^{(2)}(x)\cos^2\theta + M_{13}(x)x\cos 2\theta\right]\sin\theta \\
\tilde{G}^R_{12,2'} &= \tilde{G}^R_{21,2'} = \vartheta_1 F_1(\kappa)\left[4H_2^{(2)}(x)\cos\theta\cos 2\theta - M_{31}(x)x\sin\theta\sin 2\theta\right] \\
\tilde{G}^R_{22,2'} &= \vartheta_1 F_1(\kappa)\left[2H_1^{(2)}(x)x + 8H_2^{(2)}(x)\cos^2\theta - M_{13}(x)x\cos 2\theta\right]\sin\theta \\
\tilde{G}^R_{23,2'} &= -\vartheta_1 F_3(\kappa)\left[2H_1^{(2)}(x)\cos^2\theta + M_{02}(x)x\sin^2\theta\right] \\
\tilde{G}^R_{32,2'} &= -\vartheta_1 F_4(\kappa)\left[2H_1^{(2)}(x)\cos^2\theta + M_{02}(x)x\sin^2\theta\right] \\
\tilde{G}^R_{33,2'} &= \vartheta_1 F_5(\kappa) 2H_1^{(2)}(x)x\sin\theta \\
\tilde{G}^R_{11,3'} &= \vartheta_2 F_{1,3'}(\kappa)\left[H_0^{(2)}(x) - H_2^{(2)}(x)\cos 2\theta\right] \\
\tilde{G}^R_{12,3'} &= \tilde{G}^R_{21,3'} = -\vartheta_2 F_{1,3'}(\kappa) H_2^{(2)}(x)\sin 2\theta \\
\tilde{G}^R_{22,3'} &= \vartheta_2 F_{1,3'}(\kappa)\left[H_0^{(2)}(x) + H_2^{(2)}(x)\cos 2\theta\right] \\
\tilde{G}^R_{13,3'} &= \vartheta_2 F_{3,3'}(\kappa) H_1^{(2)}(x)\cos\theta \\
\tilde{G}^R_{23,3'} &= \vartheta_2 F_{3,3'}(\kappa) H_1^{(2)}(x)\sin\theta
\end{aligned}$$

$$\tilde{G}^R_{31,3'} = \vartheta_2 F_{4,3'}(\kappa) H_1^{(2)}(x)\cos\theta$$
$$\tilde{G}^R_{32,3'} = \vartheta_2 F_{4,3'}(\kappa) H_1^{(2)}(x)\sin\theta$$
$$\tilde{G}^R_{33,3'} = \vartheta_2 F_{5,3'}(\kappa) H_0^{(2)}(x)$$

where, $x = \kappa r$, $\vartheta_1 = -\mathrm{i}/[8\mu r R'(\kappa)]$, $\vartheta_2 = 2r\vartheta_1$, and $M_{mn}(x) = H_m^{(2)}(x) - H_n^{(2)}(x)$. Substituting these expressions into Eq. (6.6.1) yields the Rayleigh wave displacement component $\tilde{u}_i^R$ generated by a shear dislocation point source:

$$u_i^R(\boldsymbol{x},\omega) = M_{pq}(\omega) G^R_{ip,q'}(\boldsymbol{x},\boldsymbol{x}',\omega) \tag{6.6.8}$$

6.7 Static Solution at Surface for Lamb's Problem

In Sect. 6.4.1.2, during the discussion of the solution system for the P-SV case, it was mentioned that when $\omega = 0$, there is a situation where the eigenvalues are equal. This scenario requires special consideration. In fact, the equality of eigenvalues does not only occur when $\omega = 0$. When $\omega \neq 0$, $k = \omega/\alpha$ or $k = \omega/\beta$ can also result in equal eigenvalues. However, for a fixed frequency, this occurs only for isolated k values, which can be neglected in the context of integration. Therefore, it is sufficient to consider only $\omega = 0$. In this case, for the P-SV case, there is always a degeneracy regardless of the value of k, while for the SH case, no degeneracy exists (the isolated case of $k = 0$ is also neglected in the context of integration).

Below, we study the static Green's function for the $\mathcal{LP}_2$.[26]

6.7.1 Static Green's Function for $\mathcal{LP}_2$

6.7.1.1 Expansion Coefficients of Displacement in SH Case

When $\omega = 0$, for the SH case, there are no equal eigenvalues. Therefore, the analysis in Sect. 6.4.1.1 remains valid, and the corresponding results can be obtained by simply setting $\omega = 0$. Equation (6.4.22a) can be written as

$$u_m^T(k,z)\Big|_{z=0} = \frac{\mathrm{e}^{-kz_s}}{k\mu}\mathscr{F}_m^T \tag{6.7.1}$$

where the expression for $\mathscr{F}_m^T$ can be found in Eq. (6.4.15).

[26] Although the method in this section is not limited to the $\mathcal{LP}_2$, we focus on it because, for $z_s > 0$ and $z = 0$, the results have a more concise form. Interested readers can extend the method to the $\mathcal{LP}_3$ on their own.

6.7.1.2 Expansion Coefficients of Displacement in P-SV Case

When $\omega = 0$, the coefficient matrix for the P-SV case is

$$\mathbf{A} = \frac{1}{\mu}\begin{bmatrix} 0 & k\mu & 1 & 0 \\ -k\mu(\xi' - \xi) & 0 & 0 & \xi \\ 4k^2\mu^2\xi' & 0 & 0 & k\mu(\xi' - \xi) \\ 0 & 0 & -k\mu & 0 \end{bmatrix} \quad \left(\xi = \frac{\mu}{\lambda + 2\mu},\ \xi' = 1 - \xi\right)$$

In this case, $\mathbf{A}$ is not similar to a diagonal matrix but is in Jordan form $\mathbf{J}$[27]

$$\mathbf{J} = \begin{bmatrix} \mathbf{J}_1 & \mathbf{0} \\ \mathbf{0} & \mathbf{J}_2 \end{bmatrix}, \quad \mathbf{J}_1 \triangleq k\begin{bmatrix} -1 & \xi' \\ 0 & -1 \end{bmatrix}, \quad \mathbf{J}_2 \triangleq k\begin{bmatrix} 1 & \xi' \\ 0 & 1 \end{bmatrix}$$

The matrices $\mathbf{E}$ and its inverse $\mathbf{E}^{-1}$ are respectively[28]

$$\mathbf{E} = \begin{bmatrix} \mathbf{E}_{11} & \mathbf{E}_{12} \\ \mathbf{E}_{21} & \mathbf{E}_{22} \end{bmatrix}, \quad \mathbf{E}^{-1} = \begin{bmatrix} (\mathbf{E}^{-1})_{11} & (\mathbf{E}^{-1})_{12} \\ (\mathbf{E}^{-1})_{21} & (\mathbf{E}^{-1})_{22} \end{bmatrix} \tag{6.7.2}$$

where,

---

[27] In general linear algebra or advanced algebra, a Jordan matrix is defined as

$$\mathbf{J} = \begin{bmatrix} \mathbf{J}_1 & \mathbf{0} \\ \mathbf{0} & \mathbf{J}_2 \end{bmatrix}, \quad \mathbf{J}_1 \triangleq \begin{bmatrix} -k & 1 \\ 0 & -k \end{bmatrix}, \quad \mathbf{J}_2 \triangleq \begin{bmatrix} k & 1 \\ 0 & k \end{bmatrix}$$

However, in the current problem, k is a physical quantity with the dimension of wavenumber. The elements at positions (1, 2) and (3, 4) of the matrix have different dimensions compared to k, which is inconvenient. This matrix needs to be modified to make all elements dimensionally consistent. We should replace 1 with quantity with the same dimensions as k. Different choices will lead to different levels of complexity in the representation of $\mathbf{E}$. Through trial and error, it was found that choosing $k\xi'$ leads to a relatively simple expression for $\mathbf{E}$.

[28] Since $\mathbf{E}^{-1}\mathbf{A}\mathbf{E} = \mathbf{J}$, it follows that $\mathbf{A}\mathbf{E} = \mathbf{E}\mathbf{J}$. If we denote $\mathbf{E} = [\boldsymbol{e}_1, \boldsymbol{e}_2, \boldsymbol{e}_3, \boldsymbol{e}_4]$, then we have

$$\mathbf{A}[\boldsymbol{e}_1, \boldsymbol{e}_2, \boldsymbol{e}_3, \boldsymbol{e}_4] = [\boldsymbol{e}_1, \boldsymbol{e}_2, \boldsymbol{e}_3, \boldsymbol{e}_4]\mathbf{J} = [-k\boldsymbol{e}_1, k\xi'\boldsymbol{e}_1 - k\boldsymbol{e}_2, k\boldsymbol{e}_3, k\xi'\boldsymbol{e}_3 + k\boldsymbol{e}_4]$$

Given $(k\mathbf{I} + \mathbf{A})\boldsymbol{e}_1 = 0$ and $(k\mathbf{I} - \mathbf{A})\boldsymbol{e}_3 = 0$, we can choose

$$\boldsymbol{e}_1 = [-1, -1, 2k\mu, 2k\mu]^{\mathrm{T}}, \quad \boldsymbol{e}_3 = [1, -1, 2k\mu, -2k\mu]^{\mathrm{T}}$$

The vectors $\boldsymbol{e}_2$ and $\boldsymbol{e}_4$ satisfy $(k\mathbf{I} + \mathbf{A})\boldsymbol{e}_2 = k\xi'\boldsymbol{e}_1$ and $(k\mathbf{I} - \mathbf{A})\boldsymbol{e}_4 = -k\xi'\boldsymbol{e}_3$ respectively, leading to

$$\boldsymbol{e}_2 = [1, -\xi, -2k\mu\xi', 0]^{\mathrm{T}}, \quad \boldsymbol{e}_4 = [1, \xi, 2k\mu\xi', 0]^{\mathrm{T}}$$

Thus, we obtain $\mathbf{E}$.

$$\mathbf{E}_{11} \triangleq \begin{bmatrix} -1 & 1 \\ -1 & -\xi \end{bmatrix}, \qquad \mathbf{E}_{12} \triangleq \begin{bmatrix} 1 & 1 \\ -1 & \xi \end{bmatrix}$$

$$\mathbf{E}_{21} \triangleq 2k\mu \begin{bmatrix} 1 & -\xi' \\ 1 & 0 \end{bmatrix}, \quad \mathbf{E}_{22} \triangleq 2k\mu \begin{bmatrix} 1 & \xi' \\ -1 & 0 \end{bmatrix}$$

$$\left(\mathbf{E}^{-1}\right)_{11} \triangleq \frac{1}{2} \begin{bmatrix} 0 & -\xi' \\ 1 & -1 \end{bmatrix}, \left(\mathbf{E}^{-1}\right)_{12} \triangleq \frac{1}{4k\mu} \begin{bmatrix} \xi & 1 \\ -1 & 1 \end{bmatrix}$$

$$\left(\mathbf{E}^{-1}\right)_{21} \triangleq \frac{1}{2} \begin{bmatrix} 0 & -\xi' \\ 1 & 1 \end{bmatrix}, \left(\mathbf{E}^{-1}\right)_{22} \triangleq \frac{1}{4k\mu} \begin{bmatrix} \xi & -1 \\ 1 & 1 \end{bmatrix}$$

Replace $\mathbf{E}$ and $\mathbf{E}^{-1}$ in Eq. (6.3.19) with the forms in Eq. (6.7.2), and replace $\mathbf{Q}$ and $\mathbf{Q}^{-1}$ with[29]

$$\mathbf{Q}(z) = \begin{bmatrix} e^{-kz}\mathbf{Q}_1(z) & \mathbf{0} \\ \mathbf{0} & e^{kz}\mathbf{Q}_1(z) \end{bmatrix}, \quad \mathbf{Q}^{-1}(z) = \begin{bmatrix} e^{kz}\mathbf{Q}_2(z) & \mathbf{0} \\ \mathbf{0} & e^{-kz}\mathbf{Q}_2(z) \end{bmatrix} \tag{6.7.3}$$

where,

$$\mathbf{Q}_1(z) \triangleq \begin{bmatrix} 1 & k\xi' z \\ 0 & 1 \end{bmatrix}, \quad \mathbf{Q}_2(z) \triangleq \begin{bmatrix} 1 & -k\xi' z \\ 0 & 1 \end{bmatrix}$$

Therefore, according to Eq. (6.4.11), we obtain

$$\begin{aligned} \boldsymbol{F}_1(z) &= \int_{z_s^-}^{z} e^{k\eta}\mathbf{Q}_2(\eta) \left[\mathbf{E}^{-1}\right]_{12} \boldsymbol{f}_2(\eta)\, d\eta \\ &= -\frac{e^{k\eta}}{4k\mu} \int_{z_s^-}^{z} \begin{bmatrix} 1 & -k\xi' z \\ 0 & 1 \end{bmatrix} \begin{bmatrix} \xi & 1 \\ -1 & 1 \end{bmatrix} \begin{bmatrix} \mathscr{F}_m^S \\ \mathscr{F}_m^R \end{bmatrix} \delta(\eta - z_s)\, d\eta \\ &= H(z - z_s)\boldsymbol{s}_1 \end{aligned} \tag{6.7.4a}$$

---

[29] To find the general solution $\boldsymbol{y}_g$ of the homogeneous equation, we need to solve the system of ordinary differential equations

$$\frac{d}{dz}\boldsymbol{Y}(z) = \mathbf{J}\boldsymbol{Y}(z), \quad \boldsymbol{Y}(z) = \mathbf{E}^{-1}\boldsymbol{y}_g(z).$$

Note that the current $\mathbf{J}$ is not a diagonal matrix. $Y_2(z)$ and $Y_4(z)$ satisfy

$$\frac{d}{dz}Y_2(z) + kY_2(z) = 0, \quad \frac{d}{dz}Y_4(z) - kY_4(z) = 0,$$

with solutions $Y_2(z) = C_2 e^{-kz}$ and $Y_4(z) = C_4 e^{kz}$. Consequently, the ordinary differential equations for $Y_1(z)$ and $Y_3(z)$ are

$$\frac{d}{dz}Y_1(z) + kY_1(z) = C_2 k\xi' e^{-kz}, \quad \frac{d}{dz}Y_3(z) - kY_3(z) = C_4 k\xi' e^{kz},$$

which have solutions $Y_1(z) = (C_1 + C_2 k\xi' z)e^{-kz}$ and $Y_3(z) = (C_3 + C_4 k\xi' z)e^{kz}$. Writing this in matrix form gives us $\mathbf{Q}(z)$.

$$
\begin{aligned}
\boldsymbol{F}_2(z) &= \int_{z_s^-}^{z} \mathrm{e}^{-k\eta}\mathbf{Q}_2(\eta)\left[\mathbf{E}^{-1}\right]_{22} \boldsymbol{f}_2(\eta)\,\mathrm{d}\eta \\
&= -\frac{\mathrm{e}^{-k\eta}}{4k\mu}\int_{z_s^-}^{z} \begin{bmatrix} 1 & -k\xi' z \\ 0 & 1 \end{bmatrix} \begin{bmatrix} \xi & -1 \\ 1 & 1 \end{bmatrix} \begin{bmatrix} \mathscr{F}_m^S \\ \mathscr{F}_m^R \end{bmatrix} \delta(\eta - z_s)\,\mathrm{d}\eta \\
&= H(z - z_s)\boldsymbol{s}_2
\end{aligned} \tag{6.7.4b}
$$

where, the expressions for $\mathscr{F}_m^S$ and $\mathscr{F}_m^R$ can be found in Eq. (6.4.16), and

$$
\boldsymbol{s}_1 = \frac{\mathrm{e}^{kz_s}}{4k\mu}\begin{bmatrix} -\left(\xi + k\xi' z_s\right)\mathscr{F}_m^S - \left(1 - k\xi' z_s\right)\mathscr{F}_m^R \\ \mathscr{F}_m^S - \mathscr{F}_m^R \end{bmatrix} \tag{6.7.5a}
$$

$$
\boldsymbol{s}_2 = \frac{\mathrm{e}^{-kz_s}}{4k\mu}\begin{bmatrix} -\left(\xi - k\xi' z_s\right)\mathscr{F}_m^S + \left(1 + k\xi' z_s\right)\mathscr{F}_m^R \\ -\mathscr{F}_m^S - \mathscr{F}_m^R \end{bmatrix} \tag{6.7.5b}
$$

Under the condition of $\omega = 0$, substitute Eq. (6.4.10) into the boundary condition (6.1.1b),

$$
\begin{aligned}
\boldsymbol{y}_1(k, z)\Big|_{z\to\infty} &= 0 \cdot \mathbf{E}_{11}\mathbf{Q}_1(\infty)\left(\boldsymbol{C}_1 + \boldsymbol{s}_1\right) + \infty \cdot \mathbf{E}_{12}\mathbf{Q}_1(\infty)\left(\boldsymbol{C}_2 + \boldsymbol{s}_2\right) = \mathbf{0} \\
&\Rightarrow \quad \boldsymbol{C}_2 = -\boldsymbol{s}_2 \\
\boldsymbol{y}_2(k, z)\Big|_{z=0} &= \mathbf{E}_{21}\boldsymbol{C}_1 + \mathbf{E}_{22}\boldsymbol{C}_2 = \mathbf{0} \\
&\Rightarrow \boldsymbol{C}_1 = \mathbf{E}_{21}^{-1}\mathbf{E}_{22}\boldsymbol{s}_2 = -\begin{bmatrix} (\boldsymbol{s}_2)_1 \\ \frac{2}{\xi'}(\boldsymbol{s}_2)_1 + (\boldsymbol{s}_2)_2 \end{bmatrix}
\end{aligned}
$$

Substituting Eq. (6.7.5) into Eqs. (6.4.20) and (6.4.21), and then into Eq. (6.4.10), we obtain:

$$
\begin{aligned}
\left[u_m^S,\ u_m^R\right]^{\mathrm{T}} &= \mathrm{e}^{-kz}\mathbf{E}_{11}\mathbf{Q}_1(z)\left[\mathbf{E}_{21}^{-1}\mathbf{E}_{22}\boldsymbol{s}_2 + H(z - z_s)\boldsymbol{s}_1\right] \\
&\quad + \mathrm{e}^{kz}\mathbf{E}_{12}\mathbf{Q}_1(z)\left[-\boldsymbol{s}_2 + H(z - z_s)\boldsymbol{s}_2\right]
\end{aligned}
$$

For the $\mathcal{LP}_2$, the above equation has a more compact form. Let $z = 0$, $\mathbf{Q}_1(0)$ and $\mathbf{Q}_2(0)$ be identity matrices, and $H(-z_s) = 0$. Thus, we have

$$
\begin{bmatrix} u_m^S \\ u_m^R \end{bmatrix}\Bigg|_{z=0} = \frac{\mathrm{e}^{-kz_s}}{2k\mu\xi'}\begin{bmatrix} \left(1 - k\xi' z_s\right)\mathscr{F}_m^S - \left(\xi + k\xi' z_s\right)\mathscr{F}_m^R \\ \left(-\xi + k\xi' z_s\right)\mathscr{F}_m^S + \left(1 + k\xi' z_s\right)\mathscr{F}_m^R \end{bmatrix} \tag{6.7.6}
$$

6.7.1.3 Static Green's Function for $\mathcal{LP}_2$

According to Eq. (6.3.1), the static displacement components at the surface, $u_r^{\mathrm{s}}(r, \theta, 0)$, $u_\theta^{\mathrm{s}}(r, \theta, 0)$, and $u_z^{\mathrm{s}}(r, \theta, 0)$ (where the superscript "s" stands for "static"), are expressed as

$$u_r^{\mathrm{s}} = \frac{1}{2\pi}\sum_{m=-\infty}^{+\infty} \mathrm{e}^{\mathrm{i}m\theta}\int_0^{+\infty}\left[u_m^S(k,0)J_m'(kr) + \frac{\mathrm{i}m}{kr}u_m^T(k,0)J_m(kr)\right]k\,\mathrm{d}k \quad (6.7.7a)$$

$$u_\theta^{\mathrm{s}} = \frac{1}{2\pi}\sum_{m=-\infty}^{+\infty} \mathrm{e}^{\mathrm{i}m\theta}\int_0^{+\infty}\left[\frac{\mathrm{i}m}{kr}u_m^S(k,0)J_m(kr) - u_m^T(k,0)J_m'(kr)\right]k\,\mathrm{d}k \quad (6.7.7b)$$

$$u_z^{\mathrm{s}} = -\frac{1}{2\pi}\sum_{m=-\infty}^{+\infty} \mathrm{e}^{\mathrm{i}m\theta}\int_0^{+\infty} u_m^R(k,0)J_m(kr)k\,\mathrm{d}k \quad (6.7.7c)$$

Substitute Eqs. (6.4.15) and (6.4.16) into Eqs. (6.7.1) and (6.7.6), noting that

$$\sum_{m=-\infty}^{+\infty} \mathrm{e}^{\mathrm{i}m\theta}J_m'(kr)\mathscr{F}_m^S \overset{J_{-1}'=J_1'}{=\!=\!=} \frac{1}{2}J_1'(kr)\left[\mathrm{e}^{\mathrm{i}\theta}(\delta_{1j} - \mathrm{i}\delta_{2j}) + \mathrm{e}^{-\mathrm{i}\theta}(\delta_{1j} + \mathrm{i}\delta_{2j})\right]$$

$$= \left(\delta_{1j}\cos\theta + \delta_{2j}\sin\theta\right)J_1'(kr) \quad (6.7.8a)$$

$$\sum_{m=-\infty}^{+\infty} \mathrm{e}^{\mathrm{i}m\theta}J_m'(kr)\mathscr{F}_m^R \overset{J_0'=-J_1}{=\!=\!=} \delta_{3j}J_1(kr) \quad (6.7.8b)$$

$$\sum_{m=-\infty}^{+\infty} \mathrm{i}m\mathrm{e}^{\mathrm{i}m\theta}J_m(kr)\mathscr{F}_m^T \overset{J_{-1}=J_1}{=\!=\!=} \frac{1}{2}J_1(kr)\left[\mathrm{e}^{\mathrm{i}\theta}(\delta_{1j} - \mathrm{i}\delta_{2j}) + \mathrm{e}^{-\mathrm{i}\theta}(\delta_{1j} + \mathrm{i}\delta_{2j})\right]$$

$$= \left(\delta_{1j}\cos\theta + \delta_{2j}\sin\theta\right)J_1(kr) \quad (6.7.8c)$$

$$\sum_{m=-\infty}^{+\infty} \mathrm{i}m\mathrm{e}^{\mathrm{i}m\theta}J_m(kr)\mathscr{F}_m^S = \frac{1}{2}J_1(kr)\left[\mathrm{e}^{\mathrm{i}\theta}(\delta_{2j} + \mathrm{i}\delta_{2j}) + \mathrm{e}^{-\mathrm{i}\theta}(\delta_{2j} - \mathrm{i}\delta_{1j})\right]$$

$$= \left(\delta_{2j}\cos\theta - \delta_{1j}\sin\theta\right)J_1(kr) \quad (6.7.8d)$$

$$\sum_{m=-\infty}^{+\infty} \mathrm{i}m\mathrm{e}^{\mathrm{i}m\theta}J_m(kr)\mathscr{F}_m^R = 0 \quad (6.7.8e)$$

$$\sum_{m=-\infty}^{+\infty} \mathrm{e}^{\mathrm{i}m\theta}J_m'(kr)\mathscr{F}_m^T = -\frac{1}{2}J_1'(kr)\left[\mathrm{e}^{\mathrm{i}\theta}(\delta_{2j} + \mathrm{i}\delta_{1j}) + \mathrm{e}^{-\mathrm{i}\theta}(\delta_{2j} - \mathrm{i}\delta_{1j})\right]$$

$$= -\left(\delta_{2j}\cos\theta - \delta_{1j}\sin\theta\right)J_1'(kr) \quad (6.7.8f)$$

$$\sum_{m=-\infty}^{+\infty} \mathrm{e}^{\mathrm{i}m\theta}J_m(kr)\mathscr{F}_m^S = \left(\delta_{1j}\cos\theta + \delta_{2j}\sin\theta\right)J_1(kr) \quad (6.7.8g)$$

$$\sum_{m=-\infty}^{+\infty} \mathrm{e}^{\mathrm{i}m\theta}J_m(kr)\mathscr{F}_m^R = -\delta_{3j}J_0(kr) \quad (6.7.8h)$$

substituting Eqs. (6.7.1), (6.7.6), and (6.7.8) into Eq. (6.7.7), we obtain

$$\begin{aligned} u_r^{\mathrm{s}} =& \frac{1}{8\pi\mu\xi'}\Big\{\left(\delta_{1j}\cos\theta+\delta_{2j}\sin\theta\right)\left[K_{00}-K_{20}-\xi' z_s\left(K_{01}-K_{21}\right)\right] \\ & -2\delta_{3j}\left(\xi K_{10}+\xi' z_s K_{11}\right)+2\xi'\left(\delta_{1j}\cos\theta+\delta_{2j}\sin\theta\right)\left(K_{00}+K_{20}\right)\Big\} \end{aligned} \tag{6.7.9a}$$

$$\begin{aligned} u_\theta^{\mathrm{s}} =& \frac{1}{8\pi\mu\xi'}\Big\{\left(\delta_{2j}\cos\theta-\delta_{1j}\sin\theta\right)\left[K_{00}+K_{20}-\xi' z_s\left(K_{01}+K_{21}\right)\right] \\ & +2\xi'\left(\delta_{2j}\cos\theta-\delta_{1j}\sin\theta\right)\left(K_{00}-K_{20}\right)\Big\} \end{aligned} \tag{6.7.9b}$$

$$u_z^{\mathrm{s}} = \frac{1}{4\pi\mu\xi'}\Big\{\left(\delta_{1j}\cos\theta+\delta_{2j}\sin\theta\right)\left(\xi K_{10}-\xi' z_s K_{11}\right)+\delta_{3j}\left(K_{00}+\xi' z_s K_{01}\right)\Big\} \tag{6.7.9c}$$

where,

$$K_{mn} \triangleq \int_0^{+\infty} \mathrm{e}^{-k z_s} J_m(kr) k^n \,\mathrm{d}k$$

Using the relationship between the components in polar and Cartesian coordinates,

$$u_x = u_r\cos\theta - u_\theta\sin\theta, \quad u_y = u_r\sin\theta + u_\theta\cos\theta$$

we obtain from Eq. (6.7.9) that

$$\begin{aligned} u_x^{\mathrm{s}} =& \frac{1}{8\pi\mu\xi'}\Big\{\delta_{1j}(1+2\xi')K_{00}-\delta_{1j}(1-2\xi')\cos 2\theta K_{20}-\delta_{2j}\sin 2\theta(1-2\xi')K_{20} \\ & -\delta_{1j}\xi' z_s K_{01}+\xi' z_s\left(\delta_{1j}\cos 2\theta+\delta_{2j}\sin 2\theta\right)K_{21} \\ & -2\delta_{3j}\cos\theta\left(\xi K_{10}+\xi' z_s K_{11}\right)\Big\} \end{aligned} \tag{6.7.10a}$$

$$\begin{aligned} u_y^{\mathrm{s}} =& \frac{1}{8\pi\mu\xi'}\Big\{\delta_{2j}(1+2\xi')K_{00}-\delta_{1j}\sin 2\theta(1-2\xi')K_{20}+\delta_{2j}(1-2\xi')\cos 2\theta K_{20} \\ & -\delta_{2j}\xi' z_s K_{01}-\xi' z_s\left(\delta_{2j}\cos 2\theta-\delta_{1j}\sin 2\theta\right)K_{21} \\ & -2\delta_{3j}\sin\theta\left(\xi K_{10}+\xi' z_s K_{11}\right)\Big\} \end{aligned} \tag{6.7.10b}$$

By setting j to 1, 2, and 3 in Eqs. (6.7.10) and (6.7.9c), we obtain the components of the static Green's function as

$$\begin{aligned} G_{11}^{\mathrm{s}} &= \frac{1}{8\pi\mu\xi'}\left[(1+2\xi')K_{00}-(1-2\xi')\cos 2\theta K_{20}-\xi' z_s(K_{01}-\cos 2\theta K_{21})\right] \\ G_{12}^{\mathrm{s}} = G_{21}^{\mathrm{s}} &= \frac{\sin\theta\cos\theta}{4\pi\mu\xi'}\left[-(1-2\xi')K_{20}+\xi' z_s\sin 2\theta K_{21}\right] \\ G_{22}^{\mathrm{s}} &= \frac{1}{8\pi\mu\xi'}\left[(1+2\xi')K_{00}+(1-2\xi')\cos 2\theta K_{20}-\xi' z_s(K_{01}+\cos 2\theta K_{21})\right] \end{aligned}$$

$$G_{13}^{\mathrm{s}} = -\frac{\cos\theta}{4\pi\mu\xi'}\left(\xi K_{10} + \xi' z_s K_{11}\right), \quad G_{23}^{\mathrm{s}} = -\frac{\sin\theta}{4\pi\mu\xi'}\left(\xi K_{10} + \xi' z_s K_{11}\right)$$
$$G_{31}^{\mathrm{s}} = \frac{\cos\theta}{4\pi\mu\xi'}\left(\xi K_{10} - \xi' z_s K_{11}\right), \quad G_{32}^{\mathrm{s}} = \frac{\sin\theta}{4\pi\mu\xi'}\left(\xi K_{10} - \xi' z_s K_{11}\right)$$
$$G_{33}^{\mathrm{s}} = \frac{1}{4\pi\mu\xi'}\left(K_{00} + \xi' z_s K_{01}\right)$$

The integral results for K_{ij} involved are[30]:

$$K_{00} = \frac{1}{R}, \quad K_{10} = \frac{R - z_s}{rR}, \quad K_{20} = \frac{(R - z_s)^2}{r^2 R}$$
$$K_{01} = \frac{z_s}{R^3}, \quad K_{11} = \frac{r}{R^3}, \quad K_{21} = \frac{2}{r^2} - \frac{z_s\left(3r^2 + 2z_s^2\right)}{r^2 R^3}$$

In the formula, $R = \sqrt{r^2 + z_s^2}$. Substituting these results into the above equation, we finally obtain the static Green's function for the $\mathcal{LP}_2$ as follows,

$$G_{11}^{\mathrm{s}} = c\left[1 + \kappa m_1 m_3^{-1} + \cos^2\theta\left(m_3^2 - \kappa m_1^2\right)\right] \tag{6.7.11a}$$
$$G_{12}^{\mathrm{s}} = G_{21}^{\mathrm{s}} = c\left(m_3^2 - \kappa m_1^2\right)\sin\theta\cos\theta \tag{6.7.11b}$$
$$G_{22}^{\mathrm{s}} = c\left[1 + \kappa m_1 m_3^{-1} + \sin^2\theta\left(m_3^2 - \kappa m_1^2\right)\right] \tag{6.7.11c}$$
$$G_{13}^{\mathrm{s}} = -c\left(m_2 m_3 + \kappa m_1\right)\cos\theta \tag{6.7.11d}$$
$$G_{23}^{\mathrm{s}} = -c\left(m_2 m_3 + \kappa m_1\right)\sin\theta \tag{6.7.11e}$$
$$G_{31}^{\mathrm{s}} = -c\left(m_2 m_3 - \kappa m_1\right)\cos\theta \tag{6.7.11f}$$
$$G_{32}^{\mathrm{s}} = -c\left(m_2 m_3 - \kappa m_1\right)\sin\theta \tag{6.7.11g}$$
$$G_{33}^{\mathrm{s}} = c\left(1 + \kappa + m_2^2\right) \tag{6.7.11h}$$

where,

$$\kappa = \frac{\xi}{\xi'} = \frac{\beta^2}{\alpha^2 - \beta^2}, \quad m_1 = \frac{R - z_s}{r}$$
$$m_2 = \frac{z_s}{R}, \quad m_3 = \frac{r}{R}, \quad c = \frac{1}{4\pi\mu R}$$

[30] These integrals can be easily obtained using Maple. Enter the following command:

```
> m:=1; n:=1; int(e^(-k*zs)*BesselJ(m,k*r)*k^n, k=0..infinity);
```

By changing the values of m and n, you can get the results for K_{mn}.

The static problem of a concentrated force acting within an elastic half-space, with the boundary of the half-space being free, is called the *Mindlin problem*. Here, we obtained the solution to the Mindlin problem within the same framework used for the corresponding dynamic problem of the half-space, as expressed by Eq. (6.7.11).

6.7.2 Static Surface Solution to a Shear Dislocation Point Source

In Sect. 6.6.1, the dynamic displacement field caused by a shear dislocation source in a half-space was studied. In the frequency domain, this can be expressed as the product of the moment tensor and the spatial derivative of the Green's function (see Eq. (6.6.1)). For the static case, we have correspondingly:

$$u_n^{\mathrm{s}}(x_1, x_2, 0) = M_{pq} G^{\mathrm{s}}_{np,q'} \tag{6.7.12}$$

where, M_{pq} represents the static form of the seismic moment tensor listed in Eq. (3.4.8a), which is obtained by replacing $M_0(t)$ with $M_0 = \mu\overline{[u]}A$. $G^{\mathrm{s}}_{np,q'}$ is the result of differentiating with respect to the source coordinates according to Eq. (6.7.11). Note that Eq. (6.7.11) is derived for a source located on the x_3 axis, so generalization is required for differentiating with respect to the source coordinates,

$$r = \sqrt{(x_1 - x_1')^2 + (x_2 - x_2')^2}, \quad \cos\theta = \frac{x_1 - x_1'}{r}, \quad \sin\theta = \frac{x_2 - x_2'}{r}$$

Substituting the above equation into Eq. (6.7.11) and differentiating,[31] we obtain:

[31] It's a process that, while not complicated, is very tedious. This calculation can be carried out using Maple. As an example, consider the simplest case of solving $G^{\mathrm{s}}_{33,1'}$:

```
> r:=sqrt((x-xs)^2+(y-ys)^2); R:=sqrt(r^2+zs^2); m2:=zs/R;
G33:=(m2^2+kappa+1)/(4*pi*mu*R); > diff(G33,xs);
```

Maple provides the following result:

$$-\frac{1}{4}\frac{zs^2(-2x+2x_s)}{\left((x-xs)^2+(y-ys)^2+zs^2\right)^{5/2}\pi\mu} - \frac{1}{8}\frac{\left(\dfrac{zs^2}{(x-xs)^2+(y-ys)^2+zs^2}+\kappa+1\right)(-2x+2x_s)}{\pi\mu\left((x-xs)^2+(y-ys)^2+zs^2\right)^{3/2}}$$

A bit of simplification leads to the final result. For other components, the form becomes more complex, requiring further simplification of the results directly obtained from Maple.

$$G^{s}_{11,1'} = c'c_\theta\left[n + m_3^2 + \left(m_3^2c_\theta^2 - 2\,s_\theta^2\right) f(1) - 2m_2c_\theta^2 f(m_2)\right] \tag{6.7.13a}$$
$$G^{s}_{12,1'} = G^{s}_{21,1'} = c's_\theta\left[\left(m_3^2c_\theta^2 + c_{2\theta}\right) f(1) - 2m_2c_\theta^2 f(m_2)\right] \tag{6.7.13b}$$
$$G^{s}_{13,1'} = c'\left[\left(s_\theta^2 - m_3^2c_\theta^2\right) d^+ + m_2c_\theta^2h^+\right] \tag{6.7.13c}$$
$$G^{s}_{22,1'} = c'c_\theta\left[n + m_3^2 + \left(m_3^2 + 2\right) s_\theta^2 f(1) - 2m_2s_\theta^2 f(m_2)\right] \tag{6.7.13d}$$
$$G^{s}_{23,1'} = G^{s}_{13,2'} = -c's_\theta c_\theta\left[\left(1 + m_3^2\right) d^+ - m_2h^+\right] \tag{6.7.13e}$$
$$G^{s}_{31,1'} = c'\left[\left(s_\theta^2 - m_3^2c_\theta^2\right) d^- + m_2c_\theta^2h^-\right] \tag{6.7.13f}$$
$$G^{s}_{32,1'} = G^{s}_{31,2'} = -c's_\theta c_\theta\left[\left(1 + m_3^2\right) d^- m_2h^-\right] \tag{6.7.13g}$$
$$G^{s}_{33,1'} = c'c_\theta m_3^2\left(1 + \kappa + 3m_2^2\right) \tag{6.7.13h}$$
$$G^{s}_{11,2'} = c's_\theta\left[n + m_3^2 + \left(m_3^2 + 2\right) c_\theta^2 f(1) - 2m_2c_\theta^2 f(m_2)\right] \tag{6.7.13i}$$
$$G^{s}_{12,2'} = G^{s}_{21,2'} = c'c_\theta\left[\left(m_3^2\, s_\theta^2 - c_{2\theta}\right) f(1) - 2m_2s_\theta^2 f(m_2)\right] \tag{6.7.13j}$$
$$G^{s}_{22,2'} = c's_\theta\left[n + m_3^2 + \left(m_3^2\, s_\theta^2 - 2c_\theta^2\right) f(1) - 2m_2s_\theta^2 f(m_2)\right] \tag{6.7.13k}$$
$$G^{s}_{23,2'} = c'\left[\left(c_\theta^2 - m_3^2\, s_\theta^2\right) d^+ + m_2s_\theta^2h^+\right] \tag{6.7.13l}$$
$$G^{s}_{32,2'} = c'\left[\left(c_\theta^2 - m_3^2\, s_\theta^2\right) d^- + m_2s_\theta^2h^-\right] \tag{6.7.13m}$$
$$G^{s}_{33,2'} = c's_\theta m_3^2\left(1 + \kappa + 3m_2^2\right) \tag{6.7.13n}$$
$$G^{s}_{11,3'} = -c'\left[\kappa m_1 + m_2m_3\left(1 + c_\theta^2 f(1)\right) + 2m_3c_\theta^2 f(m_2)\right] \tag{6.7.13o}$$
$$G^{s}_{12,3'} = G^{s}_{21,3'} = -c's_\theta c_\theta\left[m_2m_3 f(1) + 2m_3 f(m_2)\right] \tag{6.7.13p}$$
$$G^{s}_{13,3'} = c'c_\theta\left(m_2m_3d^+ + \kappa m_1m_3 + g\right) \tag{6.7.13q}$$
$$G^{s}_{22,3'} = -c'\left[\kappa m_1 + m_2m_3\left(1 + s_\theta^2 f(1)\right) + 2m_3s_\theta^2 f(m_2)\right] \tag{6.7.13r}$$
$$G^{s}_{23,3'} = c's_\theta\left(m_2m_3d^+ + \kappa m_1m_3 + g\right) \tag{6.7.13s}$$
$$G^{s}_{31,3'} = c'c_\theta\left(m_2m_3d^- - \kappa m_1m_3 + g\right) \tag{6.7.13t}$$
$$G^{s}_{32,3'} = c's_\theta\left(m_2m_3d^- - \kappa m_1m_3 + g\right) \tag{6.7.13u}$$
$$G^{s}_{33,3'} = -c'm_2m_3\left(\kappa + 2m_2^2 - m_3^2\right) \tag{6.7.13v}$$

where, $c' = 1/(4\pi\mu Rr)$, $s_\theta = \sin\theta$, $c_\theta = \cos\theta$, $c_{2\theta} = \cos 2\theta$, $n = 2\kappa m_1/m_3 - \kappa$, and

$$\begin{aligned} d^+ &= m_2m_3 + \kappa m_1,\ h^+ = m_3(m_2^2 - m_3^2) + \kappa m_1,\ f(x) = xm_3^2 - \kappa m_1^2 \\ d^- &= m_2m_3 - \kappa m_1,\ h^- = m_3(m_2^2 - m_3^2) - \kappa m_1,\ g = m_3^2\left(m_2^2 - m_3^2\right) \end{aligned}$$

By substituting the specific expressions from Eqs. (3.4.8a) and (6.7.13) into Eq. (6.7.12), we obtain the static displacement field caused by a shear dislocation source.

Since the methods developed in Sects. 6.3 to 6.5 do not include a static solution, the static Green's function and the surface static displacement caused by a shear dislocation point source derived in this section complement the previous solutions.

In recent years, geodetic techniques such as GPS and InSAR have been increasingly used in seismology. The theoretically derived static displacement field is crucial for interpreting GPS observational data.

6.8 Summary

In this chapter, we used a frequency-domain solution method to obtain the Green's functions for the Lamb's problem and the displacement field generated by a shear dislocation source, expressed as a double integral over wavenumber and frequency. We began by constructing the basis functions, then solved the system of ordinary differential equations using matrix properties, and finally derived the Green's functions in the frequency domain in the form of wavenumber integrals. Based on this, we performed a theoretical analysis of the two components of the Green's solution: the direct wave contribution and the free surface contribution. We proved that the direct wave contribution is exactly equal to the Green's function in an infinite medium. By analyzing the free surface contribution and calculating the zero points of the Rayleigh function, we derived the excitation formula for Rayleigh wave in a half-space. For the static case when the frequency is zero, we provided special treatment to obtain the static solution. In addition to the Green's functions themselves, we also derived the displacement field generated by a shear dislocation source based on their spatial derivatives.

Although, in theory, the numerical computation of the wavenumber integral can be performed using standard functions in mathematical software like Matlab, our wavenumber integral is an oscillatory improper integral involving Bessel functions. Therefore, it would be beneficial to design a targeted approach that fully considers the properties of the integrand to improve calculation accuracy and efficiency. In the next chapter, we will explore the numerical implementation of the formulas derived in this chapter, develop several targeted numerical techniques, present the numerical results, and analyze and discuss these results.

Chapter 7
Frequency-Domain Solution of Lamb's Problem (II): Numerical Results

In Chap. 6, we derived the expressions for the displacement fields caused by the Green's function of the Lamb's problem and shear dislocation sources. These expressions can be represented as double integrals over wavenumber and frequency. The next challenge is how to compute these integrals. There are two approaches: One approach is to use various asymptotic analysis techniques to directly obtain closed-form analytical solutions, as many mathematicians, including Lamb himself, did in the first half of the 20th century. In fact, the Green's function for the two-dimensional Lamb's problem does have a closed-form solution, while the three-dimensional problem has a generalized closed-form analytical solution. However, using frequency domain analysis makes it difficult to obtain generalized closed-form solutions.[1] Clearly, from the perspective of obtaining accurate rather than asymptotic solutions, this approach is not ideal.

Another approach is to directly use numerical methods to compute these integrals. Compared to theoretical analysis, numerical methods have clear advantages: they can handle many problems that theoretical methods cannot, including complex issues that were unimaginable 60 years ago. However, when using numerical methods, there is a balance between accuracy and efficiency that must be carefully managed. Ensuring computational accuracy while maximizing efficiency is a key consideration when applying numerical methods.

Specifically, in Chap. 6, the double integrals for the Green's function and the displacement fields caused by dislocation sources are computed using the fast Fourier transform (FFT) for the frequency integrals, while the wavenumber integrals for each frequency component are calculated using numerical integration methods. For numerical integration, developing targeted strategies based on a thorough analysis of

---

[1] Using another method mentioned in the review in Chap. 5, the Cagniard-de Hoop method based on Laplace transforms, can achieve this. This will be the focus of the next volume of this book.

H. Zhang, *Lamb's Problem in Seismology (Volume 1)*,
https://doi.org/10.1007/978-981-95-3170-7_7

the integrand's properties can significantly enhance computational efficiency. This chapter first introduces several techniques related to numerical implementation,[2] including the inverse Fourier transform for frequency ω and its discrete form, which are essential for implementing the formulas from Chap. 6 in code. Before delving into the displacement fields of point forces and dislocation sources, we ensure the correctness of the formulas and programs by carefully comparing our results with those of previous studies on three types of Lamb's problems. The main focus of this chapter is on calculating the displacement fields in Lamb's problems. We examine the time-varying displacement and particle motion at three levels: point force sources, point dislocation sources, and finite-scale dislocation sources. Finally, we present specific results for the static displacement field and Rayleigh wave.

7.1 Numerical Implementation of Wavenumber Integration

The half-space Green's function (see Eqs. (6.5.1) and (6.5.2)) and its spatial derivatives (see Eqs. (6.6.3) and (6.6.4)) involve wavenumber integrals of the following form

$$I(\omega) = \int_0^{+\infty} F(k, \omega) J_m(kr) k \, \mathrm{d}k \quad (m = 0, 1, 2, 3) \tag{7.1.1}$$

Several numerical techniques for calculating the above integral will be discussed in this section. However, before that, a special treatment method related to this will be introduced.

7.1.1 *Discrete Wavenumber Method*

The discrete wavenumber method, introduced primarily by Bouchon in the late 1970s and early 1980s, is an effective technique for computing theoretical seismograms (Bouchon 1977, 1979, 1981). The main idea of this method is to transform results expressed in integral form into discrete sums by introducing periodically placed sources.

[2] This includes the discrete wavenumber method for the wavenumber k integration, the adaptive Filon's integration method, and the peak-trough averaging method. The latter two methods are primarily based on the following works: Zhang and Chen (2001), Chen and Zhang (2001), Zhang et al. (2001, 2003).

7.1.1.1 Principle of Discrete Wavenumber Method

This section uses the example of a spherical P-wave in an infinite elastic medium to illustrate the principle of the discrete wavenumber method.[3] A spherical P-wave can be represented by the Sommerfeld integral (see Eq. (6.5.5a)) as follows:

$$\frac{\mathrm{e}^{-\mathrm{i}\omega\frac{R}{\alpha}}}{R} = \int_0^{+\infty} \frac{\mathrm{e}^{-\gamma|z|}}{\gamma} J_0(kr)k\,\mathrm{d}k \tag{7.1.2}$$

where $\gamma = \sqrt{k^2 - \omega^2/\alpha^2}$ with $\mathrm{Re}(\gamma) > 0$, and α is the P-wave velocity.

Consider the Helmholtz equation in cylindrical coordinates for an infinite elastic medium

$$\frac{\partial^2\phi}{\partial r^2} + \frac{1}{r}\frac{\partial\phi}{\partial r} + \frac{1}{r^2}\frac{\partial^2\phi}{\partial\theta^2} + \frac{\partial^2\phi}{\partial z^2} + \frac{\omega^2}{\alpha^2}\phi = 0$$

Let $\phi(r, \theta, z; \omega) = R(r)\Phi(\theta)Z(z, \omega)$, where $R(r)$, $\Phi(\theta)$, and $Z(z, \omega)$ satisfy the three equations in Eq. (6.2.7). Their solutions are given in Eq. (6.2.8). Therefore, the general solution for $\phi(r, \theta, z)$ is:

$$\phi(r, \theta, z; \omega) = \sum_{m=-\infty}^{\infty} \mathrm{e}^{im\theta} \int_0^{+\infty} f_m(k, \omega) J_m(kr)\mathrm{e}^{-\gamma|z|}\,\mathrm{d}k \tag{7.1.3}$$

in which, $f_m(k, \omega)$ is an arbitrary function of k and ω. Specifically, referring to the right-hand side of Eq. (7.1.2), we take[4]

$$\begin{aligned} f_{2n} &= (-1)^n(2 - \delta_{n0})\frac{k\mathrm{e}^{-\mathrm{i}2n\theta_s}}{\pi\gamma}[H(\theta_s) - H(\theta_s - \pi)] \\ f_{-2n} &= (-1)^n(2 - \delta_{n0})\frac{k\mathrm{e}^{\mathrm{i}2n\theta_s}}{\pi\gamma}[H(\theta_s - \pi) - H(\theta_s - 2\pi)] \\ f_{2n+1} &= f_{-2n+1} = 0 \end{aligned}$$

where $n = 0, 1, 2, \ldots$, and θ_s is the angle between the line connecting the origin to the source and the x axis (counterclockwise is positive) when the source is not at the origin. Therefore, a particular solution to Eq. (7.1.3) is:

$$\phi_s(r, \theta, z; \omega) = \frac{1}{\pi}\sum_{n=0}^{\infty}(-1)^n(2 - \delta_{n0})\mathrm{e}^{\pm\mathrm{i}2n(\theta-\theta_s)} \int_0^{+\infty} \frac{\mathrm{e}^{-\gamma|z|}}{\gamma} J_{2n}(kr)k\,\mathrm{d}k \tag{7.1.4}$$

---

[3] This is the simplest example that clearly demonstrates the concept of the discrete wavenumber method. Bouchon (1981) also used this example in his paper on the discrete wavenumber method for three-dimensional cases.

[4] The reason for this choice is to facilitate converting the integral expression into a summation using the corresponding relations.

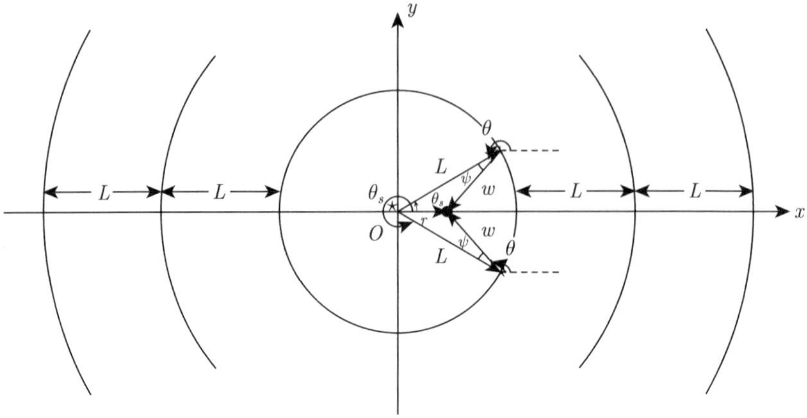

Fig. 7.1 Diagram for discrete wavenumber method. The symbol ⋆ represents the source point. The actual source is located at the origin O (slightly offset in the diagram for clarity), while the additional sources are positioned on a series of circles centered at the origin with radii mL. The projection of the observation point at depth z onto the xy plane is on the x axis (black dot), with a horizontal distance of r from the actual source. θ_s is the polar angle of the source point, w is the horizontal distance from the source point on the circle to the observation point, and ψ is the angle between the line connecting the source point on a circle of radius L to the observation point's projection and the line connecting the source point to the origin

where the "+" and "−" in the exponential are taken when θ_s is in $(0, \pi)$ and $(\pi, 2\pi)$, respectively. This represents the P-wave radiated by a source located at the origin. We choose the coordinate system $Oxyz$ such that the source is at the origin and the projection of the observation point on the xy plane is on the x-axis, as shown in Fig. 7.1.

Let's consider the following problem: if identical sources to the actual source (referred to as *additional sources*) are continuously distributed on a series of circles centered at the origin O in the xy plane, with radii mL (where m is a positive integer), what is the relationship between the displacement field caused by both the actual and additional sources and Eq. (7.1.2)?

We will study this problem in three steps:

(1) Obtain the displacement ϕ_1 caused by a single point source on a circle with radius L;
(2) Integrate the result from the first step over θ_s to find the displacement ϕ_2 caused by a continuously distributed source on a circle with radius L;
(3) Sum the contributions of all circles to obtain ϕ_3, and then add it to the actual source to get the total displacement ϕ^* caused by both the actual and additional sources.

Consider a single point source located on a circle with radius L. Based on the geometric relationships shown in Fig. 7.1, the results listed in Table 7.1 are obtained.

Table 7.1 The values of θ, θ_s, and $\theta - \theta_s$ under different conditions

	Source is above x axis	Source is below x axis
$\theta \in$	$(\pi, 2\pi)$	$(0, \pi)$
$\theta_s \in$	$(0, \pi)$	$(\pi, 2\pi)$
$\theta - \theta_s =$	$\pi + \psi$	$-\pi - \psi$

Therefore, when the source is located on a circle with a radius L, according to Table 7.1 and using Eq. (7.1.4), we have

$$\phi_1(r, z; L, \theta_s; \omega) = \frac{1}{\pi}\sum_{n=0}^{\infty}(-1)^n(2-\delta_{n0})\mathrm{e}^{\mathrm{i}2n\psi}\int_0^{+\infty}\frac{\mathrm{e}^{-\gamma|z|}}{\gamma}J_{2n}(kw)k\,\mathrm{d}k \quad (7.1.5)$$

where $w = \sqrt{r^2 + L^2 - 2rL\cos\theta_s}$ represents the horizontal distance between the source and observation points, calculated using the cosine rule. According to Graf's addition theorem (Abramowitz and Stegun 1964, p. 363),

$$J_{2n}(kw)\mathrm{e}^{\mathrm{i}2n\psi} = \sum_{m=-\infty}^{\infty} J_m(kr)J_{2n+m}(kL)\mathrm{e}^{\mathrm{i}m\theta_s}$$

Equation (7.1.5) can be rewritten as

$$\phi_1 = \frac{1}{\pi}\sum_{n=0}^{+\infty}(-1)^n(2-\delta_{n0})\int_0^{+\infty}\frac{\mathrm{e}^{-\gamma|z|}}{\gamma}\left[\sum_{m=-\infty}^{\infty} J_m(kr)J_{2n+m}(kL)\mathrm{e}^{\mathrm{i}m\theta_s}\right]k\,\mathrm{d}k \quad (7.1.6)$$

Equation (7.1.6) is for the case of a single additional source. For sources continuously distributed on a circle with radius L, integrate over θ_s from 0 to 2π. Notice that

$$\int_0^{2\pi}\mathrm{e}^{\mathrm{i}m\theta_s}\,\mathrm{d}\theta_s = 2\pi\delta_{m0}$$

Therefore, we get

$$\begin{aligned}\phi_2(r, z; L; \omega) &= \int_0^{2\pi}\phi_1(r, z; L, \theta_s; \omega)\,\mathrm{d}\theta_s \\ &= 2\sum_{n=0}^{\infty}(-1)^n(2-\delta_{n0})\int_0^{+\infty}\frac{\mathrm{e}^{-\gamma|z|}}{\gamma}J_0(kr)J_{2n}(kL)k\,\mathrm{d}k\end{aligned}$$

Summing the contributions of all circles with radii mL:

$$\begin{aligned}\phi_3(r,z;\omega)=&\sum_{m=1}^{\infty}\phi_2(r,z;mL;\omega)\\=&2\sum_{m=1}^{\infty}\sum_{n=0}^{\infty}(-1)^n(2-\delta_{n0})\int_0^{+\infty}\frac{\mathrm{e}^{-\gamma|z|}}{\gamma}J_0(kr)J_{2n}(mkL)k\,\mathrm{d}k\end{aligned}$$

Combining this with the right-hand side of Eq. (7.1.2), the displacement caused by the actual source and all additional sources is

$$\phi^*(r,z;\omega)=\int_0^{+\infty}\frac{\mathrm{e}^{-\gamma|z|}}{\gamma}J_0(kr)\left[1+2\sum_{m=1}^{\infty}\sum_{n=0}^{\infty}(-1)^n(2-\delta_{n0})J_{2n}(mkL)\right]k\,\mathrm{d}k \tag{7.1.7}$$

Note the Jacobi expansion with Bessel function coefficients (Watson 1922, p. 22):

$$\mathrm{e}^{\pm\mathrm{i}x\sin\theta}=J_0(x)+2\sum_{n=1}^{\infty}J_{2n}(x)\cos 2n\theta\pm 2\mathrm{i}\sum_{n=0}^{\infty}J_{2n+1}(x)\sin(2n+1)\theta$$

Setting $\theta=\pi/2$, we get

$$\mathrm{e}^{\pm\mathrm{i}x}=\sum_{n=0}^{\infty}(-1)^n(2-\delta_{n0})J_{2n}(x)\pm 2\mathrm{i}\sum_{n=0}^{\infty}(-1)^nJ_{2n+1}(x)$$

Adding these two expressions, we obtain

$$\mathrm{e}^{\mathrm{i}x}+\mathrm{e}^{-\mathrm{i}x}=2\sum_{n=0}^{\infty}(-1)^n(2-\delta_{n0})J_{2n}(x)$$

Replacing x with mkL and summing over m from 1 to ∞, while also including the result when $m=0$ which is 1, we have

$$\sum_{m=-\infty}^{\infty}\mathrm{e}^{\mathrm{i}mkL}=1+2\sum_{m=1}^{\infty}\sum_{n=0}^{\infty}(-1)^n(2-\delta_{n0})J_{2n}(mkL) \tag{7.1.8}$$

Note that the Fourier series expansion of the Dirac δ function is

$$\delta(x)=\frac{1}{2\pi}\sum_{m=-\infty}^{\infty}\mathrm{e}^{\mathrm{i}mx}$$

Given that the right-hand side of this equation has a period of $2n\pi$ (where n is an integer), this implies

$$\sum_{m=-\infty}^{\infty} \mathrm{e}^{imkL} = 2\pi \sum_{n=-\infty}^{\infty} \delta(kL - 2n\pi) \tag{7.1.9}$$

Combining this with Eq. (7.1.8) and substituting into Eq. (7.1.7), we get

$$\phi^*(r, z; \omega) = 2\pi \int_0^{+\infty} \frac{\mathrm{e}^{-\gamma|z|}}{\gamma} J_0(kr) \sum_{n=-\infty}^{\infty} \delta(kL - 2n\pi) k\, \mathrm{d}k \tag{7.1.10}$$

Note that

$$\delta(kL - 2n\pi) = \frac{1}{L}\delta(k - n\Delta k), \quad \Delta k = \frac{2\pi}{L}$$

Thus, Eq. (7.1.10) can be rewritten as

$$\phi^*(r, z; \omega) = \sum_{n=1}^{\infty} \frac{\mathrm{e}^{-\gamma_n|z|}}{\gamma_n} J_0(k_n r) k_n \Delta k \tag{7.1.11}$$

where $k_n = n\Delta k$ and $\gamma_n = \sqrt{k_n^2 - \omega^2/\alpha^2}$.

Now we can address the question posed earlier. By comparing Eqs. (7.1.11) with (7.1.2), it is clear that the former is a discrete representation of the latter. This is because the continuous transverse wavenumber k has been discretized into distinct values k_n, leading to the term *discrete wavenumber method.* It is important to note that while Eq. (7.1.11) might appear to be a numerical discretization of the integral in Eq. (7.1.2), there is a significant difference between them. The result given by the discrete wavenumber method (Eq. (7.1.11)) is an exact expression for the wave field generated by the actual source and all additional sources, whereas the numerical discretization of the integral in Eq. (7.1.2) is merely an approximation.[5]

The discrete wavenumber method offers a way to rigorously convert wavenumber integrals into summations by introducing additional sources. However, to ensure accurate results without altering the original problem, certain conditions must be met. In practical applications, we typically calculate signals within a finite time window T. This means that the earliest signals from the nearest additional sources haven't reached the observation point within this period, ensuring that the signals during T are solely from the real source. The nearest additional sources are located at

[5] This distinction is clearly due to the introduction of additional sources. The inclusion of these sources transforms the originally finite-duration time signal from the real source into an infinitely continuous time signal. This happens because the time signals from the additional sources on circles of different radii continuously reach the receiving point. In Sect. 7.1.1.2, we will examine the side effects caused by this factor and address them by introducing imaginary frequencies.

the intersection of a circle with radius L and the positive x axis. Thus, the condition can be expressed as

$$\sqrt{(L-r)^2+z^2} > \alpha T \quad \Rightarrow \quad L > r + \sqrt{\alpha^2 T^2 - z^2}$$

In practice, choosing an appropriate T is crucial. It should be large enough to encompass all wave components emitted by the real source within the time window, typically 2 to 3 times the arrival time of the S-wave. L must be "sufficiently large", which also implies that Δk should be "sufficiently small". Essentially, the discrete wavenumber method is equivalent to the numerical algorithm for integrating the displacement produced by only the real source.[6]

7.1.1.2 Complex Frequency

The discrete wavenumber method converts integrals into sums by introducing additional sources. However, this introduction comes with a significant side effect that must be carefully managed in practical computations. To illustrate this, we must first consider the characteristics of the discrete Fourier transform (which will be detailed in Sect. 7.2). Unlike the integral form of the Fourier transform, the discrete Fourier transform requires the function to be periodic. Because computers can only perform discrete operations, the mathematical inverse Fourier transform is implemented on a computer through the Fourier series of a periodic function. This difference introduces some numerical implementation issues that need careful attention.

Let's consider the time interval of interest as $[0, T]$. As illustrated in Fig. 7.2, the discrete Fourier transform effectively shifts the signal within the $[0, T]$ interval to the left (time series labeled (1) and (2) in the figure) and to the right (time series labeled (4) and (5) in the figure) by nT (where n is a positive integer), and then adds these shifted signals to the original signal (time series labeled (3) in the figure). This process forms an infinitely long periodic sequence with period T (time series labeled (6) in the figure), from which we then extract the portion within the $[0, T]$ interval. If this interval encompasses all non-zero signals,[7] the discrete Fourier transform can accurately reconstruct the signal. The signals within the top and bottom boxes in Fig. 7.2 are identical.

---

[6] It is important to note that in Chap. 6, we derived the integral expression for the Green's function as Eq. (6.5.1). From the example process of the discrete wavenumber method in this section, it is evident that this method heavily relies on the use of the Graf addition theorem and the Jacobi expansion. Without summing over the Bessel function orders, the discrete wavenumber method cannot be applied. This means that for the expression in Eq. (6.5.1), we cannot use the discrete wavenumber method to convert the integral strictly into a sum and must resort to numerical methods to perform the integration. As previously mentioned, this is essentially equivalent to the discrete wavenumber method. If we wish to use the discrete wavenumber method, it must be applied at the initial stage of the solution process that involves summation over m.

[7] In other words, if T is increased, the additional portions of the signal will always be zero.

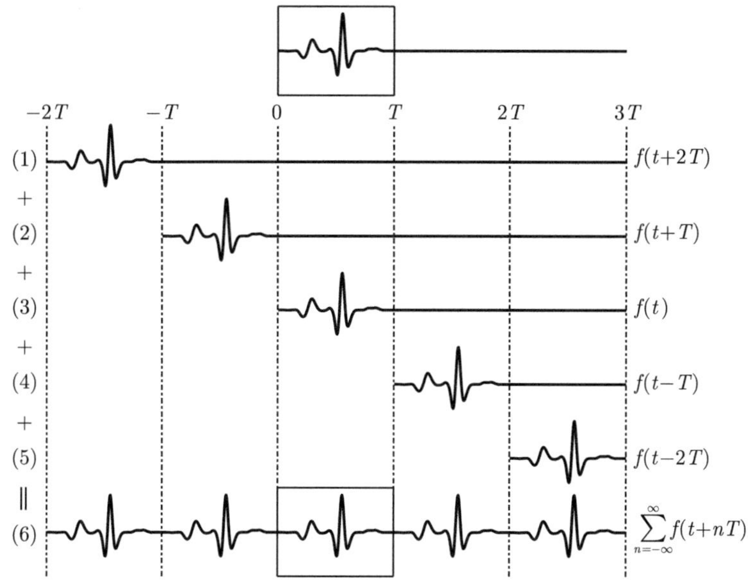

Fig. 7.2 Diagram illustrating discrete Fourier transform on Computer (1). At the top is the actual signal where all non-zero values are within the $[0, T]$ range. (1) through (5) are the signals shifted left and right by multiples of T. (6) represents their superposition, with the boxed area showing the signal output by the computer within the $[0, T]$ range

However, if the actual signal has non-zero values outside the $[0, T]$ range, the signal provided by the discrete Fourier transform will differ from the real signal. Figure 7.3 shows an example. The actual signal has non-zero components within the $[T, 2T]$ range, but since we only compute the result for the $[0, T]$ interval, after shifting left and right and summing, the portion of the time series labeled (6) within the $[0, T]$ range significantly differs from the real signal in this interval. The sequence shifted left by one period (2) "mixes" into the actual signal within the $[0, T]$ range. This phenomenon is called *aliasing*.

For the Lamb's problem we are currently considering, if additional sources are not introduced, it is sufficient to choose a large enough T such that all non-zero signals are contained within it. However, if the discrete wavenumber method is adopted, signals from some additional sources further away will always be outside of T, regardless of how large T is. In this case, the portions of the different time series shifted to the left within the $[0, T]$ interval will all be superimposed on the real signal, and more critically, because some of these signals have amplitudes comparable to that of the real signal, the discrete Fourier transform will yield severely distorted results. This is an inherent issue of the discrete wavenumber method. Without addressing this aliasing problem, the discrete wavenumber method cannot be applied in practical calculations.

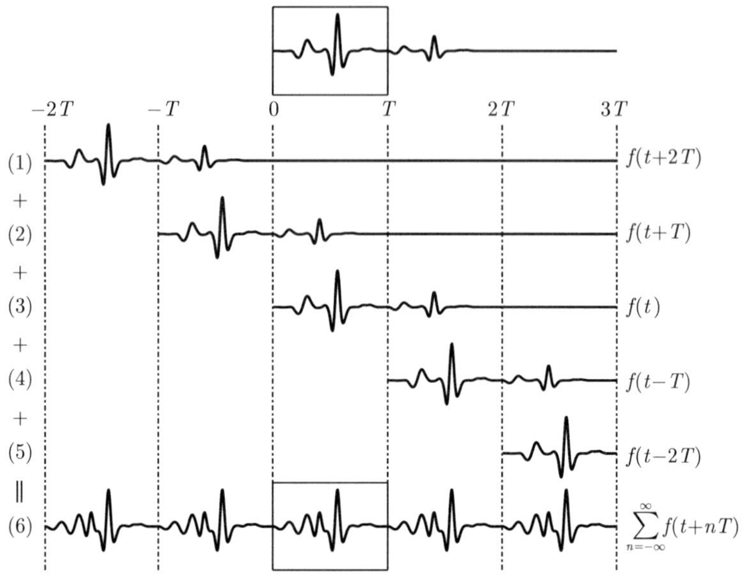

Fig. 7.3 Diagram illustrating discrete Fourier transform on Computer (2). Unlike in Fig. 7.2, the actual signal here has non-zero values outside the [0, T] range. This situation will result in aliasing, causing the portion within the box in the time series (6) to differ from the real signal

To address this issue, an effective method is to introduce an imaginary frequency $\omega_{\rm I}$, replacing the original real frequency ω with the complex frequency ω^*:

$$\omega^* = \omega - \mathrm{i}\omega_{\rm I} \tag{7.1.12}$$

Notice that the inverse Fourier transform defined by

$$f(t) = \frac{1}{2\pi} \int_{-\infty}^{+\infty} F(\omega) \mathrm{e}^{\mathrm{i}\omega t}\, \mathrm{d}\omega$$

becomes, in numerical implementation,

$$\sum_{n=-\infty}^{\infty} f(t + nT) = \frac{\Delta\omega}{2\pi} \sum_{n=-\infty}^{\infty} F(n\Delta\omega) \mathrm{e}^{\mathrm{i}n\Delta\omega t} \tag{7.1.13}$$

where $\Delta\omega$ is the step size of angular frequency. If we replace $f(t)$ in the Fourier transform with $f(t)\mathrm{e}^{-\omega_{\rm I} t}$, the corresponding spectrum $F(\omega)$ changes to $F(\omega^*)$

$$F(\omega^*) = \int_{-\infty}^{+\infty} f(t)\mathrm{e}^{-\omega_{\rm I} t} \mathrm{e}^{-\mathrm{i}\omega t}\, \mathrm{d}t = \int_{-\infty}^{+\infty} f(t)\mathrm{e}^{-\mathrm{i}(\omega - \mathrm{i}\omega_{\rm I})t}\, \mathrm{d}t$$

Thus, Eq. (7.1.13) becomes

$$\sum_{n=-\infty}^{\infty} f(t+nT)e^{-\omega_I(t+nT)} = \frac{\Delta\omega}{2\pi}\sum_{n=-\infty}^{\infty} F(n\Delta\omega - i\omega_I)e^{in\Delta\omega t}$$

which means

$$\sum_{n=-\infty}^{\infty} f(t+nT)e^{-n\omega_I T} = \frac{\Delta\omega}{2\pi}\sum_{n=-\infty}^{\infty} F(n\Delta\omega - i\omega_I)e^{in(\Delta\omega - i\omega_I)t} \tag{7.1.14}$$

A typical value for ω_I is

$$\omega_I = \frac{\pi}{T}$$

Equation (7.1.14) shows that if the angular frequency ω is replaced by the complex frequency ω^* defined by Eq. (7.1.13), the time series produced by the discrete Fourier transform is the result of shifting the signal by $\pm nT$ (where n is a positive integer) and multiplying by $e^{-n\pi}$ and $e^{n\pi}$, respectively. This means that the left-shifted signal is attenuated.

Figure 7.4 illustrates the effect of introducing the complex frequency. The actual signal is the same as in Fig. 7.3. Due to the significant attenuation of the left-shifted signal, its amplitude within the $[0, T]$ interval is almost negligible compared to the real signal. As a result, the summed sequence labeled (6) (corresponding to the left side of Eq. (7.1.14)) within the $[0, T]$ interval closely matches the real signal.[8]

It is worth mentioning that the introduction of complex frequency is essential for using the discrete wavenumber method but not for the equivalent direct numerical calculation of Eq. (6.5.2). The latter only requires a sufficiently large T to ensure the non-zero signal is within the $[0, T]$ interval. However, in the direct numerical integration method, if we want to include the zero-frequency case, complex frequency must still be used. This is because:

(1) With the introduction of complex frequency, even for $\omega = 0$, there is still a non-zero ω^* due to the presence of imaginary frequency. In the solution developed in Chap. 6, it is noted that the eigenvalue corresponding to $\omega = 0$ is degenerate. In this case, the coefficient matrix is not similar to a diagonal matrix, and the result does not include this special case. Note that for the P-SV case, the denominator of the matrix $\mathbf{E}$ in Eq. (6.4.6) contains ω, which means $\omega \neq 0$. Once imaginary frequency is introduced, the complex frequency is always non-zero, allowing it to cover the zero-frequency case.

[8] Clearly, the larger the imaginary frequency ω_I, the more it suppresses the left-shifted signal. However, note that ω_I appears in the exponent on the right side of Eq. (7.1.14), leading to exponential growth as t increases. Excessive values can cause numerical instability during the inverse Fourier transform. Therefore, a balance must be struck between these factors. In Sect. 7.2, we will present the numerical results for different values of ω_I.

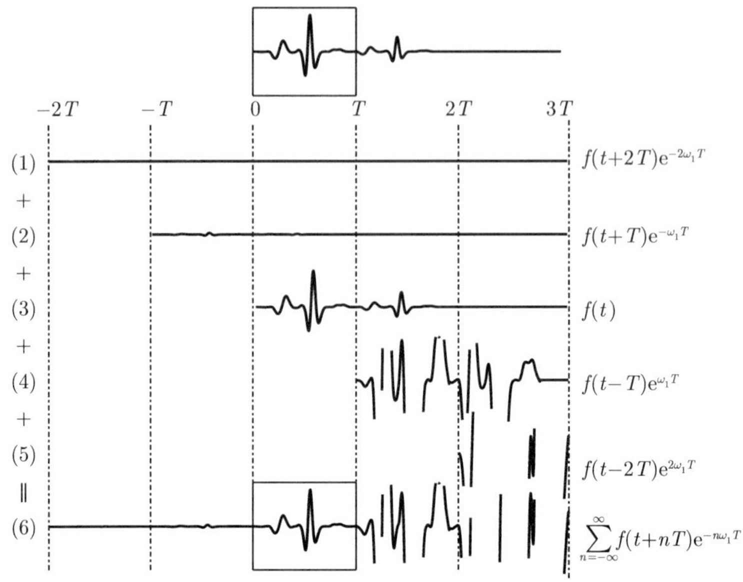

Fig. 7.4 Diagram of discrete fourier transform with introduction of complex frequency. After introducing complex frequency, the signals shifted to the left and right by nT are multiplied by $e^{-n\omega_I}$ and $e^{n\omega_I}$, respectively. This causes the left-shifted signal to shrink, significantly reducing the aliasing effect. The portion within the box in the synthesized time series (6) is nearly identical to the real signal. For sequences (4) to (6), the amplification leads to large amplitudes, and truncation is applied in the diagram

(2) Including the zero-frequency component means that the non-zero signal exists outside the $[0, T]$ interval. To accurately reproduce the results, complex frequency must be used.

In summary, complex frequency not only includes the zero-frequency component in the solution but is also necessary to achieve this effect. Incidentally, in the theoretical analysis of frequency-domain Rayleigh wave in Sect. 6.5.3.2, the branch points and Rayleigh poles are located in the second and fourth quadrants of the complex plane of k, rather than on the real axis of k, which is also a result of using complex frequency.

7.1.2 Adaptive Filon's Integration Method

The discrete wavenumber method introduced in Sect. 7.1.1 cannot be directly used for the calculation of Eq. (7.1.1); therefore, we need to employ numerical integration techniques to handle it.

7.1.2.1 Problem Statement

The wavenumber integrals in the Green's function and its spatial derivative components are similar, so we take I_5 from Eq. (6.5.2e) as an example. Here, in Eq. (7.1.1), $m = 0$,

$$F(k) = -\frac{\beta^2}{\omega^2}\left[\frac{R^*(k)}{\nu R(k)}\left(\nu\gamma e^{PP} + k^2 e^{SS}\right) - \frac{4k^2\gamma\chi}{R(k)}\left(e^{SP} + e^{PS}\right)\right] \tag{7.1.15}$$

Note that in Sect. 6.5.2, it has been proven that the direct wave components are the same as the solution for an infinite medium. Therefore, the direct wave components have been removed from the above equation, and their contributions in the final result are replaced by the solutions from the infinite-space Green's function discussed in Chap. 4.

Figure 7.5 shows the variation of $F(k)$ with k at a frequency of $f = 1$ Hz. Panels (a) and (b) depict the cases without and with the introduction of complex frequency, respectively. It is evident that the curve becomes smoother when the complex frequency is introduced.[9] In Fig. 7.5a, several sharp peaks can be seen, which correspond exactly to the two poles of Eq. (7.1.15). Both ν and $R(k)$ appear in the denominator, thus they correspond to the roots $\nu = 0$ at $k_1^* = \frac{\omega}{\beta}$ and $R(k) = 0$ at $k_2^* = \sqrt{x_R}k_1^*$, where x_R is the root of the Rayleigh function $R(x)$ when it equals zero. For a Poisson medium, $x_R \approx 1.183$.

The variation of the entire integrand $F(k)J_0(kr)k$ with k exhibits new characteristics. Figure 7.6 shows the curves for different epicentral distances r. Due to the presence of the Bessel function $J_0(kr)$ in the integrand, the overall behavior is oscillatory. Another notable feature is that the frequency of oscillation significantly increases with larger r. For an oscillatory integrand, standard numerical integration methods require selecting a sufficient number of integration points within one period to ensure accuracy. Consequently, the computational workload increases significantly as r grows.

7.1.2.2 Filon's Method Based on Adaptive Selection of Integration Points

Noting that $F(k)$ is independent of r, if we can adaptively select integration points based on the variation of $F(k)$ with k, then approximate $F(k)$ with an interpolation polynomial and replace $J_0(kr)$ with its asymptotic expression in the interval between adjacent integration points, we can obtain a closed-form expression for the integral

---

[9] At the end of Sect. 7.1.1.2, the benefits of introducing complex frequency were mentioned, hinting at another advantage: the smoothed curve is more conducive to numerical integration. As we will see later, our integration method adapts the integration points based on the variation of $F(k)$. If the curve is too sharp, it results in an overly dense distribution of local integration points. In practice, the smoothed curve helps improve the efficiency of numerical integration.

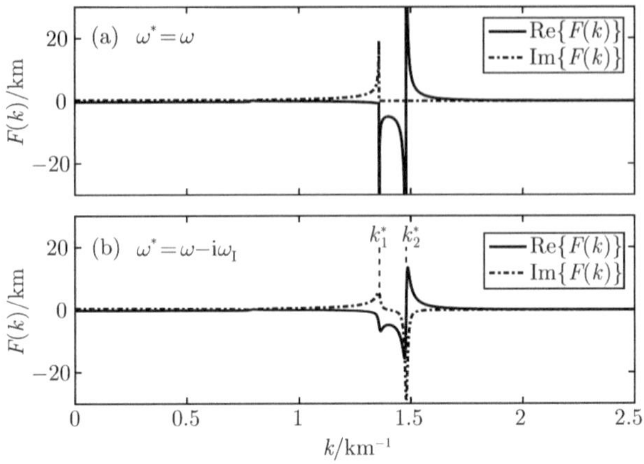

Fig. 7.5 Curves of the function $F(k)$ with k for real and complex frequencies. **a** The case of real frequencies; **b** the case of complex frequencies. The frequency $f = 1$ Hz, and $\omega_I = 0.01\pi$. The solid and dashed lines represent the real part $\text{Re}\{F(k)\}$ and the imaginary part $\text{Im}\{F(k)\}$ of $F(k)$, respectively. $k_1^* = \omega/\beta$, $k_2^* = \sqrt{x_R}k_1^*$, where x_R is the root of the Rayleigh function $R(x)$ when it equals zero

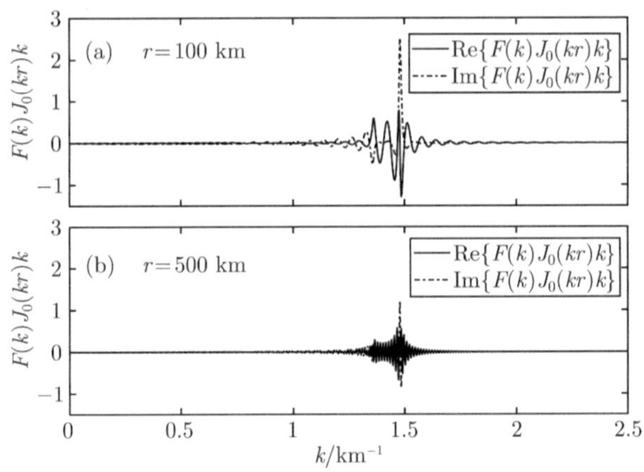

Fig. 7.6 Curves of the integrand $F(k)J_0(kr)k$ with k for different epicentral distances r. **a** $r = 100$ km; **b** $r = 500$ km. The frequency $f = 1$ Hz, and $\omega_I = 0.01\pi$. The solid and dashed lines represent the real part $\text{Re}\{F(k)J_0(kr)k\}$ and the imaginary part $\text{Im}\{F(k)J_0(kr)k\}$ of the integrand $F(k)J_0(kr)k$, respectively

in that interval. This approach is known as the *adaptive Filon's integration method.*[10] Simply put, this method involves two key points: first, adaptively selecting integration points based on $F(k)$; and second, approximating the integrand in each subinterval with interpolation polynomials and asymptotic expressions to achieve an analytical expression for the integral. The final result is obtained by summing the contributions from all subintervals.

Before explaining the adaptive Filon's integration method, it's important to note that this approach is based on the premise that the Bessel function $J_0(x)$ can be approximated by its asymptotic expression in small integration intervals. This approximation is valid only when the argument x of $J_0(x)$ meets certain conditions. The Hankel asymptotic expression for $J_0(x)$ is given by Abramowitz and Stegun (1964, p. 364):

$$J_0^{(i)}(x) \sim \sqrt{\frac{2}{\pi x}}\left[P^{(i)}(x)\cos\left(x-\frac{1}{4}\pi\right) - Q^{(i)}(x)\sin\left(x-\frac{1}{4}\pi\right)\right]$$
$$= \sqrt{\frac{1}{\pi x}}\left[\left(P^{(i)}(x)+Q^{(i)}(x)\right)\cos x + \left(P^{(i)}(x)-Q^{(i)}(x)\right)\sin x\right]$$

where,

$$P^{(1)}(x)=1, \quad P^{(2)}(x)=1-\frac{9}{128x^2}, \quad P^{(3)}(x)=1-\frac{9}{128x^2}+\frac{11025}{98304x^4}$$
$$Q^{(1)}(x)=0, \quad Q^{(2)}(x)=-\frac{1}{8x}, \quad Q^{(3)}(x)=-\frac{1}{8x}+\frac{225}{3072x^3}$$

The approximations $J_0^{(1)}(x)$, $J_0^{(2)}(x)$, and $J_0^{(3)}(x)$ increasingly improve their fit to $J_0(x)$. Figure 7.7a, b illustrate the variation of $J_0(x)$ with x and compare the relative errors $\varepsilon_i(x)$ of the different asymptotic solutions $J_0^{(i)}(x)$ $(i=1,2,3)$. For an allowable relative error of 10^{-6}, the asymptotic expression $J_0^{(3)}(x)$ meets the requirement when $x>10$. Therefore, we will use $J_0^{(3)}(x)$ as the asymptotic approximation for $J_0(x)$.

Since $J_0^{(3)}(x)$ does not approximate $J_0(x)$ well when $x<10$, $J_0(x)$ must be calculated exactly in this range. To address this, we split the integral in Eq. (7.1.1) into two parts:

$$I(\omega)=\int_0^{k^*}F(k,\omega)J_0(kr)k\,\mathrm{d}k+\int_{k^*}^{+\infty}F(k,\omega)J_0(kr)k\,\mathrm{d}k \quad \left(k^*=\frac{10}{r}\right)$$

---

[10] "Adaptive" means automatically selecting integration points based on the variation of the function. Where the function changes rapidly, the integration points are chosen more densely, while in regions where the function changes slowly, the points are more sparsely distributed. This approach ensures maximum efficiency in numerical integration while maintaining the desired accuracy.

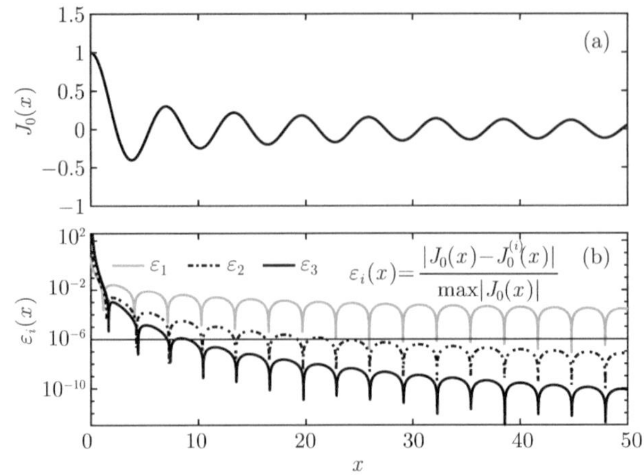

Fig. 7.7 Comparison of Bessel function $J_0(x)$ with different asymptotic solutions $J_0^{(i)}(x)$. **a** $J_0(x)$; **b** The relative errors of different asymptotic solutions $\varepsilon_i(x)$ $(i = 1, 2, 3)$. The fine horizontal line in the figure represents a relative error of 10^{-6}

Our goal is to adaptively divide the integration range $[k^*, k_{\max}]$[11] into several small intervals $[k_i, k_{i+1}]$ $(i = 1, 2, \ldots)$ based on the variations in $\text{Re}\{F(k)\}$[12] for the second integral (denoted as $I^{(2)}(\omega)$) on the right-hand side of the equation. Within each small interval $[k_i, k_{i+1}]$, we approximate $J_0(x)$ with its asymptotic expression $J_0^{(3)}(x)$ and replace $\text{Re}\{F(k)\}$ with an interpolation polynomial. This allows us to obtain a closed-form solution for the integral over each mall interval. Finally, we sum the solutions from all the sub-intervals.

Let's start by adaptively selecting the integration points. First, divide $[k^*, k_{\max}]$ into four equal segments to form k_i and compute the corresponding $F_i = \text{Re}\{F(k_i)\}$ for $i = 1, 2, \ldots, 5$. Let $P^{(lmn)}(k)$ be the quadratic polynomial formed by selecting (k_l, F_l), (k_m, F_m), and (k_n, F_n) as interpolation points:

$$P^{(lmn)}(k) = a^{(lmn)}k^2 + b^{(lmn)}k + c^{(lmn)}$$

where,

[11] Here, the upper limit of the integral is truncated, and a sufficiently large $k_{\max}$ is chosen as the upper limit. In Sect. 7.1.3, we will investigate the selection of the upper limit and related issues.

[12] The real and imaginary parts of $F(k)$ vary similarly with k, as shown in Fig. 7.5. Therefore, for the purpose of selecting integration points, it is sufficient to consider only one of them. Here, we take the real part as an example.

$$a^{(lmn)} = -\frac{(k_l - k_m)F_n + (k_m - k_n)F_l + (k_n - k_l)F_m}{(k_l - k_m)(k_m - k_n)(k_n - k_l)}$$
$$b^{(lmn)} = \frac{(k_l^2 - k_m^2)F_n + (k_m^2 - k_n^2)F_l + (k_n^2 - k_l^2)F_m}{(k_l - k_m)(k_m - k_n)(k_n - k_l)}$$
$$c^{(lmn)} = -\frac{(k_l - k_m)k_l k_m F_n + (k_m - k_n)k_m k_n F_l + (k_n - k_l)k_n k_l F_m}{(k_l - k_m)(k_m - k_n)(k_n - k_l)}$$

Figure 7.8 illustrates the formation of the three quadratic polynomials $P^{(135)}(k)$, $P^{(123)}(k)$, and $P^{(345)}(k)$.

Let

$$S_{11} = \int_{k_1}^{k_3} P^{(135)}(k)\,\mathrm{d}k = \frac{a^{(135)}}{3}\left(k_3^3 - k_1^3\right) + \frac{b^{(135)}}{2}\left(k_3^2 - k_1^2\right) + c^{(135)}\,(k_3 - k_1)$$
$$S_{12} = \int_{k_3}^{k_5} P^{(135)}(k)\,\mathrm{d}k = \frac{a^{(135)}}{3}\left(k_5^3 - k_3^3\right) + \frac{b^{(135)}}{2}\left(k_5^2 - k_3^2\right) + c^{(135)}\,(k_5 - k_3)$$
$$S_{21} = \int_{k_1}^{k_3} P^{(123)}(k)\,\mathrm{d}k = \frac{a^{(123)}}{3}\left(k_3^3 - k_1^3\right) + \frac{b^{(123)}}{2}\left(k_3^2 - k_1^2\right) + c^{(123)}\,(k_3 - k_1)$$
$$S_{22} = \int_{k_3}^{k_5} P^{(345)}(k)\,\mathrm{d}k = \frac{a^{(345)}}{3}\left(k_5^3 - k_3^3\right) + \frac{b^{(345)}}{2}\left(k_5^2 - k_3^2\right) + c^{(345)}\,(k_5 - k_3)$$

For a pre-set allowable error ε (typically 10^{-3}), if the following three inequalities are

$$\left|\frac{S_{11} + S_{12} - S_{21} - S_{22}}{S_{11} + S_{12} + S_{21} + S_{22}}\right| < \varepsilon, \quad \left|\frac{S_{11} - S_{21}}{S_{11} + S_{21}}\right| < \varepsilon, \quad \left|\frac{S_{12} - S_{22}}{S_{12} + S_{22}}\right| < \varepsilon$$

then the quadratic polynomial $P^{(135)}(k)$ can be considered an acceptable approximation of $F(k)$ within the interval $[k_1, k_5]$. If not, the interval $[k_1, k_3]$ is further

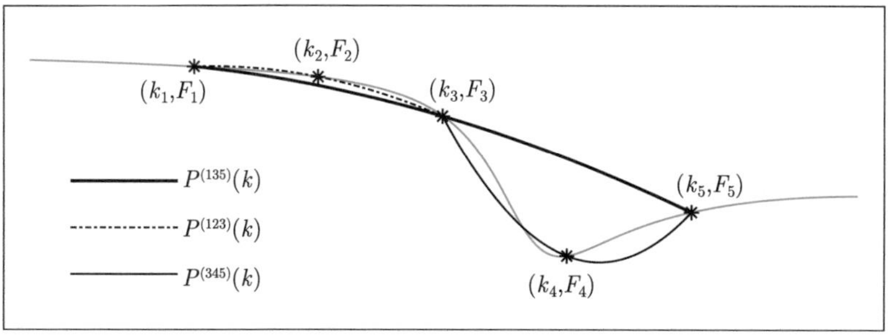

Fig. 7.8 Diagram of adaptive selection process for integration points. The gray line represents $\mathrm{Re}\{F(k)\}$, and $*$ marks the five interpolation points (k_i, F_i) $(i = 1, 2, \ldots, 5)$. The thick solid line, the dashed line, and the thin solid line represent the quadratic polynomials $P^{(135)}(k)$, $P^{(123)}(k)$, and $P^{(345)}(k)$ obtained from interpolation, respectively

subdivided into four segments, updating k_i $(i = 1, 2, ..., 5)$, and the criteria are re-evaluated. This process is repeated until the criteria are satisfied. The points k_i $(i = 1, 3, 5)$ and the coefficients of the corresponding quadratic polynomial $P^{(135)}(k)$ are recorded, and the interval $[k_1, k_5]$ is removed from the intervals to be evaluated. This procedure is repeated, reducing the intervals to be evaluated until all satisfy the criteria. All recorded points that meet the criteria form the adaptively selected set of integration points k_i $(i = 1, 2, ..., N)$.

Figure 7.9 illustrates an example with an allowable relative error $\varepsilon = 10^{-3}$. In Fig. 7.9a, the distribution of integration points across the entire $[k^*, k_{\max}]$ interval is represented by thin vertical lines. To provide a clearer view of densely populated regions, Fig. 7.9b, c zoom in on two specific intervals. It is evident that in areas where $\mathrm{Re}\{F(k)\}$ changes rapidly, the integration points are more densely distributed, whereas in areas with slower changes, the points are more sparsely distributed. This is the characteristic feature of an adaptive integration point distribution.

For the set of integration points k_i $(i = 1, 2, \ldots, N)$ selected according to the adaptive algorithm described above, the integral $I^{(2)}(\omega)$ can be expressed as

$$I^{(2)}(\omega) = \sum_{m=1}^{\frac{N-1}{2}} \int_{k_{2m-1}}^{k_{2m+1}} F(k, \omega) J_0(kr) k \, \mathrm{d}k \triangleq \sum_{m=1}^{\frac{N-1}{2}} I_m$$

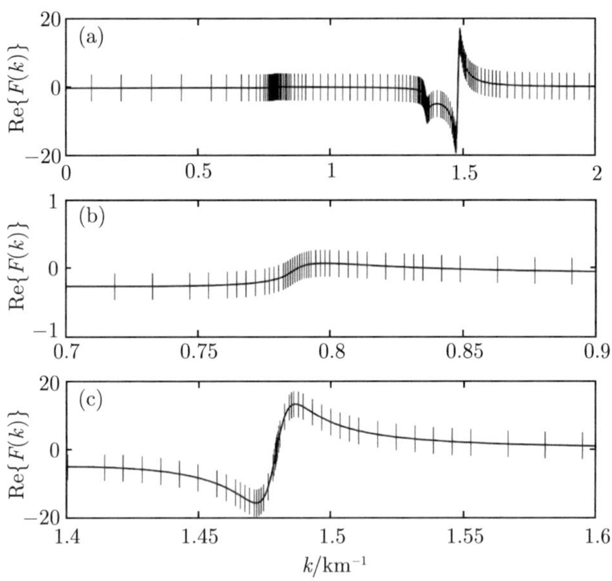

Fig. 7.9 Example of adaptive distribution of integration points. **a** The interval [0, 2]; **b** The zoomed-in interval [0.7, 0.9]; **c** The zoomed-in interval [1.4, 1.6]. The selected integration points are shown as short vertical lines. Where the function changes rapidly, the integration points are densely distributed, and where it changes slowly, they are more sparse

In each subinterval $[k_{2m-1}, k_{2m+1}]$ of the integration variable, we replace $F(k,\omega)$ and $J_0(kr)$ with the quadratic polynomial $P^{(135)}(k)$ and the asymptotic expression $J_0^{(1)}(kr)$, respectively, to obtain a closed-form expression for I_m:

$$I_m = \sqrt{\frac{1}{\pi r}} \int_{k_{2m-1}}^{k_{2m+1}} (a_m k^{\frac{5}{2}} + b_m k^{\frac{3}{2}} + c_m k^{\frac{1}{2}}) \left[M_1(k)\cos(kr) + M_2(k)\sin(kr)\right] \mathrm{d}k$$

where a_m, b_m, and c_m are the coefficients of the quadratic interpolation polynomial in the interval $[k_{2m-1}, k_{2m+1}]$,[13]

$$M_\alpha(k) = 1 + \frac{(-1)^\alpha}{8kr} - \frac{9}{128k^2r^2} - \frac{(-1)^\alpha 225}{3072k^3r^3} + \frac{11025}{98304k^4r^4} \quad (\alpha = 1, 2)$$

The integrals I_m involve terms of the form:

$$\begin{bmatrix} I_{mn}^c \\ I_{mn}^s \end{bmatrix} \triangleq \int_{k_{2m-1}}^{k_{2m+1}} k^{\frac{2n-9}{2}} \begin{bmatrix} \cos(kr) \\ \sin(kr) \end{bmatrix} \mathrm{d}k \quad (n = 1, 2, \cdots, 7)$$

specifically,[14]

$$I_{m1}^c = \frac{2\sqrt{r^5}}{15} \left[4\sqrt{2\pi}S(cx) + 2K_{-3}^s(x) + (4x^4 - 3)K_{-5}^c(x)\right]\Big|_{k_{2m-1}}^{k_{2m+1}}$$

$$I_{m1}^s = -\frac{2\sqrt{r^5}}{15} \left[4\sqrt{2\pi}C(cx) + 2K_{-3}^c(x) - (4x^4 - 3)K_{-5}^s(x)\right]\Big|_{k_{2m-1}}^{k_{2m+1}}$$

$$I_{m2}^c = -\frac{2\sqrt{r^3}}{3} \left[2\sqrt{2\pi}C(cx) - 2K_{-1}^s(x) + K_{-3}^c(x)\right]\Big|_{k_{2m-1}}^{k_{2m+1}}$$

$$I_{m2}^s = -\frac{2\sqrt{r^3}}{3} \left[2\sqrt{2\pi}S(cx) + 2K_{-1}^c(x) + K_{-3}^s(x)\right]\Big|_{k_{2m-1}}^{k_{2m+1}}$$

$$I_{m3}^c = -2\sqrt{r} \left[\sqrt{2\pi}S(cx) + K_{-1}^c(x)\right]\Big|_{k_{2m-1}}^{k_{2m+1}}$$

---

[13] During the process of selecting adaptive integration points, these coefficients are calculated. Therefore, it is necessary to record these coefficients alongside the integration points for future use.

[14] As an example of using Maple to solve the indefinite integral corresponding to I_{m3}^c, input the following commands:

```
> f:=sin(k*r)/k^(3/2);
> simplify(int(f,k));
```

The direct result given is

$$-\frac{2\left(\sqrt{r}\sqrt{2}\sqrt{\pi}\mathrm{FresnelS}\left(\frac{\sqrt{2}\sqrt{k}\sqrt{r}}{\sqrt{\pi}}\right)\sqrt{k} + \cos(kr)\right)}{\sqrt{k}}.$$

$$
\begin{aligned}
I_{m3}^{s} &= 2\sqrt{r}\left[\sqrt{2\pi}C(cx) - K_{-1}^{s}(x)\right]\Big|_{k_{2m-1}}^{k_{2m+1}} \\
I_{m4}^{c} &= \sqrt{\frac{2\pi}{r}}C(cx)\Big|_{k_{2m-1}}^{k_{2m+1}}, \quad I_{m4}^{s} = \sqrt{\frac{2\pi}{r}}S(cx)\Big|_{k_{2m-1}}^{k_{2m+1}} \\
I_{m5}^{c} &= -\frac{1}{2\sqrt{r^3}}\left[\sqrt{2\pi}S(cx) - 2K_{1}^{s}(x)\right]\Big|_{k_{2m-1}}^{k_{2m+1}} \\
I_{m5}^{s} &= \frac{1}{2\sqrt{r^3}}\left[\sqrt{2\pi}C(cx) - 2K_{1}^{c}(x)\right]\Big|_{k_{2m-1}}^{k_{2m+1}} \\
I_{m6}^{c} &= -\frac{1}{4\sqrt{r^5}}\left[3\sqrt{2\pi}C(cx) - 4K_{3}^{s}(x) - 6K_{1}^{c}(x)\right]\Big|_{k_{2m-1}}^{k_{2m+1}} \\
I_{m6}^{s} &= -\frac{1}{4\sqrt{r^5}}\left[3\sqrt{2\pi}S(cx) + 4K_{3}^{c}(x) - 6K_{1}^{s}(x)\right]\Big|_{k_{2m-1}}^{k_{2m+1}} \\
I_{m7}^{c} &= \frac{1}{8\sqrt{r^7}}\left[15\sqrt{2\pi}S(cx) + 20K_{3}^{c}(x) + 2(4x^4 - 15)K_{1}^{s}(x)\right]\Big|_{k_{2m-1}}^{k_{2m+1}} \\
I_{m7}^{s} &= -\frac{1}{8\sqrt{r^7}}\left[15\sqrt{2\pi}C(cx) - 20K_{3}^{s}(x) + 2(4x^4 - 15)K_{1}^{c}(x)\right]\Big|_{k_{2m-1}}^{k_{2m+1}}
\end{aligned}
$$

where,

$$
x = \sqrt{kr}, \quad c = \sqrt{2/\pi}, \quad K_i^c(x) = x^i \cos\left(x^2\right), \quad K_i^s(x) = x^i \sin\left(x^2\right)
$$

$C(x)$ and $S(x)$ are two transcendental functions known as *Fresnel integrals* (Abramowitz and Stegun 1964, p. 300):

$$
C(x) \triangleq \int_0^x \cos\left(\frac{\pi}{2}t^2\right) \mathrm{d}t, \quad S(x) \triangleq \int_0^x \sin\left(\frac{\pi}{2}t^2\right) \mathrm{d}t
$$

Most commonly used mathematical software has dedicated functions for calculating them; for example, in Matlab, the corresponding functions are `fresnelc(x)` and `fresnels(x)`.

In summary, for the numerical calculation of the integral in Eq. (7.1.1), the entire integration interval is adaptively divided into several small subintervals based solely on the variation of $F(k, \omega)$. In each subinterval, $F(k, \omega)$ is approximated by a quadratic polynomial, and the Bessel function is replaced by its asymptotic expression. This approach allows for a closed-form solution of the integral in each subinterval. It is worth noting that during the adaptive selection of integration points, additional computational effort is required to verify the criteria. Therefore, this adaptive Filon's integration method does not offer an advantage over the standard integration scheme in the near-field case, but its benefits become increasingly significant as the epicentral distance increases.

7.1.3 *Peak and Trough Averaging Method*

The integral in Eq. (7.1.1) is an improper integral with an upper limit of $+\infty$, and numerical integration requires truncation. For oscillatory integrals, especially those with slow convergence,[15] a very large upper limit is often needed to ensure accuracy, resulting in low computational efficiency.

7.1.3.1 Problem Statement

Using I_5 from Eq. (6.5.2e) as an example, the real and imaginary parts of the integrand in Eq. (7.1.1) exhibit similar behavior. For simplicity and clarity in showing the effect of the truncation limit, only the real part is considered here. Based on Eq. (7.1.1), the integral function is defined as:

$$P(k) = \int_0^k \mathrm{Re}\{F(k')\} J_m(k'r)k' \,\mathrm{d}k'$$

Clearly, $P(+\infty) = I(\omega)$.

Figures 7.10 and 7.11 show the curves of the real part of the integrand $\mathrm{Re}\{F(k)\}J_0(kr)k$ and the integral function $P(k)$ with k for both fast convergence ($z_s = 5$ km, $z = 2$ km) and slow convergence ($z_s = z = 0$ km). In these figures, k_2^* corresponds to the root of the Rayleigh function, as seen in Fig. 7.5b. When $k > k_2^*$, the integrand oscillates around zero with a monotonically decreasing amplitude. The difference between the two cases is that in the case of $z_s = z = 0$ km (Fig. 7.11), the oscillation amplitude decays very slowly. This is because the phase factors e^{PP}, e^{SS}, e^{PS}, and e^{SP} in $F(k)$ (see Eq. (7.1.15)) lose their damping effect on the oscillation amplitude of the integrand when $z_s = z = 0$ km. For $F(k)$ containing only the contribution from the free surface, the situation is similar when both z_s and z are small. In such cases, without special treatment, a very large upper integration limit is required, resulting in low computational efficiency.

7.1.3.2 Peak and Trough Averaging Method Based on Repeated Averaging Method

Notably, in Fig. 7.11, the integral function $P(k)$ oscillates around a stable value when $k > k_2^*$. If an algorithm can be developed to quickly determine the central value of these oscillations based on the characteristics of the oscillating function, computational efficiency would be greatly improved. This goal can be achieved using the repeated averaging method in numerical computations (Dahlquist and Björck 1974, p. 278).

---

[15] This typically occurs when the observation point and the source point are at similar or identical depths, causing the integrand to converge very slowly.

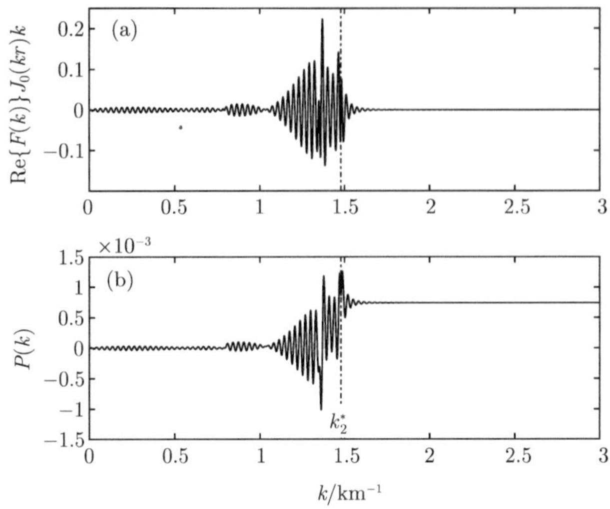

Fig. 7.10 Curves of $\mathrm{Re}\{F(k)\}J_0(kr)k$ and corresponding integral function $P(k)$ with k (1). **a** $\mathrm{Re}\{F(k,\omega)\}\, J_0(kr)k$, **b** $P(k)$. k_2^* is the root of Rayleigh function. $z_s = 5$ km, $z = 2$ km

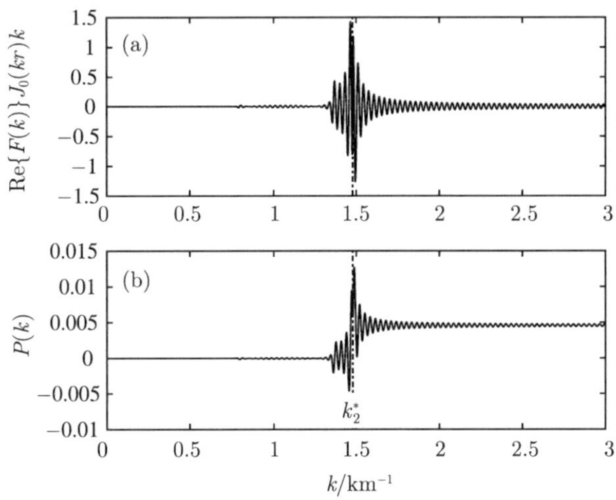

Fig. 7.11 Curves of $\mathrm{Re}\{F(k)\}J_0(kr)k$ and corresponding integral function $P(k)$ with k (2). The same as in Fig. 7.10, except that $z_s = z = 0$ km

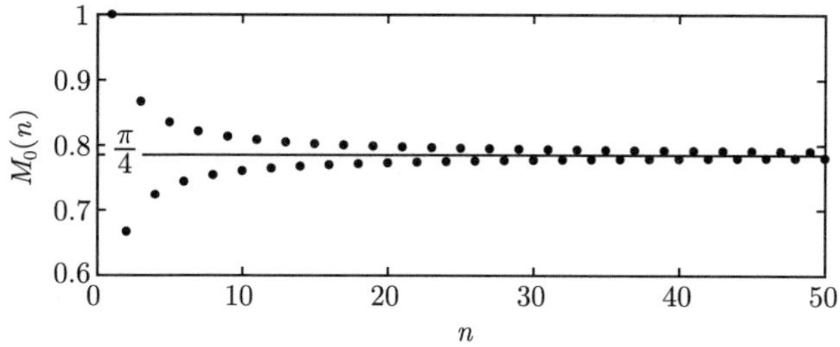

Fig. 7.12 Variation of the Leibniz series $M_0(n)$ with n. The black dots represent the values of the series, and the horizontal line represents $\pi/4$. As n increases, the series oscillates around $\pi/4$ and slowly converges

Using the Leibniz series for π as an example, we illustrate the method of repeated averaging. If we define

$$M_0(n) = \sum_{i=1}^{n} \frac{(-1)^{i-1}}{2i-1}$$

then

$$\frac{\pi}{4} = \lim_{n\to+\infty} \{M_0(n)\}$$

Figure 7.12 shows how $M_0(n)$ varies with n. As n increases, $M_0(n)$ oscillates around $\pi/4$ and gradually approaches this value. This is a typical example of a *slowly converging sequence*. For such sequences, we define a *reduced sequence* as follows:

$$M_i(n) = \frac{1}{2}\left[M_{i-1}(n) + M_{i-1}(n+1)\right] \quad (i = 1, 2, \cdots)$$

This means that the n-th element of the ith reduced sequence is the arithmetic mean of the nth and $(n+1)$th elements of the $(i-1)$th reduced sequence. Clearly, the reduced sequences defined in this way have the following properties:

(1) The length of the i-th reduced sequence is one less than that of the $(i-1)$-th sequence.
(2) They share the same limit, i.e., $M_i(+\infty) = M_0(+\infty)$ $(i = 1, 2, \ldots)$.

Figure 7.13 shows the first six reduced sequences for the 6–12th elements of $M_0(n)$. With each reduction, the sequence length decreases by one, so by the 6th reduced sequence, the length is reduced to one. This single value represents the final convergence value of the slowly converging sequence.

The construction of reduced sequences involves repeatedly calculating arithmetic means, hence the method is called the *repeated averaging method*. As shown in Fig. 7.13, this method involves minimal computation, making the algorithm both simple and highly efficient. To demonstrate the efficiency of the repeated averaging

n	$M_0(n)$	$M_1(n)$	$M_2(n)$	$M_3(n)$	$M_4(n)$	$M_5(n)$	$M_6(n)$
6	0.744012	0.782474	0.785038	0.785340	0.785387	0.785396	0.785398
7	0.820935	0.787602	0.785641	0.785434	0.785405	0.785400	
8	0.754268	0.783680	0.785228	0.785376	0.785395		
9	0.813092	0.786776	0.785523	0.785414			
10	0.760460	0.784270	0.785305				
11	0.808079	0.786340					
12	0.764501						

Fig. 7.13 Execution diagram of the repeated averaging method. $M_0(n)$ is the original sequence, and $M_i(n)$ $(i = 1, 2, \ldots, 6)$ are the reduced sequences. n represents the number of terms in the sequence. The arrows indicate that the element in the direction they point to is the arithmetic mean of the two corresponding elements from the previous reduced sequence

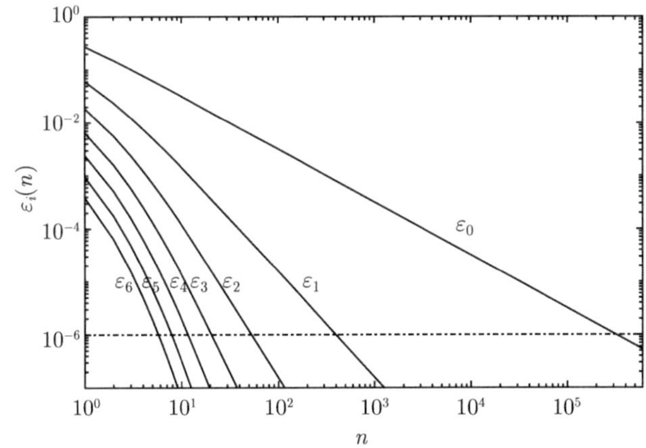

Fig. 7.14 Relative error ε_i of $M_0(n)$ and the reduced sequences $M_i(n)$. The horizontal dashed line represents a relative error of 10^{-6}

method, we define the relative error sequence $\varepsilon_i(n)$ corresponding to the original sequence $M_0(n)$ and the ith reduced sequence as follows:

$$\varepsilon_i(n) = \frac{|M_i(n) - M_i(+\infty)|}{|M_i(+\infty)|} \quad (i = 0, 1, 2, \cdots)$$

Figure 7.14 displays the variations of $\varepsilon_i(n)$ $(i = 0, 1, \cdots, 6)$ with n on a double logarithmic scale. If the allowable relative error is 10^{-6}, then for $M_0(n)$, n must be greater than 31,800, whereas for $M_6(n)$, n only needs to be greater than 5. The reduction in computational effort is remarkable.

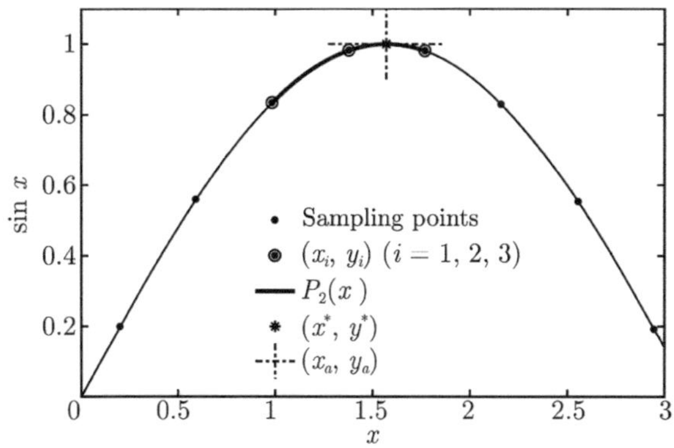

Fig. 7.15 Diagram of precise peak value calculation: Using $\sin x$ as an example. The • represents sampling points, • represents the three sampling points used to determine the precise value, $P_2(x)$ represents the quadratic polynomial obtained by interpolating these three points, ⊙ represents the precise peak position (x^*, y^*) obtained by setting the derivative of the quadratic polynomial to zero, and ∗ represents the actual precise peak position $(\pi/2, 1)$

From the above description, it is evident that the repeated averaging method is simple and efficient for finding the convergence value of slowly converging sequences. Referring back to the integral function $P(k)$ shown in Fig. 7.11, it has been mentioned that when $k > k_2^*$, the function oscillates around its convergence value with a monotonically decreasing amplitude. If we take the peak and trough values of this oscillation to form a sequence, we get a standard slowly converging sequence. The repeated averaging method can be applied to quickly determine its convergence value, and this approach can be called the *peak and trough averaging method.*

The key to implementing the peak and trough averaging method lies in accurately identifying the peak and trough values. In numerical calculations, a finite Δk is always chosen to compute the integrand values. Using a finite Δk as the step size does not guarantee the precise identification of peak and trough values. Therefore, a refined algorithm is necessary to locate these values accurately. Figure 7.15 demonstrates the process of accurately determining the peak value using a simple sine function $\sin x$ as an example, where the dots represent sampling points.

By taking three consecutive sampling points (x_i, f_i) and checking if $f_2 \geq f_1$ and $f_2 \geq f_3$, it can be concluded that an accurate peak lies within the interval (x_1, x_3) if the condition is met. If not, the process is repeated with $\Delta x = \pi/8$ as the step size. The refinement involves using (x_i, f_i) $(i = 1, 2, 3)$ as interpolation points to form a quadratic polynomial:

$$P_2(x) = ax^2 + bx + c$$

The coefficients are:

$$a = -\frac{f_1(x_2 - x_3) + f_2(x_3 - x_1) + f_3(x_1 - x_2)}{(x_1 - x_2)(x_2 - x_3)(x_3 - x_1)}$$
$$b = \frac{f_1(x_2^2 - x_3^2) + f_2(x_3^2 - x_1^2) + f_3(x_1^2 - x_2^2)}{(x_1 - x_2)(x_2 - x_3)(x_3 - x_1)}$$
$$c = -\frac{f_1 x_2 x_3(x_2 - x_3) + f_2 x_3 x_1(x_3 - x_1) + f_3 x_1 x_2(x_1 - x_2)}{(x_1 - x_2)(x_2 - x_3)(x_3 - x_1)}$$

The extremum is found at $P_2'(x) = 0$, which gives $x^* = -b/2a$. The refined peak value is then $y^* = P_2(x^*)$. For the example shown in Fig. 7.15, the error of the refined peak value relative to the exact value is approximately 0.05%.

By refining the process described above, a series of peak and trough values are found and combined into a slowly converging sequence $M_0(n)$. Using the repeated averaging method introduced earlier, the convergence value of the integral function $P(k)$ can be quickly obtained. Figure 7.16 illustrates the results of the peak and trough averaging method, using $P(k)$ from Fig. 7.11b as an example. The starting point k^* for finding peak and trough values is crucial and must meet the condition that $P(k)$ oscillates monotonically decreasing when $k > k^*$. Therefore, it can be conservatively set to $k^* = 1.1k_2^*$. The enlarged view in Fig. 7.16b clearly shows

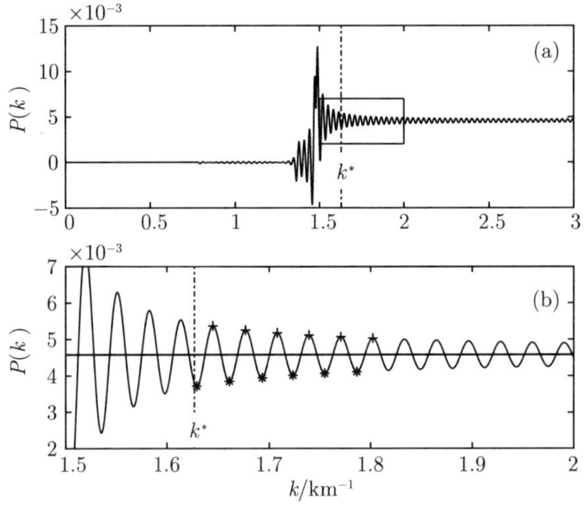

Fig. 7.16 Results of the peak and trough average method. (**a**) shows the same as in Fig. 7.11b with $k^* = 1.1k_2^*$. The boxed area is the region displayed in (**b**). (**b**) is a zoomed-in view of the boxed area in (**a**). The symbols + represent peaks, and $*$ represent troughs. The horizontal line represents the results obtained by applying the repeated averaging method to the sequence of peak and trough values

the convergence value obtained from a sequence composed of six peaks and six troughs. Its accuracy can be visually assessed. Despite being based on very limited calculations, the efficiency improvement is significant.

7.2 Discrete Fourier Transform and Source Time Function

Section 7.1 addresses issues related to the computation of wavenumber integrals. Once we obtain the results of these integrals through numerical methods, we achieve displacement in the frequency domain. To eventually get results in the time domain, we need to perform an inverse Fourier transform. In practice, the inverse Fourier transform is implemented using standard fast Fourier transform (FFT) programs, which is a fast algorithm for the discrete Fourier transform (DFT). To effectively use FFT programs, one must have a clear understanding of DFT. This section will illustrate the characteristics of DFT with specific examples.

7.2.1 A Few Basic Fourier Transform Pairs

Before delving into specific problems, it's essential to understand a few basic Fourier transform pairs. For convenience, we use "$\Longleftrightarrow$" to represent a Fourier transform pair. Here, $f(t)$ denotes a function in the time domain, while $F(f)$ or $F(\omega)$ represents a function in the frequency domain, where $\omega = 2\pi f$. Therefore,

$$F(\omega) = \int_{-\infty}^{+\infty} f(t)\mathrm{e}^{-\mathrm{i}\omega t}\,\mathrm{d}t \quad \Longleftrightarrow \quad f(t) = \frac{1}{2\pi}\int_{-\infty}^{+\infty} F(\omega)\mathrm{e}^{\mathrm{i}\omega t}\,\mathrm{d}\omega$$

To convert a continuous-time function into a discrete-time series, a process called *sampling* is required. Mathematically, this is represented using a discrete δ function.[16] The sampling function in the time domain, $\Delta(t)$, is given by

$$\Delta(t) = \sum_{n=-\infty}^{+\infty} \delta(t - n\Delta t) \tag{7.2.1}$$

as shown in Fig. 7.17a. The corresponding sampling function in the frequency domain, $\Delta(f)$, is

---

[16] The discrete δ function differs significantly from the continuous Dirac δ function, on which integration can be performed. The discrete δ function $\delta(t - t_0)$ takes the value 1 when $t = t_0$. Therefore, $f(t)\delta(t - t_0) = f(t_0)$.

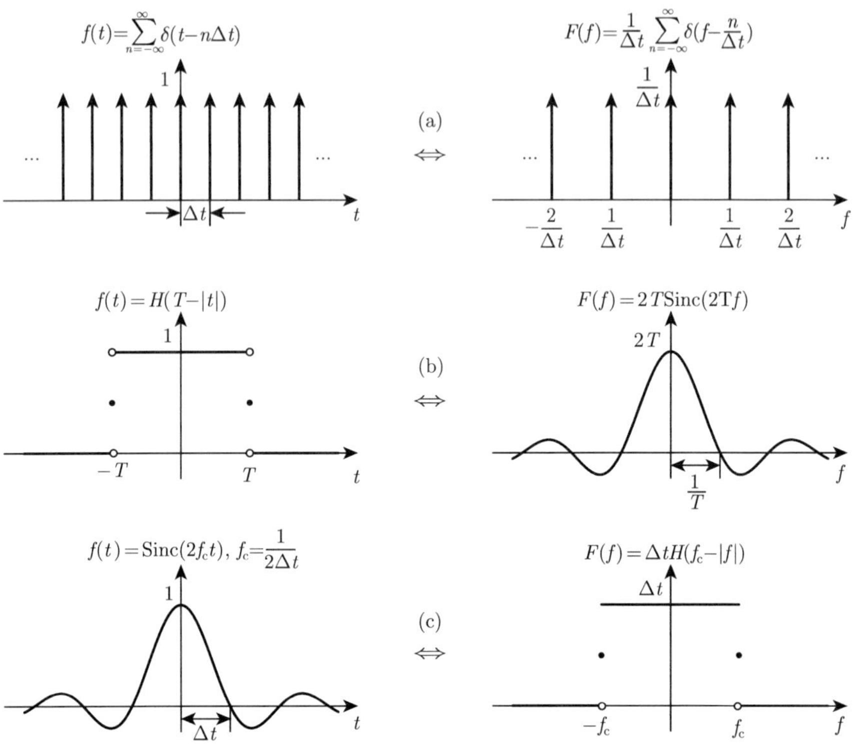

Fig. 7.17 A few basic Fourier transform pairs. The function in the time domain is denoted as $f(t)$, while the corresponding function in the frequency domain is denoted as $F(f)$. Δt represents the sampling interval in the time domain, T is the half-width of a rectangular function in the time domain, and $f_c = 1/2\Delta t$ is the Nyquist frequency

$$\Delta(f) = \sum_{n=-\infty}^{+\infty} \int_{-\infty}^{+\infty} \delta(t - n\Delta t)e^{-i2\pi f t}\,dt = \sum_{n=-\infty}^{+\infty} e^{-i2n\pi f \Delta t}$$

$$\overset{(7.1.9)}{=\!=\!=} 2\pi \sum_{n=-\infty}^{+\infty} \delta(2\pi f \Delta t - 2n\pi) = \frac{1}{\Delta t} \sum_{n=-\infty}^{+\infty} \delta\left(f - \frac{n}{\Delta t}\right).$$

Figure 7.17a shows that the spectrum of the picket fence-like sampling function in the time domain also resembles a picket fence, although with different spacing and height.

In addition to sampling, it is common to use a rectangular window function in the time or frequency domain to extract a portion of the signal. For example, the window function in the time domain can be represented as $f(t) = H(T - |t|)$,

$$F(f) = \int_{-T}^{T} e^{-i2\pi f t}\,dt = \frac{e^{-i2\pi f T} - e^{i2\pi f T}}{-i2\pi f} = \frac{2\sin(2\pi f T)}{2\pi f} = 2T\,\mathrm{Sinc}(2fT)$$

where the Sinc function is defined as

$$\mathrm{Sinc}(x) \triangleq \frac{\sin(\pi x)}{\pi x}$$

Similarly, we have

$$f(t) = \mathrm{Sinc}(2 f_c t) \iff F(f) = \Delta t H(f_c - |f|) \tag{7.2.2}$$

where $f_c = 1/2\Delta t$ is the Nyquist frequency (detailed in Sect. 7.2.2). Figure 7.17b, c illustrate these Fourier transform pairs. From the figures, we can see that truncation in the time domain leads to sidelobes in the frequency domain, known as *spectral leakage*, and vice versa. This is a significant property of the relationship between a time-domain signal and its spectrum.

7.2.2 Discretization of Continuous Waveforms and Sampling Theorem

Unlike the theoretical analysis done through Fourier transforms and their inverses, the Fourier transforms and inverses performed on a computer are discrete, namely the DFT and IDFT. The primary question to address before performing discrete Fourier transforms is under what conditions a continuous waveform can be accurately reconstructed from its discretized signal. From the signal processing perspective, discretization equates to sampling. Thus, understanding the effects of sampling a continuous waveform is fundamental to performing discrete Fourier transforms and their inverses.

Consider a time-domain function $f(t)$ with a finite bandwidth and a highest frequency of f_c.[17] Figure 7.18a shows the graphs of $f(t)$ and its spectrum $F(f)$, where $F(f)$ is non-zero only for $|f| < f_c$. Sampling $f(t)$ mathematically corresponds to multiplying it by the time-domain sampling function $\Delta(t)$ from Eq. (7.2.1), as shown in Fig. 7.18b. The product of $f(t)$ and $\Delta(t)$ results in the discrete signal shown in Fig. 7.18c. According to the property "multiplication in the time domain corresponds to convolution in the frequency domain", the convolution of $F(f)$ and $\Delta(f)$ effectively replicates the spectrum at positions centered around $i/\Delta t$ for $i = \pm 1, \pm 2, \ldots$, because

$$F(f) * \Delta(f) = \int_{-\infty}^{+\infty} F(f-\tau)\frac{1}{\Delta t}\sum_{n=-\infty}^{+\infty}\delta\left(\tau - \frac{n}{\Delta t}\right)\mathrm{d}\tau = \frac{1}{\Delta t}\sum_{n=-\infty}^{+\infty} F\left(f - \frac{n}{\Delta t}\right)$$

---

[17] Bandwidth refers to the difference between the maximum and minimum frequencies in the time signal spectrum. For example, the signal in Fig. 7.18a has a bandwidth of $2 f_c$.

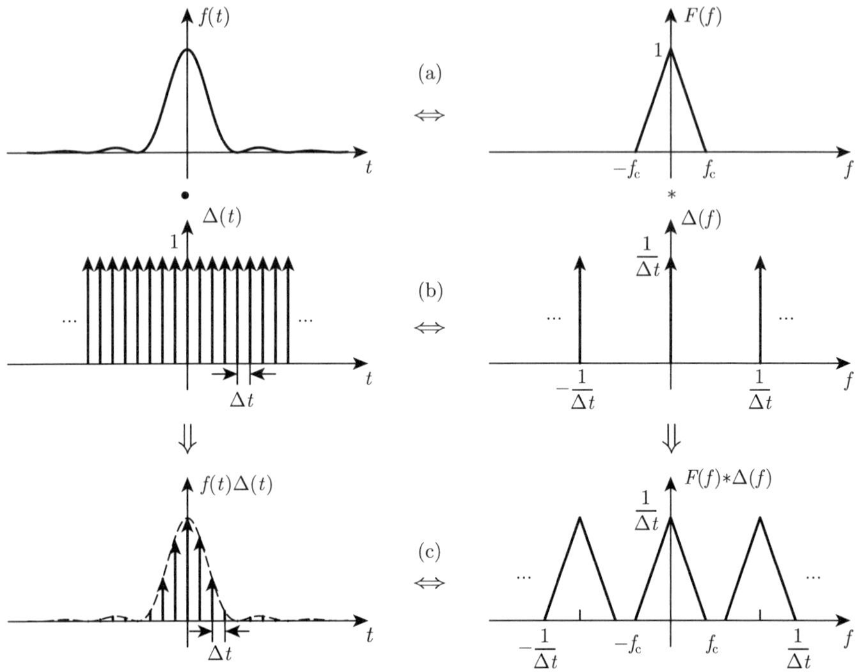

Fig. 7.18 How time-domain sampling leads to periodicity in frequency domain (1). $\Delta(t)$ and $\Delta(f)$ are the sampling functions in the time domain and frequency domain, respectively. Δt is the sampling interval in the time domain, and f_c is the highest frequency of the signal. The symbol $\cdot$ denotes multiplication, and $*$ denotes convolution

Figure 7.18 clearly illustrates this process. It shows that discretizing a continuous waveform in the time domain with an interval of Δt causes the corresponding frequency spectrum to exhibit periodic repetitions with a period of $1/\Delta t$. This demonstrates an important property of the relationship between a time function and its spectrum: discreteness in the time domain corresponds to periodicity in the frequency domain, and similarly, discreteness in the frequency domain corresponds to periodicity in the time domain.

For the case we are considering, with the highest frequency f_c, if Δt is sufficiently small (making $1/\Delta t$ large enough), the replicated spectra will not overlap, as shown in Fig. 7.18c. However, if Δt is not small enough, resulting in the spacing $1/\Delta t$ in the frequency domain not being large enough, a phenomenon called *aliasing* occurs, where the replicated spectra overlap, as illustrated in Fig. 7.19. Aliasing means that the time signal cannot be accurately reconstructed from the spectrum in this scenario.

From the spectrum shown in Fig. 7.18c, it is clear that the critical condition for the time-domain sampling interval Δt is met when

$$f_c = \frac{1}{\Delta t} - f_c \quad \Longrightarrow \quad \Delta t = \frac{1}{2f_c}$$

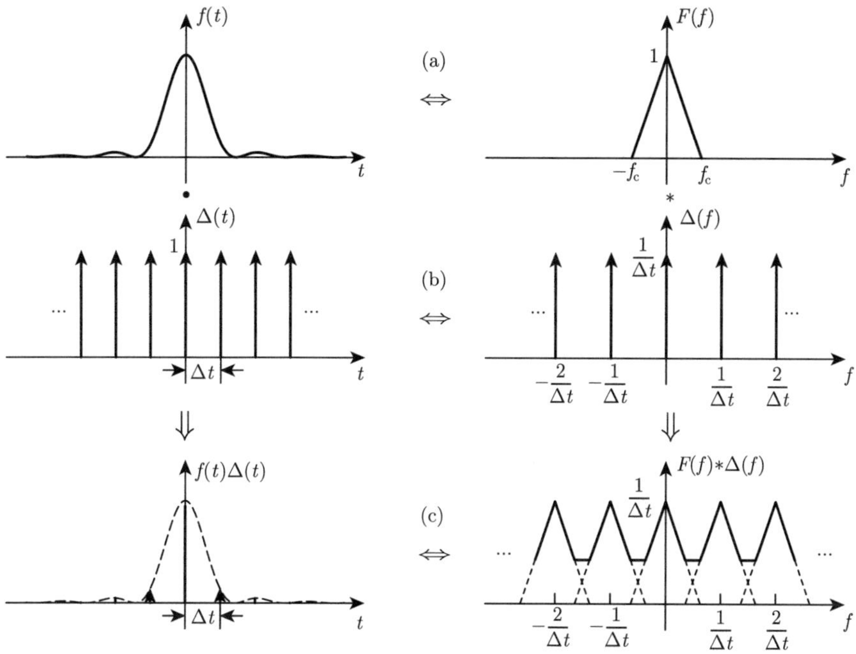

Fig. 7.19 How time-domain sampling leads to periodicity in frequency domain (2). The same as in Fig. 7.18, but the difference is that the sampling interval Δt in the time domain has increased, causing the sampling interval $1/\Delta t$ in the frequency domain to decrease, which leads to aliasing in the frequency domain waveform

This means that for a time signal $f(t)$ with a highest frequency f_c, to avoid aliasing, the maximum sampling interval for the continuous signal must be $\Delta t = 1/2f_c$. In other words, the minimum sampling frequency should be twice the highest frequency, i.e., $1/\Delta t = 2f_c$. Alternatively, for a time signal with a sampling interval Δt, the highest resolvable frequency f_c is $1/2\Delta t$, known as the *Nyquist frequency*.

Based on the above facts, the following sampling theorem[18] can be derived: *For a continuous signal $f(t)$ with a highest frequency f_c in its spectrum, the continuous function $f(t)$ can be uniquely determined by its sampled values with a sampling interval $\Delta t = 1/2f_c$. Specifically, $f(t)$ can be expressed as*

$$f(t) = \sum_{n=-\infty}^{+\infty} h(n\Delta t)\text{Sinc}\,[2f_c(t - n\Delta t)] \qquad (7.2.3)$$

[18] The sampling theorem was first proposed by the Swedish-American physicist H. Nyquist in 1928 and is thus known as the *Nyquist sampling theorem*. In 1933, Soviet radio physicist and cryptographic communications pioneer V. A. Kotelnikov provided a rigorous formulation of the theorem, which is known as the *Kotelnikov sampling theorem* in Soviet literature. In 1948, American mathematician, electrical engineer, and cryptographer C. E. Shannon clearly explained and formally cited the theorem, leading to its reference as the *Shannon sampling theorem* in many texts.

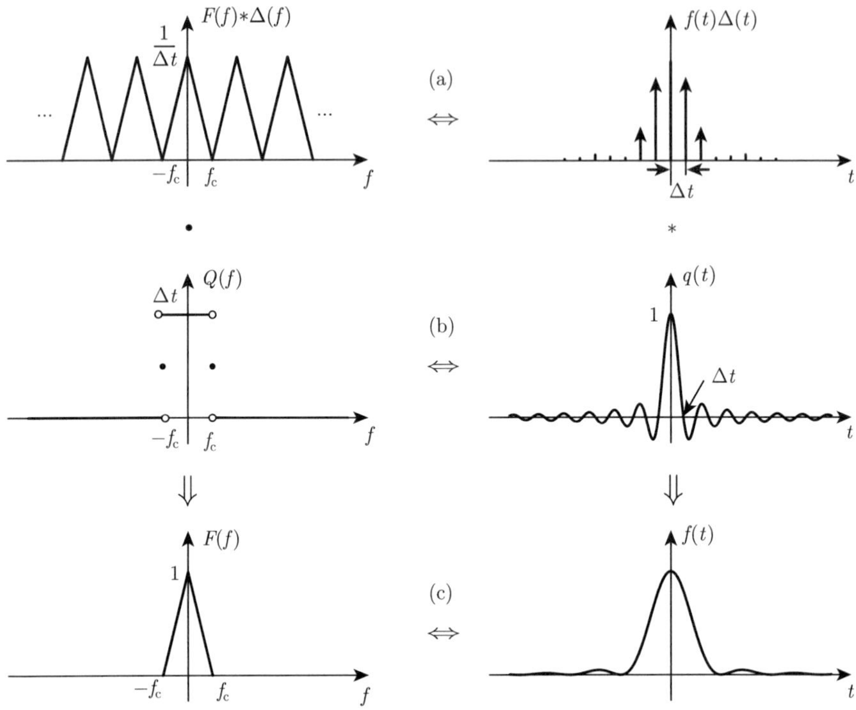

Fig. 7.20 Diagram of the sampling theorem. Illustrates how to reconstruct a continuous signal from a discretized signal. **a** Discretized time-domain signal and its corresponding periodic spectrum; **b** Window function in the frequency domain and its corresponding time-domain function; **c** Non-periodic spectrum and continuous time-domain function

The diagram in Fig. 7.20 illustrates this process: Windowing (multiplying by a window function $Q(f)$) the periodically distributed spectrum $F(f) * \Delta(f)$ truncates it to obtain the non-periodic spectrum $F(f)$. This corresponds to the convolution of the discrete-time signal $f(t)\Delta(t)$ with the time-domain function $q(t)$, related to the window function in the frequency domain, to produce the continuous-time function $f(t)$. Therefore,

$$
\begin{aligned}
f(t) = [f(t)\Delta(t)] * q(t) &= \sum_{n=-\infty}^{+\infty} [f(n\Delta t)\delta(t - n\Delta t)] * q(t) \\
&= \sum_{n=-\infty}^{+\infty} f(n\Delta t) q(t - n\Delta t) \overset{(7.2.2)}{=\!=\!=} \sum_{n=-\infty}^{+\infty} h(n\Delta t)\text{Sinc}\,[2f_c(t - n\Delta t)]
\end{aligned}
$$

The sampling theorem can be seen as a bridge between continuous and discrete domains, as well as between theoretical analysis and practical computer implementation. Computers can only handle discrete time signals, and using these discrete points to represent a continuous time function introduces errors. The question arises: can the continuous signal be accurately reconstructed from these sampled points? The sampling theorem provides the answer: by finding the signal's maximum frequency f_c and sampling at a frequency twice this maximum, the continuous signal can be accurately reconstructed using Eq. (7.2.3).

7.2.3 *From Fourier Transform to Discrete Fourier Transform*

The previously introduced sampling theorem bridges the gap between continuous time functions and discrete time signals. How do we transition from theoretical analysis to numerical implementation? Simply put, it involves two steps: sampling in the time domain and sampling in the frequency domain. The following explanation, along with Fig. 7.21, illustrates this process in detail.

7.2.3.1 Time-Domain Sampling

Sampling the continuous function $f(t)$ in Fig. 7.21a involves multiplying it by the sampling function $\Delta_1(t)$:

$$f(t)\Delta_1(t) = f(t)\sum_{k=-\infty}^{+\infty}\delta(t-k\Delta t) = \sum_{k=-\infty}^{+\infty} f(k\Delta t)\delta(t-k\Delta t)$$

This process is illustrated in Fig. 7.21b, c. As shown in Sect. 7.2.2, time-domain sampling results in periodic distribution in the frequency domain. It is important to note that in general, the spectrum is not band-limited, so aliasing effects are inevitable.[19]

7.2.3.2 Frequency-Domain Sampling

Sampling the spectrum in Fig. 7.21c yields the periodic discrete signal $\tilde{f}(t)$ in the time domain. This process is depicted in Fig. 7.21d, e. Sampling in frequency domain corresponds to convolution in time domain, resulting in the periodic discrete signal $\tilde{f}(t)$:

[19] One way to mitigate this issue is to apply low-pass filtering (discussed in detail in Sect. 7.2.5), which retains only the frequency components below f_c and filters out those above f_c. The trade-off is that this alters the original signal's spectrum, leading to a loss of high-frequency components in the time signal obtained through inverse discrete Fourier transform. However, high-frequency components typically do not significantly alter the overall waveform but rather reflect local details.

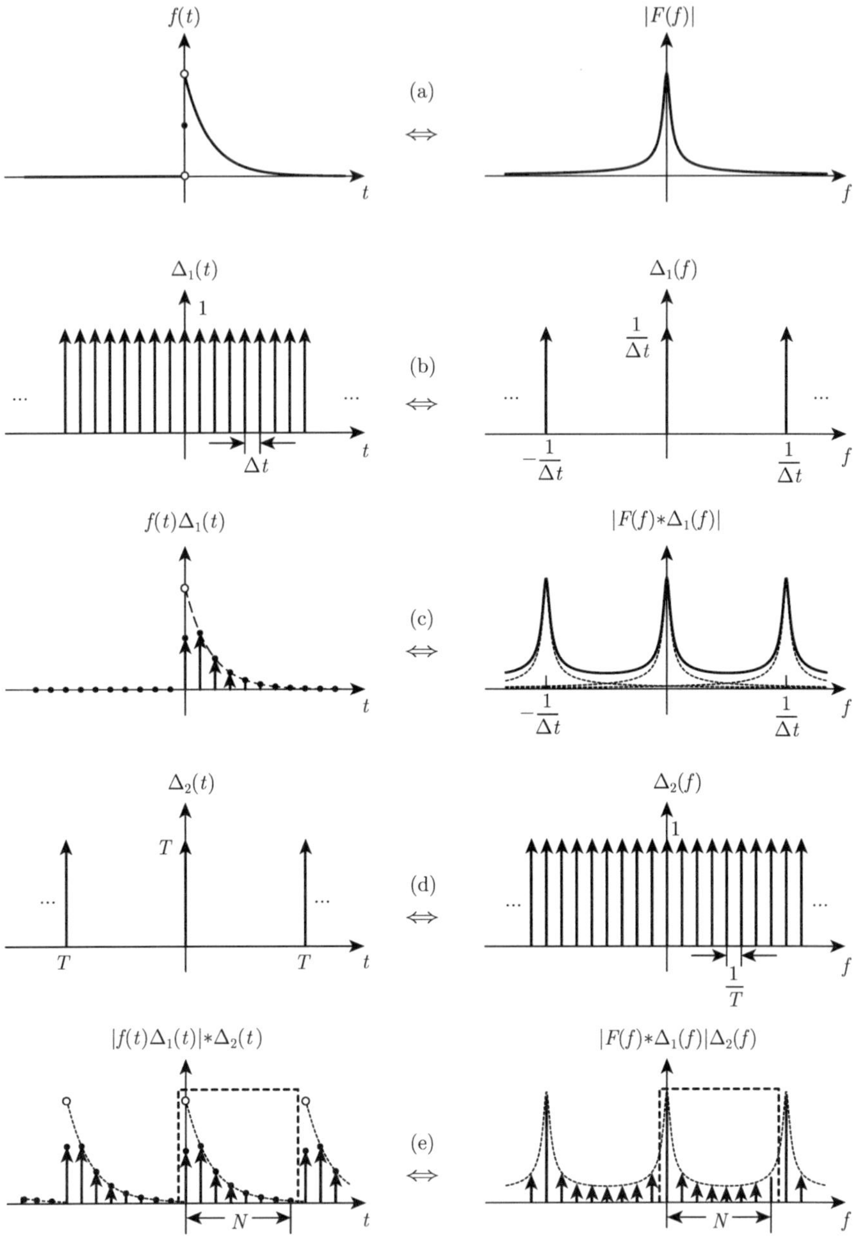

Fig. 7.21 Diagram of transition from continuous Fourier transform to discrete Fourier transform. Perform two operations on the continuous function $f(t)$ in (**a**): time-domain sampling in (**b**) and (**c**), and frequency-domain sampling in (**d**) and (**e**)

$$\begin{aligned}\tilde{f}(t) &= [f(t)\Delta_1(t)] * \Delta_2(t) \\ &= \int_{-\infty}^{+\infty} \sum_{k=-\infty}^{+\infty} f(k\Delta t)\delta(\tau - k\Delta t) T \sum_{m=-\infty}^{+\infty} \delta(t - \tau - mT)\, \mathrm{d}\tau \\ &= T \sum_{m=-\infty}^{+\infty} \sum_{k=-\infty}^{+\infty} f(k\Delta t)\delta(t - k\Delta t - mT)\end{aligned}$$

It's important to note that because sampling the spectrum results in a periodic distribution of the discrete signal, if all non-zero time signals are concentrated within the interval $[0, T]$,[20] the above expression simplifies to

$$\tilde{f}(t) = T \sum_{m=-\infty}^{+\infty} \sum_{k=0}^{N-1} f(k\Delta t)\delta(t - k\Delta t - mT) \tag{7.2.4}$$

7.2.3.3 Discrete Fourier Transform and Inverse Transform

The periodicity in the time domain corresponds to discreteness in the frequency domain, and time-domain sampling also leads to periodicity in the frequency domain. Therefore, the spectrum of $\tilde{f}(t)$, denoted as $\tilde{F}(f)$, can be expressed as a Fourier series:

$$\tilde{F}(f) = \sum_{n=-\infty}^{+\infty} \alpha_n \delta\left(f - \frac{n}{T}\right) \tag{7.2.5}$$

where

$$\begin{aligned}\alpha_n &= \frac{1}{T} \int_{-\frac{\Delta t}{2}}^{T-\frac{\Delta t}{2}} \tilde{f}(t) \mathrm{e}^{-\mathrm{i}2\pi \frac{n}{T} t}\, \mathrm{d}t \\ &\overset{(7.2.4)}{=\!=} \sum_{m=-\infty}^{+\infty} \sum_{k=0}^{N-1} \int_{-\frac{\Delta t}{2}}^{T-\frac{\Delta t}{2}} f(k\Delta t)\delta(t - k\Delta t - mT) \mathrm{e}^{-\mathrm{i}2\pi \frac{n}{T} t}\, \mathrm{d}t \\ &= \sum_{k=0}^{N-1} f(k\Delta t) \mathrm{e}^{-\mathrm{i}2\pi \frac{nk}{N}}\end{aligned}$$

[20] If not, problems like those described in Fig. 7.3 will occur, where signals beyond T fold into the $[0, T]$ interval. This can be mitigated by introducing complex frequencies.

Substituting this into Eq. (7.2.5), the spectrum within one period, as shown in the dashed box in Fig. 7.21e, is given by

$$F(n) = \sum_{k=0}^{N-1} f(k) W_N^{nk}, \quad W_N \triangleq e^{-i\frac{2\pi}{N}} \tag{7.2.6}$$

In the above expression, $f(k)$ is used instead of $f(k\Delta t)$ to represent the discrete point at index k, and correspondingly, $F(n)$ is used instead of $F(n\Delta f)$. Notice that[21]

$$\sum_{n=0}^{N-1} e^{i\beta n\frac{2\pi}{N}} = N\delta_{\beta,jN} \quad (j = 0, \pm 1, \pm 2, \cdots)$$

$$\sum_{n=0}^{N-1} F(n) e^{i\frac{2\pi nk}{N}} = \sum_{n=0}^{N-1}\sum_{m=0}^{N-1} f(m\Delta t) e^{i(k-m)n\frac{2\pi}{N}} = N\sum_{m=0}^{N-1} f(m\Delta t)\delta_{km} = Nf(k)$$

Thus, we have

$$f(k) = \frac{1}{N}\sum_{n=0}^{N-1} F(n) W_N^{-nk} \tag{7.2.7}$$

Equations (7.2.6) and (7.2.7) form a pair of discrete Fourier transform and its inverse. According to the above analysis, we can see that discretizing a continuous signal naturally leads to periodicity in the spectrum, and discretizing the spectrum in turn leads to periodicity in the time series. Thus, the original continuous-time function, after several sampling and windowing processes, results in a time signal and spectrum that are both periodic and discrete.

7.2.3.4 Fast Fourier Transform

Equations (7.2.6) and (7.2.7) provide a clear computational method. It is evident that directly solving them requires $O(N^2)$ operations: there are N outputs for $f(k)$, and each output requires N summations. For larger N, this results in a significant computational load. In 1965, Cooley and Tukey introduced the Fast Fourier Transform (FFT)

[21] When $\beta = jN$, each term in the series on the left side of the equation is 1, so the result is N. When $\beta \neq jN$, the left side is a geometric series, so

$$\sum_{n=0}^{N-1} e^{i\beta n\frac{2\pi}{N}} = \frac{1 - \left(e^{i\beta\frac{2\pi}{N}}\right)^N}{1 - e^{i\beta\frac{2\pi}{N}}} = 0$$

.

algorithm.[22] This algorithm, known as the Cooley-Tukey algorithm, decomposes a long sequence of length N into a series of shorter sequences. It leverages the symmetry and periodicity in the DFT computation to calculate the DFT of these shorter sequences and appropriately combines them, eliminating redundant calculations and reducing the number of multiplications. This compression reduces the computational complexity to $O(N \log_2 N)$. For example, with $N = 2^{10} = 1024$, direct computation requires 1,048,576 operations, whereas the FFT reduces this to 10,240 operations, just 1% of the direct method. Consequently, once introduced, the FFT was rapidly adopted in various fields of scientific and engineering computation.[23]

Since the FFT is a fast algorithm for the DFT, it provides the same results as the DFT but with significantly improved efficiency. As a commonly used algorithm in numerical computation, corresponding functions are available in various mathematical software (e.g., the `fft` command in Matlab). Therefore, it will not be elaborated on here, and interested readers can refer to relevant signal processing literature for more information.

7.2.4 Examples of Applications of Discrete Fourier Transform

After understanding the specifics of the discrete Fourier transform operations, in this section we explore its characteristics through a concrete example, examining the forward transform, inverse transform, and filtering. Consider the following Fourier transform pair:

$$f(t) = e^{-t} H(t) \quad \Longleftrightarrow \quad F(f) = \frac{1}{1 + i2\pi f}$$

The real and imaginary parts of the spectrum are given by

$$\mathrm{Re}\,\{F(f)\} = \frac{1}{1 + (2\pi f)^2}, \quad \mathrm{Im}\,\{F(f)\} = -\frac{2\pi f}{1 + (2\pi f)^2} \tag{7.2.8}$$

[22] It was later discovered that these authors had merely reinvented an algorithm that the "Prince of Mathematicians", Gauss, had originally proposed in 1805. This algorithm has been repeatedly rediscovered in various forms throughout history. It wasn't until the last 60 years that the FFT algorithm gained significant attention, likely due to the advent of computers. After all, the benefits of FFT become more evident with larger N. The magic of the FFT algorithm lies in the fact that, on one hand, it is simply an algorithm used to accelerate the DFT, appearing unremarkable at first glance; but on the other hand, due to the quantitative change leading to qualitative change, its application is extremely wide-ranging, covering almost all fields of science and engineering.

[23] It was named one of the top ten algorithms of the 20th century by the IEEE journal on scientific and engineering computation. American mathematician G. Strang even hailed the FFT as "the most important numerical algorithm of our lifetime".

It can be seen that the real part of the spectral function $F(f)$ is symmetric about $f = 0$, while the imaginary part is antisymmetric about $f = 0$.

7.2.4.1 Disrete Fourier Transform

Given $T = 7.75$ s and $N = 2^5 = 32$, we have $\Delta t = T/(N-1) = 0.25$ s and $f_c = 1/2\Delta t = 2$ Hz. Figure 7.22a shows the function $f(t)$ and the sampled discrete sequence. According to the sampling theorem discussed in Sect. 7.2.2, sampling the time signal at intervals of 0.25 s allows us to resolve frequencies up 2 Hz. Figure 7.22b, c display the real and imaginary parts of the spectrum obtained after performing the DFT, respectively. For comparison, the solid line represents the spectrum of the continuous signal calculated using Eq. (7.2.8).[24]

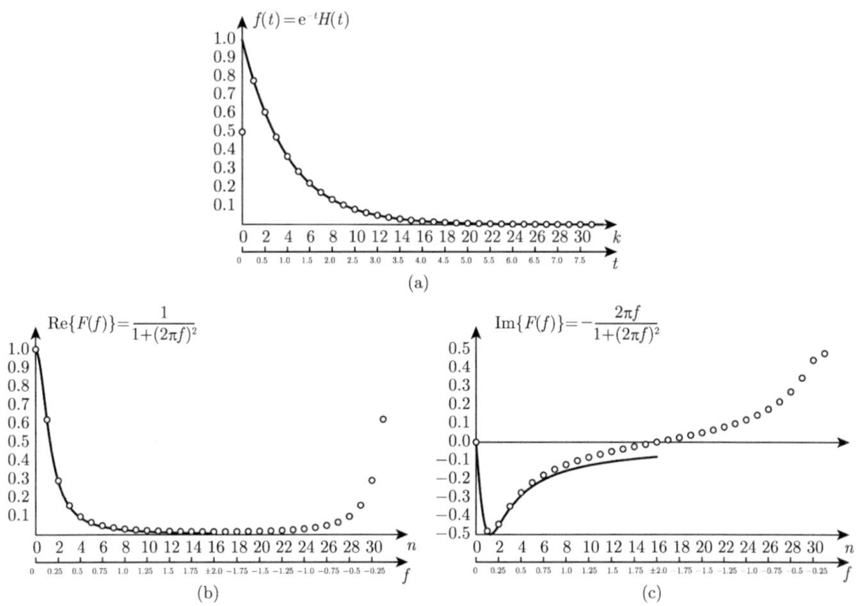

Fig. 7.22 An example of applications of the discrete Fourier transform. **a** The time-domain signal $f(t) = H(t)e^{-t}$; **b** The real part of the frequency-domain signal $\mathrm{Re}\{F(f)\} = 1/[1 + (2\pi f)^2]$; **c** The imaginary part of the frequency-domain signal $\mathrm{Im}\{F(f)\} = -2\pi f/[1 + (2\pi f)^2]$. The bold solid line represents the analytical results, and the hollow circles represent the results corresponding to the discrete Fourier transform

[24] It is worth noting that the DFT in Eq. (7.2.6) corresponds to the integrand of the Fourier transform. To compare with the result directly obtained from the Fourier transform in Eq. (7.2.8), it must be multiplied by Δt. For instance, when using the Matlab command `fft` for DFT, if the discretized time series is `ft`, the comparison with the result of Eq. (7.2.8) should be made with `fft(ft)*dt`.

If we represent the array storing the spectrum with $F(n)$ ($n = 0, 1, \cdots, 31$), based on Fig. 7.22b, c, the key points are: $F(0)$ stores the spectrum value at $f = 0$, $F(1)$ to $F(16)$ store the spectrum values for $f = \Delta f$ to $16\Delta f$, and $F(31)$ to $F(17)$ store the spectrum values for $f = -\Delta f$ to $-16\Delta f$ in reverse order.[25] In other words, the spectrum values in the frequency range from $-f_c$ to 0 are shifted to the latter half of the array. Figure 7.22b, c clearly show the relationship between the array index n and the corresponding frequency f using a dual horizontal axis. Understanding this is crucial for using the inverse discrete Fourier transform to reconstruct the time signal from the given spectrum values. The figures illustrate that the spectrum obtained through FFT significantly deviates from the continuous signal's spectrum at higher frequencies, especially the imaginary part. This discrepancy arises from the large sampling interval in the time domain, which can be mitigated by reducing Δt and increasing the number of samples N. Additionally, it is noteworthy that the number of sampling points for the spectrum is half that of the time domain signal, i.e., $N/2$, and the maximum frequency is f_c. Thus, the frequency interval is $\Delta f = 2f_c/N = 0.125$ Hz.

7.2.4.2 Inverse Discrete Fourier Transform

Inverse discrete Fourier transform is much more complex than discrete Fourier transform because users must prepare the spectral array for the inverse transform based on the described spectral characteristics. For the current example, if the number of time-domain signal samples $N = 32$, $\Delta t = 0.25$ s, then the preparation of the spectral array `F2` and the MATLAB code segment for the inverse discrete Fourier transform are as follows:

```
% Assign values to N and dt
   N=32; dt=0.25;
% Calculate the Nyquist frequency
   fc=1/ 2*dt);
% Calculate the sampling interval in frequency domain
   df=2*fc/N;
% Calculate the values of each frequency
   f=df*[0:1:N/2];
% Calculate the spectrum values corresponding to each frequency
   F1=1./(1+i*2*pi*f)*df*N;
% Assign a value to the spectrum at zero frequenc
   F2(1)=F1(1);
% Assign a value to the spectrum at the maximum frequency
   F2(N/2+1)=real(F1(N/2+1));
   for j=2:N/2
% Assign values to the spectrum array for indices from 2 to N/2
      F2(j)=F1(j);
```

[25] $F(16)$ stores the spectrum value for both $f = \pm 16\Delta f$.

```
% Assign values to the spectrum array for indices from N/2+2 to N
        F2(N-j+2)=conj(F1(j));
    end
% Perform the inverse discrete Fourier transform on F2
    ft=ifft(F2);
```

In the code snippet above, `F1` represents the spectrum calculated based on the analytical expression of the spectrum, and during the assignment, it is multiplied by `df` and `N`. The former is because the inverse discrete Fourier transform only corresponds to the integrand in the Fourier transform integral expression. Therefore, to obtain the actual $f(t)$, it must be multiplied by `df`. The latter multiplication by `N` cancels out the coefficient in Eq. (7.2.7). The zero-frequency spectrum component is stored separately and thus needs to be assigned separately. The spectra at $f = \pm f_c$ are stored at the `N/2+1` index.[26] Due to the spectrum's real part being symmetric about zero and the imaginary part being antisymmetric, the real part is directly assigned, while the imaginary part is forcibly set to zero. The loop effectively centers at the `N/2+1` position and mirrors the spectrum (excluding the zero-frequency component) according to the symmetry of the real part and antisymmetry of the imaginary part.

Figure 7.23 shows the spectrum array prepared according to the described process and the result of applying the inverse discrete Fourier transform to it. The time series obtained through the inverse discrete Fourier transform in Fig. 7.23c closely matches the exact continuous function, but there is significant error towards the end. Recall from Sect. 7.2.1 that basic Fourier transform pairs indicate that a window function in the frequency domain corresponds to noticeable side lobes in the time domain, as seen in Fig. 7.17c. Therefore, the time series obtained through the inverse discrete Fourier transform is actually a convolution of the exact time function with a Sinc function, and these oscillations are a reflection of the side lobes.

7.2.5 *Filtering: Low-Pass Filtering and Band-Pass Filtering*

How can we mitigate the impact of these sidelobes? An effective method is to filter the spectrum, known as *filtering*. There are many types of filters; here, we introduce a very simple and effective frequency domain window function $W(f)$, defined as:

$$W(f) = \begin{cases} 0, & f \le f_1 \text{ or } f \ge f_4 \\ 1, & f_2 \le f \le f_3 \\ \dfrac{1}{2}\left[1 - \cos\left(\dfrac{f - f_1}{f_2 - f_1}\right)\pi\right], & f_1 < f < f_2 \\ \dfrac{1}{2}\left[1 + \cos\left(\dfrac{f - f_3}{f_4 - f_3}\right)\pi\right], & f_3 < f < f_4 \end{cases}$$

---

[26] Note that in Matlab, array indexing starts at 1, which differs from the numbering in Fig. 7.22.

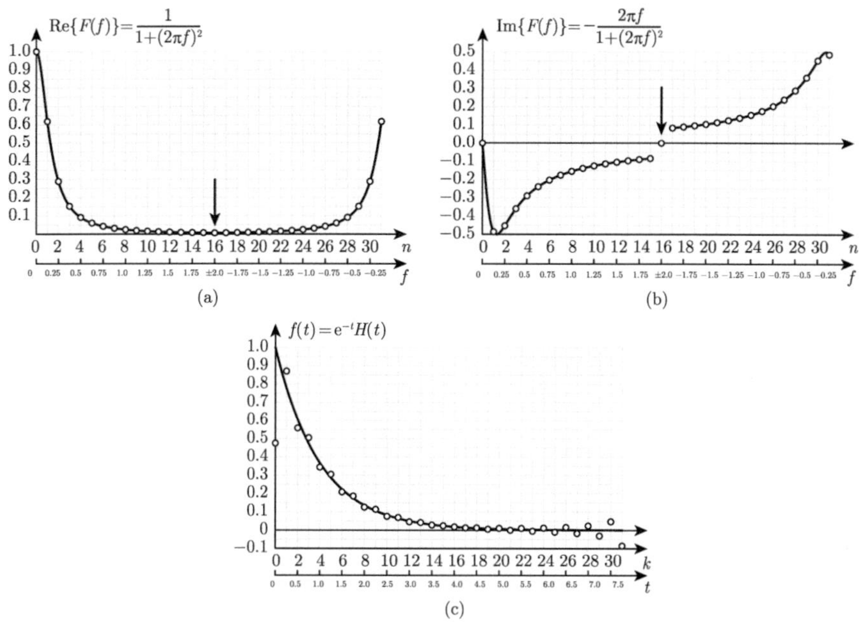

Fig. 7.23 Examples of applying the inverse discrete Fourier transform. **a** The real part of the frequency-domain signal $\mathrm{Re}\{F(f)\} = 1/[1 + (2\pi f)^2]$; **b** The imaginary part of the frequency-domain signal $\mathrm{Im}\{F(f)\} = -2\pi f/[1 + (2\pi f)^2]$; **c** Time-domain signal $f(t) = H(t)\mathrm{e}^{-t}$. The thick solid line represents the analytical result, and the hollow circles represent the result corresponding to the discrete Fourier transform. The arrows in (**a**) and (**b**) indicate the folding points in the spectrum

where f_i $(i = 1, 2, \cdots, 4)$ are control frequencies that satisfy $f_1 < f_2 < f_3 < f_4 \leq f_c$.[27] Figure 7.24 shows the variation of $W(f)$ with frequency f. By choosing different control frequencies and multiplying the spectrum by $W(f)$, certain frequency components can be filtered out. The smoothness of the curve in the intervals $f \in (f_1, f_2)$ or (f_3, f_4) results in significantly reduced sidelobes in the time domain compared to a steep rectangular function.

If $f_1 = f_2 = 0$ and $f_4 < f_c$, $W(f)$ becomes a *low-pass filter*. As the name suggests, it retains the lower frequency components while filtering out the higher frequencies. Considering the example in Fig. 7.23, let's take $f_1 = f_2 = 0$, $f_3 = 1.0$ Hz, and $f_4 = 2.0$ Hz. Multiplying $W(f)$ with the spectrum $F(f)$ and performing the inverse discrete Fourier transform gives the results shown in Fig. 7.25. It can be observed that after applying the low-pass filter, the discontinuities in the high-frequency part of the spectrum's imaginary component near the symmetry point disappear, replaced by a waveform that gradually decreases to zero. Consequently, the oscillations in the time

[27] When $f_1 = f_2$ or $f_3 = f_4$, the third or fourth line in the definition does not apply, so there is no issue of division by zero.

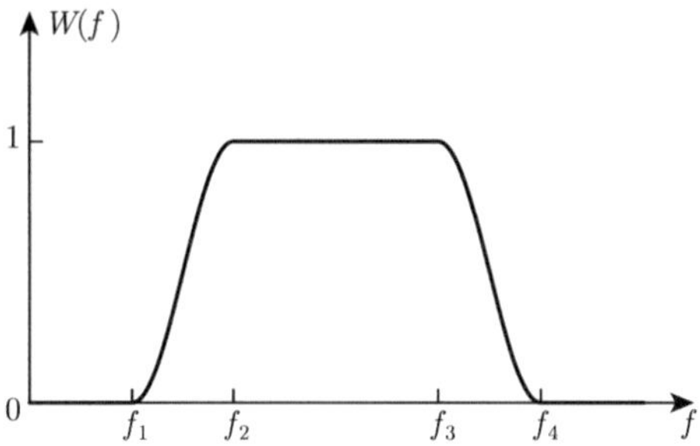

Fig. 7.24 Common band-pass filter $W(f)$. f_i $(i = 1, 2, \ldots, 4)$ are control frequencies. When $f_1 = f_2 = 0$, $W(f)$ is a low-pass filter

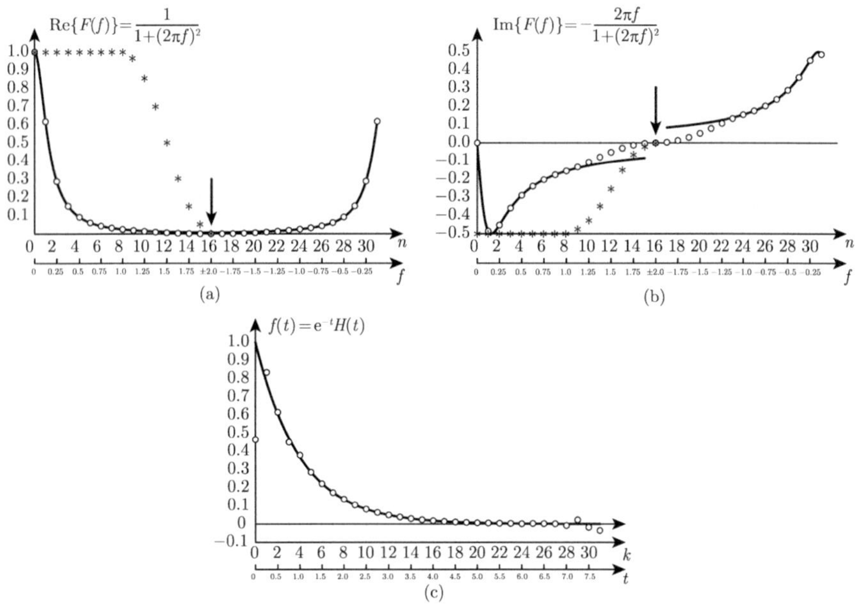

Fig. 7.25 Example of applying low-pass filter in inverse discrete Fourier transform. The legend is the same as in Fig. 7.23. For reference, the values of the low-pass filter are marked with $*$ in (**a**) and (**b**). In (**b**), the result is multiplied by -0.5 for clarity. Here, $f_1 = f_2 = 0$, $f_3 = 1.0$ Hz, and $f_4 = 2.0$ Hz

series obtained from the inverse discrete Fourier transform are significantly reduced compared to the case without the low-pass filter, as seen by comparing Figs. 7.23c and 7.25c.

If $0 \le f_1 < f_2$ and $f_3 < f_4 \le f_c$, $W(f)$ becomes a band-pass filter, meaning it retains the middle frequency components while filtering out the high and low frequencies. Figure 7.26 shows the corresponding results, where $f_1 = 0$ Hz, $f_2 = 0.5$ Hz, $f_3 = 1.0$ Hz, and $f_4 = 2.0$ Hz. In this case, because part of the low-frequency

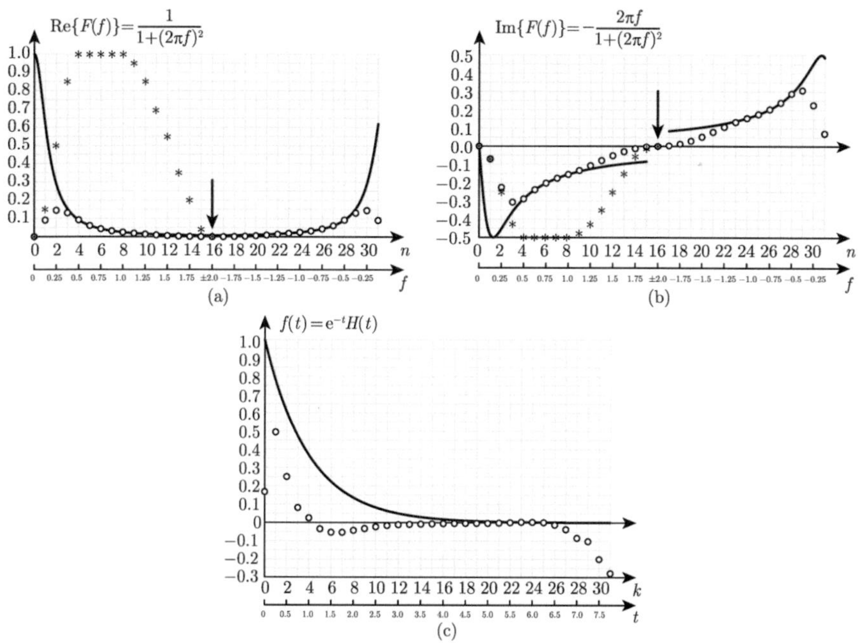

Fig. 7.26 An Example of applying the inverse discrete Fourier transform with band-pass filtering. The same as in Fig. 7.25, except that $f_1 = 0$ Hz, $f_2 = 0.5$ Hz, $f_3 = 1.0$ Hz, and $f_4 = 2.0$ Hz

spectrum components are filtered out, there is a significant change in the spectrum, as seen in Fig. 7.26a, b. Since the low-frequency part of the spectrum determines the main shape of the corresponding time-domain function, the time signal obtained from the inverse discrete Fourier transform deviates significantly from the original function.

The instruments that record seismic signals often have a certain frequency range, acting as natural filters for the recorded seismic motion. Therefore, to compare synthetic seismograms with seismic records for structural inversion or source rupture process inversion to understand the Earth's internal structure or the behavior of the source, it is necessary to filter the theoretical seismograms according to the frequency range of the recorded data.

7.2.6 Several Common Source Time Functions

According to the source representation theorem in Chap. 3, Eq. (3.2.5), the displacement field at a point in an elastic medium can be represented as the weighted convolution of the dislocation on fault and the spatial derivatives of Green's function. In kinematic forward modeling, the source time functions (STFs), such as ramp functions, describe how the displacement changes over time. These commonly used STFs have analytical expressions for their spectra. As described in Sect. 7.2.4.2, one can calculate the analytical form of the spectra, multiply it with the Green's function spectra in the frequency domain, and then use IFFT to transform it back to the time domain. This process corresponds to convolution in the time domain. However, the spectra of these time functions are often proportional to ω^{-1} or ω^{-2}, so to include the DC component (zero frequency), complex frequencies need to be introduced.

7.2.6.1 Introduction of Complex Frequencies: An Example of Step Function

The Fourier transform of the step function is given by

$$f(t) = H(t) \quad \Longleftrightarrow \quad F(\omega) = \pi\delta(\omega) + \frac{1}{\mathrm{i}\omega} \tag{7.2.9}$$

At $\omega = 0$, the spectrum is singular. Therefore, even if the Green's function itself can include the zero-frequency component, computing the time function remains problematic.

In Sect. 7.1.1.2, complex frequencies were introduced to suppress the effects of additional sources in the discrete wavenumber method. It was previously concluded that if $f(t)\mathrm{e}^{-\omega_\mathrm{I} t}$ is used instead of $f(t)$ in the Fourier transform, the corresponding spectrum $F(\omega)$ becomes $F(\omega^*)$, that is,

$$f(t)\mathrm{e}^{-\omega_\mathrm{I} t} \quad \Longleftrightarrow \quad F(\omega^*) \quad (\omega^* = \omega - \mathrm{i}\omega_\mathrm{I}) \tag{7.2.10}$$

Using the complex frequency ω^*, the spectrum in Eq. (7.2.9) is no longer singular, that is,

$$f'(t) = f(t)\mathrm{e}^{-\omega_\mathrm{I} t} \quad \Longleftrightarrow \quad F(\omega^*) = \frac{1}{\mathrm{i}\omega^*}$$

Thus, according to Eq. (7.2.10), the formula for obtaining the step function using the IDFT with complex frequencies is

$$H(t) = \mathrm{IDFT}\left\{\frac{1}{\mathrm{i}\omega^*}\right\}\mathrm{e}^{\omega_\mathrm{I} t} \tag{7.2.11}$$

Equation (7.2.11) shows that by directly replacing ω with ω^* in the spectral function and using the conventional IFFT algorithm, the resulting time-domain result multiplied by $\mathrm{e}^{\omega_\mathrm{I} t}$ yields the desired $f(t)$.

If we set

$$\omega_\mathrm{I} = \frac{\varsigma\pi}{T}$$

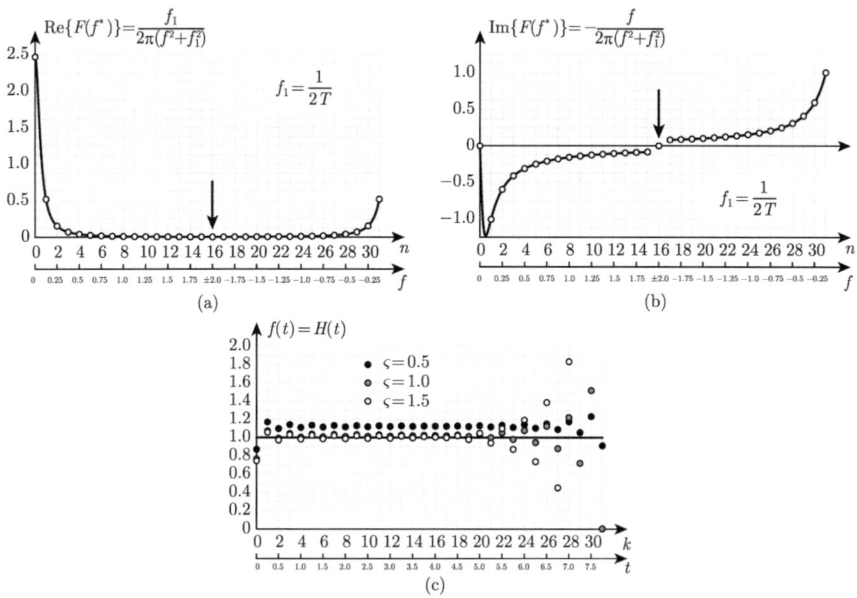

Fig. 7.27 Commonly used source time functions (1): Step function $H(t)$. **a** and **b** are the real and imaginary parts of the spectrum for $\varsigma = 1$, respectively; **c** is the time signal obtained via IFFT for different values of ς

where T is the length of the time window and ς is the coefficient of the imaginary frequency, typically valued at 1. Figure 7.27 shows the spectrum based on complex frequencies for $\varsigma = 1$ and the time domain results for different values of ς obtained via IFFT. From Fig. 7.27c, it is clear that when $\varsigma = 0.5$, the time signal deviates significantly from the accurate value of 1. In contrast, for $\varsigma = 1$ and 1.5, the results match the accurate value well. However, as ς increases, the oscillations in the latter part of the time signal become more pronounced.

This phenomenon can be explained by Figs. 7.3 or 7.21. The step function remains at a constant value of 1 for $t > 0$, thus it is still 1 after the time window T. When transitioning from the continuous Fourier transform to the discrete Fourier transform, frequency-domain sampling causes periodicity in the time-domain signal, as shown in Fig. 7.3. The tail of the shifted signal extends into the $[0, T]$ interval, causing signal distortion. Figure 7.4 illustrates how introducing complex frequencies can suppress this part of the signal. It is evident that the smaller the imaginary frequency, the less effective the suppression. However, according to Eq. (7.2.11), the signal obtained by performing an IDFT on the spectrum with complex frequencies needs to be multiplied by the factor $e^{\omega_I t}$ to retrieve the actual signal. This factor grows exponentially with time, which means that if the imaginary frequency is large, numerical errors in the latter part of the signal will be significantly amplified, as shown in Fig. 7.27c. Therefore, ω_I should neither be too small nor too large. An empirical value of ς is 1.

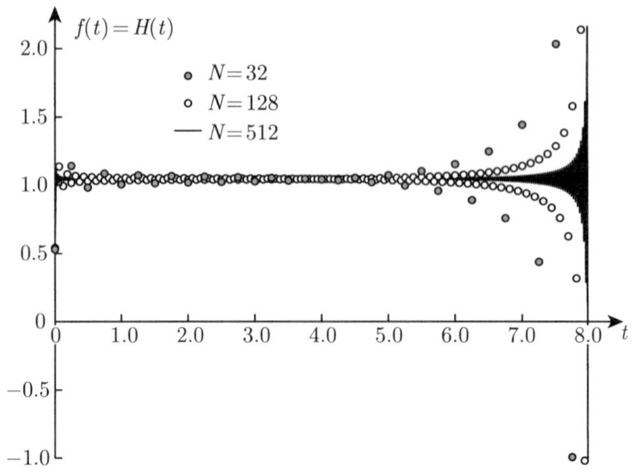

Fig. 7.28 The step function $H(t)$ ($\varsigma = 1$) with different numbers of sampling points N

It is worth noting that the phenomenon of amplified errors in the recovered signal's tail cannot be mitigated by increasing the number of sampling points. Figure 7.28 shows the results for different numbers of sampling points N. As the number of sampling points increases, the region of amplified numerical errors shrinks, but the magnitude of these errors does not decrease with increasing N.

Although the step function was often used by early seismologists studying the Lamb's problem, for frequency domain solutions, its spectrum decays slowly with increasing frequency (ω^{-1}), making the truncation effects in the frequency domain quite evident. From a numerical computation perspective, this inevitably leads to distortion in the time-domain signal. Noting that the actual seismic source rupture process usually occurs over a period rather than instantaneously, various source time functions that more closely approximate the actual situation have been proposed based on the Heaviside function.

7.2.6.2 Ramp Function $R(t)$

The ramp function is one of the most commonly used source time functions in practice, as introduced in Sect. 4.5.2.2. Here, we focus on its relationship with the spectrum. The Fourier transform pair for the ramp function is

$$R(t) = \frac{t}{t_0} H(t) + \left(1 - \frac{t}{t_0}\right) H(t - t_0) \quad \Longleftrightarrow \quad R(\omega) = \frac{e^{-i\omega t_0} - 1}{\omega^2 t_0}$$

where t_0 is the rise time, the period during which the amplitude increases from 0 to 1. The graph of $R(t)$ is shown as the solid line in Fig. 7.29c.

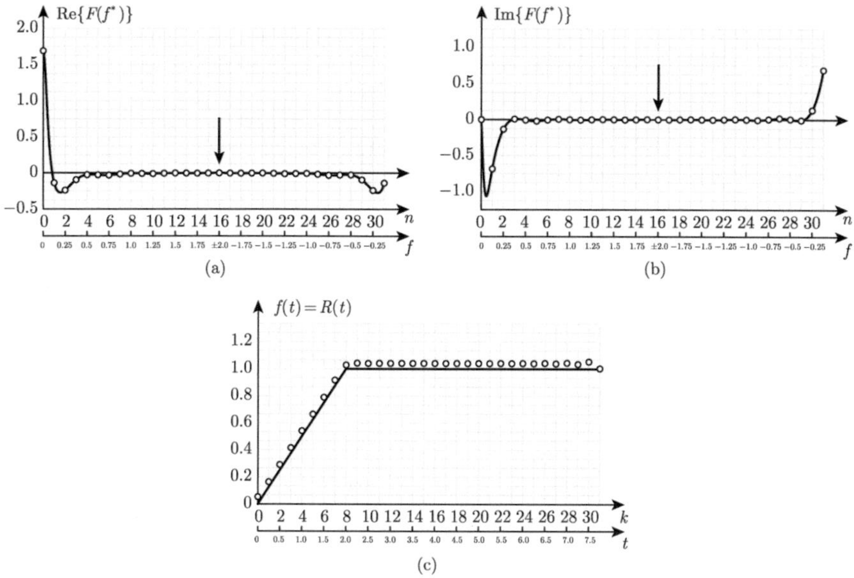

Fig. 7.29 Commonly used source time functions (2): Ramp function $R(t)$ ($t_0 = 2$ s)

The spectrum of the ramp function, $R(\omega) \propto \omega^{-2}$, indicates that its spectrum decays faster with ω compared to the step function, making truncation effects less significant and resulting in weaker oscillations in the time signal. By replacing ω in the spectrum with the complex frequency ω^* and following the previously described steps, the recovered time series is shown in Fig. 7.29c. It is worth noting that the recovered time signal has a small systematic offset compared to the accurate ramp function, caused by the extension of the shifted signal into the $[0, T]$ interval. Increasing the coefficient of the imaginary frequency ς can reduce this effect. The results demonstrate that by introducing complex frequencies, the recovered signal effectively reproduces the zero-frequency information.

7.2.6.3 Smoothed Ramp Function $Ex(t)$

Another time function that describes the time history of a seismic source is represented by an exponential function, which can be regarded as a special smoothed case of the ramp function. Its Fourier transform pair is

$$Ex(t) = \left(1 - \mathrm{e}^{-\frac{t}{t_0}}\right) H(t) \quad \Longleftrightarrow \quad Ex(\omega) = \frac{1}{\omega^2(\mathrm{i} - \omega t_0)}$$

where, t_0 is also the rise time, but unlike the ramp function, the time function approaches 1 only as $t \to \infty$. The parameter t_0 simply controls the rate of rise of

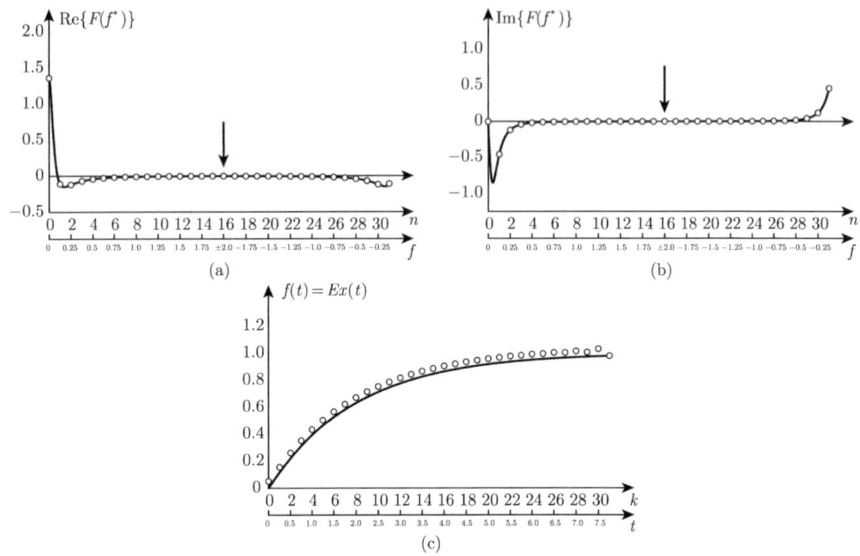

Fig. 7.30 Commonly used source time functions (3): Smoothed ramp function $Ex(t)$ ($t_0 = 2$ s)

the time function. Figure 7.30 shows the complex spectrum and the recovered time function. The characteristics are similar to those of the ramp function, so they are not repeated here.

7.2.6.4 Trapezoidal Function $Tr(t)$

We know that the Green's function is caused by an impulsive force source, and sometimes we need to study the displacement field caused by similar impulsive source time functions. However, in practical problems, the time behavior of the force source is not as sharp as an impulse function. Thus, we use a smoothed impulsive function–a trapezoid function–to describe it. Its Fourier transform pair is

$$Tr(t) = \begin{cases} 0, & t \le 0 \text{ or } t \ge t_3 \\ \dfrac{t}{t_1}, & 0 < t < t_1 \\ 1, & t_1 \le t \le t_2 \\ \dfrac{t_3 - t}{t_3 - t_2}, & t_2 < t < t_3 \end{cases} \iff Tr(\omega) = \frac{e^{-i\omega t_1} - 1}{\omega^2 t_1} + \frac{e^{-i\omega t_2} - e^{-i\omega t_3}}{\omega^2 (t_3 - t_2)}$$

where t_i ($i = 1, 2, 3$) are control parameters. In the interval $t \in [0,\ t_1]$, the function value increases linearly from 0 to 1; in the interval $t \in [t_1,\ t_2]$, the function value remains constant at 1; and in the interval $t \in [t_2,\ t_3]$, the function value decreases linearly from 1 to 0, and then stays at 0. The variation of $Tr(t)$ over time is shown

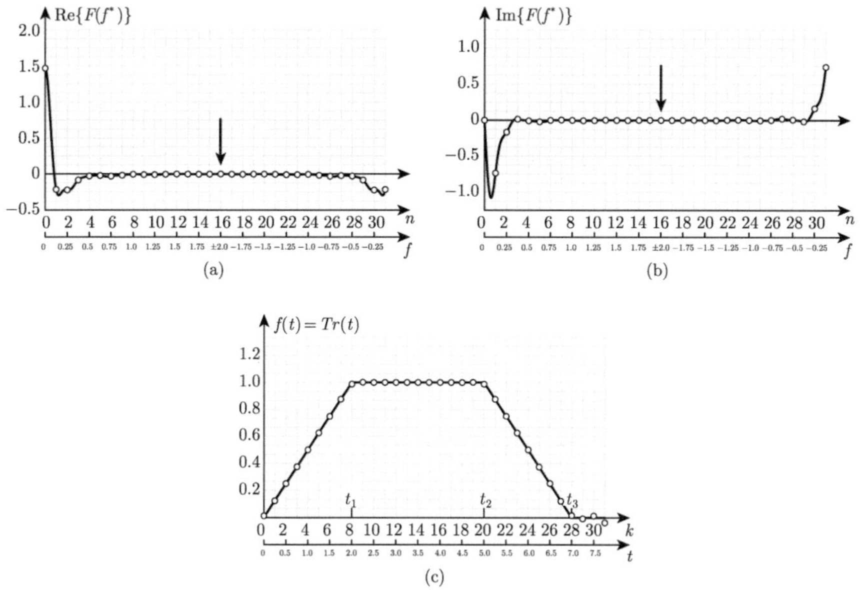

Fig. 7.31 Commonly used source time functions (4): Trapezoid function $Tr(t)$. The three parameters involved are $t_1 = 2$ s, $t_2 = 5$ s, $t_3 = 7$ s

by the solid line in Fig. 7.31c. For the trapezoid function, the signal recovered from the complex frequency spectrum matches the exact time function very well. Unlike the ramp function and the smooth ramp function, although the spectrum of the trapezoid function is also proportional to ω^{-2}, the recovered time signal does not have a systematic shift. This is because the trapezoid function itself is 0 for $t > T$, and all non-zero signals are within $[0, T]$, so the shifted signal does not affect the signal in the $[0, T]$ interval.

In Sect. 4.6.1, we have obtained the following conclusion: the time function of the far-field displacement generated by a shear dislocation source is the derivative of the time functions of near-field and intermediate-field displacements. It is noted that the integral of a trapezoid function is a smooth ramp function, and therefore it is often used to characterize the temporal behavior of far-field displacements.

7.2.6.5 Ricker Wavelet $Ri(t)$

Another source time function used to simulate an impulsive force source is the Ricker wavelet. Its Fourier transform pair is

$$Ri(t) = \left[1 - 2\pi^2 f_0^2 (t - t_0)^2\right] \mathrm{e}^{-\pi^2 f_0^2 (t - t_0)^2} \quad \Longleftrightarrow \quad Ri(f) = \frac{2f^2}{\sqrt{\pi} f_0^3} \mathrm{e}^{-\frac{f^2}{f_0^2} - \mathrm{i}2\pi f t_0}$$

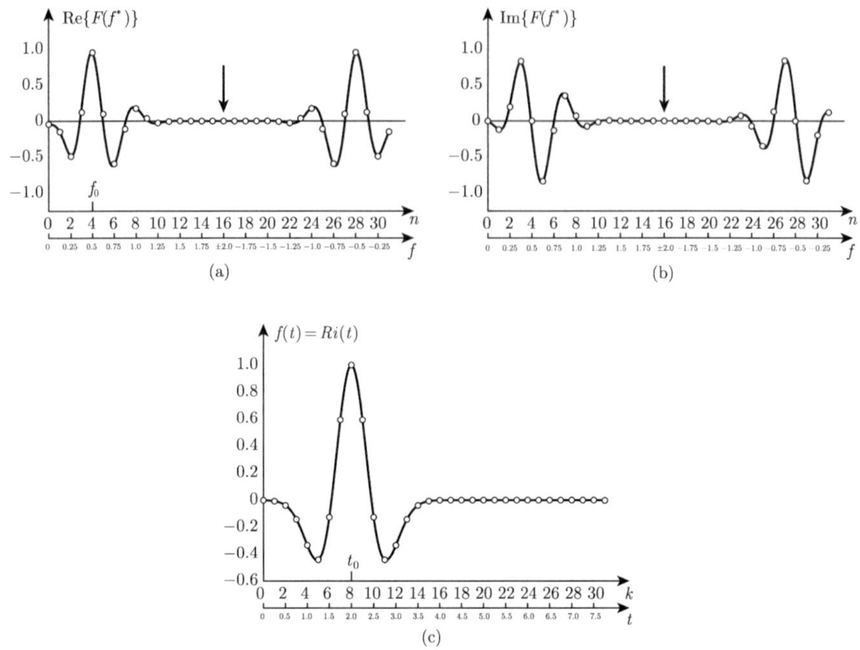

Fig. 7.32 Commonly used source time functions (5): Ricker wavelet $Ri(t)$ ($t_0 = 2$ s, $f_0 = 0.5$ Hz)

where t_0 and f_0 represent the time shift and central frequency, respectively, as shown by the solid line in Fig. 7.32. Unlike the other time functions mentioned earlier, the spectrum of the Ricker wavelet decays as e^{-f^2} with f, allowing the time series recovered from the complex frequency spectrum to match the exact time function perfectly.

Compared to the trapezoid function, the shape of the Ricker wavelet is more similar to a pulse, which is why it is widely used in seismic wavefield simulations. Another important reason is that the spectrum of the Ricker wavelet decays very rapidly with increasing frequency, allowing for very smooth waveforms to be obtained without the need to compute very high frequencies. This is particularly advantageous for numerical computation methods such as finite element and finite difference methods that rely on extensive calculations.

7.3 Correctness Verification

With the foundation laid in the previous two sections, we can now numerically compute wavenumber integrals in the frequency domain using the expressions derived in Chap. 6. By applying the FFT algorithm, we obtain the displacement field in the time domain through discrete inverse Fourier transforms. Before analyzing specific problems, we need to verify the correctness of our formulas and programs. In this section, we compare the numerical results of the Green's functions for three types of Lamb's problems,[28] their spatial derivatives, and the displacement field caused by a dislocation source with previous studies to ensure their accuracy.

7.3.1 *Green's Function for $\mathcal{LP}_1$*

For the $\mathcal{LP}_1$, where both the source and observation points are on the surface, the geometry appears quite simple. However, using the current method, the displacement integral expressed in cylindrical coordinates converges very slowly as the transverse wavenumber k increases. This is because the exponential decay term in the integrand ceases to be effective. In such cases, the peak and trough averaging method described in Sect. 7.1.3 becomes necessary.

7.3.1.1 Comparison with Results of Pekeris (1955a) and Chao (1960)

Consider a source located at the origin (0, 0, 0) and an observation point at (10, 0, 0) km. The velocities of the P-wave and S-wave are $\alpha = 8.00$ km/s and $\beta = 4.62$ km/s, respectively, and the density is $\rho = 3.30$ g/cm^3.[29]

Figures 7.33 and 7.34 display the displacement components over time caused by vertical and horizontal forces applied at the origin, respectively. The horizontal axis represents time non-dimensionalized by the S-wave arrival time $\tau = \beta t/r$, and the vertical axis represents the non-dimensionalized Green's function components $\bar{G}_{ij} = \pi\mu r G_{ij}$ with $1/\pi\mu r$.[30] For the chosen observation point, $G_{12} = G_{21} = G_{23} = G_{32} = 0$. Since $G_{31} = -G_{13}$, G_{31} is not shown in Fig. 7.34. In the review of Lamb's problems in Chap. 5, Pekeris (1955a) and Chao (1960) calculated the displacement

---

[28] Strictly speaking, the examples shown in this section are not exact Green's functions but rather convolutions of Green's functions with a step function. In the frequency domain, this corresponds to multiplying their spectra. This approach serves two purposes: it facilitates comparison with previous results and mitigates severe oscillations around the arrival times of P-wave and S-wave that would occur if we directly transformed the Green's function spectra into the time domain. This oscillation is known as the Gibbs phenomenon. The step function spectrum decays as ω^{-1}, effectively filtering the Green's function spectrum. Although the Gibbs phenomenon is still significant, it is considerably reduced compared to the unfiltered Green's function. For convenience, this section presents results for a step-function source time function, without distinguishing it from the Green's function.

[29] The values for P-wave and S-wave velocities, as well as density, are taken from Johnson (1974). These are typical values for considering the Earth's medium as an elastic half-space Poisson medium. Unless otherwise specified, these values are used for calculations throughout this book.

[30] Based on the meaning of the components of the Green's function, G_{33} and G_{13} represent the vertical and horizontal displacements caused by vertical forces, respectively. G_{31} and G_{11} correspond to the vertical and horizontal displacements along the x_1 direction caused by horizontal forces in the x_1 direction. Finally, G_{22} represents the horizontal displacement along the x_2 axis caused by horizontal forces in the x_2 direction.

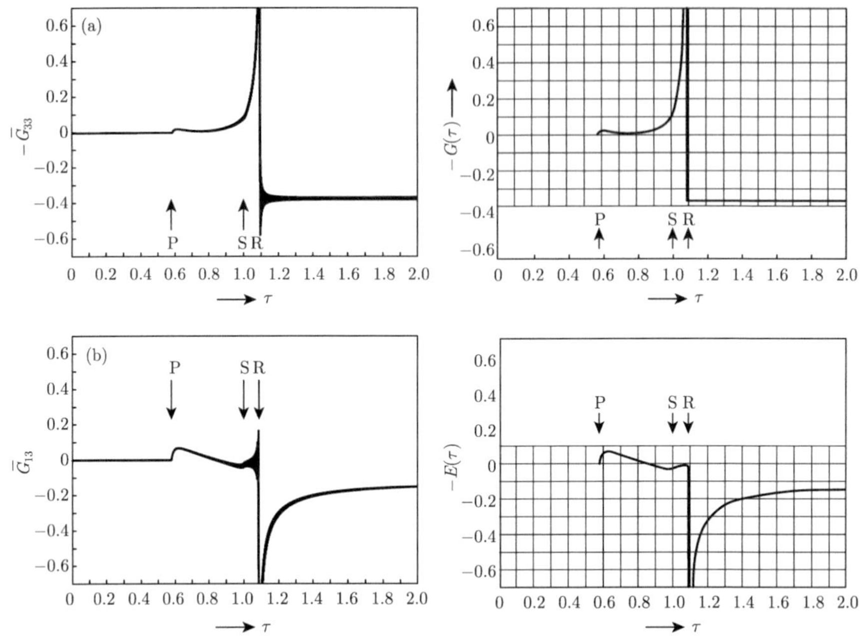

Fig. 7.33 Comparison of displacements due to vertical force (left) and with Pekeris (1955a) (right). **a** Dimensionless vertical displacement $\bar{G}_{33} = \pi\mu r G_{33}$; (b) Dimensionless horizontal displacement $\bar{G}_{13} = \pi\mu r G_{13}$. The horizontal axis $\tau = \beta t/r$ represents time scaled by the arrival time of the S-wave. In the figure, "P", "S", and "R" denote the arrival times of the P-wave, S-wave, and Rayleigh wave, respectively. To compare with Pekeris (1955a), the vertical displacement is taken as the negative value

fields due to vertical and horizontal forces and presented the corresponding numerical results in Figs. 5.7 and 5.9. Together, their results form the components of the Green's function.[31] According to Figs. 7.33 and 7.34, the current calculations are generally consistent with the results of Pekeris (1955a) and Chao (1960).

It's worth noting that a very prominent phenomenon appears in the numerical results obtained here–noticeable oscillations at the points of abrupt jumps in the time-domain signals. This is a common occurrence when obtaining time signals from sharp time signals based on their spectrum and is known as the *Gibbs phenomenon*.[32]

[31] In Pekeris (1955a), the vertical and horizontal displacements correspond to G_{33} and G_{13}, respectively. While in Chao (1960), the displacement components u_r, u_θ, and u_z correspond to G_{11}, G_{22}, and G_{31}, respectively.

[32] We know that any periodic signal $x(t)$ can be represented by an infinite series of sine and cosine functions. In particular, does this hold true for functions in the time domain that have corners (such as discontinuities)? In the early 19th century, French mathematicians J. B. J. Fourier and J. L. Lagrange had a dispute about whether "sine curves can be combined to form a signal with corners". Later, it was found out that both were "right": indeed, it is not possible to compose a signal with corners using sine curves, but we can approximate it with sine curves to the extent that there is no energy difference. In 1898, American physicist A. Michelson observed fluctuations near the

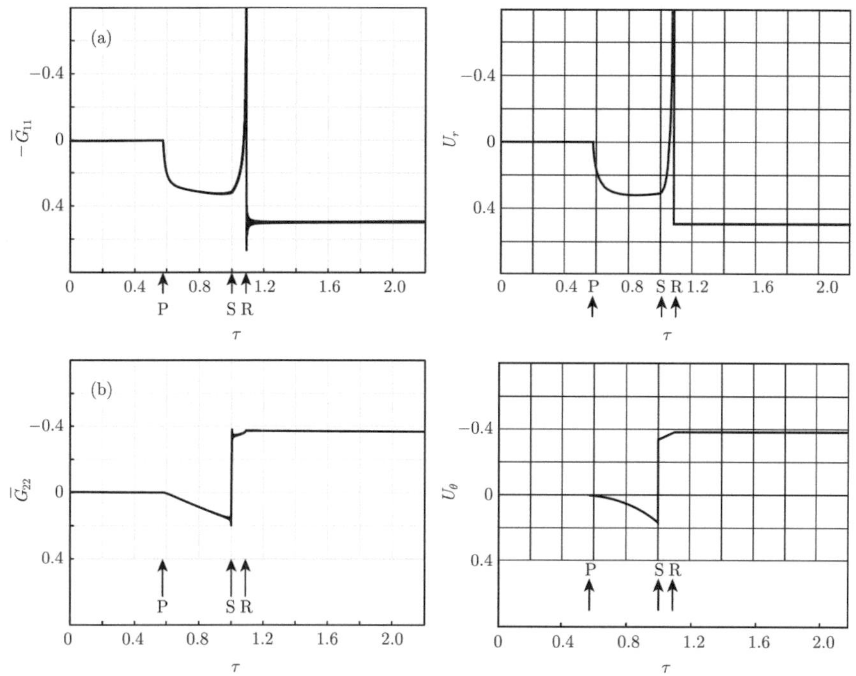

Fig. 7.34 Comparison of displacements due to horizontal force (left) and with Chao (1960) (right). **a** Dimensionless horizontal displacement $\bar{G}_{11} = \pi\mu r G_{11}$; **b** Dimensionless horizontal displacement $\bar{G}_{22} = \pi\mu r G_{22}$. The horizontal axis $\tau = \beta t/r$ represents time made dimensionless by the arrival time of the S-wave. In the figure, "P", "S", and "R" denote the arrival times of the P-wave, S-wave, and Rayleigh wave, respectively. To compare with Chao (1960), the horizontal displacement in (**a**) is taken as the negative value

Specifically, the G_{22} component, primarily consisting of SH wave, shows significant oscillations at the arrival of the S-wave, while the G_{11}, G_{13}, and G_{33} components, primarily consisting of P-SV waves, exhibit notable oscillations at the arrival of the Rayleigh wave. This is related to the choice of the source time function; for a step function, due to its discontinuity at $t = 0$, the displacement waveform reflects this by showing sharp peaks at the phase arrivals, making the Gibbs effect quite pronounced.

discontinuities of a square wave using a harmonic analyzer, and the magnitude of these fluctuations did not seem to decrease as the number of terms N in the Fourier series increased. He was surprised and even suspected there was an issue with his instrument. He wrote to the mathematical physicist J. Gibbs about this problem. Gibbs studied the issue and published a paper in *Nature* in 1899, where he theoretically proved that as N increases, the fluctuations of the partial sums contract towards the discontinuities, but for any finite value of N, the magnitude of the fluctuations remains constant. This phenomenon was later named the *Gibbs phenomenon* or *Gibbs effect*. A Fourier truncation approximation of a discontinuous signal will show high-frequency fluctuations near the discontinuities, and it is necessary to choose a sufficiently large N to ensure that the total energy of these fluctuations can be neglected. In the limit as $N \to \infty$, the energy of the approximation error is zero, and the Fourier series of a discontinuous signal (such as a square wave) converges.

Increasing the number of sample points N will only make the oscillatory region narrower and sharper, but this oscillatory phenomenon will not disappear. We will demonstrate this with examples later and show that if the source time function is chosen as a ramp function, especially with a longer rise time, the Gibbs phenomenon becomes less pronounced.

7.3.1.2 Comparison with Results of Kausel (2012)

Kausel (2012) obtained the closed-form solution for the $\mathcal{LP}_1$ and presented results for the aforementioned issues. He also provided results for different values of Poisson's ratio ν (0, 0.05, 0.10, 0.15, 0.25, 0.33, 0.40, 0.45, 0.50).

Figure 7.35 shows a comparison between our results and Kausel's results (right column). Apart from the issues caused by the Gibbs effect, the results are consistent. Note that

$$\alpha = \sqrt{\frac{\lambda + 2\mu}{\rho}} = \sqrt{\frac{(1-\nu)E}{(1-2\nu)(1+\nu)\rho}}, \quad \beta = \sqrt{\frac{\mu}{\rho}} = \sqrt{\frac{E}{2(1+\nu)\rho}}$$

where λ and μ are the Lamé's coefficients, E is the Young's modulus, and ν is Poisson's ratio. Therefore, the ratio of P-wave arrival time to S-wave arrival time η is

$$\eta = \frac{\beta}{\alpha} = \sqrt{\frac{1-2\nu}{2(1-\nu)}} = \sqrt{1 - \frac{1}{2(1-\nu)}}$$

It can be seen that η decreases monotonically with the increase of ν. Considering two extreme cases: when $\nu = 0$, $\eta = \sqrt{2}/2 \approx 0.707$, the phase of P-wave is very sharp; and when $\nu = 0.5$, $\eta = 0$. A Poisson's ratio of 0.5 corresponds to an incompressible elastic body, where for compressive P-wave, the elastic body behaves like a rigid body, and the effect of the applied force is transmitted instantaneously.

7.3.2 *Green's Function for $\mathcal{LP}_2$*

In practical seismological problems, the seismic source is located underground, and the majority of seismic records are obtained at the surface. Therefore, the $\mathcal{LP}_2$ has widespread practical significance in seismology. Compared to the $\mathcal{LP}_1$, solving the $\mathcal{LP}_2$ is considerably more complex.

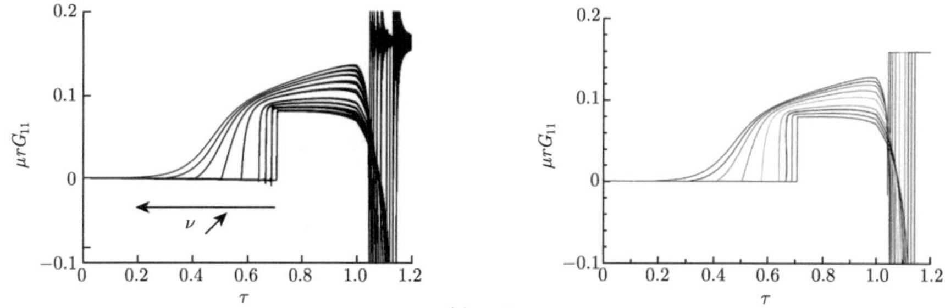

(a) $\mu r G_{11}$

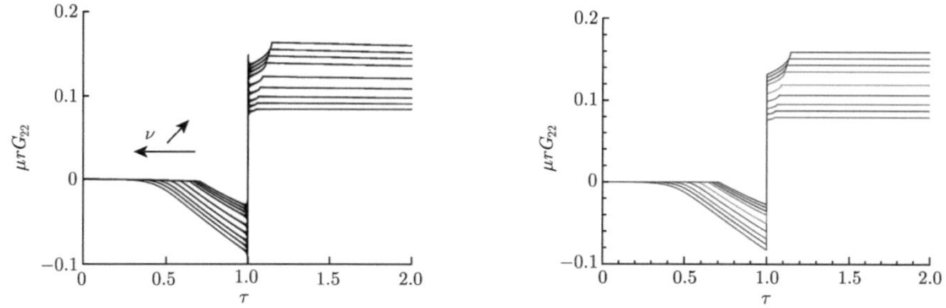

(b) $\mu r G_{22}$

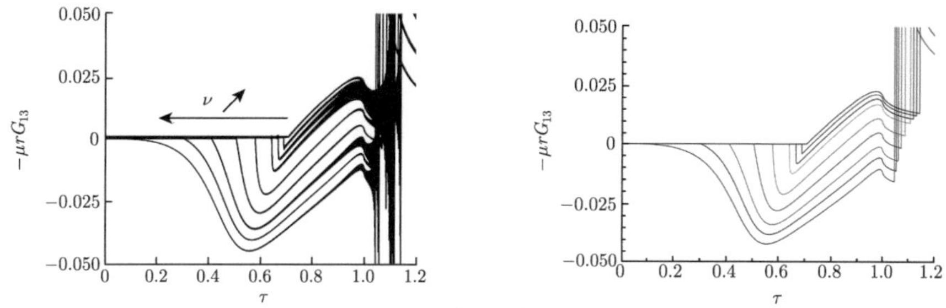

(c) $-\mu r G_{13}$

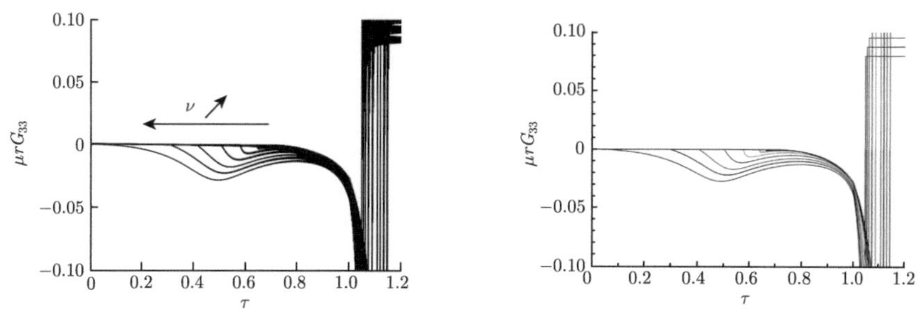

(d) $\mu r G_{33}$

Fig. 7.35 Comparison of solutions in this book (left) with those in Kausel (2012) (right). **a** $\mu r G_{11}$; **b** $\mu r G_{22}$; **c** $-\mu r G_{13}$; **d** $\mu r G_{33}$. The direction of increasing ν is indicated in the figure, with ν values of 0, 0.05, 0.10, 0.15, 0.25, 0.33, 0.40, 0.45, and 0.50

7.3.2.1 Comparison with Results of Johnson (1974)

Johnson (1974) used the Cagniard-de Hoop method to derive a compact integral expression for the Green's function of the $\mathcal{LP}_2$. He also provided results for the Green's function for several specific source-receiver configurations, as shown in Fig. 5.12. Figures 7.36, 7.37 and 7.38 compare the Green's function components G_{ij}^{H} (where "H" denotes a step function time dependency) for three scenarios: (1). the source at (0, 0, 10) km and the receiver at (2, 0, 0) km; (2). the source at (0, 0, 2) km and the receiver at (10, 0, 0) km; (3). the source at (0, 0, 0.2) km and the receiver at (10, 0, 0) km. The comparison shows a high degree of consistency between the results.

If we denote the ratio of epicentral distance r to source depth H (i.e., x_3' or z_s) by ζ: $\zeta = r/H$, it is clear that the values of ζ corresponding to Figs. 7.36, 7.37 and 7.38 are 0.2, 5, and 50, respectively. For $\zeta = 0.2$, the waveforms in Fig. 7.36 show only two seismic phases around $t = 1.27$ s and 2.21 s, corresponding to the P-wave and S-wave, respectively. Comparing with the waveforms of the infinite space Green's function generated by a step-function force (Fig. 4.9), we find that they are very similar. This indicates that, in this situation, the main components of the seismic waves observed at the surface are direct waves. All components of the Green's function exhibit significant permanent displacement, which is characteristic of a step-function time dependency. Compared to the results for the $\mathcal{LP}_1$, the Gibbs phenomenon in the Green's function for the $\mathcal{LP}_2$ is significantly reduced. The Gibbs phenomenon is related to the sharpness of the time-domain signal, with sharper signals exhibiting

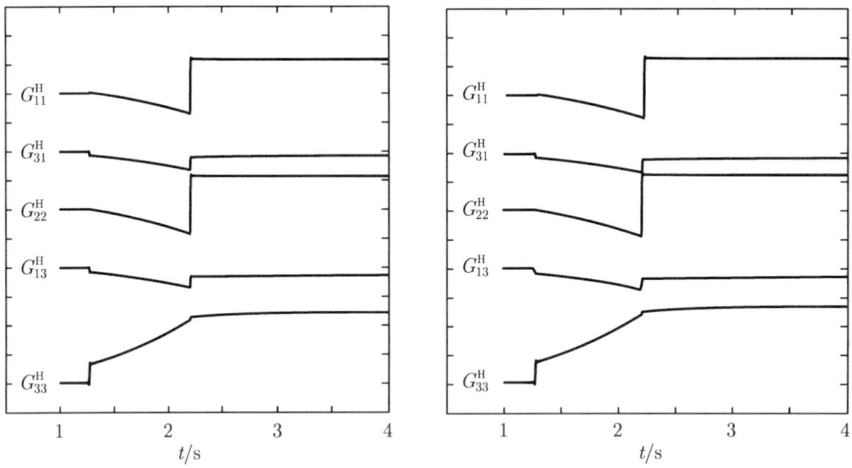

Fig. 7.36 Comparison of G_{ij}^{H} in this book (left) with those in Johnson (1974) (right) (1). Only the non-zero Green's function components are shown. The source position is $\boldsymbol{\xi} = (0,\ 0,\ 10)$ km and the receiver position is $\boldsymbol{x} = (2,\ 0,\ 0)$ km

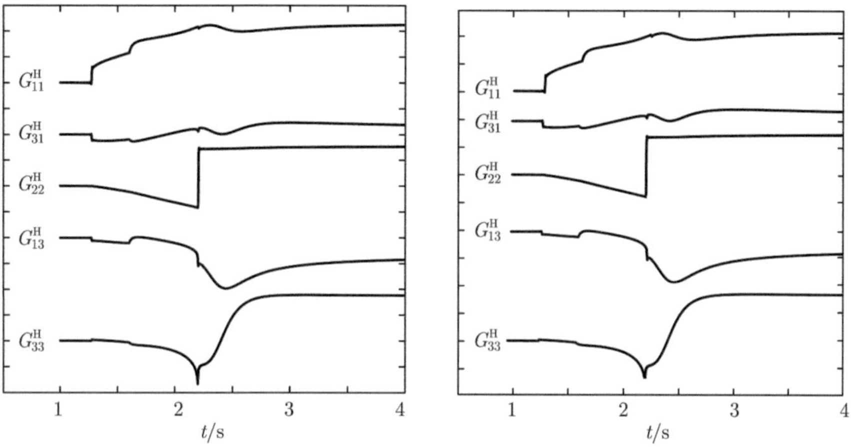

Fig. 7.37 Comparison of G_{ij}^{H} in this book (left) with those in Johnson (1974) (right) (2). The same as in Fig. 7.36, except that $\boldsymbol{\xi} = (0,\ 0,\ 2)$ km, and $\boldsymbol{x} = (10,\ 0,\ 0)$ km

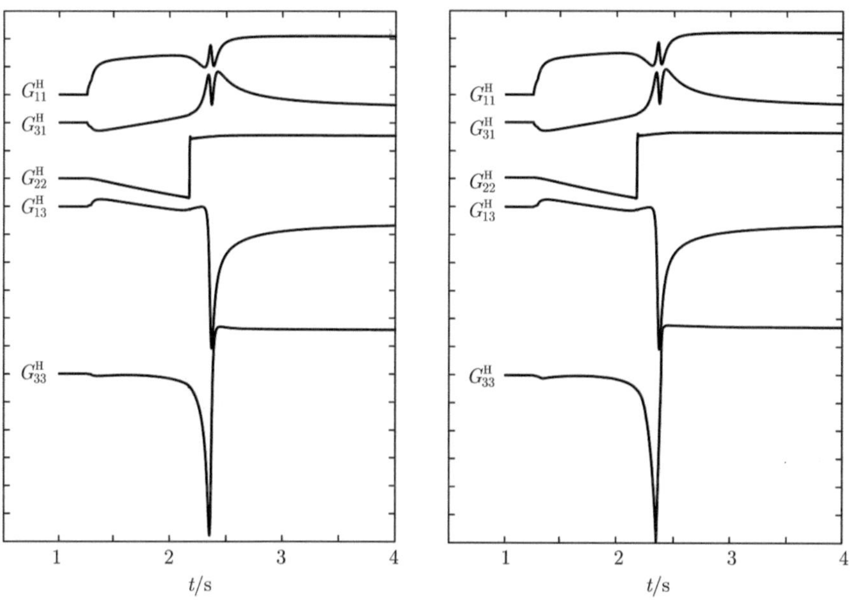

Fig. 7.38 Comparison of G_{ij}^{H} in this book (left) with those in Johnson (1974) (right) (3). The same as in Fig. 7.36, except that $\boldsymbol{\xi} = (0,\ 0,\ 0.2)$ km, and $\boldsymbol{x} = (10,\ 0,\ 0)$ km

more pronounced Gibbs effects.[33] The waveforms at the arrival times of the seismic phases in the Green's function for the $\mathcal{LP}_2$ are noticeably smoother, and the highest frequency components contained are much lower than those in the $\mathcal{LP}_1$. Therefore, the Gibbs phenomenon is greatly diminished.

When $\zeta = 5$, as shown in Fig. 7.37, noticeable changes occur in the Green's function components except for G_{22}^{H}. For the current field-source configuration, with the field point coordinates $\boldsymbol{x} = (10, 0, 0)$ km and the source point coordinates $\boldsymbol{\xi} = (0, 0, 2)$ km, the G_{22}^{H} component corresponds to SH wave, while the other non-zero components correspond to coupled P-SV waves. Comparing this to Fig. 7.36, it is evident that the free surface of the half-space medium model significantly impacts the P-SV wave system. A particularly notable feature is the appearance of a new phase around $t = 1.60$ s, corresponding to an S wave incident at the critical angle $\theta_c = \arcsin(\beta/\alpha)$, converting to a P wave and propagating along the surface.[34] Another distinct feature is a very smooth yet clearly identifiable phase appearing around $t = 2.4$ s, shortly after the arrival of the S wave at $t = 2.21$ s. This is the Rayleigh surface wave. When ζ increases further to 50, as shown in Fig. 7.38, the Rayleigh wave becomes the dominant phase, especially in the G_{13}^{H} and G_{33}^{H} components induced by a vertical force.

7.3.2.2 Comparison with Results of Pekeris and Lifson (1957)

Pekeris and Lifson (1957) used numerical calculations to demonstrate the time variation of Green's function components for different ζ values, as shown in Fig. 5.6. Figures 7.39 and 7.40 illustrate the vertical and horizontal Green's function components induced by a vertical force for different ζ values (left column), compared with the results from Pekeris and Lifson (1957) (right column). The horizontal axis represents the dimensionless time $\tau = \beta t/r$ scaled by the S-wave arrival time, and the vertical axis represents the dimensionless components $-\bar{G}_{33}^{\mathrm{H}} = -\pi\mu r G_{33}^{\mathrm{H}}$ and $\bar{G}_{13}^{\mathrm{H}} = \pi\mu r G_{13}^{\mathrm{H}}$, scaled by $1/\pi\mu r$. The arrival times of P-waves, S-waves, SP-waves, and Rayleigh waves are marked with "P", "S", "SP", and "R", respectively. Except for the oscillations at the Rayleigh wave arrival times due to the Gibbs effect when $\zeta = 1000$ and $\zeta = \infty$ (i.e., $H = 0$), our results match well with those of Pekeris and Lifson (1957).

[33] Essentially, the Gibbs phenomenon arises from truncation in the frequency domain. The sharper a waveform is in the time domain, the wider its corresponding frequency band. An extreme example is the delta function $\delta(t)$ in the time domain, which has a spectrum of 1. No matter how densely it is sampled, it cannot cover all frequency components. As a result, reconstructing the time-domain signal from the delta function's spectrum inevitably leads to a severe Gibbs phenomenon.

[34] Pekeris and Lifson (1957) referred to this as the "SP" phase, as shown in Figs. 7.39 and 7.40. Note that the current frequency-domain method, based on calculating the spectrum and performing a discrete inverse Fourier transform, is not convenient for analyzing phase arrivals. In the second volume of this book, we will use a time-domain Green's function approach, which is more suitable for phase analysis, to provide a detailed examination of the various phases in the results.

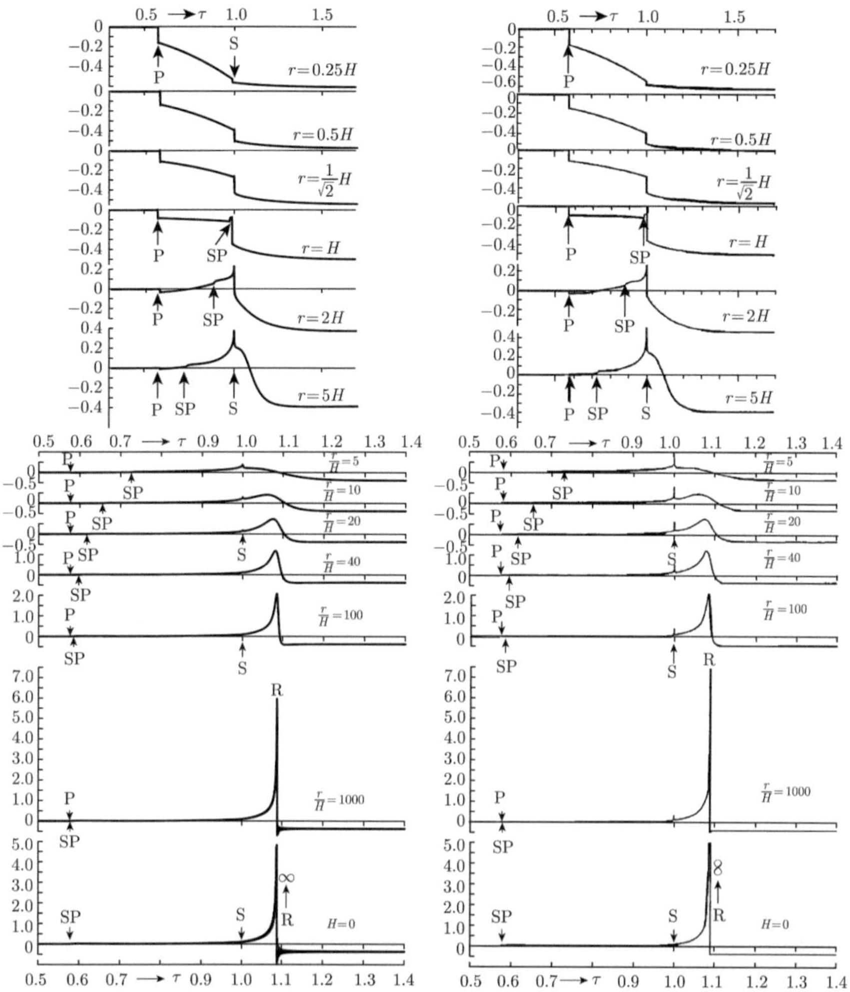

Fig. 7.39 Comparison of $-\bar{G}_{33}^{\mathrm{H}}$ with (left) and those in Pekeris and Lifson (1957) (right). $\zeta = r/H = 0.25, 0.5, 1/\sqrt{2}, 1, 2, 5, 10, 20, 40, 100, 1000$ and ∞ (H=0). "P", "SP", "S", and "R" represent P-waves, SP-waves, S-waves, and Rayleigh waves, respectively

Figure 7.39 shows the time variation of the vertical displacement $-\bar{G}_{33}$ caused by a vertical force, using a consistent scale for ζ values of 0.25, 0.5, $1/\sqrt{2}$, 1, 2, and 5, as well as 5, 10, 20, 40, 100, 1000, and ∞. Several key features are clearly visible:

(1) When $\zeta > 1/\sqrt{2}$, the SP phase appears, and as ζ increases, its arrival time decreases from the S-wave arrival time to the P-wave arrival time.
(2) For $\zeta > 5$, the Rayleigh wave emerges and becomes more pronounced as ζ increases. As ζ approaches infinity, the Rayleigh wave amplitude tends to infinity.

(3) Despite the complex intermediate variations, the final dimensionless permanent displacement is roughly the same.
(4) The Gibbs phenomenon is notably evident near the Rayleigh wave arrival for very large ζ values.

Figure 7.40 presents the results for the horizontal displacement $\bar{G}_{13}$ caused by a vertical force. The features observed are similar to those in Fig. 7.39.

It is worth mentioning that Pekeris and Lifson (1957) were the first to provide detailed numerical results for the Green's functions of the $\mathcal{LP}_2$. Their findings illus-

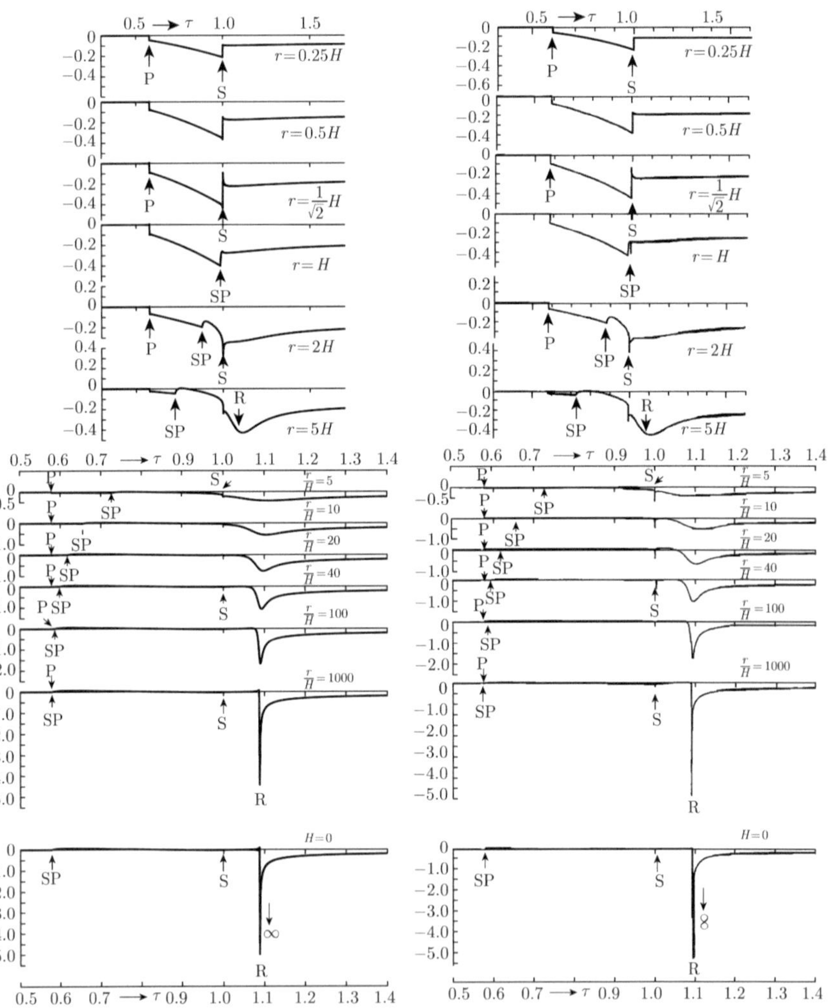

Fig. 7.40 Comparison of $\bar{G}^{\mathrm{H}}_{13}$ with (left) and those in Pekeris and Lifson (1957) (right). The same as in Fig. 7.39, except that here is for a different component of Green's function

trated how different ratios of epicentral distance to source depth affect the results,[35] thus becoming a benchmark for verifying the accuracy of subsequent studies on Lamb's problem.

7.3.3 Green's Function for $\mathcal{LP}_3$

Compared to the $\mathcal{LP}_2$, the application scope of the $\mathcal{LP}_3$ is much narrower. It primarily appears in studies of source dynamics, where both the source and the field points are located underground. Due to the reflection and conversion of elastic waves at the free surface, the waveforms are expected to be more complex.

7.3.3.1 Comparison with Results of Johnson (1974)

Using the Cagniard-de Hoop method, Johnson (1974) derived integral expressions and provided numerical results for reflected and converted waves. For a source point located at $\boldsymbol{\xi} = (0,\ 0,\ 2)$ km, Johnson (1974) calculated the Green's functions at two field points, $\boldsymbol{x} = (10,\ 0,\ 1)$ km and $\boldsymbol{x} = (10,\ 0,\ 4)$ km. Figures 7.41 and 7.42 show a comparison of our results with those of Johnson (1974) on the right.

For an observation point located 1 km below the surface, Fig. 7.41 shows a good agreement between the two results. Compared to the surface observation (Fig. 7.37), the waveform is much more complex, including direct P and S-waves, reflected P-wave (PP), reflected S-wave (SS), converted P-wave (PS), converted S-wave (SP),[36] and Rayleigh wave. The Rayleigh wave phase, which is smoother than the others, appears around $t = 2.46$ s and is more noticeable in the G_{31}^{H} component. In the second volume of this book, we will use the time-domain Cagniard-de Hoop method to derive and analyze the waveforms and arrival times of these phases in detail. Figure 7.42 presents the results for an observation point located 4 km below the surface, showing general consistency with Fig. 7.41 but without a clear Rayleigh wave. Notably, our results show an additional phase around $t = 1.71$ s compared to Johnson (1974). Based on theoretical arrival times, this is identified as the SP wave.

[35] For instance, the phenomenon of Rayleigh wave becoming more pronounced with increasing ζ values, and the occurrence of SP wave only within a certain range, with their arrival time shifting from S-wave to P-wave arrival time as ζ increases, are all clearly shown for the first time. This is of great significance for understanding the behavior of seismic waves.

[36] The "SP" phase mentioned here differs from the "SP" phase described by Pekeris and Lifson (1957). Their "SP" phase refers to an S wave incident at the critical angle $\arcsin(\beta/\alpha)$, which converts to a P wave at the surface and then propagates along the surface to the observation point. In contrast, the "SP" phase referred to here is an S wave that reaches the surface, converts to a P wave, and then reflects back to be received by an underground observation point.

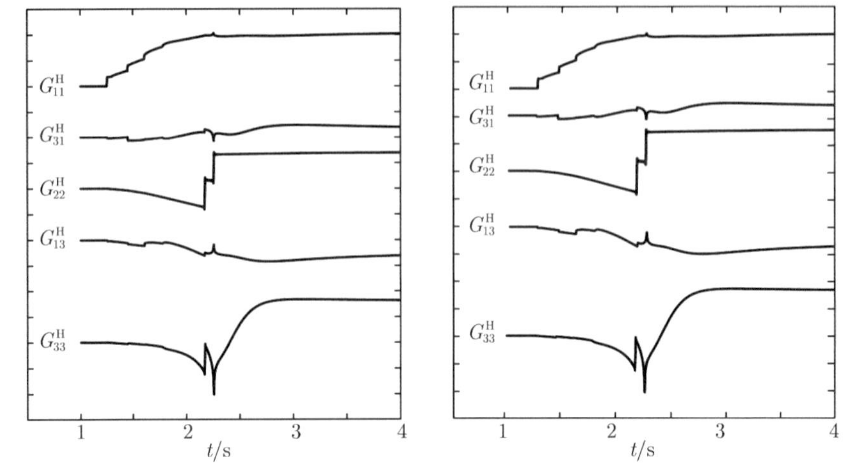

Fig. 7.41 Comparison of G_{ij}^{H} in this book (left) with those in Johnson (1974) (right) (4). The same as in Fig. 7.36, except that $\boldsymbol{\xi} = (0,\ 0,\ 2)$ km, and $\boldsymbol{x} = (10,\ 0,\ 1)$ km

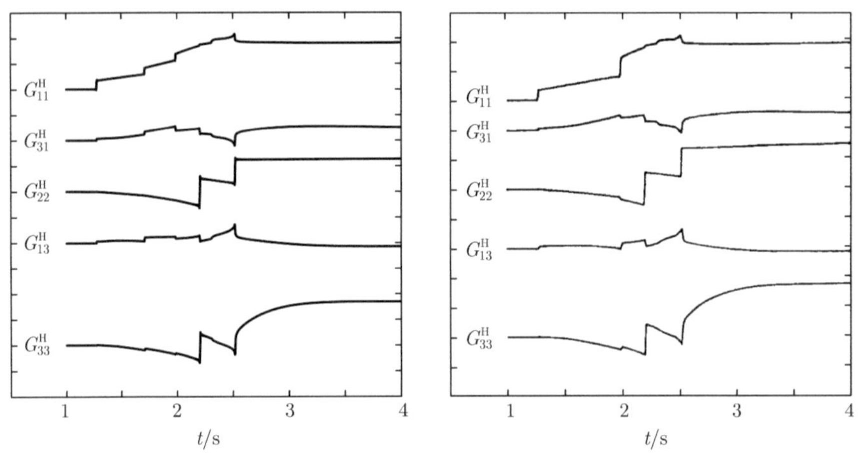

Fig. 7.42 Comparison of G_{ij}^{H} in this book (left) with those in Johnson (1974) (right) (5). The same as in Fig. 7.36, except that $\boldsymbol{\xi} = (0,\ 0,\ 2)$ km, and $\boldsymbol{x} = (10,\ 0,\ 4)$ km

7.3.3.2 Indirect Verification and Analysis of Differences

To verify the existence of this phase, we used an indirect method. The source was placed at $\boldsymbol{\xi} = (0,\ 0,\ 2)$ km, and the observation point depth was varied from the surface to a depth of 5 km in 0.1 km steps. Figure 7.43 shows the results for the horizontal displacement component G_{11}^{H} and the vertical component G_{31}^{H} caused by horizontal force in the x_1 direction. The thick lines correspond to the respective components in Figs. 7.37, 7.41, and 7.42. The waveforms in the figure reveal continuous

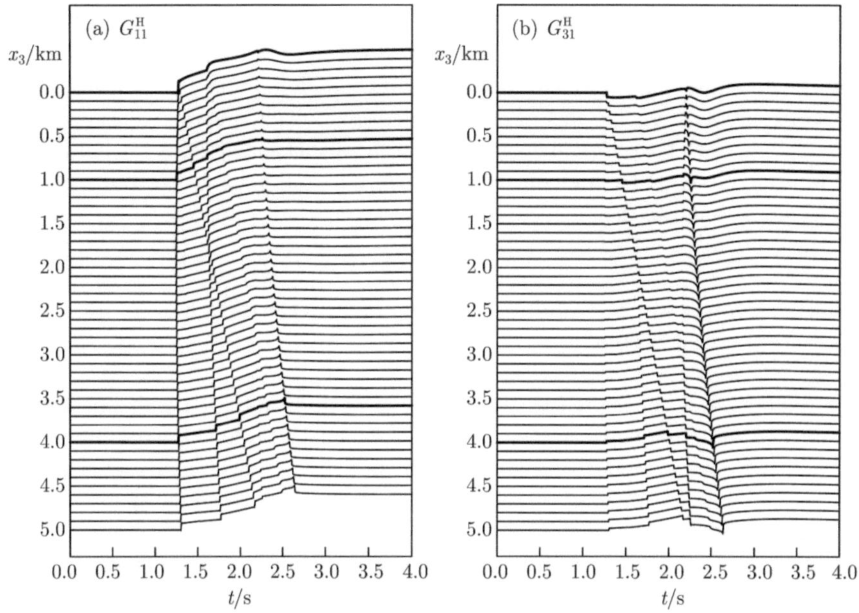

Fig. 7.43 Green's functions at different depths caused by horizontal force at a depth of 2 km. **a** $G^{H}_{11}(10, 0, x_3; 0, 0, 2, 0)$; **b** $G^{H}_{31}(10, 0, x_3; 0, 0, 2, 0)$. The depth x_3 of the observation point ranges from 0 to 5 km, with intervals of 0.1 km. The results for $x_3 = 0$, 1.0, and 4.0 km are indicated with thick lines

changes in each phase with depth, indicating that the additional phase mentioned earlier does indeed exist. Since the current results are based on frequency-domain solutions, the time-domain signals are obtained through the inverse discrete Fourier transform, which is a global transform, so local differences in the time domain will affect all frequency components in the frequency domain. Therefore, the likelihood of an error in a local phase in the time domain due to errors in the frequency domain calculations of all frequency components is very small. Johnson (1974), however, performed direct time-domain solutions and solved for each phase independently. The final waveform was the sum of contributions from each phase, which could have led to the omission of a phase.

7.3.4 Spatial Derivatives of Green's Function and Displacement Due to Shear Dislocation Point Source

The accuracy of the Green's functions for several types of Lamb's problems has been verified through comparisons with different studies. However, in seismology, we are more concerned with the displacement field generated by a shear dislocation source. As concluded in Chap. 3, this can be obtained by convolving the seismic moment tensor with the spatial derivatives of the Green's function. In the frequency domain, this corresponds to the product of the spectra of the seismic moment tensor and the Green's function. Therefore, the validation of the displacement field generated by a shear dislocation point source involves two steps: first, the spatial derivatives of the Green's function, and then its convolution with the seismic moment tensor.

7.3.4.1 Comparison with Johnson (1974): Spatial Derivatives of Green's Function

Johnson (1974) is considered a comprehensive work on solving Lamb's problems, not only because it provides complete results for the Green's functions of three types of Lamb's problems, but also because it offers numerical results for the spatial derivatives of the Green's functions. In his paper, Johnson presented all the spatial derivatives of the Green's function $G^{\mathrm{H}}_{ij,k'}$ for the field-source position combination corresponding to Fig. 7.37: $\boldsymbol{x} = (10,\ 0,\ 0)$ km and $\boldsymbol{\xi} = (0,\ 0,\ 2)$ km. Figures 7.44, 7.45

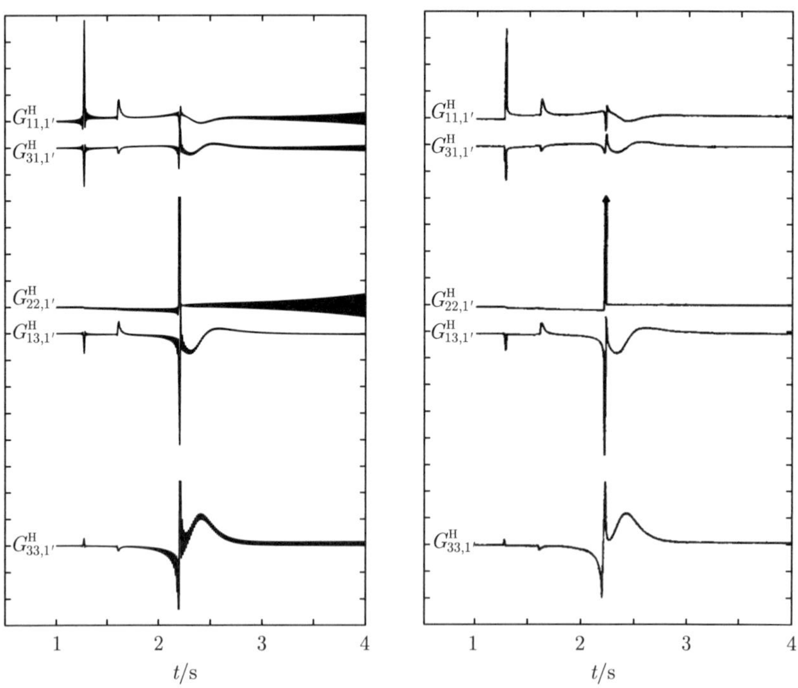

Fig. 7.44 Comparison of $G^{\mathrm{H}}_{ij,1'}$ in this book (left) with those in Johnson (1974) (right). Only the non-zero Green's function components are shown. The source position is $\boldsymbol{\xi} = (0,\ 0,\ 2)$ km and the receiver position is $\boldsymbol{x} = (10,\ 0,\ 0)$ km

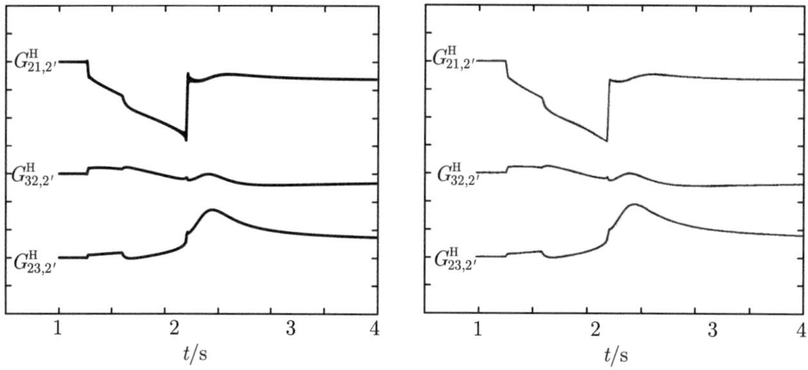

Fig. 7.45 Comparison of $G^{\mathrm{H}}_{ij,2'}$ in this book (left) with those in Johnson (1974) (right). The source and receiver positions are the same as in Fig. 7.44

and 7.46 respectively show the non-zero components of $G^{\mathrm{H}}_{ij,1'}(10, 0, 0, t; 0, 0, 2, 0)$, $G^{\mathrm{H}}_{ij,2'}(10, 0, 0, t; 0, 0, 2, 0)$, and $G^{\mathrm{H}}_{ij,3'}(10, 0, 0, t; 0, 0, 2, 0)$ (left) and their correspondences in Johnson's (1974) (right).

From these comparison figures, it can be seen that our calculations are correct. However, unlike the Green's function itself (Fig. 7.37), the waveforms of the Green function's spatial derivatives are very sharp at the arrival of seismic phases, resulting in significant Gibbs effects. Unlike previous instances where Gibbs effects appeared only near the arrival times in the time domain, here they affect the entire subsequent waveform, causing noticeable oscillations. Examples of this can be seen in $G^{\mathrm{H}}_{11,1'}$ and $G^{\mathrm{H}}_{22,1'}$ in Fig. 7.44, and $G^{\mathrm{H}}_{11,3'}$ and $G^{\mathrm{H}}_{22,3'}$ in Fig. 7.46. Since this causes significant deviations in the overall waveform, it warrants a more in-depth analysis.

As mentioned earlier, the Gibbs effect is related to the truncation in the frequency domain of sharp time signals. The sharper the time-domain signal, the more pronounced the truncation effect in the frequency domain, resulting in a more noticeable Gibbs effect when transformed back to the time domain using the inverse discrete Fourier transform. The reason for the intense oscillations following the sharp signal is the use of complex frequencies in our calculations. After introducing complex frequencies, the original time signal is recovered by multiplying the inverse discrete Fourier transform result by $e^{\omega_{\mathrm{I}} t}$ (refer to Eq. (7.2.11)), effectively amplifying the signal over time. Thus, the time-domain signal distortion caused by the Gibbs effect becomes more evident after this amplification.

The results shown in Figs. 7.44, 7.45 and 7.46 are clearly unsatisfactory. Firstly, increasing the number of sampling points N does not fundamentally improve the effect. Figure 7.47a uses $G^{\mathrm{H}}_{11,1'}(10, 0, 0, t; 0, 0, 2, 0)$ as an example to show results for different N values. It can be seen that more sampling points result in more intense local oscillations due to the Gibbs effect, rather than eliminating it. So, how can this effect be mitigated? The clue lies in addressing the root cause of the Gibbs effect. It arises from sharp features in the time-domain signal; if these features are

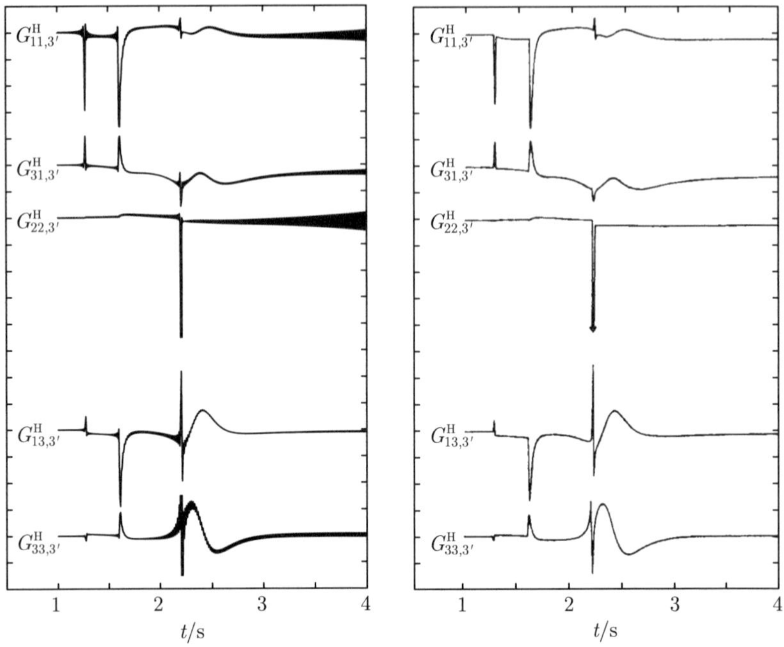

Fig. 7.46 Comparison of $G^{H}_{ij,3'}$ in this book (left) with those in Johnson (1974) (right). The source and receiver positions are the same as in Fig. 7.44

smoothed, the Gibbs effect will naturally diminish. So far, our results are based on a step function as the source time function. Based on the discussion in Sect. 7.2.6, we know that the spectrum of the step function is proportional to $1/\omega$. If we use a time function that decays faster with increasing ω, such as a ramp function, it is expected that the sharpness of the time-domain signal will be reduced. Figure 7.47b shows the results for a ramp function with a rise time $t_0 = 0.5$ s. As seen, except for $N = 256$ where the resolution is low due to too few sampling points, the results for $N = 512$, 1024, and 2048 show almost no difference, and the latter part of the waveform is very clean. Of course, for step functions, using frequency domain filtering can also reduce the Gibbs effect, but the trade-off is that the time function is no longer a strict step function after filtering.

7.3.4.2 Comparison with Apsel (1979): Displacement Due to Shear Dislocation Point Source

By multiplying the spatial derivatives of the frequency-domain Green's functions with the spectra of the seismic moment tensor and then performing an inverse discrete Fourier transform, we obtain the displacement field generated by a shear dislocation point source. In Chap. 5 of his Ph.D. dissertation (Apsel 1979), Apsel showed the

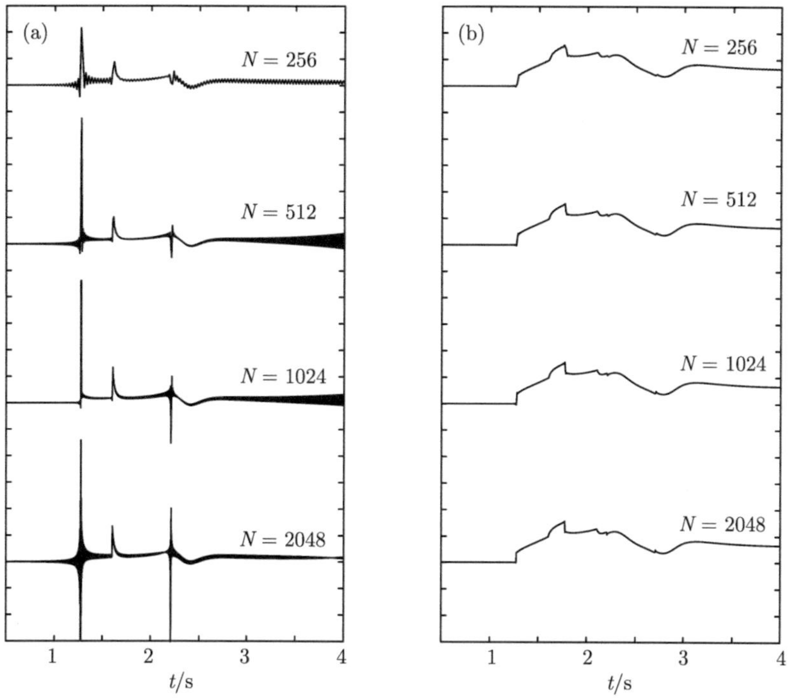

Fig. 7.47 Spatial derivatives of Green's function with different sampling points N. **a** The Green function $G^{\mathrm{H}}_{11,1'}(10, 0, 0, t; 0, 0, 2, 0)$ with a step function as the time function; **b** The Green function $G^{\mathrm{R}}_{11,1'}(10, 0, 0, t; 0, 0, 2, 0)$ with a ramp function as the time function, $t_0 = 0.5$ s. From top to bottom, the N values are 256, 512, 1024, and 2048. For clarity, the amplitude of the fluctuations in (**b**) has been magnified by a factor of 5

three-component displacements produced by a shear dislocation point source for a ramp function with a rise time of $t_0 = 8$ s, for a strike angle $\phi_s = 67.5°$,[37] and various dip δ and rake λ angles. The source is located at $\boldsymbol{\xi} = (0,\ 0,\ 5)$ km, and the observation point is at $\boldsymbol{x} = (20,\ 0,\ 0)$ km. Table 7.2 summarizes the fault models and corresponding fault angles considered by Apsel (1979). Below, we will compare our results with those of Apsel (1979).

Figures 7.48, 7.49 and 7.50 illustrate the displacement fields generated by faults with different dip and rake angles, and compare these with the results of Apsel (1979) shown on the right. Apsel (1979) also presents a comparison between his calculated results (solid lines) and those obtained using Johnson's (1974) method (dashed lines), showing some noticeable discrepancies, particularly in the latter part of the waveforms.

---

[37] In Apsel's (1979) original text, it is stated that "the observation point makes an angle of 22.5° with the fault strike direction". According to our definition in Fig. 3.6, the strike should be the angle between the fault trace and the north direction, which should be taken as 67.5°.

Table 7.2 Fault models and corresponding fault angles considered in Apsel (1979)

Figure no.	Fault type	$\phi_s/(°)$	$\delta/(°)$	$\lambda/(°)$
7.48	Vertical strike/Thrust	67.5	90	0/90
7.49	Oblique strike/Thurst	67.5	45	0/90
7.50	Horizontal strike	67.5	0	0/90

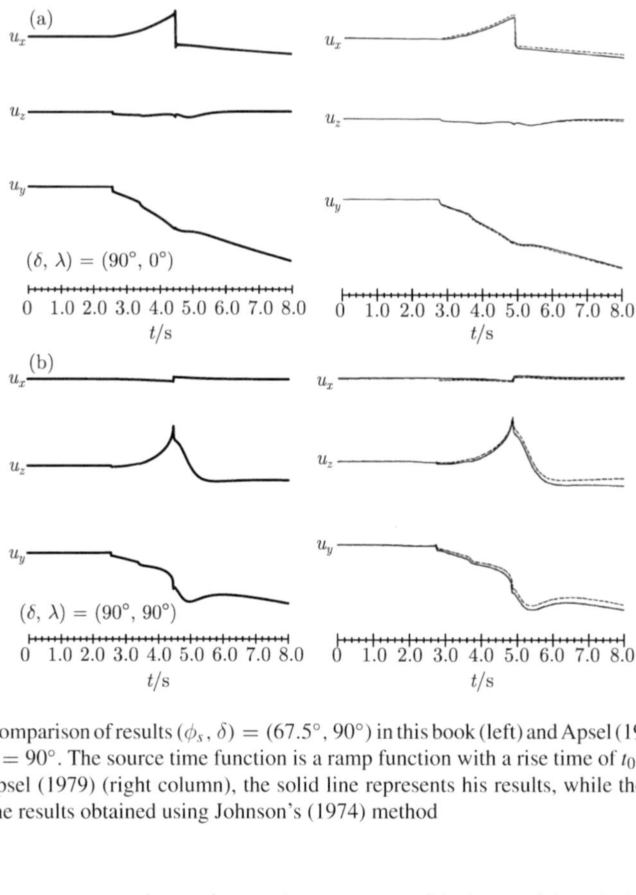

Fig. 7.48 Comparison of results $(\phi_s, \delta) = (67.5°, 90°)$ in this book (left) and Apsel (1979) (right). **a** $\lambda = 0°$; **b** $\lambda = 90°$. The source time function is a ramp function with a rise time of $t_0 = 8$ s. In the results of Apsel (1979) (right column), the solid line represents his results, while the dashed line represents the results obtained using Johnson's (1974) method

For all cases, our results are in good agreement with those of Apsel. As discussed in Sect. 7.3.4.1 regarding the Gibbs effect, since the current time function is a ramp function with a rise time equal to the entire time window, $t_0 = 8$ s, the signal in the time domain is relatively smooth, and the Gibbs effect is not pronounced. Additionally, due to the long rise time, some components, such as u_y in Fig. 7.48a, exhibit a clear linear trend.

It is noted that for the current source and field point positions, where the ratio of epicentral distance to source depth $\zeta = 4$, Rayleigh waves can be identified from the waveforms of u_z and u_y. Furthermore, similar to the results for a single force, the waveforms clearly show P, S, and SP-phases.

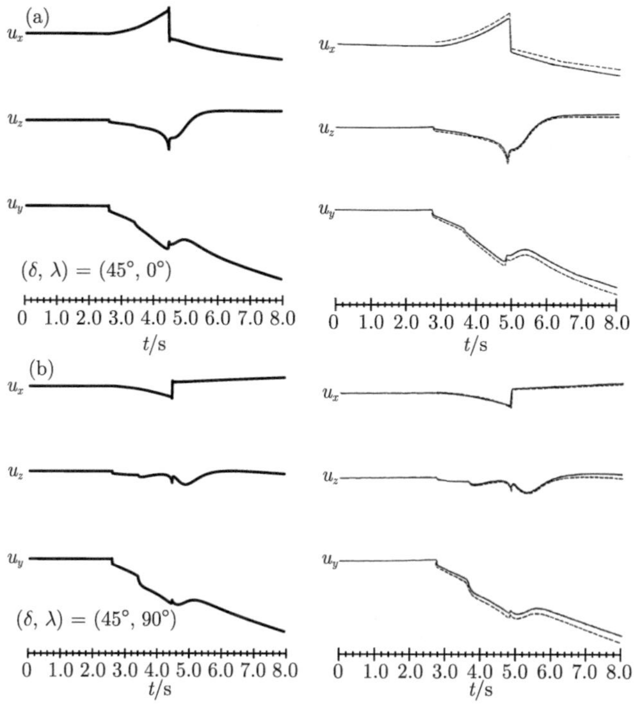

Fig. 7.49 Comparison of results $(\phi_s, \delta) = (67.5°, 45°)$ in this book (left) and Apsel (1979) (right). The same as in Fig. 7.48

7.4 Theoretical Seismogram of Lamb's Problem

In Sect. 7.3, we thoroughly validated the formulas established in Chap. 6, as well as the numerical algorithms for wavenumber integration introduced at the beginning of this chapter. This was done by comparing the Green's functions for three types of Lamb's problems, their spatial derivatives, and the displacement fields from shear dislocation sources. To gain a deeper understanding of the characteristics of seismic wave fields in half-space, in this section, we will continue to use the developed methods and programs to examine the effects of time functions on displacement and the resulting particle motion, the displacement fields caused by point and finite-sized dislocation sources, static solutions, and the properties of Rayleigh waves. The displacement (or velocity, acceleration) time histories obtained theoretically or through various numerical methods are commonly referred to as *theoretical seismograms* or *synthetic seismogram*.

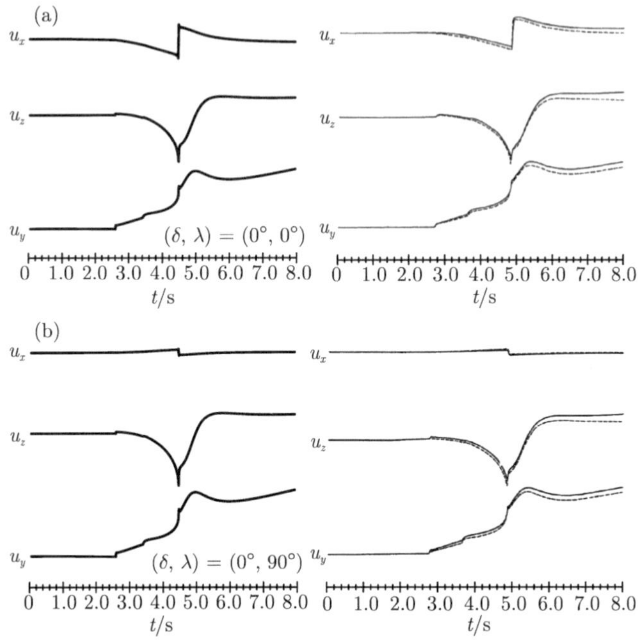

Fig. 7.50 Comparison of results $(\phi_s, \delta) = (67.5°, 0°)$ in this book (left) and Apsel (1979) (right). The same as in Fig. 7.48

7.4.1 Displacement Due to Single Force with General Time Function

Recalling the discussion in Chap. 4 about the Green's function in infinite space and the displacement fields generated by dislocation sources, we studied the properties of the displacement fields from both temporal and spatial perspectives. This was feasible because the Green's function for infinite space is relatively simple, with explicit separation of spatial and temporal variations, allowing for independent examination. However, for the Lamb's problem, the method involves first computing in the frequency domain and then transforming to the time domain, so there is no explicit separation of spatial and temporal components. Nevertheless, for point sources, the frequency-domain displacement field induced by a source time function $S(t)$ is the product of $S(\omega)$ and the Green's function spectrum. Therefore, different $S(t)$ will result in different displacement fields. In the following, we will examine the properties of displacement fields corresponding to several typical source time functions (see Sect. 7.2.6), with special attention paid on the particle motion trajectory, which is a visual representation of ground motion caused by seismic waves.

7.4.1.1 Particle Motion Trajectory Induced by Step-Function Force

Since (Lapwood 1949), studies on the Lamb's problem have almost universally used step functions for the time function. The simplest model to describe the time process of a seismic source involves applying a force from a certain moment and maintaining it constant. In the examples examined in Sect. 7.3, except for the one compared with Apsel (1979), all other examples use this function. Thus, we are familiar with the displacement-time variation images it produces, especially characterized by sharp seismic phases and permanent displacements.

Taking the two examples of the $\mathcal{LP}_2$ compared with Johnson (1974) in Sect. 7.3.2.1 as an example, Figs. 7.37 and 7.38 show the Green's functions for a step function force received at $\boldsymbol{x} = (10, 0, 0)$ km from sources located at $\boldsymbol{\xi} = (0, 0, 2)$ km and $(0, 0, 0.2)$ km, respectively. In this specific source-receiver setup, the non-zero displacement components generated by a horizontal force along the x_1 axis are G_{11}^{H} and G_{31}^{H}, while those generated by a vertical force along the x_3 axis are G_{13}^{H} and G_{33}^{H}. None of these forces generate horizontal displacements along the x_2 axis.

Plotting the particle's motion trajectory with horizontal displacement on the x_1 axis and vertical displacement on the x_3 axis provides a clear visualization of the particle's movement. Figures 7.51 and 7.52 illustrate the particle trajectories in the first quadrant of the x_1x_3 plane for concentrated point forces (both horizontal and vertical) with a step function time profile at depths of 2 and 0.2 km, respectively. The circles mark the initial positions of the particles, and each black dot represents the particle's position at equal time intervals.

For a force applied at a depth of 2 km, as shown in Fig. 7.51, horizontal and vertical forces generate significant permanent displacements in their respective directions. The magnitude of these permanent displacements decreases markedly with distance from the point source. Before reaching their final positions, the particles follow complex motion paths. When seismic phases arrive, there are noticeable jumps, characterized by abrupt changes in the direction of particle motion and a sudden increase in velocity.

When the depth of the concentrated force is 0.2 km, as shown in Fig. 7.52, an additional distinct feature emerges compared to Fig. 7.51: the motion trajectory at the surface includes a clear elliptical path. This part of the trajectory corresponds to Rayleigh wave. Displacements caused by vertical and horizontal forces are predominant in their respective directions. However, due to the presence of the free surface, vertical motion is less constrained than horizontal motion, making Rayleigh wave generated by vertical forces more pronounced than those by horizontal forces. Notably, the larger distances between adjacent black dots on the Rayleigh wave part of the trajectory indicate that this part of the path forms in a very short time. Unlike P-wave and S-wave, Rayleigh wave does not have sharp seismic phase arrival signal. In the direction of wave propagation, the motion of particles in Rayleigh wave approximates a retrograde ellipse, consistent with the conclusion drawn from analyses based on the plane wave assumption.

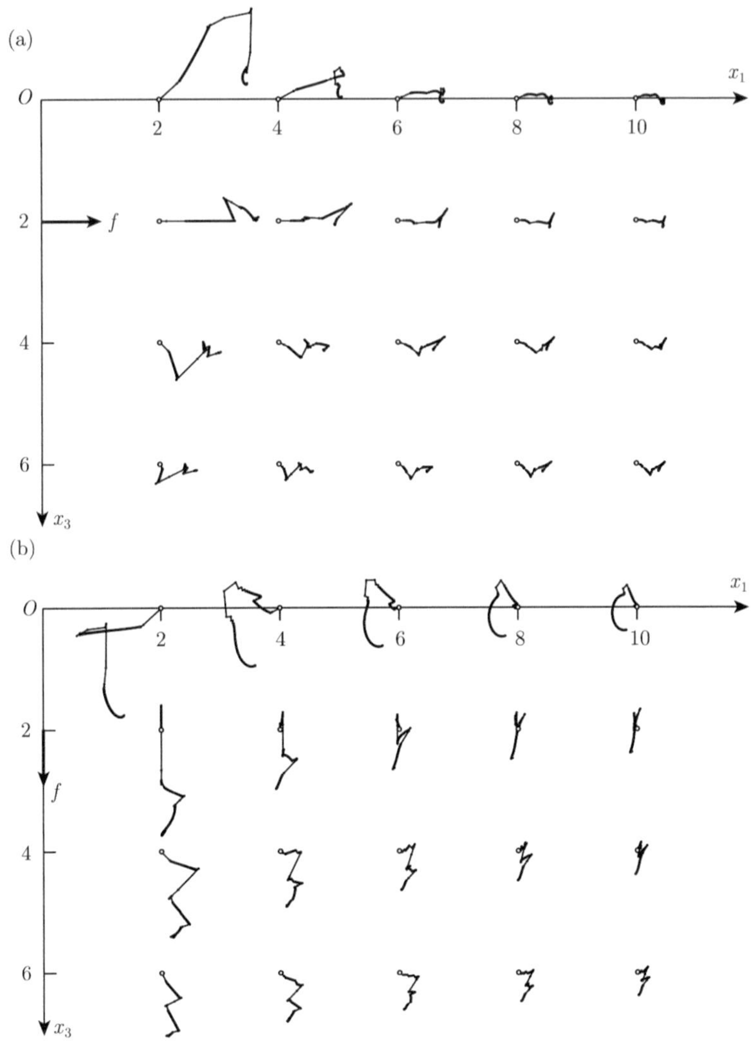

Fig. 7.51 Particle motion trajectory induced by step-function point force at depth of 2 km. **a** Results for horizontal force; **b** Results for vertical force. The initial position of the particle is indicated by a circle, and the thick line arrows represent the force f

7.4.1.2 Displacement and Particle Motion Trajectory Induced by Ramp-Function Force

The ramp function, which incorporates a rise time t_0, better represents the actual situation of a seismic source compared to the step function. Therefore, it is one of the most frequently used source time functions in practical work. In Sect. 4.5.2.2, we showed the displacement field in an infinite space produced by a point force

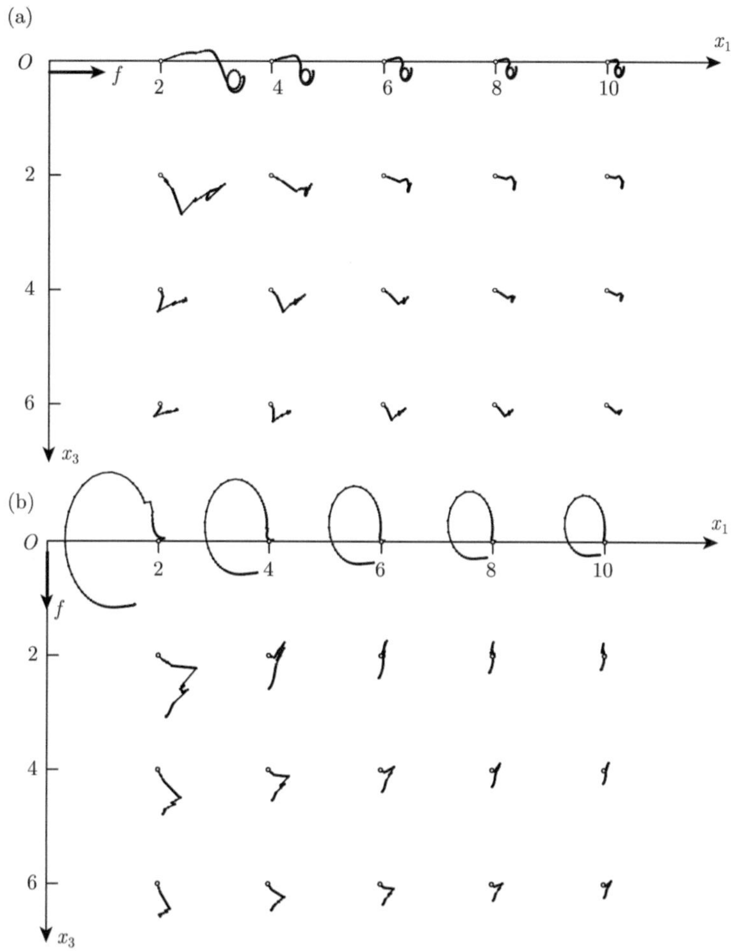

Fig. 7.52 Particle motion trajectory induced by step-function point force at depth of 0.2 km. The same as in Fig. 7.51

using the ramp function (see Fig. 4.12). To examine the changes in displacement when a free surface is present, Fig. 7.53 presents the corresponding results in a half-space model. The difference here is the presence of a free surface, with the source located at the origin on the surface and the observation point at (2, 1, 3) km within the half-space. Compared to Fig. 4.12, the overall shape remains consistent, but a notable difference is that after the S-wave arrival, the waveform transitions slowly to a permanent displacement in the presence of a free surface. In contrast, without a free surface, the waveform reaches its final displacement value immediately after t_0.

To illustrate the changes introduced by the rise time when a free interface is present, Fig. 7.54 shows the results for the ramp function with a time constant of $t_0 = 0.3$ s, corresponding to the time functions in Figs. 7.41 and 7.42. Unlike the

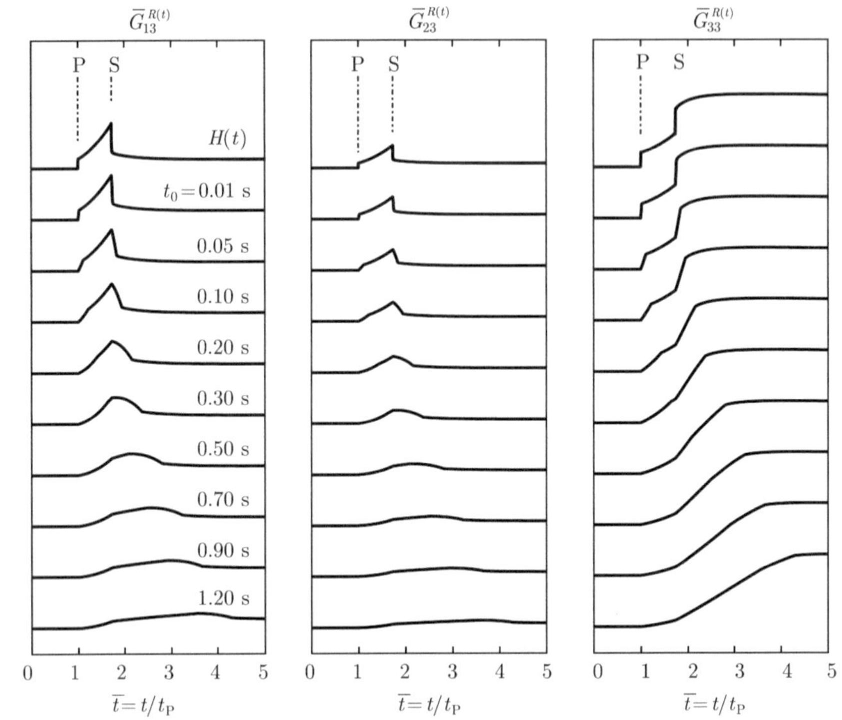

Fig. 7.53 $\bar{G}_{i3}^{R(t)}$ generated by vertical ramp-function force with different rise times t_0. $\bar{G}_{i3}^{R(t)} = G_{i3}^{R(t)} \cdot 4\pi\mu r$. The source is located at the origin (0, 0, 0), while the observation point is at (2, 1, 3) km. Other details are the same as in Fig. 4.12

step function case, where the components of the Green's function show distinct jumps at the arrival of seismic phases (as seen in Figs. 7.41 and 7.42), the Green's function for the ramp function with $t_0 = 0.3$ s exhibits a smooth variation in the entire displacement curve, with no noticeable jumps at the seismic phase arrival times.

Figures 7.55 and 7.56 show the particle motion trajectories induced by a ramp-function point force, corresponding to Figs. 7.51 and 7.52, respectively. A comparison reveals two distinct features: first, the permanent displacement caused by the point force is consistent. This is understandable because the difference between a ramp function and a step function lies only in the transition process; the final value is the same. Therefore, the displacement they cause only differs in the process, not the permanent displacement. Second, the particle motion trajectory induced by the ramp function is noticeably smoother compared to the step function. This is evident in the simplified motion path, the significantly reduced amplitude, and the lack of sudden changes in particle velocity. The abrupt changes in motion direction at the arrival of seismic phases are also much less pronounced. This smoothing effect significantly reduces the amplitude of Rayleigh waves, as seen in Fig. 7.56b.

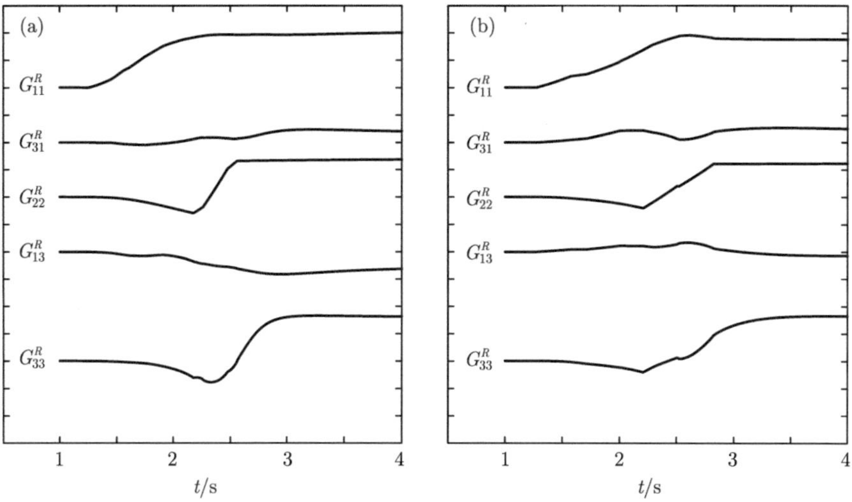

Fig. 7.54 Results for ramp-function force corresponding to those in Figs. 7.41 and 7.42. **a** $\boldsymbol{x} = (10, 0, 1)$ km; **b** $\boldsymbol{x} = (10, 0, 4)$ km. In G^R_{ij}, "R" stands for "Ramp function". $t_0 = 0.3$ s

According to the discussion in Sect. 7.2.6, from a perspective of spectral analysis, the spectrum of a ramp function decays much faster than that of a step function. This essentially means that the results of a ramp function can be seen as a low-pass filtered version of the step function results, leading to a smoother waveform in the time domain.

7.4.1.3 Displacements and Particle Motion Trajectories Due to Trapezoidal Function and Ricker-Wavelet Forces

In Sect. 7.2.6, we introduced two types of time functions used to characterize the behavior of impulsive seismic sources: the trapezoidal function and the Ricker wavelet. These can be considered as smoothed versions of impulsive functions. Unlike the step function and the ramp function, these two time functions return to zero for large values of t, and therefore do not cause permanent displacement.

Figures 7.57 and 7.58 show the results for the trapezoidal function and Ricker wavelet, respectively, corresponding to Figs. 7.41 and 7.42. The overall waveform characteristics are similar, but the Ricker wavelet produces a smoother waveform. Compared to the results for the step function and ramp function, it is evident that different time functions lead to significant differences in the displacement field over time.

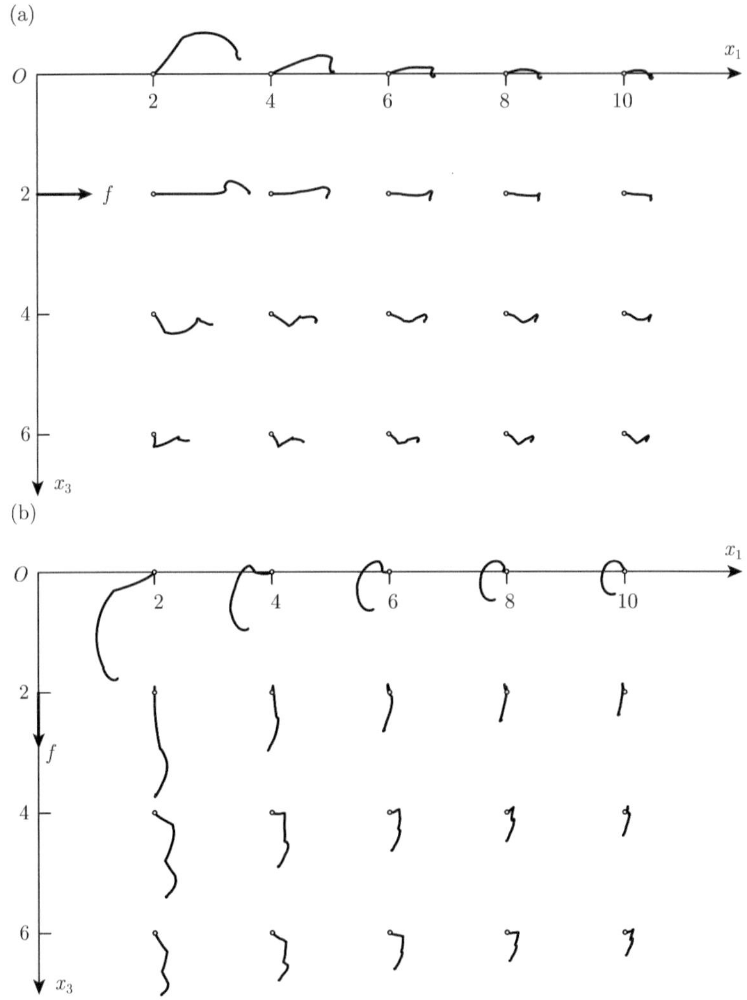

Fig. 7.55 Particle motion trajectory induced by ramp-function point force at depth of 2 km. The same as in Fig. 7.51

Figures 7.59, 7.60, 7.61 and 7.62 show the particle motion trajectories induced by a point force with a trapezoidal function and a Ricker wavelet at depths of 2 and 0.2 km, respectively. Unlike the previously discussed step and ramp functions, since these two time functions are used to imitate a delta-type time dependence, which do not cause permanent displacement, so the particles return to their starting points after each cycle. The smoother time variation of displacement induced by the Ricker wavelet is reflected in the smoother particle motion trajectories. While the particle motion trajectories induced by the trapezoidal function point force are relatively less smooth. In the near-field region, some trajectories somewhat resembles rectangles.

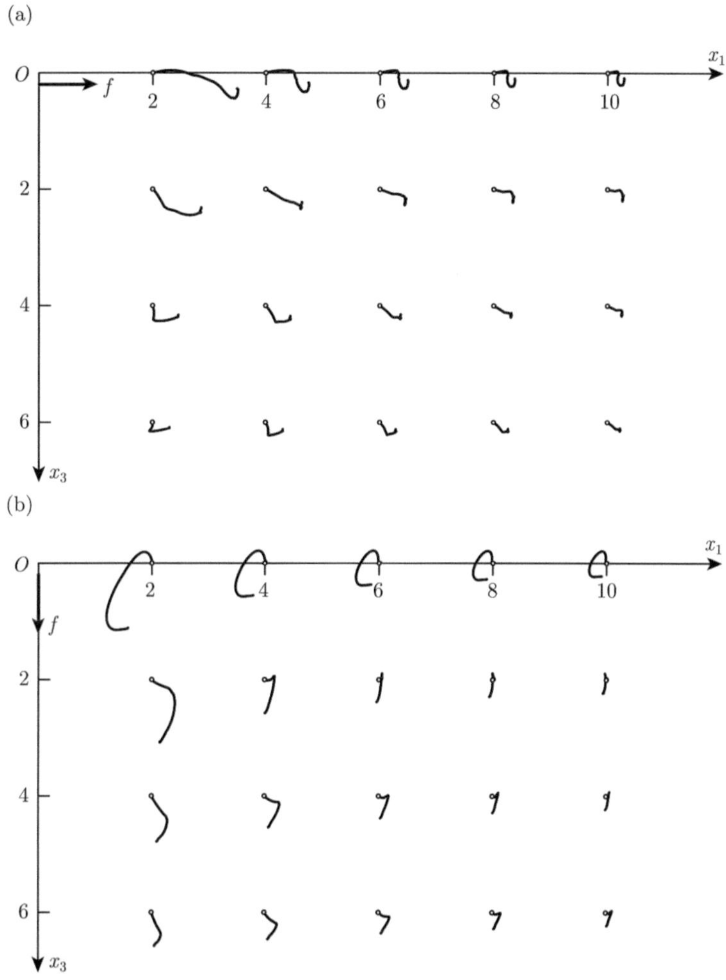

Fig. 7.56 Particle motion trajectory induced by ramp-function point force at depth of 0.2 km. The same as in Fig. 7.51

And whether the force is horizontal or vertical, the resulting Rayleigh waves exhibit nearly retrograde elliptical motion in the direction of wave propagation.

Notably, in Fig. 7.62, the motion trajectories of Rayleigh waves near the surface close to the point force are nearly oblique elliptical, and as the distance from the point force increases, the ellipses become more upright. This interesting characteristic caused by the presence of the source will be further explored in Sect. 7.5.2, where we discuss the properties of Rayleigh waves in the frequency domain.

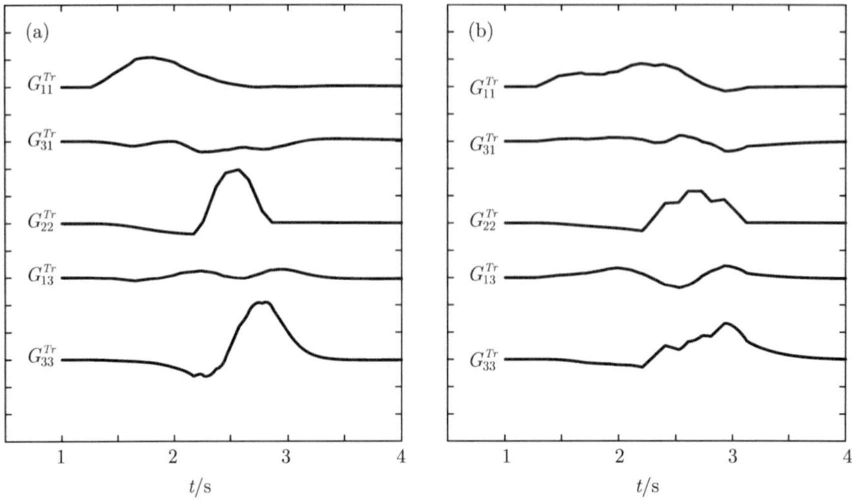

Fig. 7.57 Results for trapezoidal function force corresponding to those in Figs. 7.41 and 7.42. In G_{ij}^{Tr}, "*Tr*" stands for "Trapezoid function". $t_1 = 0.2$ s, $t_2 = 0.4$ s, t_3=0.6 s

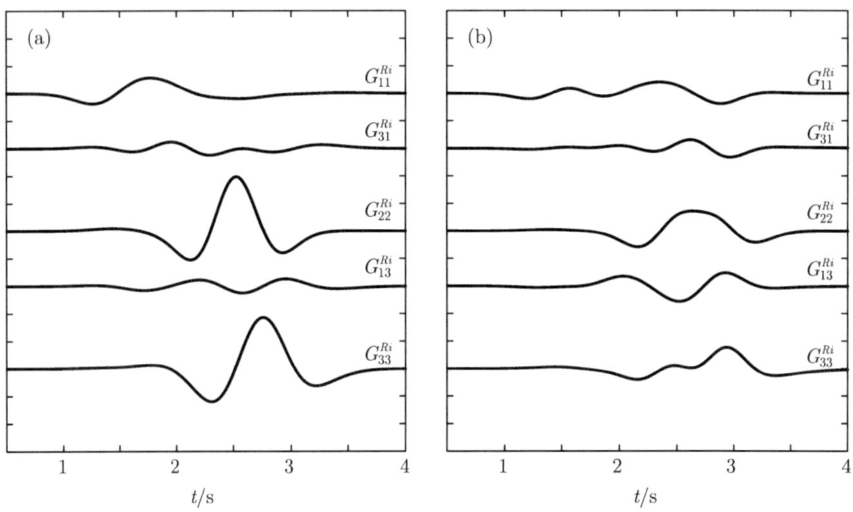

Fig. 7.58 Results for Ricker-wavelet force corresponding to those in Figs. 7.41 and 7.42. In G_{ij}^{Ri}, "*Ri*" stands for "Ricker wavelet". $f_0 = 1$ Hz, $t_0 = 0.3$ s

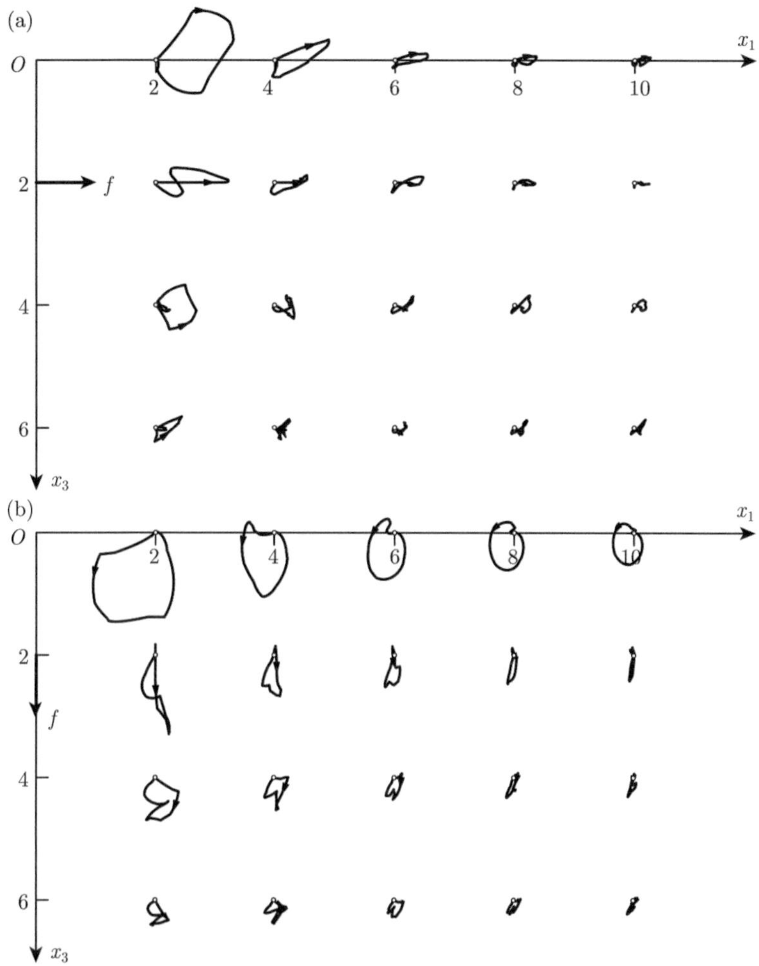

Fig. 7.59 Particle motion trajectory induced by trapezoidal function point force at depth of 2 km. The same as in Fig. 7.51

7.4.2 Displacement Induced by Dislocation Point Source

For seismological issues, dislocation sources are more practically significant. A finite-sized dislocation source can be approximated by dividing the fault plane into small patches, each of which behaves like a point source. By summing the contributions of these point sources, the displacement field can be computed. Thus, calculating the displacement field induced by a dislocation point source is fundamental.

In this section, we compute the displacement field in a half-space model for the examples studied in Sect. 4.7.1 and compare it with the case of an unbounded space, demonstrating the impact of a free surface on the seismic displacement field. To make

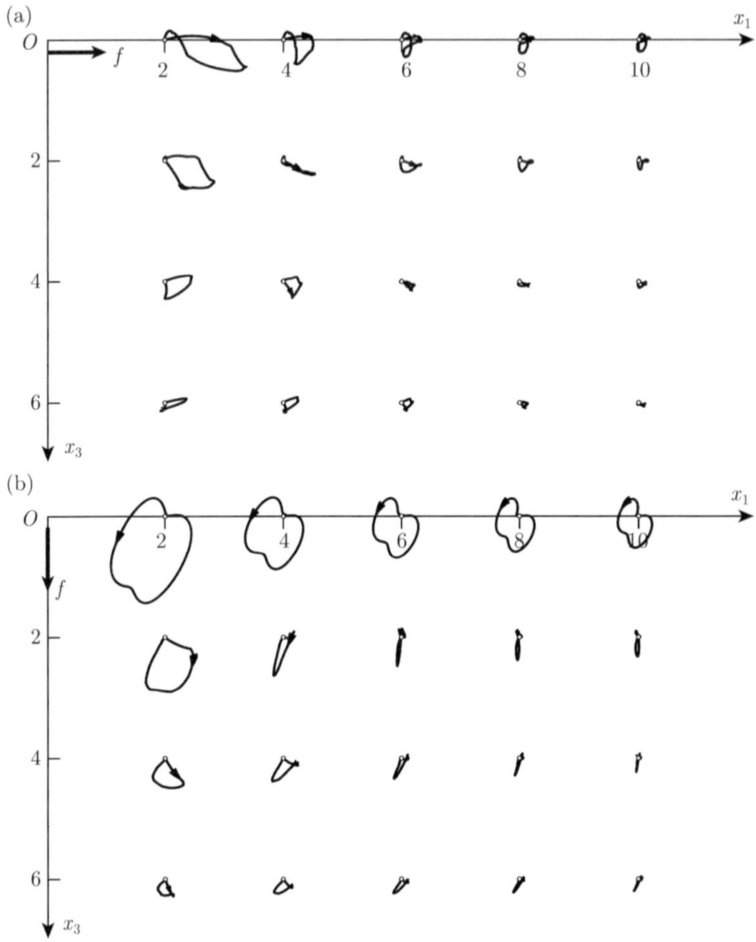

Fig. 7.60 Particle motion trajectory induced by trpeziod-function point force at depth of 0.2 km. The same as in Fig. 7.51

the results more relevant, we will focus on the $\mathcal{LP}_2$, where the observation point is at the surface and the seismic source is buried underground. We consider two typical seismic sources: a left-lateral strike-slip fault and a thrust fault.

7.4.2.1 Left-Lateral Strike-Slip Fault

We first study the example previously calculated in Sect. 4.7.1 using an unbounded space model, except now the medium model is a half-space. All other parameters remain the same as in Sect. 4.7.1. To investigate the effect of the surface on the displacement field, we consider two depths for the fault: $h = 10$ km and $h = 1$ km.

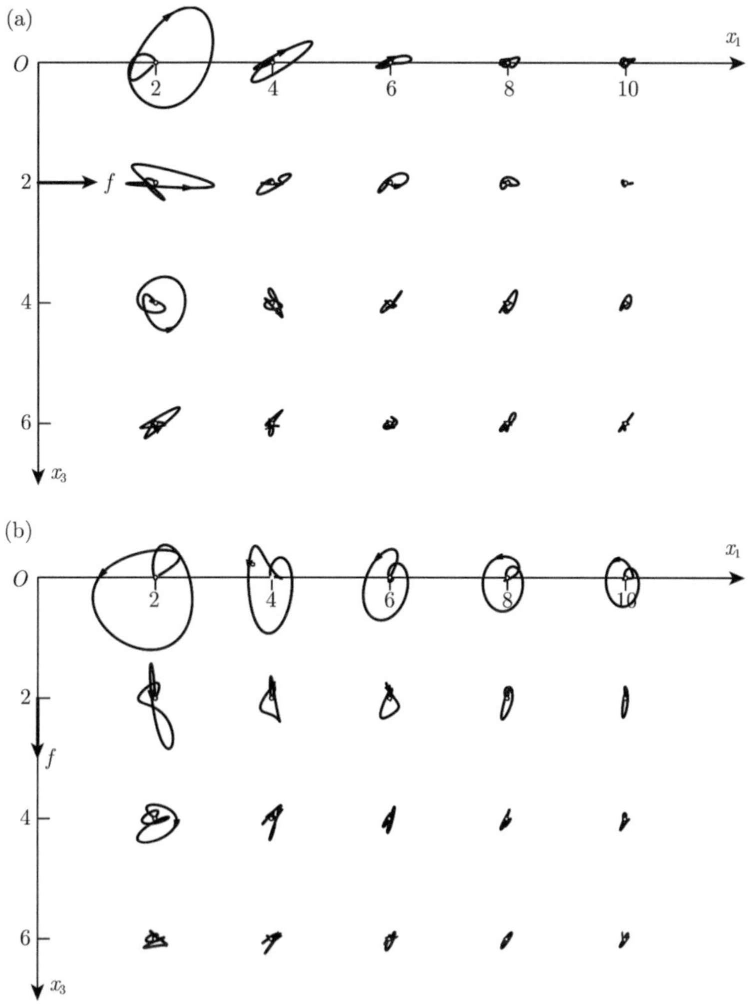

Fig. 7.61 Particle motion trajectory induced by Ricker-wavelet point force at depth of 2 km. The same as in Fig. 7.51

The angle parameters for the left-lateral strike-slip fault are: $\phi_s = 90°$, $\delta = 90°$, and $\lambda = 0°$. We still consider a fault size of 1 km×1 km with an average dislocation of $\overline{[u]} = 1$ m, and a ramp time function with $t_0 = 3$ s. Similarly, to show the azimuthal effect, we compute the displacements at several points A_i and B_i ($i = 1, 2, \ldots, 11$) on two survey lines on the surface, as shown in Fig. 4.29.

Figures 7.63 and 7.64 show the displacement fields at various measurement points induced by a left-lateral strike-slip fault at $x_3' = 10$ km and 1 km, respectively (bold lines). For comparison, the figures also display the corresponding results for an

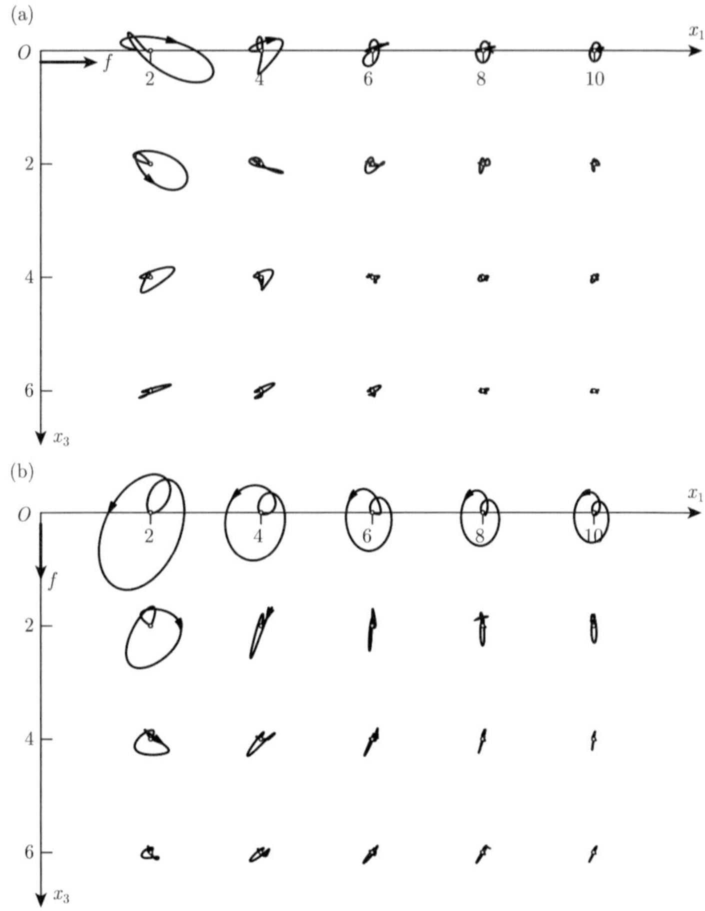

Fig. 7.62 Particle motion trajectory induced by Ricker-wavelet point force at depth of 0.2 km. The same as in Fig. 7.51

infinite space model (thin lines).[38] In Sect. 4.7.1, we discussed the results for the infinite space model in detail, and these characteristics still hold for the current half-space model. Here, we focus on the differences between the two models' results.

First, the amplitude of the static displacement field in the half-space model is larger than the corresponding value in the infinite space model. This is understandable because the free surface in the half-space model allows free vertical movement. Second, compared to the infinite space model, which includes only simple P and S-waves, the half-space model contains more complex wave phases, and the overall

[38] These results are calculated by omitting the terms related to the free surface in the half-space model, as referenced in Eq. (6.4.23). Therefore, the results are identical to those in Figs. 4.30 and 4.31, which again verifies the accuracy of the method and code.

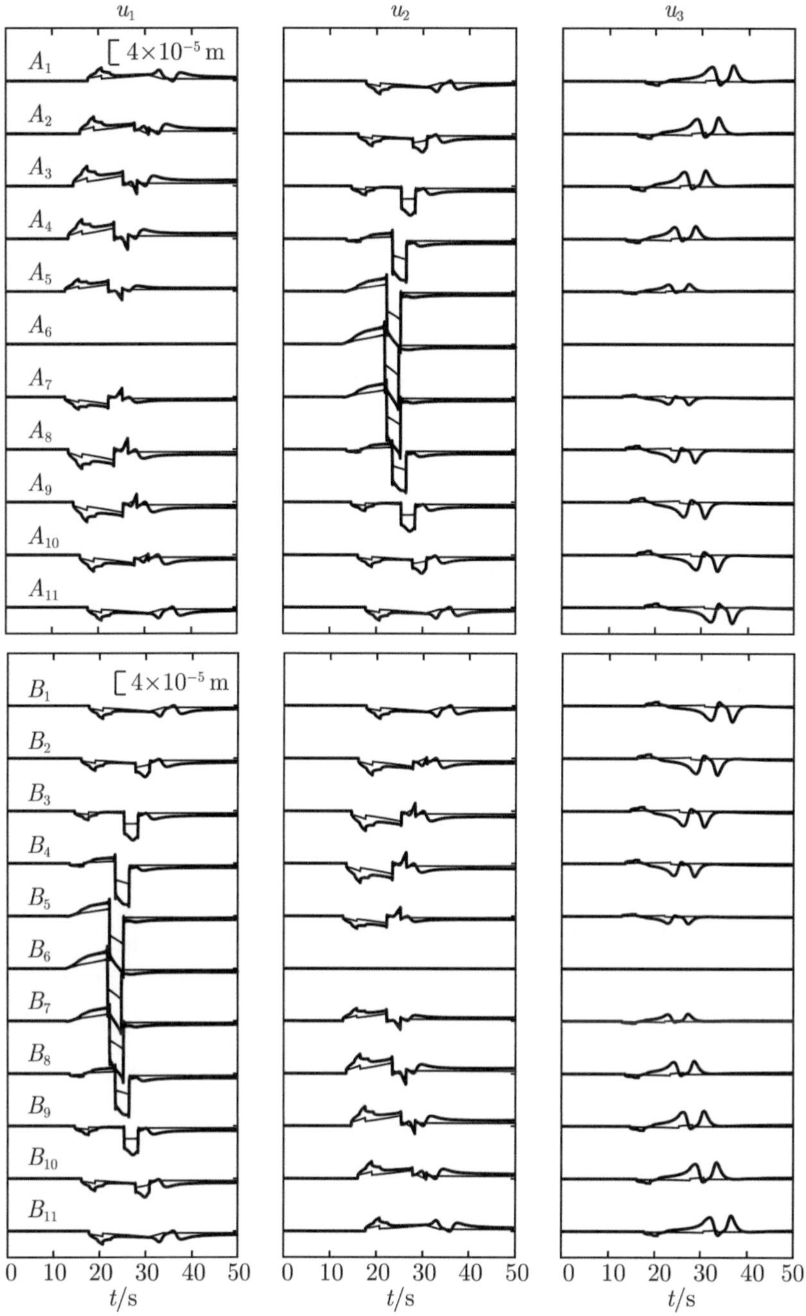

Fig. 7.63 Displacements at A_i and B_i induced by left-lateral strike-slip source at $x_3' = 10$ km. The bold and thin lines represent the results of the half-space and infinite space model, respectively. $(\phi_s, \delta, \lambda) = (90°, 90°, 0°)$. For the distribution of A_i and B_i $(i = 1, 2, \ldots, 11)$, see Fig. 4.29

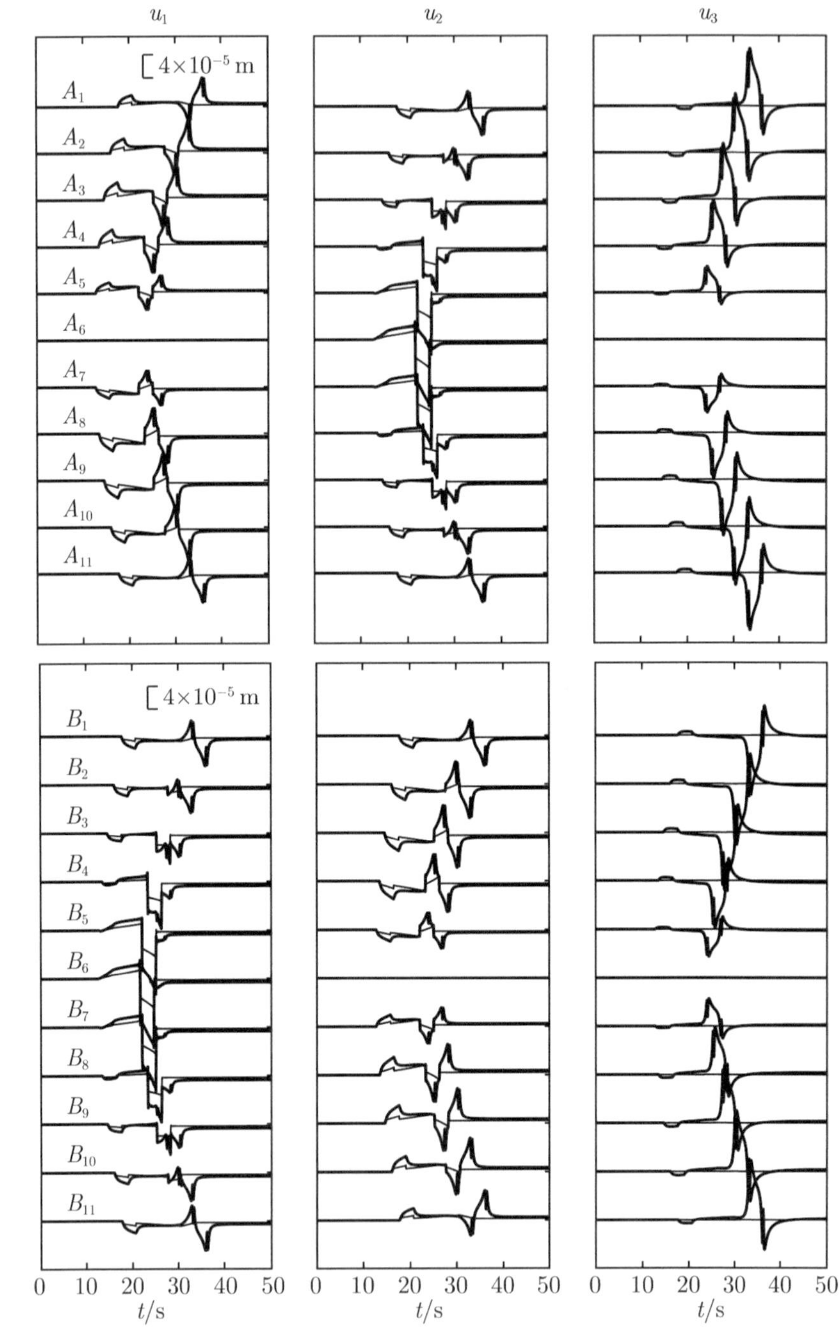

Fig. 7.64 Displacements at A_i and B_i induced by left-lateral strike-slip source at $x_3' = 1$ km. The same as in Fig. 7.63

waveform amplitude is significantly increased, almost doubling that of the infinite space model. This indicates that the presence of the free surface can substantially amplify the seismic amplitude. In particular, the vertical displacement u_3 has a much larger amplitude than its counterpart in the infinite space model.

In the infinite space model, the difference between $x_3' = 10$ km and 1 km is only in the direction of the change. However, in the half-space model, the impact is much greater. The closer the seismic source is to the surface, the more intense the interactions become near the surface. This is particularly evident in the significant increase in Rayleigh wave. Comparing the two figures clearly shows this effect.

To gain a more intuitive understanding of particle motion, Fig. 7.65 displays the trajectories of particles at points A_1, A_{11} (B_1), and B_{11} for a left-lateral strike-slip fault at different source depths in two medium models. Different magnifications are used for the particle motion trajectories (see the caption for details). The equivalent body forces of the fault are also marked. The three observation points are symmetrically located about the origin or coordinate axes and lie precisely on the P and T axes corresponding to the equivalent body force,[39] resulting in symmetric particle motion trajectories. As reflected in the waveforms, for the infinite space model (Fig. 7.65c, d), the impact of changing source depth is solely due to differences in the source-field spatial positions. Specifically, when $x_3' = 1$ km, the resulting u_3 is significantly smaller than that for $x_3' = 10$ km. However, for the half-space model (Fig. 7.65a, b), there are not only significant differences in the waveforms, but also a rapid increase in the amplitude of Rayleigh wave motion as the source becomes shallower (note the magnification in Fig. 7.65a is five times that in (b)).

7.4.2.2 Thrust Fault

In addition to the common strike-slip faults, many earthquakes in nature occur on thrust faults due to tectonic compressive forces. Consider a low-angle thrust fault with parameters: $\phi_s = 90°$, $\delta = 30°$, and $\lambda = 90°$. The other calculation parameters are the same as those for the left-lateral strike-slip fault discussed in Sect. 7.4.2.1.

Figures 7.66 and 7.67 show the displacement fields (thick lines) at various measurement points for thrust faults at depths of $x_3' = 10$ km and 1 km, respectively. Aside from some details, the overall characteristics are similar to those of strike-slip faults. An interesting feature is that for the thrust fault at $x_3' = 1$ km, shown in Fig. 7.67, there are two prominent pulse-like signals between 20 and 40 s. These signals also appear in the case of the left-lateral strike-slip fault, although they are less pronounced (see Fig. 7.64). Based on the particle motion in Fig. 7.65b, these are identified as Rayleigh waves.

[39] The P axis runs northeast-southwest, while the T axis runs northwest-southeast.

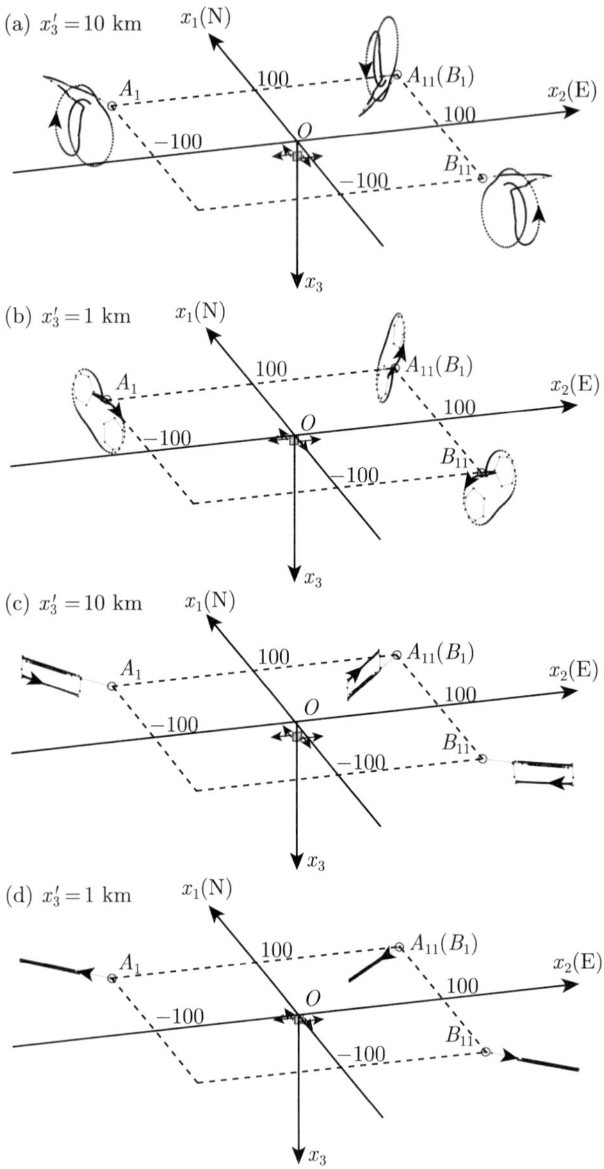

Fig. 7.65 Particle motion trajectories at A_1, A_{11}, and B_{11} caused by left-lateral strike-slip source. **a** Half-space model, source depth $x_3' = 10$ km; **b** Half-space model, source depth $x_3' = 1$ km; **c** Infinite space model, source depth $x_3' = 10$ km; **d** Infinite space model, source depth $x_3' = 1$ km. The fault parameters are $(\phi_s, \delta, \lambda) = (90°, 90°, 0°)$. The particle motion amplitude in (**a**) is magnified five times that in (**b**), and the particle motion amplitudes in (**c**) and (**d**) are magnified three times that in (**a**)

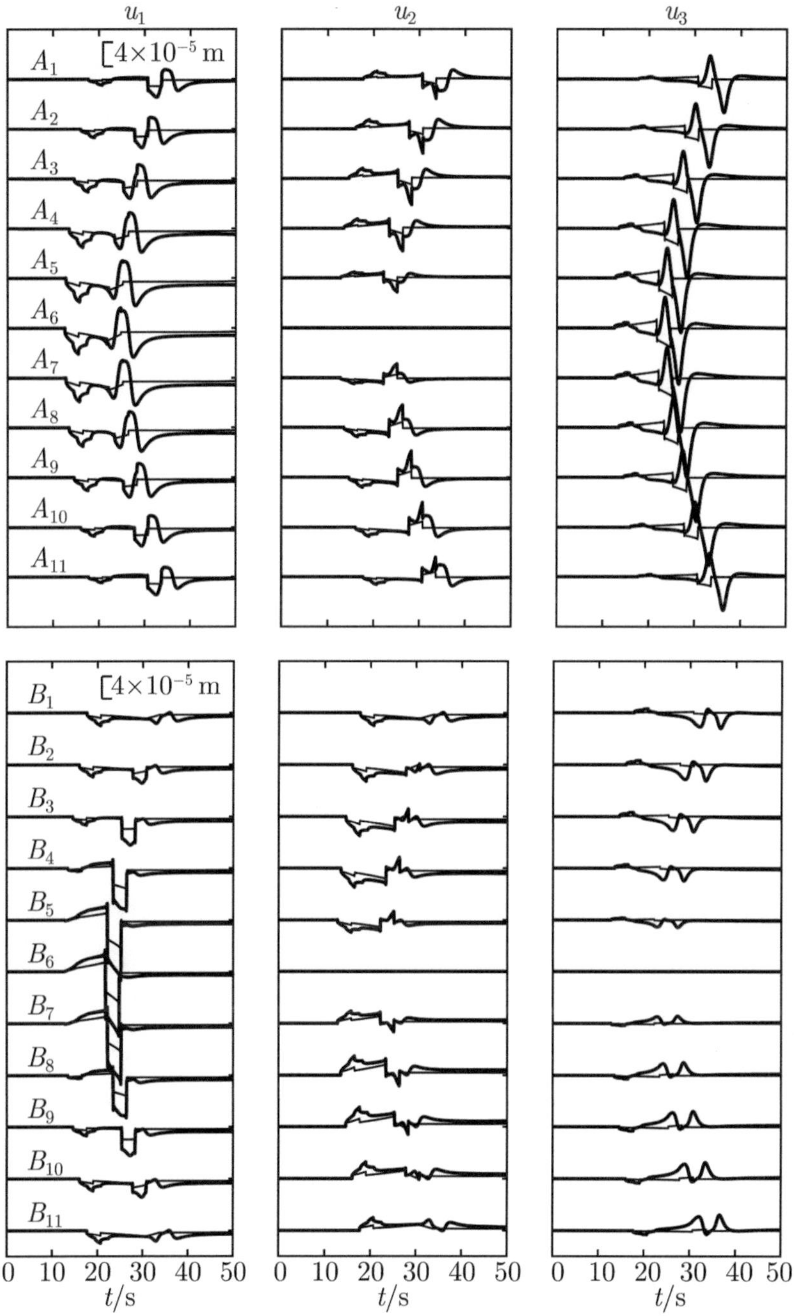

Fig. 7.66 Displacements at A_i and B_i induced by thrust source at $x_3' = 10$ km. The bold and thin lines represent the results of the half-space and infinite space model, respectively. $(\phi_s, \delta, \lambda) = (90°, 30°, 90°)$. For the distribution of A_i and B_i $(i = 1, 2, \ldots, 11)$, see Fig. 4.29

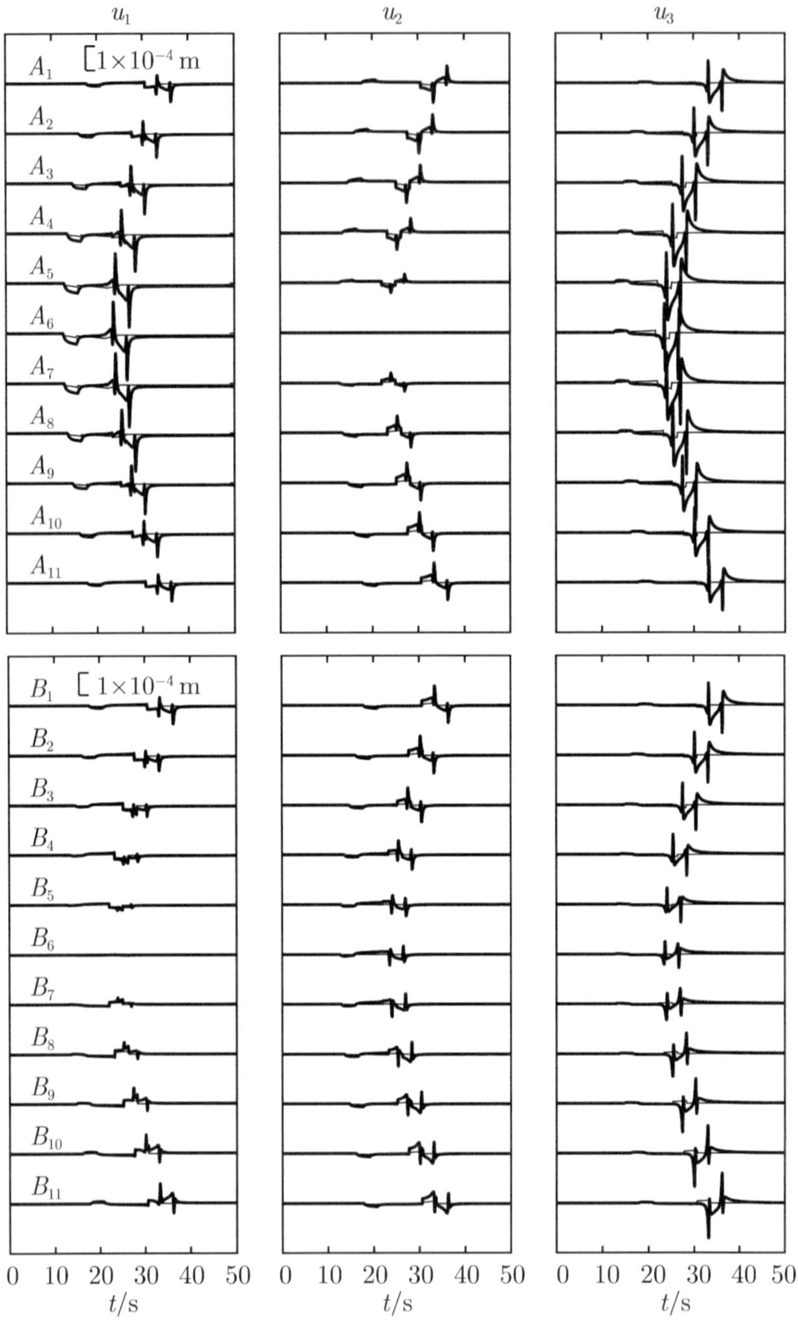

Fig. 7.67 Displacements at A_i and B_i induced by thrust source at $x_3' = 1$ km. The same as in Fig. 7.66

Figure 7.68 shows the particle motion trajectories at points A_1, A_{11} (B_1), and B_{11} for thrust faults at different source depths in two medium models. Aside from some details, the overall characteristics are similar to those of strike-slip faults. Note that for thrust faults, the equivalent body force is in the x_1x_3 plane, so the P and T axes are also in this plane. The two prominent circles in Fig. 7.68b (indicated by the diagonal arrows) correspond to the pulse signals mentioned above. Due to their short duration, the large distance between two adjacent black points in the particle motion trajectory indicates that the particle passed through these circles in a very short time.

By examining the displacement fields and particle motion trajectories for left-lateral strike-slip and thrust faults, we can understand that surface seismic activity caused by the source is complex in its transient behavior and has a much larger amplitude compared to static displacement. Furthermore, the influence of the surface is significant, especially at shallow source depths, and this is prominently reflected in the Rayleigh waves observed at fixed surface points.

7.4.3 Displacement Generated by Finite-Sized Dislocation Source

If the fault extends over a sufficiently large area and the observation point is not too far from the fault, the finite size effects of the fault cannot be ignored. In Chap. 4, we discussed the displacement field generated by a finite-sized dislocation source in an infinite space in the context of seismic waves (see Sect. 4.7.2). As a counterpart to the infinite space model, this section examines the seismic waves generated by faults of the same size in a half-space model. The presence of a free surface breaks the spatial symmetry, making the half-space fault problem dependent on the burial depth h[40] and the direction of slip. Faults that extend to the surface can cause significant surface rupture, so studying the surface displacement generated by faults directly intersecting the surface is of particular importance. Therefore, this section considers the four fault models shown in Fig. 7.69: left-lateral strike-slip and thrust faults with burial depths of $h = 10$ km and 0 km.

Additionally, the issue of finite-sized fault rupture inevitably involves a crucial parameter: the rupture propagation speed v_r. In most cases, we study a unilateral rupture with a constant propagation speed of $v_r = 3$ km/s. For a strike-slip fault with a burial depth of $h = 0$ km, we will also examine an interesting scenario where the rupture speed abruptly jumps from 3 to 5 km/s. Since the shear wave speed is set at $\beta = 4.62$ km/s, a propagation speed of 5 km/s exceeds the shear wave speed, which

[40] Here, we define the *burial depth* of the fault as the vertical distance from the upper edge of the fault to the surface.

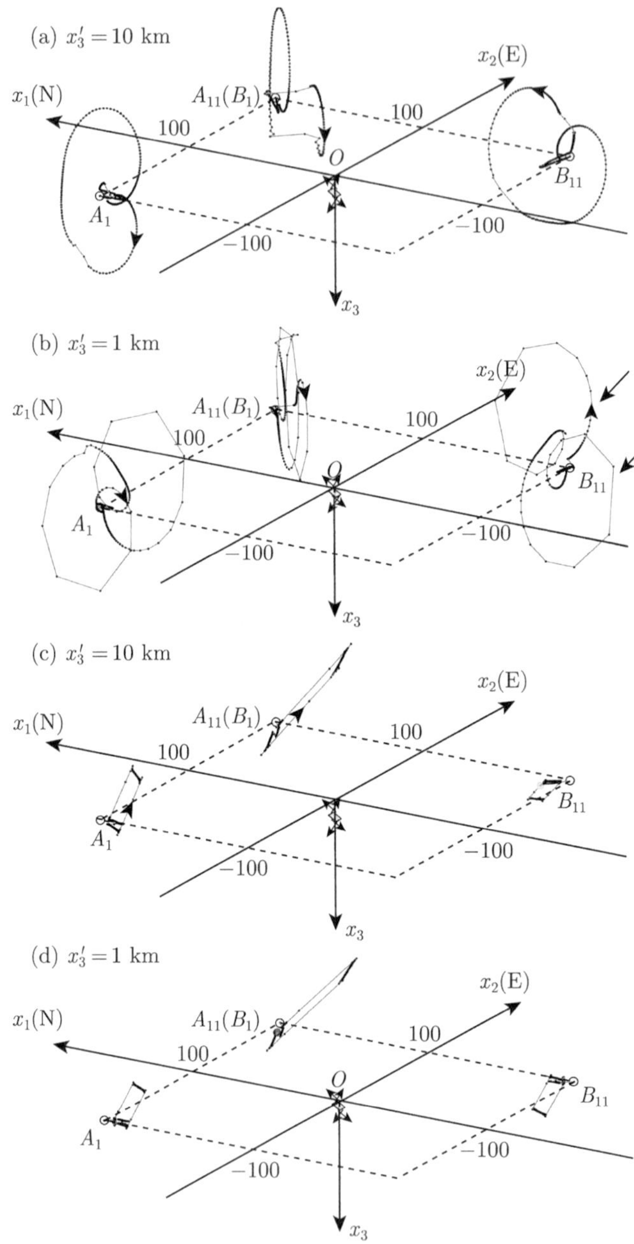

Fig. 7.68 Particle motion trajectories at A_1, A_{11}, and B_{11} caused by thrust source. The same as in Fig. 7.65, except that here, for a thrust point source, $(\phi_s, \delta, \lambda) = (90°, 30°, 90°)$

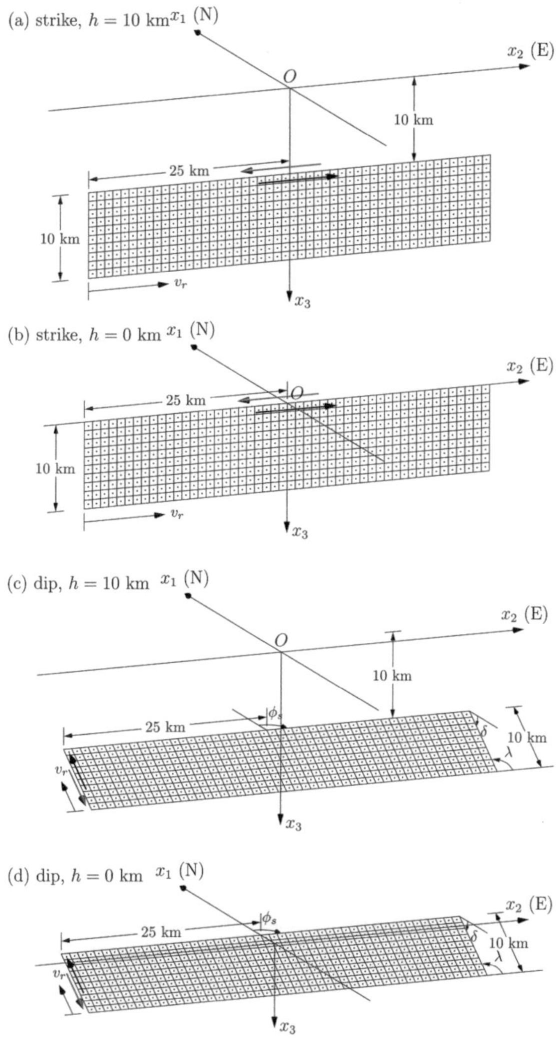

Fig. 7.69 Diagram of finite-sized fault model. **a** Left-lateral strike-slip fault at a depth of $h = 10$ km; **b** Left-lateral strike-slip fault at the surface ($h = 0$ km); **c** Thrust fault at a depth of $h = 10$ km; **d** Thrust fault at the surface ($h = 0$ km). For the strike-slip faults, the angles (ϕ_s, δ, λ) are ($90°, 90°, 0°$), and for the thrust faults, they are ($90°, 30°, 90°$). The diagrams indicate the direction of fault slip, and for the thrust faults, they also show ϕ_s, δ, and λ

is referred to as *supershear rupture*.[41] We will now explore the effects of supershear rupture on the seismic wave field. Table 7.3 outlines the various fault models and associated parameters to be studied.

[41] One key question in dynamic fracture mechanics is the speed at which a rupture propagates in an elastic solid. Initially, it was believed that rupture speed would not exceed the shear wave speed. In 1979, R. J. Archuleta found evidence of supershear rupture during the M_W 6.5 Imperial Valley earthquake. This discovery faced skepticism due to limited data. In the late 1990s, the Rosakis group at Caltech showed that rupture speeds could surpass the shear wave speed and reach compressional wave speeds. Since the 1999 M_W 7.6 Izmit earthquake, numerous studies have confirmed supershear ruptures, now widely accepted by seismologists.

Table 7.3 Several fault models to be calculated and their parameters

Figure no.	Fault type	Depth h/km	$\phi_s/(°)$	$\delta/(°)$	$\lambda/(°)$	v_r/(km/s)
7.69 (a)	Strike-slip	10	90	90	0	3
7.69 (b)	Strike-slip	0	90	90	0	3
7.69 (b)	Strike-slip	0	90	90	0	$3 \rightarrow 5$
7.69 (c)	Thrust	10	90	30	90	3
7.69 (d)	Thrust	0	90	30	90	3

7.4.3.1 Left-Lateral Strike-Slip Fault

For a left-lateral strike-slip fault of 50 km × 10 km, it is divided into 50 × 10 sub-faults, each with a side length of 1 km. The point source of each sub-fault is represented by its center point, and the final displacement is the sum of the displacement fields produced by all sub-faults. Assuming the rupture starts from the leftmost column of sub-faults and propagates at a constant speed of $v_r = 3$ km/s. The displacement fields at points A_i and B_i ($i = 1, 2, \ldots, 11$), shown in Figs. 7.70 and 7.71, are provided for the left-lateral strike-slip fault with a depth of $h = 10$ km (Fig. 7.69a) and $h = 0$ km (Fig. 7.69b). For comparison, the results for an infinite space model are also plotted in thin lines to highlight the effects of the free surface.

In Fig. 7.70, it is evident that the displacement components in the half-space model with a fault depth of $h = 10$ km are significantly larger compared to the infinite space model. The horizontal displacements u_1 and u_2 in the half-space model behave similarly to the infinite space model but with approximately double the amplitude. The vertical displacement u_3 shows a much more substantial increase in amplitude. This is because the free surface reduces the resistance to the motion of particles in the vertical direction in the half-space model.

Figure 7.71 compares the displacements for $h = 0$ km (thick lines) and $h = 10$ km (thin lines). As the fault depth becomes shallower, the horizontal displacements u_1 and u_2 slightly increase at observation points A_i ($i = 7, 8, \ldots, 11$) and B_i ($i = 2, 3, \ldots, 11$). The vertical displacement u_3 increases significantly. Even for points A_i ($i = 1, 2, \ldots, 6$) parallel to the fault, where horizontal displacements show negligible change, the u_3 displacement amplitude roughly doubles. This indicates that a strike-slip fault intersecting the surface generates significantly greater vertical displacements compared to a buried fault. From a practical perspective, this type of earthquake rupture is more hazardous and requires special attention.

It's worth noting that a noticeable "jitter" in the displacement components was observed at specific time intervals for certain measurement points. For instance, the u_2 component at points A_4 to A_8 between 20 and 40 s in Fig. 7.71 shows such jitter. Similar phenomena were discussed in Sect. 4.7.2 regarding the seismic wave field generated by a finite-scale fault in an infinite space. It was pointed out that this is due to insufficient fault grid resolution, and examples showed that refining the fault grid further reduces the jitter.

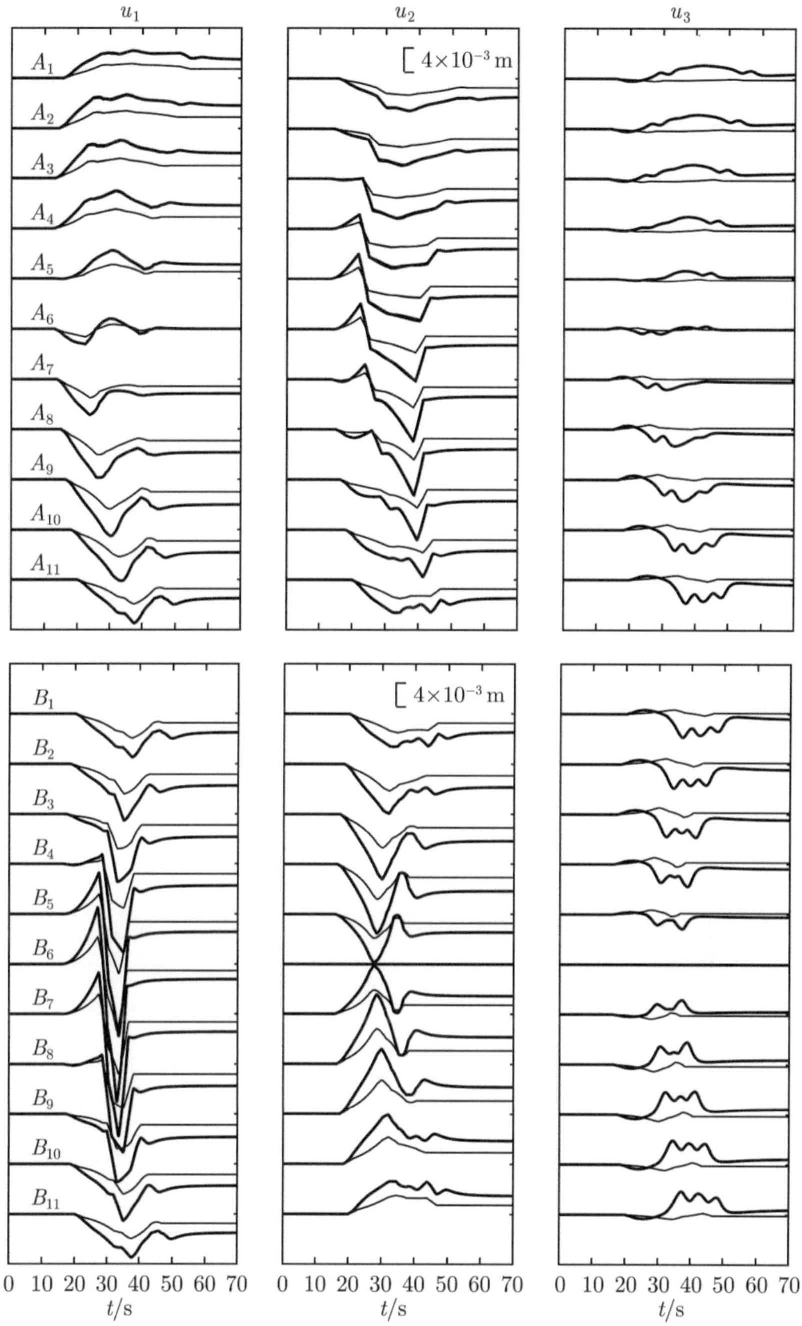

Fig. 7.70 Displacement caused by a left-lateral strike-slip fault at depth of $h = 10$ km. The bold lines represent the displacement components at points A_i and B_i; the fine lines show the corresponding results for the infinite space model. The fault dimensions are 50 km × 10 km. The rupture initiates on the left side and propagates to the right at a speed of $v_r = 3$ km/s

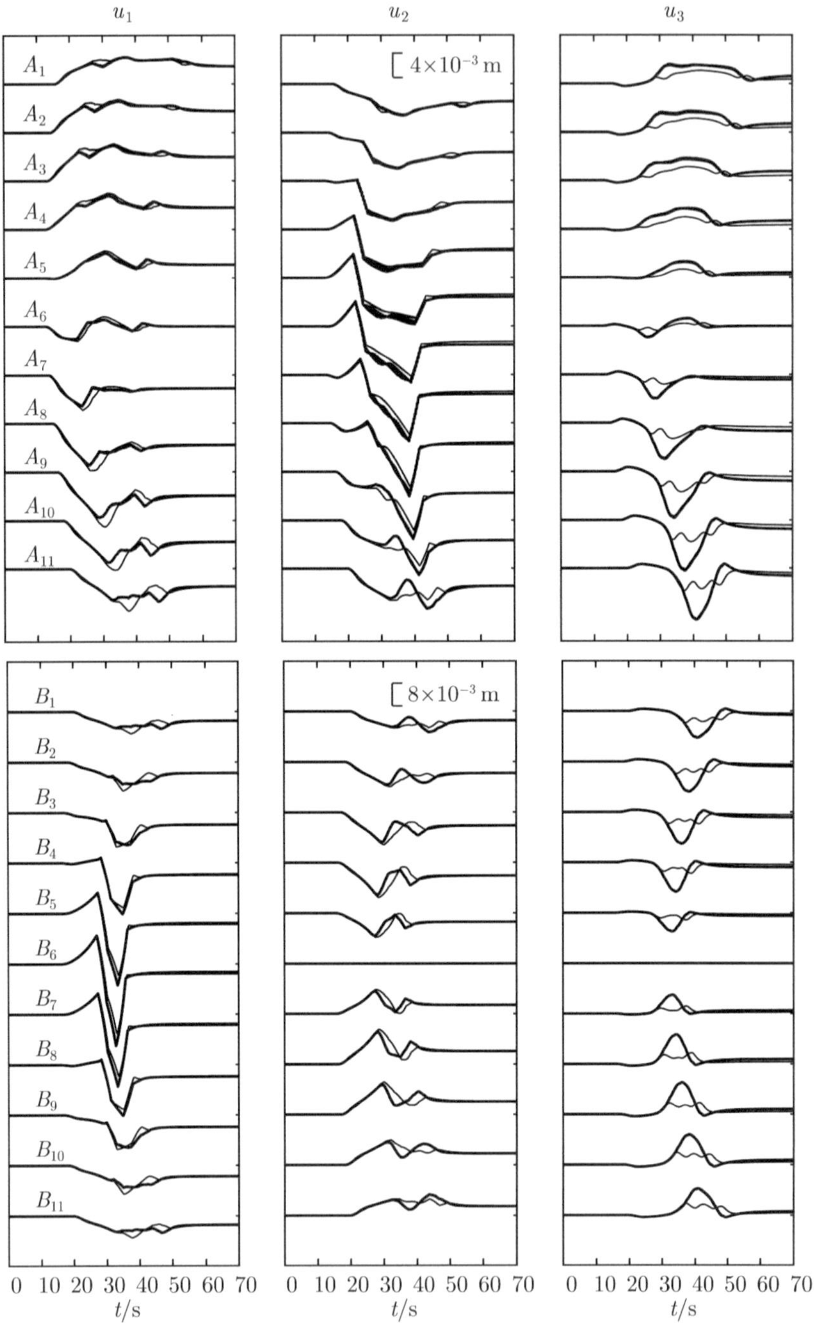

Fig. 7.71 Displacement caused by a left-lateral strike-slip fault at depth of $h = 0$ km (1). The bold lines represent the displacement components at points A_i and B_i; The fine lines represent the corresponding results for a fault at a depth of $h = 10$ km in a half-space medium model. The fault dimensions are 50 km × 10 km. The rupture initiates on the left side and propagates to the right at a speed of $v_r = 3$ km/s

Figure 7.72 vividly illustrates the particle motion trajectories at A_1, A_{11} (B_1), and B_{11} due to left-lateral strike-slip faults with different burial depths in the half-space model, all using the same magnification. In the infinite space model, changing the fault depth h merely shifts the spatial location of the source points, causing only minor waveform changes, as seen in Fig. 7.72c, d. However, in the half-space model, the impact of varying fault depths is more significant. The presence of the free surface notably amplifies the displacement components, and reducing the fault depth further increases their amplitude.

The above examples were calculated with a constant rupture speed of $v_r = 3$ km/s. However, both kinematic inversions of actual earthquakes and dynamic simulations of spontaneous rupture indicate that rupture speeds can vary during an earthquake. Particularly, when the rupture speed transitions from below the Rayleigh wave speed to above the shear wave speed, it can generate high-frequency seismic waves. Thus, it is necessary to study the impact of changing rupture speeds on the seismic wave field. We continue by considering a model with varying rupture speeds. As shown in Fig. 7.69b, the rupture propagates unilaterally from west to east. In the first 25 km, the rupture speed is $v_r = 3$ km/s, while in the next 25 km, the rupture speed increases to $v_r = 5$ km/s.

Figure 7.73 shows the displacement over time at various measurement points (bold lines) compared with the results for a constant rupture velocity of $v_r = 3$ km/s (thin lines, see Fig. 7.71). It is evident from the figure that the occurrence of supershear rupture velocity transformation makes the displacement curves sharper, indicating an increase in high-frequency components. This effect is particularly noticeable in the u_2 component observed in the forward rupture direction at points A_6 to A_{10}, and in the u_1 component observed directly ahead of the rupture at points B_3 to B_9, where very sharp peaks appear in the waveforms. Figure 7.74 shows the particle motion trajectories at A_1, A_{11} (or B_1), and B_{11} when there is a sudden change in rupture velocity. In comparison with the particle motion trajectories for a constant rupture velocity of $v_r = 3$ km/s (Fig. 7.72b), it is clear that several new abrupt changes in motion direction appear in the trajectories. These changes are caused by the sudden jump in rupture velocity along the fault plane.

The results indicate that changes in rupture velocity on the fault plane during the rupture process can lead to high-frequency components in surface displacement. Since these high-frequency components are a major cause of damage to certain buildings, further research on this topic is of particular importance for earthquake mitigation efforts.

7.4.3.2 Thrust Fault

The rocks inside the Earth are under compressive stress, and under conditions where the maximum principal stress is horizontal and the minimum principal stress is vertical, thrust faults can form. This makes thrust faults another common type of earthquake fault, especially at plate boundaries with convergent motion and within

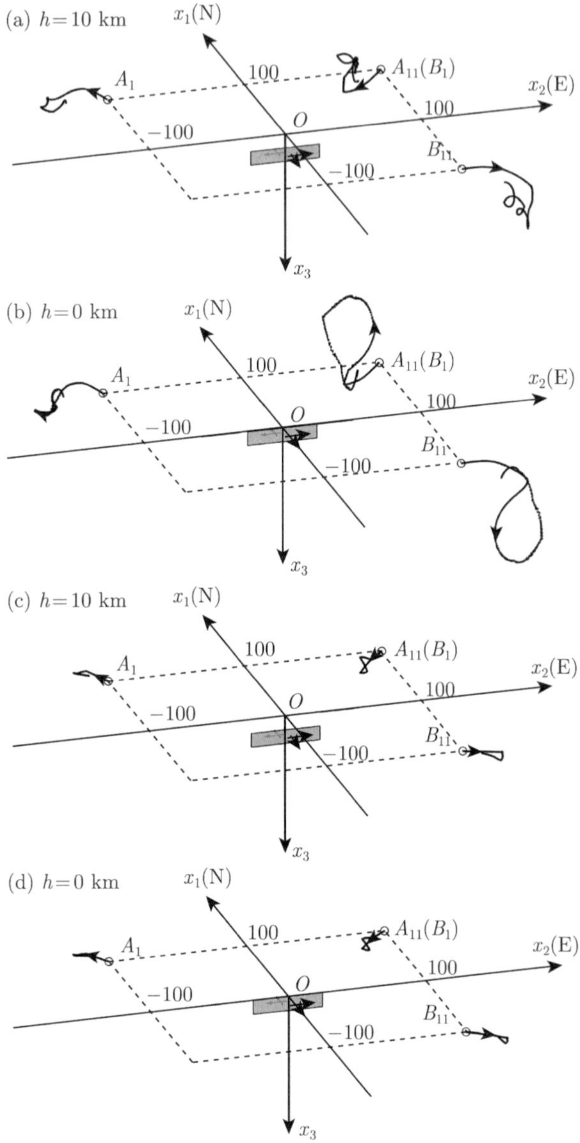

Fig. 7.72 Particle motion trajectories at A_1, A_{11}, and B_{11} caused by left-lateral strike-slip fault. **a** Half-space model with a fault depth of $h = 10$ km; **b** Half-space model with a fault depth of $h = 0$ km; **c** Full-space model with a fault depth of $h = 10$ km; **d** Full-space model with a fault depth of $h = 0$ km. The fault dimensions are 50 km × 10 km. The rupture initiates on the left side and propagates to the right at a speed of $v_r = 3$ km/s. The particle motion amplitude is magnified by the same factor in all figures

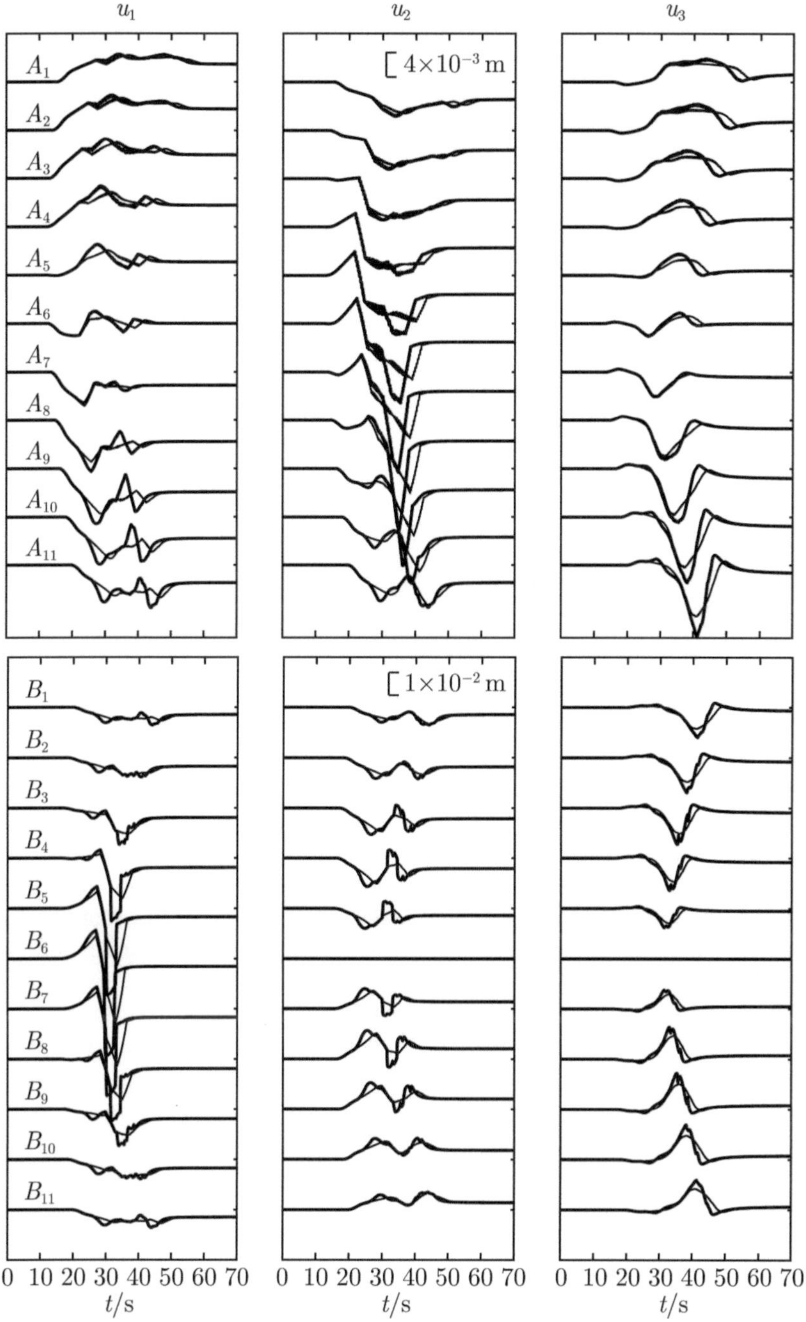

Fig. 7.73 Displacement caused by a left-lateral strike-slip fault at depth of $h = 0$ km (2). The same as in Fig. 7.71, except the fault rupture velocity v_r jumps from 3 to 5 km/s as it crosses the x_3 axis. The thin lines represent the results for a constant rupture velocity of 3 km/s

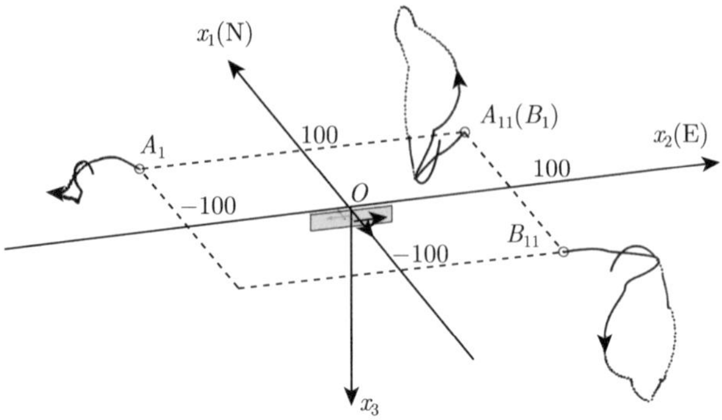

Fig. 7.74 Particle motion trajectories at A_1, A_{11}, and B_{11} caused by left-lateral strike-slip fault. The same as in Fig. 7.72b, except the rupture velocity v_r jumps from 3 to 5 km/s as it crosses the x_3 axis

intracontinental mountain belt regions. Consider the thrust faults at depths of 10 and 0 km shown in Fig. 7.69c, d, respectively. The rupture propagates upward with $v_r = 3$ km/s.

Figure 7.75 compares the displacement (thick lines) at A_i and B_i ($i = 1, 2, \ldots, 11$) caused by the thrust fault at $h = 10$ km with the corresponding results from an infinite space model (thin lines). It can be seen that the amplitude of each displacement component in the half-space model is significantly larger than in the full-space model due to the presence of the free surface, especially for the u_1 and u_3 components at measurement points A_4 to A_8 in front of the thrust fault. Once again, we see the influence of the Earth's surface on the seismic wave field.

To investigate the impact of fault depth, Fig. 7.76 compares the results for $h = 0$ km (thick lines) with those for $h = 10$ km in the half-space model (thin lines). As the fault depth decreases, there are noticeable changes in the displacement waveforms, with a significant increase in amplitude and the waveforms becoming sharper. These effects are the result of enhanced interactions between the free surface and fault rupture.

Figure 7.77 shows the particle motion trajectories at points A_1, A_{11} (B_1), and B_{11} caused by the thrust fault, comparing the results for different fault depths in both the half-space and infinite space models. Similar to strike-slip faults, comparisons between (a) and (c), as well as (b) and (d), demonstrate that the presence of the free surface significantly alters the amplitude and shape of the particle motion trajectories. Additionally, comparisons between (a) and (b) show that the shallower the fault depth, the greater the amplitude of particle vibration. It is worth noting that the "jitter" in the motion trajectories in (b) can be suppressed by increasing the number of grid divisions in the fault model.

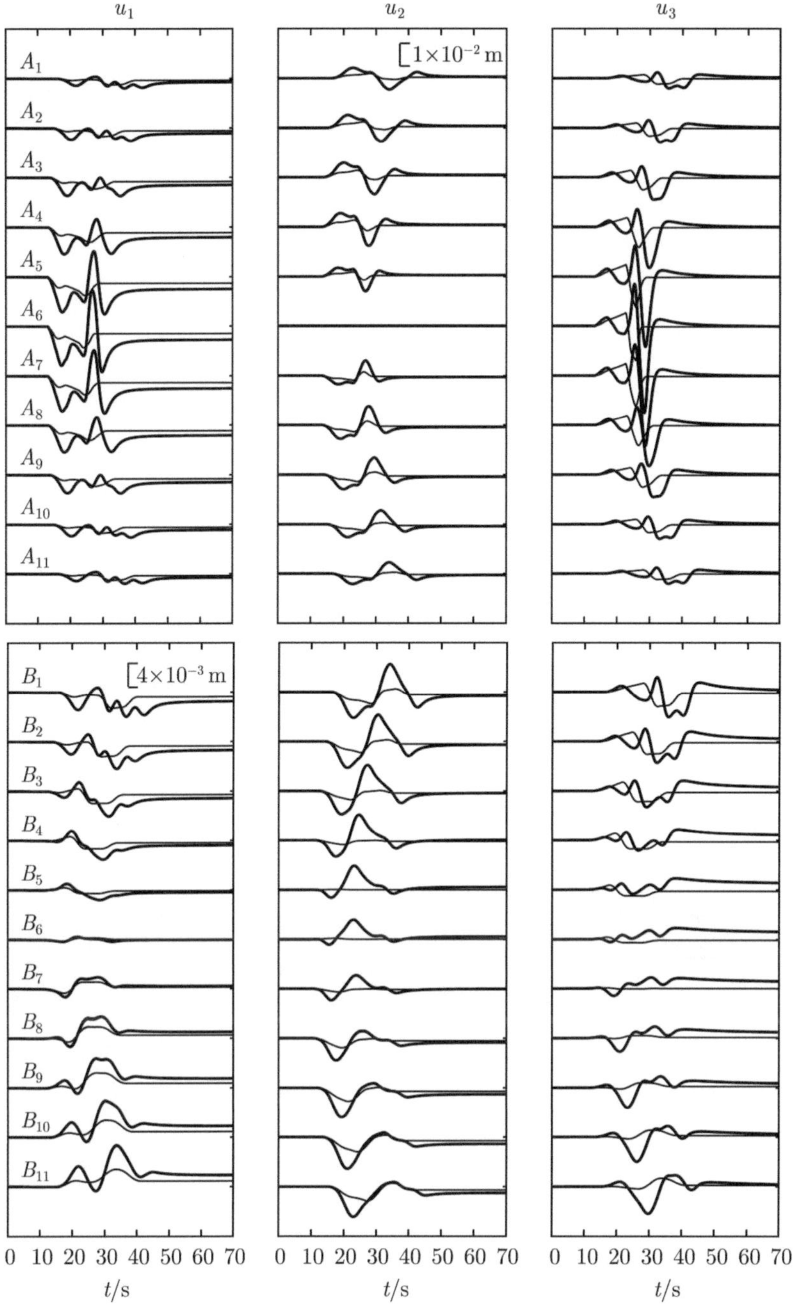

Fig. 7.75 Displacement caused by a thrust fault at depth of $h = 10$ km. The bold lines represent the displacement components at points A_i and B_i; the fine lines show the corresponding results for the infinite space model. The fault dimensions are 50 km×10 km. The rupture initiates on the left side and propagates to the right at a speed of $v_r = 3$ km/s

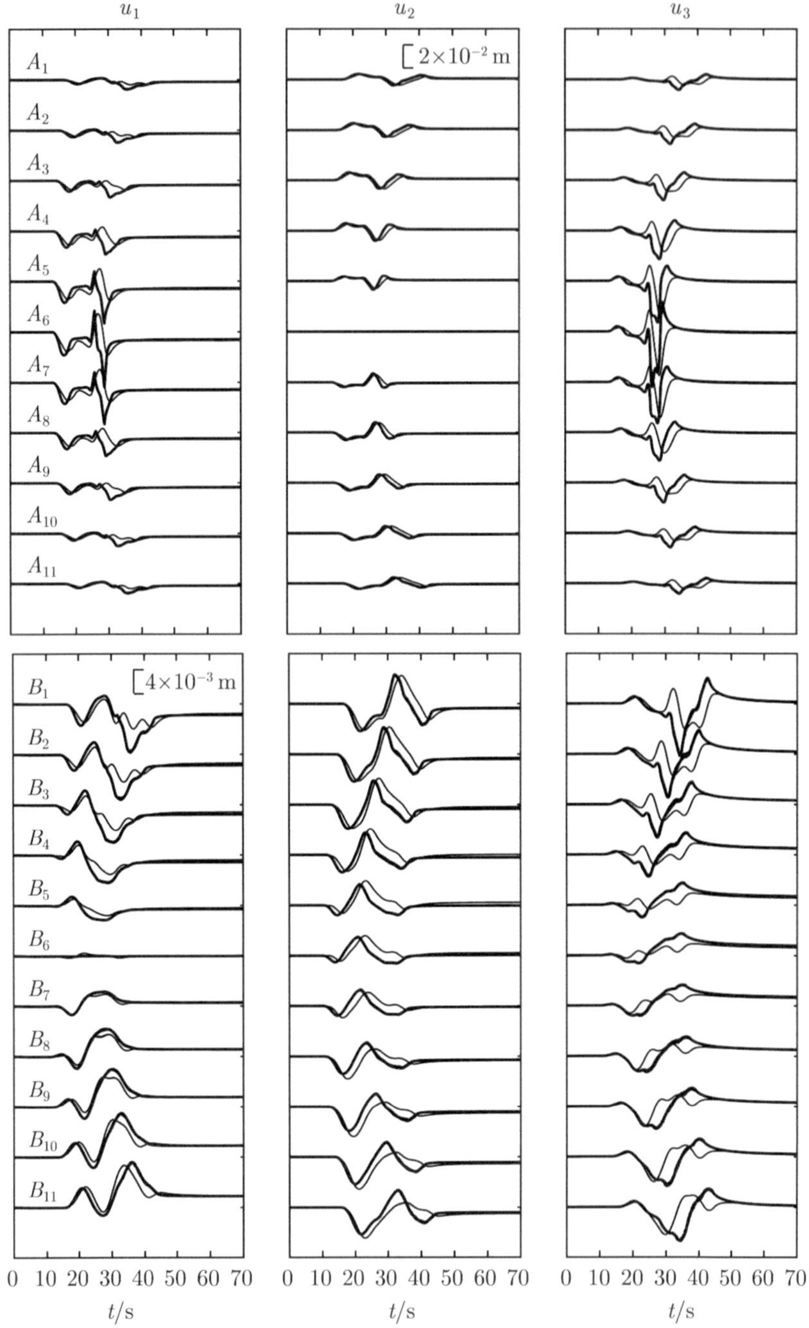

Fig. 7.76 Displacement caused by a thrust fault at depth of $h = 0$ km. The thin lines represent the corresponding results for a fault at a depth of $h = 10$ km in the half-space medium model, as shown in Fig. 7.75. The fault, which is 50 km×10 km in size, ruptures from bottom to top with a rupture velocity of $v_r = 3$ km/s

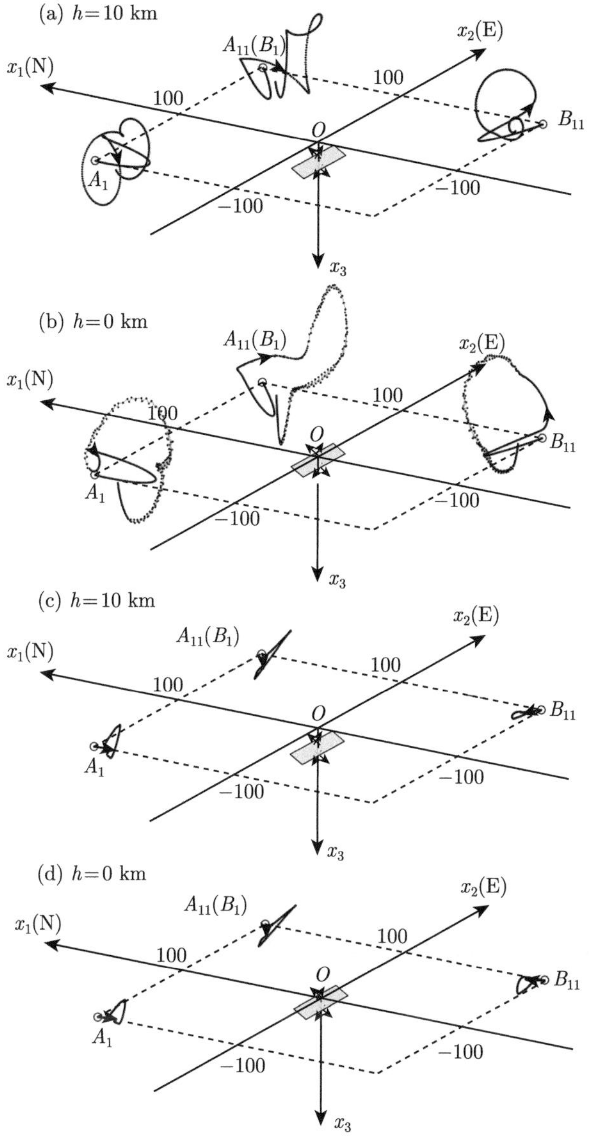

Fig. 7.77 Particle motion trajectories at A_1, A_{11}, and B_{11} caused by thrust fault. **a** Half-space model with $h = 10$ km; **b** Half-space model with $h = 0$ km; **c** Full-space model with $h = 10$ km; **d** Full-space model with $h = 0$ km. The fault size is 50 km×10 km. The fault ruptures unilaterally from bottom to top with a rupture velocity $v_r = 3$ km/s. The particle motion amplitudes in each figure are magnified by the same factor

7.5 Static Displacement and Rayleigh Wave in Lamb's Problem

In Sect. 7.4, the methods developed in Chap. 6 are applied to the numerical techniques introduced in Sect. 7.1 to compute results for various types of Lamb's problems. Specifically, in addition to the Green's function for the Lamb's problem, the study focuses on seismologically significant issues related to the seismic wave field generated by shear dislocation sources, including both point sources and finite-size sources. In Chap. 6, we addressed the special case of eigenvalue degeneracy using the $\mathcal{LP}_2$ as an example and obtained the static displacement fields generated by single force point sources and shear dislocation point sources. Additionally, we examined the Rayleigh wave problem induced by single force and dislocation sources in the frequency domain and derived the corresponding excitation formulas. This section presents the related numerical computation results. For Rayleigh wave in the frequency domain, the corresponding time-domain result can be obtained via inverse discrete Fourier transform. Since the excitation formulas only account for the contribution of the Rayleigh pole, the resulting time-domain Rayleigh wave represent the surface wave component isolated from the complete signal.

7.5.1 Static Displacement in Lamb's Problem

7.5.1.1 Static Displacement Caused by a Single Force Point Source

Equation (6.7.11) provides the static Green's function for the $\mathcal{LP}_2$. Based on this, we can determine the static surface displacement field caused by a single force in different directions.

Figure 7.78 illustrates an example of this scenario. A single force is applied at a depth of 1 km in the directions of the x_1, x_2, and x_3 axes, showing the surface deformation within a 2 km×2 km area centered at the origin.

(1) When the force is applied along the x_1 direction (see Fig. 7.78a), particles in the initial area primarily displace along the x_1 direction. In the region where $x_1 > 0$, the surface experiences expansion in the x_2 direction and elevation. Conversely, in the region where $x_1 < 0$, contraction occurs in the x_2 direction, causing surface subsidence.
(2) Due to the symmetry between the x_1 and x_2 directions, the result for a force applied along the x_2 direction is similar, as shown in Fig. 7.78b.
(3) When the force is applied along the x_3 direction (see Fig. 7.78c), particles in the initial area primarily displace downward along the x_3 direction, leading to significant surface subsidence. Additionally, due to Poisson's ratio, contraction occurs in both the x_1 and x_2 directions, reducing the initial area to a smaller region.

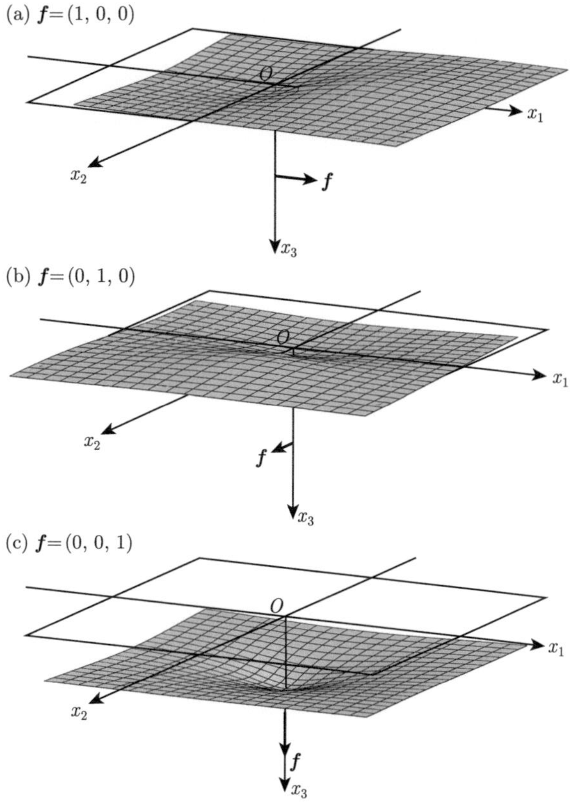

Fig. 7.78 Static surface displacement caused by single force point sources in different directions. **a** Single force in the x_1 direction; **b** Single force in the x_2 direction; **c** Single force in the x_3 direction. The force is applied at the location (0, 0, 1) km. The initial position is a 2 km×2 km square in the x_1x_2 plane

7.5.1.2 Static Displacement Caused by Dislocation Point Source

Equations (6.7.12) and (6.7.13) provide the formulas for calculating the static surface solution caused by a shear dislocation point source. Using these, one can compute the static displacement field at the surface generated by a shear dislocation point source within a half-space medium. We will again consider the examples of left-lateral strike-slip and thrust faults discussed in Sect. 7.4.2.

Examining an initial 10 km×10 km square area in the x_1x_2 plane, Fig. 7.79 illustrates the surface displacement fields caused by a left-lateral strike-slip dislocation point source at different depths. For the case with a depth of $x_3' = 1$ km (Fig. 7.79a), the fault is close to the surface, causing a significant static displacement mainly concentrated within 3 km of the origin. Horizontally, the deformation expands in the northwest and southeast quadrants and contracts in the northeast and southwest quadrants. Vertically, on a broader scale, the ground subsides in the northwest and southeast quadrants, while it uplifts in the northeast and southwest quadrants, consistent with the results of an infinite space model (Fig. 4.32). However, near the origin, there are distinct "bulges" in the northwest and southeast directions and noticeable

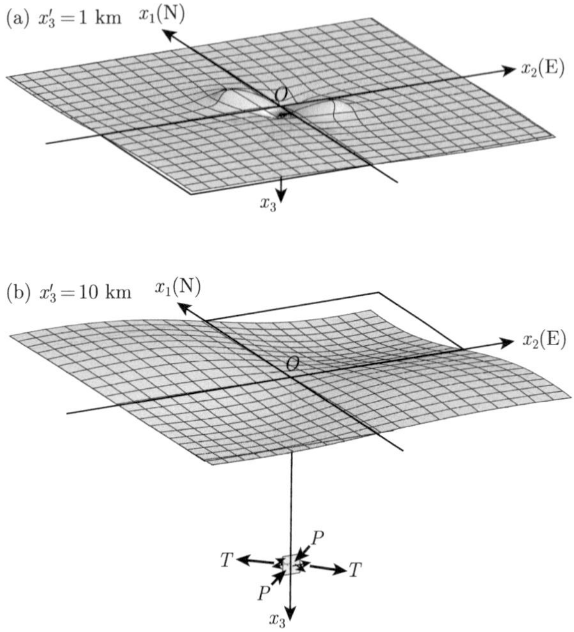

Fig. 7.79 Static surface displacement caused by left-lateral strike-slip sources at different depths. **a** Result of $x_3' = 1$ km; **b** Result of $x_3' = 10$ km. The initial position is a 10 km × 10 km square in the x_1x_2-plane. The figure indicates the equivalent body forces and the P and T axes corresponding to the left-lateral strike-slip fault. $(\phi_s, \delta, \lambda) = (90°, 90°, 0°)$. To enhance clarity, the static displacement field in (**b**) is magnified by a factor of 10 compared to (**a**)

local depressions in the northeast and southwest directions. This characteristic is due to the presence of a free surface and does not exist in the infinite space medium (Fig. 4.32). For the case shown in Fig. 7.79b with $x_3' = 10$ km, the surface motion caused by the left-lateral shear dislocation is significantly weakened (note that the surface deformation in Fig. 7.79b is magnified ten times compared to (a)), with similar overall features but a considerably larger affected area. Overall, the T and P axes of the left-lateral strike-slip fault lie in the horizontal plane parallel to the surface, mainly causing horizontal deformation.

The surface displacement fields caused by thrust dislocation point sources at different depths, shown in Fig. 7.80, are distinctly different from those caused by strike-slip dislocation point sources. In Fig. 7.80a, a thrust dislocation point source at $x_3' = 1$ km results in significant vertical displacement at the surface. The current model considers a thrust fault with a dip angle $\delta = 30°$, where the T axis slightly tilts northward. Consequently, the uplifted surface area also shifts northward, creating a noticeable depression on the southern side. For deeper point sources, as illustrated in Fig. 7.80b with $x_3' = 10$ km, the affected area is much broader, with a significantly reduced magnitude of displacement (the surface deformation in Fig. 7.80b is magnified tenfold compared to (a)), and the features become much more subdued.

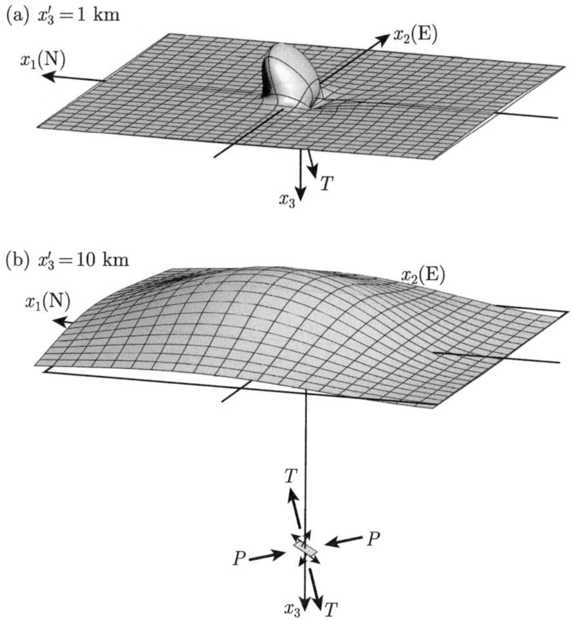

Fig. 7.80 Static surface displacement caused by thrust sources at different depths. The same as in Fig. 7.79, except $(\phi_s, \delta, \lambda) = (90°, 30°, 90°)$

7.5.1.3 Static Displacement Caused by Finite-Sized Dislocation Sources

Building on the results for the dislocation point sources mentioned above, we can further investigate the characteristics of the static displacement field caused by finite-sized dislocation sources. We will examine how the finite size effects influence the characteristics of the static displacement field. Using the example of a 50 km × 10 km vertical left-lateral strike-slip fault and a thrust fault from Sect. 7.4.3, we study the static displacement produced in an initial area of 100 km × 100 km centered at the origin under the influence of finite-sized faults.

Figure 7.81 shows the surface displacement fields caused by left-lateral strike-slip faults at different burial depths ($h = 0$ km and 10 km). Unlike the results for point sources presented in Sect. 7.5.1.2, the results for different fault burial depths h use the same magnification factor. Clearly, the area of significant surface deformation caused by finite-sized faults at different depths is much less variable than in the case of point sources. For faults that intersect directly with the surface, as seen in Fig. 7.81a, the horizontal deformation near the fault is more pronounced and features sharper characteristics compared to the case with a burial depth of $h = 10$ km. The results for thrust faults shown in Fig. 7.82 exhibit similar characteristics, with the difference being that thrust faults primarily cause vertical displacement rather than horizontal displacement.

To summarize the discussion on the static displacement field in a half-space medium, although the static displacement field information is included in the dynamic displacement field calculations of the Lamb's problem, there are situations where

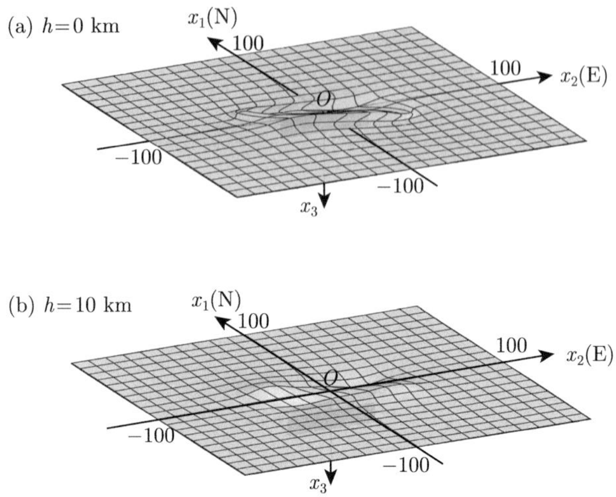

Fig. 7.81 Surface displacement caused by 50 km × 10 km left-lateral strike-slip faults. **a** Result of $h = 0$ km; **b** Result of $h = 10$ km. The initial position is a 100 km × 100 km square located in the x_1x_2-plane. $(\phi_s, \delta, \lambda) = (90°, 90°, 0°)$. The magnification factors for the static displacements in (**a**) and (**b**) are the same

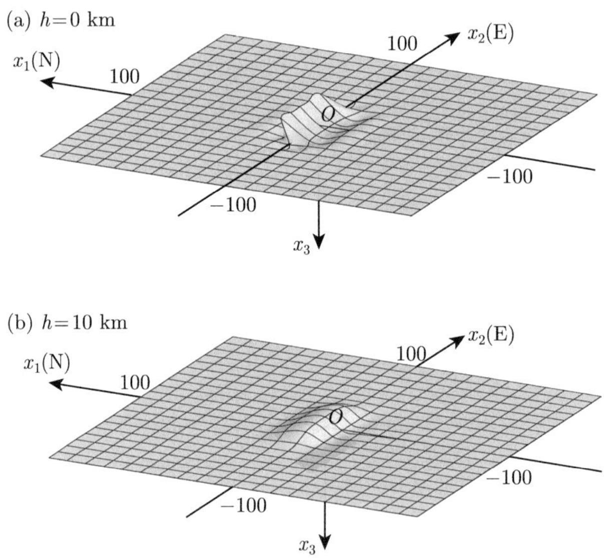

Fig. 7.82 Surface displacement caused by 50 km × 10 km thrust faults. The same as in Fig. 7.81, except $(\phi_s, \delta, \lambda) = (90°, 30°, 90°)$

static information is particularly important. For instance, when using GPS to study seismological issues, we can directly calculate the static displacement field using Eqs. (6.7.11) to (6.7.13) from Chap. 6. Exploring the characteristics of static displacement fields generated by different types of fundamental faults will help us better utilize relevant data to obtain information about the seismic source.

7.5.2 Rayleigh Wave in Lamb's Problem: A Frequency-Domain Analysis

In Sect. 6.5.3, we previously separated the contribution of Rayleigh wave by analyzing the integral over the complex k based on the Green's function in the frequency domain, as shown in Eq. (6.5.16). From this result, we derived compact expressions for the far-field approximation (Eqs. (6.5.21) to (6.5.23)) and the near-field result (Eq. (6.5.24)). Additionally, by performing spatial derivatives, we further obtained the expression for Rayleigh wave caused by a shear dislocation source (Eq. (6.6.8)). In this section, we will use these formulas to demonstrate the results through numerical calculations and provide corresponding analyses. First, we will study the far-field and near-field Rayleigh waves of a single force point source in the frequency domain, as well as the Rayleigh wave of a shear dislocation point source. Then, we will transform the frequency domain results into the time domain to examine the characteristics of Rayleigh wave in the time domain.

7.5.2.1 Far-Field Rayleigh Wave Induced by Single Force in Frequency Domain

Equation (6.5.21) provides the solution for Rayleigh wave in the far field when the source is on the surface and the observation point varies with depth. Using this equation, we can calculate the particle motion of Rayleigh wave caused by horizontal and vertical forces and analyze their characteristics.

Figures 7.83 and 7.84 illustrate the particle motion trajectories of far-field Rayleigh waves at different depths, resulting from horizontal and vertical forces applied at the surface. The numbers in the figures represent the depth x_3 (km). Except for the slightly smaller vertical and horizontal displacements caused by the horizontal force, the results are generally consistent with those of the vertical force. Near the surface, the particle motion exhibits the classic retrograde elliptical pattern. As depth increases, the minor axis of the ellipse decreases, and around 0.8 km, the horizontal displacement reverses its sign. At depths greater than 0.8 km, the particle motion changes to a prograde elliptical pattern. With further depth, both the major and minor axes of the ellipse continue to diminish. Figure 7.85a shows the variation with depth of several relevant displacement components $\tilde{G}_{11}^R$ and $\tilde{G}_{31}^R$, as well as $\tilde{G}_{13}^R$ and $\tilde{G}_{33}^R$, with their magnitudes normalized to their maximum values. Depth is normalized by

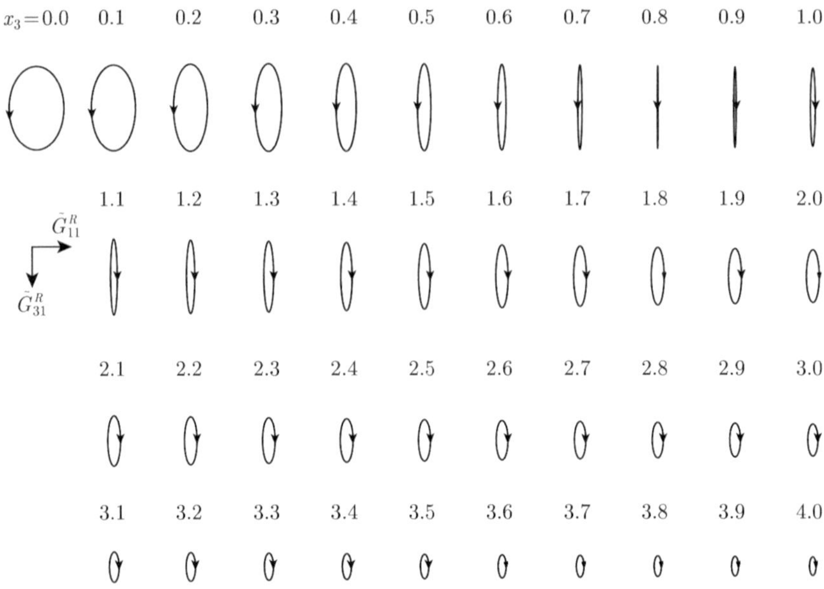

Fig. 7.83 Particle motion trajectory of far-field Rayleigh waves induced by a horizontal force. The source is located at the surface, and the numbers represent the depth of the observation point x_3 (km)

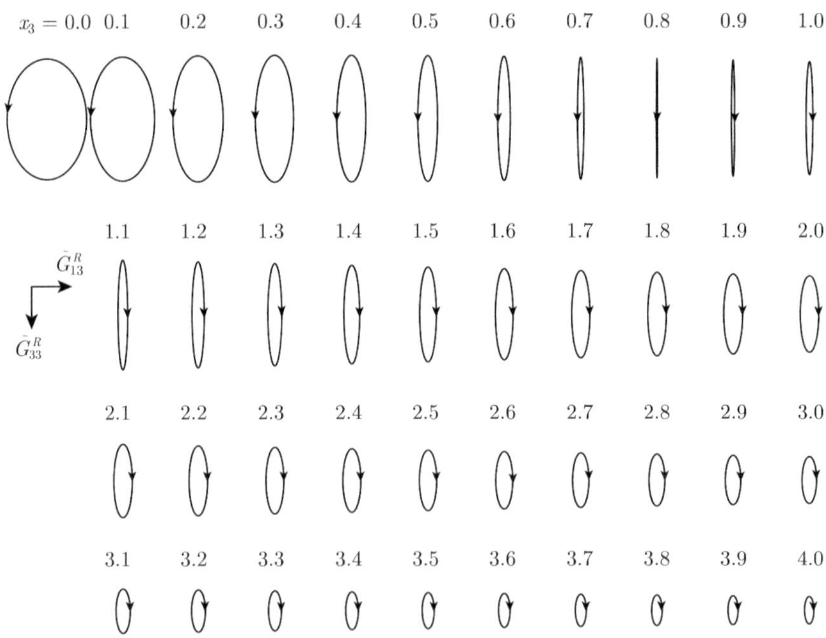

Fig. 7.84 Particle motion trajectory of far-field Rayleigh waves induced by a vertical force. The same as in Fig. 7.83

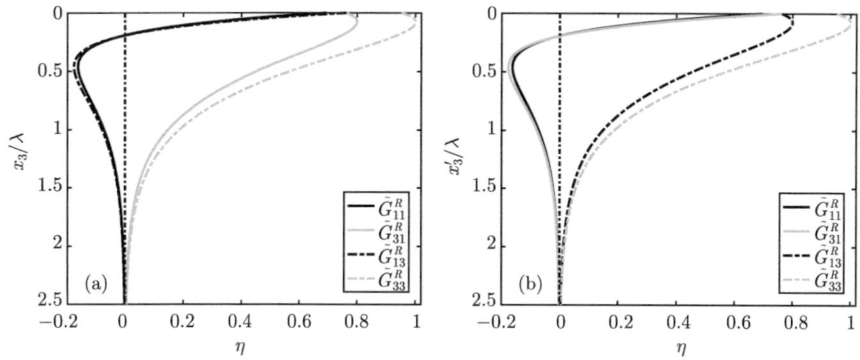

Fig. 7.85 Variation of Rayleigh wave displacement components with depth. **a** The variation of Rayleigh wave displacement components with depth, caused by horizontal and vertical forces applied at the surface; **b** The variation of surface Rayleigh wave displacement components with depth, resulting from horizontal and vertical forces applied at different depths. The horizontal axis represents the relative amplitude η (normalized by the maximum amplitudes of $\tilde{G}_{11}^R$, $\tilde{G}_{31}^R$, $\tilde{G}_{13}^R$, and $\tilde{G}_{33}^R$), and the vertical axis shows the depth normalized by the wavelength λ

the wavelength $\lambda = 2\pi/\kappa$. From Fig. 7.85a, the variation of each displacement component with depth is clearly observable. The vertical component increases initially with depth, reaches a maximum, and then decreases, while the horizontal component monotonically decreases to zero, reverses direction, and then increases before decreasing again.

The characteristics revealed by Figs. 7.83, 7.84 and 7.85a are consistent with the classical features of plane Rayleigh wave in seismology, based on the assumption of plane waves. An interesting feature is the transition from retrograde to prograde elliptical motion with increasing depth. The position of this transition point can be theoretically calculated. At the transition point, $\tilde{G}_{11}^R$ and $\tilde{G}_{13}^R$ are both zero. From Eqs. (6.5.20a) and (6.5.20c), we have

$$2\kappa^2 \mathrm{e}^{-\gamma x_3} = \chi \mathrm{e}^{-\nu x_3}, \quad \text{and} \quad \chi \mathrm{e}^{-\gamma x_3} = 2\nu\gamma \mathrm{e}^{-\nu x_3}$$

Let x_R be the real root greater than 1 derived from the Rayleigh function in Eq. (B.1). The depth normalized by the wavelength is given by

$$\bar{x}_3 = \frac{x_3}{\lambda} = \frac{x_3\kappa}{2\pi} = \frac{x_3\omega\sqrt{x_R}}{2\pi\beta}$$

From the above two equations, we obtain

$$\bar{x}_3 = \frac{\sqrt{x_R}\ln\dfrac{2x_R}{2x_R - 1}}{2\pi\left(\sqrt{x_R - m} - \sqrt{x_R - 1}\right)}, \quad \text{and} \quad \bar{x}_3 = \frac{\sqrt{x_R}\ln\dfrac{2x_R - 1}{2\sqrt{x_R - m}\sqrt{x_R - 1}}}{2\pi\left(\sqrt{x_R - m} - \sqrt{x_R - 1}\right)}$$

where $m = \beta^2/\alpha^2$. Since x_R is a root of the Rayleigh function (Eq. (B.1)), we have

$$(2x_R - 1)^2 - 4x_R\sqrt{x_R - m}\sqrt{x_R - 1} = 0$$

The two $\bar{x}_3$ values are equal. For example, for a Poisson medium where $m = 1/3$ and $x_R \approx 1.183$, we calculate $\bar{x}_3 \approx 0.19$. This indicates that at approximately 0.19 times the wavelength depth, $\tilde{G}^R_{11}$ and $\tilde{G}^R_{13}$ change sign (see Fig. 7.85a), causing the direction of elliptical motion to reverse (corresponding to $x_3 \approx 0.82$ km, as seen in Figs. 7.83 and 7.84).

Equation (6.5.22) provides the frequency-domain far-field Rayleigh wave expression for observation points at the surface, with the force applied at varying depths. This allows us to similarly consider the particle motion trajectories of far-field Rayleigh wave at the surface, induced by horizontal and vertical forces at different depths.

Figures 7.86 and 7.87 respectively show the corresponding results, with the numbers in the figures representing different source depths x_3' (km). Unlike the previous case where the source was fixed at the surface and the observation point varied with depth, here the observation point is fixed at the surface while the source depth varies. The Rayleigh wave particle motion trajectories are retrograde ellipses for all cases. The surface displacement caused by horizontal forces at different depths shows that the shape of the ellipses remains largely unchanged, but their size varies significantly. As the depth of the source increases from the surface, the ellipses gradually shrink to a point at a depth of about 0.8 km (becoming stationary), then enlarge before shrinking again as the source depth continues to increase. In contrast, for vertical forces at different depths, the surface displacement ellipses initially experience a brief enlargement before decreasing monotonically with increasing source depth. Figure 7.85b illustrates the relative amplitude variations of the displacement components normalized by wavelength as depth increases. Both the displacement components $\tilde{G}^R_{11}$ and $\tilde{G}^R_{31}$ due to horizontal forces, and $\tilde{G}^R_{13}$ and $\tilde{G}^R_{33}$ due to vertical forces, have the same signs, so the direction of the elliptical motion does not change. However, the displacement components due to horizontal forces decrease monotonically to zero, then increase and decrease again, while those due to vertical forces increase initially and then gradually decrease. These patterns explain the characteristics observed in Figs. 7.86 and 7.87.

In addition to the scenarios where either the source or the observation point is fixed while the other varies, Eq. (6.5.23) also provides solutions for cases where both the source and observation point positions can vary simultaneously. Using this equation, we can study the far-field Rayleigh wave particle motion trajectories caused by horizontal and vertical forces applied at different source depths and observed at varying depths. When the source is fixed, the far-field Rayleigh wave particle motion trajectories at different depths resemble the images shown in Figs. 7.83 and 7.84. Conversely, for a fixed observation point, the effects of sources at varying depths are similar to the results in Figs. 7.86 and 7.87, differing only in magnitude. Therefore, those cases are not displayed again here.

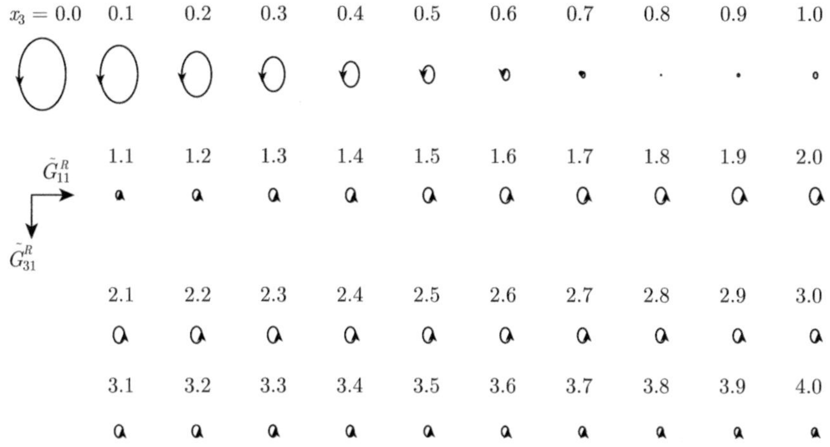

Fig. 7.86 Particle motion trajectories of far-field surface Rayleigh wave caused by horizontal forces. Observation point at the surface, numbers represent the source depth x_3' (km)

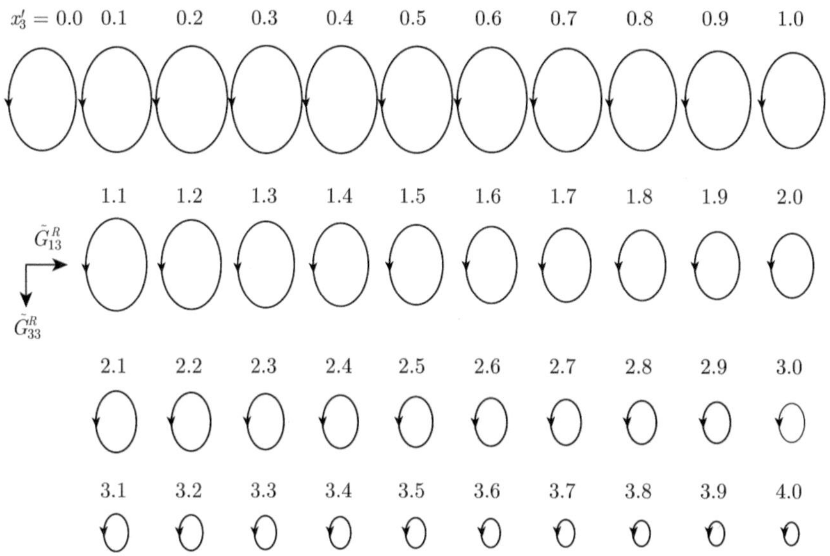

Fig. 7.87 Particle motion trajectories of far-field surface Rayleigh wave caused by vertical forces. The same as in Fig. 7.86

7.5.2.2 Near-Field Rayleigh Waves in Frequency Domain Induced by Point Force Source

After all, the far-field case is just a special instance where the asymptotic expression of the Hankel function can provide a relatively compact form. However, for the near-field case, the asymptotic expression is no longer valid. Equation (6.5.24) offers a general form, allowing for precise calculation of near-field Rayleigh wave solutions.

Figures 7.88 and 7.89 illustrate the particle motion trajectories of Rayleigh waves at different points in the near field caused by a horizontal force applied at the surface and at $x_3' = 0.25$ km, respectively. A very noticeable feature is that near the force application point, the motion trajectories are oblique ellipses with large amplitudes. As the horizontal distance from the source increases, the major axis of the ellipses becomes more vertical and the amplitude gradually decreases. Near the surface, the direction of the ellipse's travel is retrograde, while below a depth of 0.2 wavelengths, it becomes prograde. When the horizontal force is applied underground (Fig. 7.89), the amplitude of the ellipses significantly decreases. The major axis of the ellipses is inclined in the direction of wave propagation above a depth of 0.2 wavelengths and in the opposite direction below this depth.

Figures 7.90 and 7.91 show the particle motion trajectories of Rayleigh waves at different points in the near field caused by a vertical force applied at the surface and at $x_3' = 0.25$ km, respectively. The overall features are similar to those resulting from the horizontal force shown in Figs. 7.88 and 7.89 and will not be reiterated here. However, one notable difference is that the amplitude of the particle motion trajectories caused by the vertical force decreases much more slowly with increasing

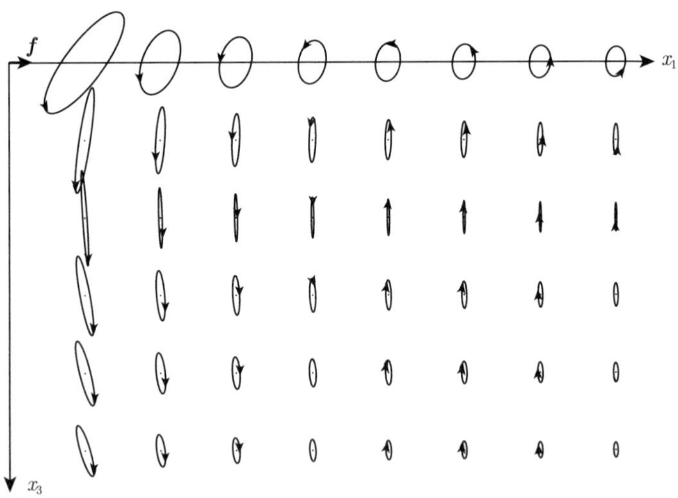

Fig. 7.88 Particle motion trajectory of near-field Rayleigh waves caused by a horizontal force (1). The force is applied at the surface. The distance between two adjacent field points is 0.5 km

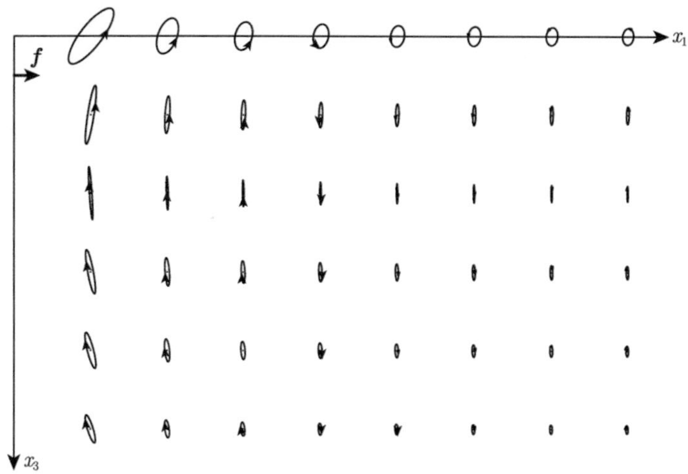

Fig. 7.89 Particle motion trajectory of near-field Rayleigh waves caused by a horizontal force (2). The force is applied at the depth of $x_3' = 0.25$ km. The distance between two adjacent field points is 0.5 km

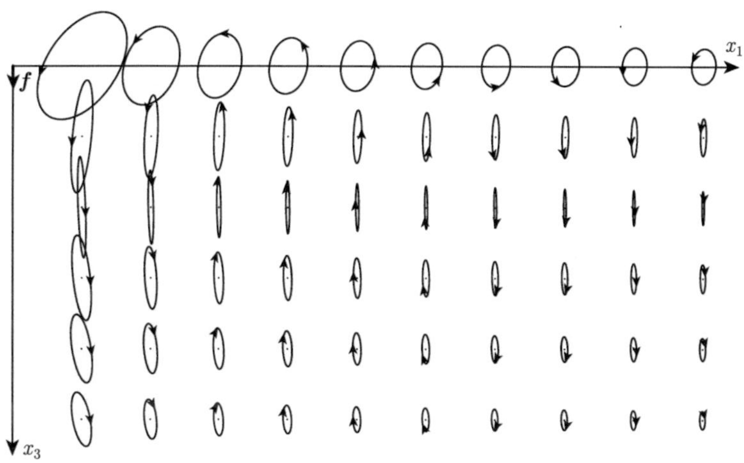

Fig. 7.90 Particle motion trajectory of near-field Rayleigh waves caused by a vertical force (1). The force is applied at the surface. The distance between two adjacent field points is 0.5 km

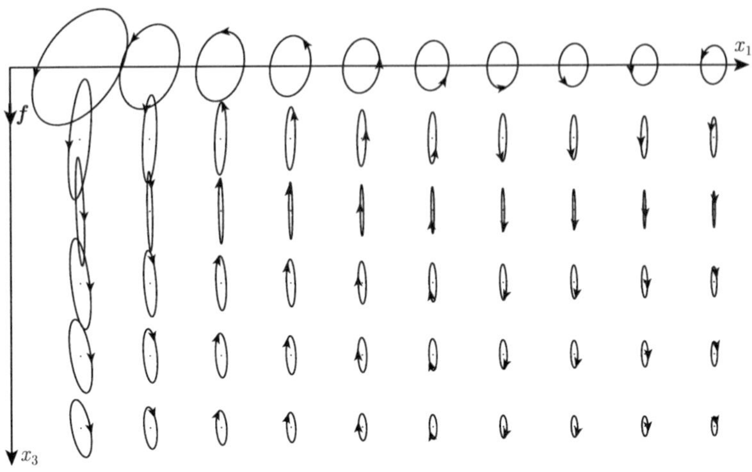

Fig. 7.91 Particle motion trajectory of near-field Rayleigh waves caused by a vertical force (2). The force is applied at the depth of $x_3' = 0.25$ km. The distance between two adjacent field points is 0.5 km

depth of the force application point.[42] Even when the source is at a depth of 0.25 km, the amplitude of the ellipses is slightly larger than when the source is at the surface. This is because, as shown in Fig. 7.85b, both components of the surface displacement caused by the vertical force increase with source depth up to 0.4 km (0.1 wavelength). Below this depth, the amplitude of the ellipses gradually decreases.

7.5.2.3 Rayleigh Waves in Frequency Domain Caused by Shear Dislocation Point Source

Compared to the simpler case of a single force, a shear dislocation point source is more seismologically significant. Section 6.6.2 explored the Rayleigh wave expression for a shear dislocation source. By taking the spatial derivatives of the Rayleigh wave Green's function and summing the products with the components of the seismic moment tensor, we derived the frequency domain displacement expression for Rayleigh wave caused by a shear dislocation point source, see Eq. (6.6.8).

[42] This difference between horizontal and vertical forces can be attributed to spatial asymmetry. At any point in space along the direction of the applied force, the medium's resistance must be overcome in two directions: one in the direction of the force and the other in the opposite direction. For example, a horizontal force applied along the positive x_1 axis must overcome the resistance of the medium in both the forward and backward directions. Similarly, a vertical force applied along the positive x_3 axis must do the same. However, unlike the horizontal force, the resistance exerted by the medium in the opposite direction for the vertical force is much smaller. Therefore, the amplitude of the particle motion trajectories decreases faster with depth for the horizontal force than for the vertical force.

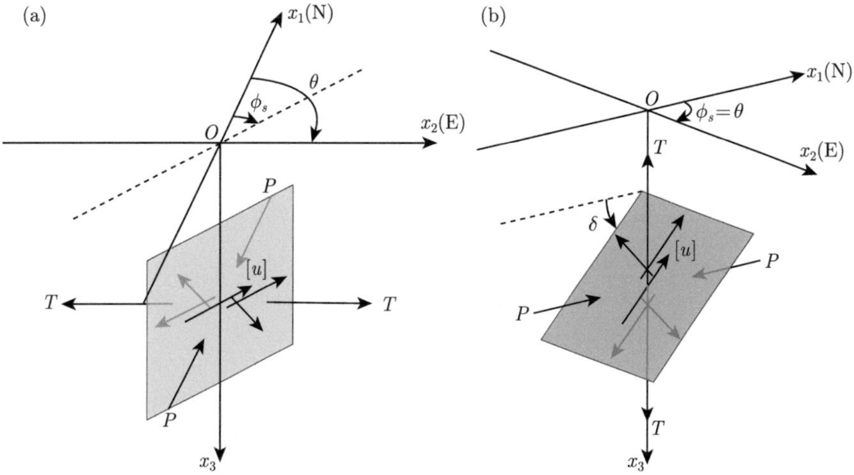

Fig. 7.92 Diagram of two typical shear dislocation models. **a** Vertical left-lateral strike-slip fault model ($\delta = 90°$, $\lambda = 0°$, $\phi_s = 45°$); **b** Thrust fault model ($\delta = 45°$, $\lambda = 90°$, $\phi_s = 90°$). The thick arrows on the faults indicate the direction of slip. The figure also shows the distribution of equivalent body forces, as well as the P and T axes

Next, we will analyze the characteristics of Rayleigh waves generated by two specific, typical dislocation sources. As shown in Fig. 7.92, we select the x_1 axis as the north direction and the x_2 axis as the east direction. We will study the Rayleigh wave generated by the vertical left-lateral strike-slip fault and the thrust fault illustrated in the figure.

For a vertical left-lateral strike-slip fault oriented along the northeast-southwest direction, $\delta = 90°$, $\lambda = 0°$, and $\phi_s = 45°$, as shown in Fig. 7.92a. By substituting these parameters into Eqs. (3.4.8a, 3.4.8b, 3.4.8c, 3.4.8d, 3.4.8e and 3.4.8f), we get

$$\mathbf{M}(\omega) = -M_0 S(\omega) \begin{bmatrix} 1 & 0 & 0 \\ 0 & -1 & 0 \\ 0 & 0 & 0 \end{bmatrix}$$

Substituting this into Eq. (6.6.8) and considering an observation point in the due east direction, $\theta = 90°$, we obtain

$$\begin{aligned}
\tilde{u}_1^R &= -M_0 S(\omega) \left(\tilde{G}_{11,1'} - \tilde{G}_{12,2'} \right) = 0 \\
\tilde{u}_2^R &= -M_0 S(\omega) \left(\tilde{G}_{21,1'} - \tilde{G}_{22,2'} \right) \\
&= \vartheta_1 M_0 S(\omega) F_1(\kappa) \left[3x H_1^{(2)}(x) - 4H_2^{(2)}(x) - x H_3^{(2)}(x) \right] \\
\tilde{u}_3^R &= -M_0 S(\omega) \left(\tilde{G}_{31,1'} - \tilde{G}_{32,2'} \right) \\
&= \vartheta_1 M_0 S(\omega) F_4(\kappa) \left[2H_1^{(2)}(x) - x H_0^{(2)}(x) + x H_2^{(2)}(x) \right]
\end{aligned}$$

From this, we see that there is no displacement component in the x_1 direction. The particle displacement caused by the Rayleigh wave is confined to the x_2x_3 plane. For a pulse force source, $S(\omega) = 1$, hence we get

$$\begin{aligned}\operatorname{Re}\left[\tilde{u}_2^R \mathrm{e}^{\mathrm{i}\omega t}\right] = & \frac{M_0 F_1(\kappa)}{8\mu r R'(\kappa)}\left\{\left[3\kappa r J_1(\kappa r) - 4J_2(\kappa r) - \kappa r J_3(\kappa r)\right]\sin\omega t\right. \\ & \left. - \left[3\kappa r Y_1(\kappa r) - 4Y_2(\kappa r) - \kappa r Y_3(\kappa r)\right]\cos\omega t\right\} \end{aligned} \quad (7.5.1a)$$

$$\begin{aligned}\operatorname{Re}\left[\tilde{u}_3^R \mathrm{e}^{\mathrm{i}\omega t}\right] = & \frac{M_0 F_4(\kappa)}{8\mu r R'(\kappa)}\left\{\left[2J_1(\kappa r) - \kappa r J_0(\kappa r) + \kappa r J_2(\kappa r)\right]\sin\omega t\right. \\ & \left. - \left[2Y_1(\kappa r) - \kappa r Y_0(\kappa r) + \kappa r Y_2(\kappa r)\right]\cos\omega t\right\} \end{aligned} \quad (7.5.1b)$$

According to Eq. (7.5.1), the particle motion trajectories of Rayleigh waves generated by the vertical left-lateral strike-slip fault in Fig. 7.92a can be plotted for both near-field and far-field regions, as shown in Figs. 7.93 and 7.94, respectively. The strike-slip fault is located at a depth of 0.25 km, with adjacent field points spaced 0.5 km apart. The figures illustrate that the particle motion trajectories are elliptical, resembling those caused by a single force. Near the vertical fault, the ellipses are inclined, and as the horizontal distance increases, the major axes of the ellipses become more upright. Near the surface, the motion is retrograde, but below 0.2 wavelengths in depth, it becomes prograde. In the far-field region (beyond 50 km), the shape of the ellipses changes very little with the horizontal distance r.

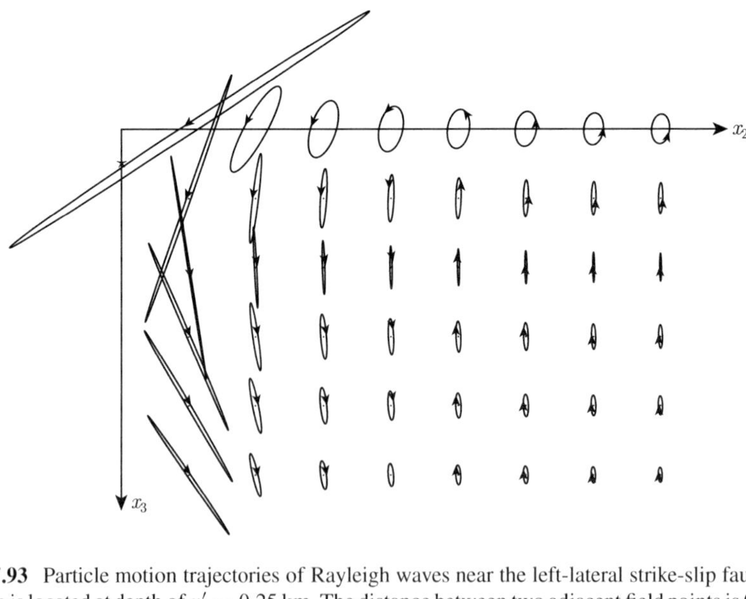

Fig. 7.93 Particle motion trajectories of Rayleigh waves near the left-lateral strike-slip fault. The source is located at depth of $x_3' = 0.25$ km. The distance between two adjacent field points is 0.5 km. The ⋆ represents the left-lateral strike-slip fault, as shown in Fig. 7.92a

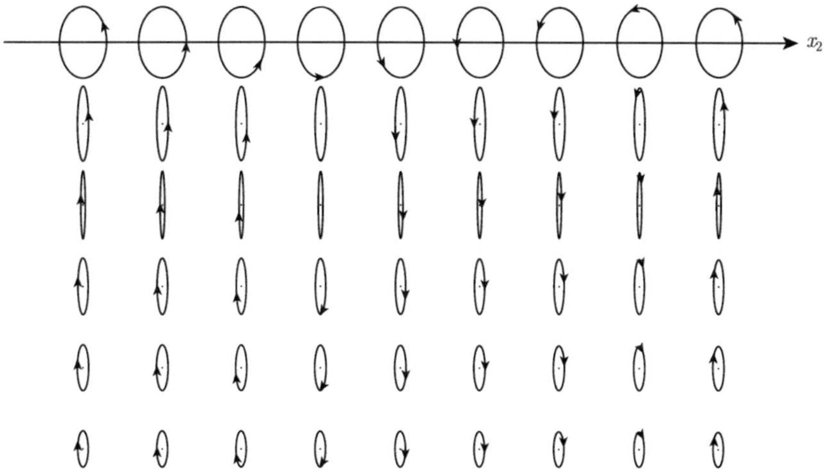

Fig. 7.94 Particle motion trajectories of Rayleigh waves in far-field region. The same as in Fig. 7.93. To clearly illustrate the motion trajectories, their amplitude has been magnified 7.5 times compared to Fig. 7.93. The left-lateral strike-slip fault is located at a horizontal distance of $r = 50$ km from the display area

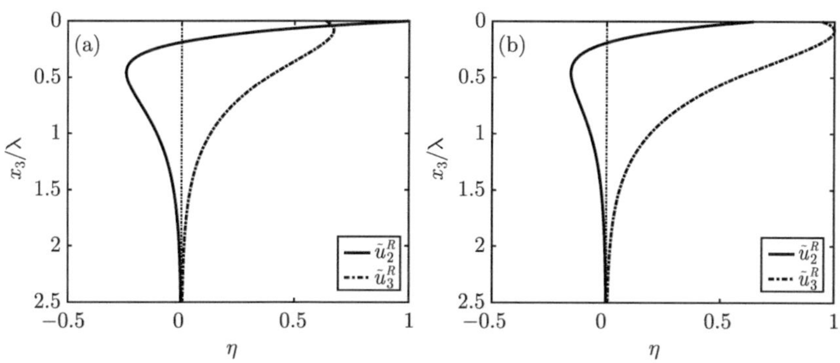

Fig. 7.95 $\tilde{u}_2^R$ and $\tilde{u}_3^R$ generated by left-lateral strike-slip fault vary with depth x_3. **a** Near-field case ($r = 0.5$ km); **b** Far-field case ($r = 50.0$ km). $\tilde{u}_2^R$ and $\tilde{u}_3^R$ are the relative amplitudes of the Rayleigh wave components. The vertical axis has been non-dimensionalized using wavelength λ

Figure 7.95 illustrates the relative amplitudes of the Rayleigh wave components $\tilde{u}_2^R$ and $\tilde{u}_3^R$ generated by a vertical strike-slip fault at horizontal distances of $r = 0.5$ km and 50 km, representing near-field and far-field conditions, respectively, as a function of depth x_3. Both displacement components show similar trends with depth: $\tilde{u}_3^R$ slightly increases with depth before monotonically decreasing, while $\tilde{u}_2^R$ significantly decreases with depth, reaching zero around 0.2 wavelengths deep, then reverses direction, peaking around 0.5 wavelengths deep, and finally monotonically decreases. A notable difference between the near-field and far-field conditions is that near the surface, $\tilde{u}_2^R$ is greater than $\tilde{u}_3^R$, causing the particle motion trajectory at the surface closest to the source (top left in Fig. 7.93) to be an ellipse with its major axis more horizontal.

For a thrust fault with a dip angle of 45° along the east-west direction, $\delta = 45°$, $\lambda = 90°$, and $\phi_s = 90°$ (see Fig. 7.92b),

$$\mathbf{M}(\omega) = -M_0 S(\omega) \begin{bmatrix} 1 & 0 & 0 \\ 0 & 0 & 0 \\ 0 & 0 & -1 \end{bmatrix}$$

Substituting into Eq. (6.6.8), for an observation point due east, where $\theta = 90°$, we have

$$\begin{aligned} \tilde{u}_1^R &= -M_0 S(\omega)\left(\tilde{G}_{11,1'} - \tilde{G}_{13,3'}\right) = 0 \\ \tilde{u}_2^R &= -M_0 S(\omega)\left(\tilde{G}_{21,1'} - \tilde{G}_{23,3'}\right) \\ &= -M_0 S(\omega)\left[4\vartheta_1 F_1(\kappa) H_2^{(2)}(x) - \vartheta_2 F_{3,3'}(\kappa) H_1^{(2)}(x)\right] \\ \tilde{u}_3^R &= -M_0 S(\omega)\left(\tilde{G}_{31,1'} - \tilde{G}_{33,3'}\right) \\ &= M_0 S(\omega)\left[2\vartheta_1 F_4(\kappa) H_1^{(2)}(x) + \vartheta_2 F_{5,3'}(\kappa) H_0^{(2)}(x)\right] \end{aligned}$$

Similarly, for a pulse force source, where $S(\omega) = 1$, we obtain

$$\begin{aligned} \mathrm{Re}\left[\tilde{u}_2^R \mathrm{e}^{\mathrm{i}\omega t}\right] = &-\frac{M_0}{4\mu r R'(\kappa)}\left\{\left[2F_1(\kappa)J_2(\kappa r) - rF_{3,3'}(\kappa)J_1(\kappa r)\right]\sin\omega t\right. \\ &\left. - \left[2F_1(\kappa)Y_2(\kappa r) - rF_{3,3'}(\kappa)Y_1(\kappa r)\right]\cos\omega t\right\} \end{aligned} \tag{7.5.2a}$$

$$\begin{aligned} \mathrm{Re}\left[\tilde{u}_3^R \mathrm{e}^{\mathrm{i}\omega t}\right] = &\frac{M_0}{4\mu r R'(\kappa)}\left\{\left[F_4(\kappa)J_1(\kappa r) + rF_{5,3'}(\kappa)J_0(\kappa r)\right]\sin\omega t\right. \\ &\left. - \left[F_4(\kappa)Y_1(\kappa r) + rF_{5,3'}(\kappa)Y_0(\kappa r)\right]\cos\omega t\right\} \end{aligned} \tag{7.5.2b}$$

Figures 7.96 and 7.97 show the particle motion trajectories of Rayleigh waves caused by a reverse fault located at a depth of 0.25 km in both the near-field and far-field regions, respectively. Except for the difference in the ratio of the major to minor axes compared to strike-slip faults, the characteristics are consistent with those of a left-lateral strike-slip fault. Figure 7.98 illustrates the relative amplitudes of the Rayleigh wave components $\tilde{u}_2^R$ and $\tilde{u}_3^R$ produced by the reverse fault as a function of depth x_3 at horizontal distances of $r = 0.5$ km and 50 km, representing near-field and far-field conditions, respectively. Except for details, the overall behavior is similar to that of a left-lateral strike-slip fault (see Fig. 7.95).

Finally, note that as $r \to \infty$, $J_m(\kappa r) \propto r^{-\frac{1}{2}}$ and $Y_m(\kappa r) \propto r^{-\frac{1}{2}}$. Thus, according to Eqs. (7.5.1) and (7.5.2), $\tilde{u}_i^R(\boldsymbol{x}, \omega) \propto r^{-\frac{1}{2}}$ for $i = 2, 3$. This indicates that the displacement components of the Rayleigh waves decay horizontally as $r^{-\frac{1}{2}}$. Figure 7.99

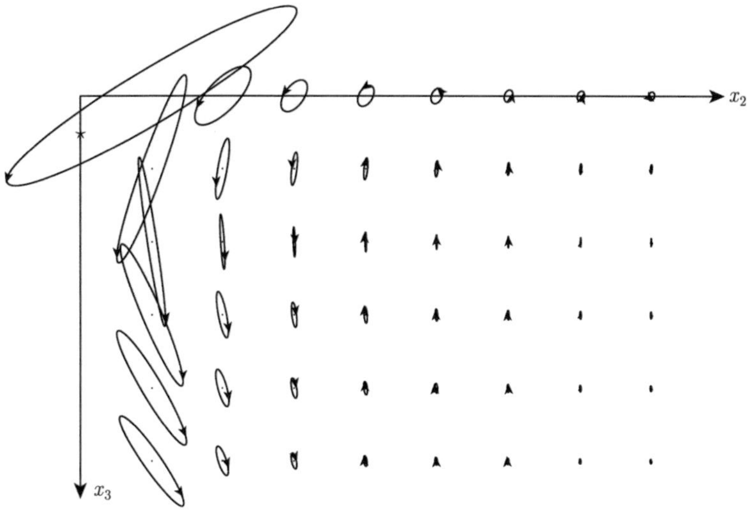

Fig. 7.96 Particle motion trajectories of Rayleigh waves near thrust fault. The source is located at depth of $x_3' = 0.25$ km. The distance between two adjacent field points is 0.5 km. The ⋆ represents the thrust fault, as shown in Fig. 7.92a. The amplitude of the motion trajectory is magnified 2.5 times compared to Fig. 7.93

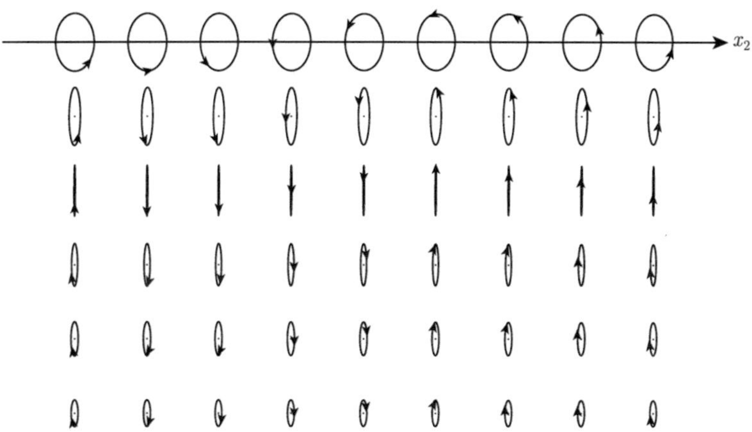

Fig. 7.97 Particle motion trajectories of Rayleigh waves in far-field region. The same as in Fig. 7.96. The amplitude has been magnified 40 times compared to Fig. 7.96. The thrust fault is located at a horizontal distance of $r = 50$ km from the display area

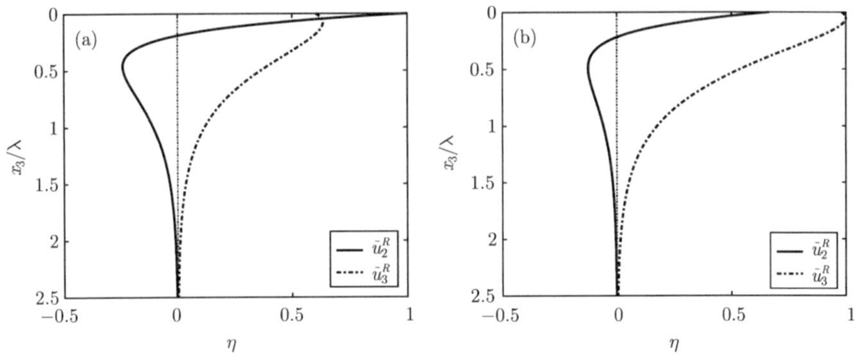

Fig. 7.98 $\tilde{u}_2^R$ and $\tilde{u}_3^R$ generated by thrust fault vary with depth x_3. The same as in Fig. 7.95

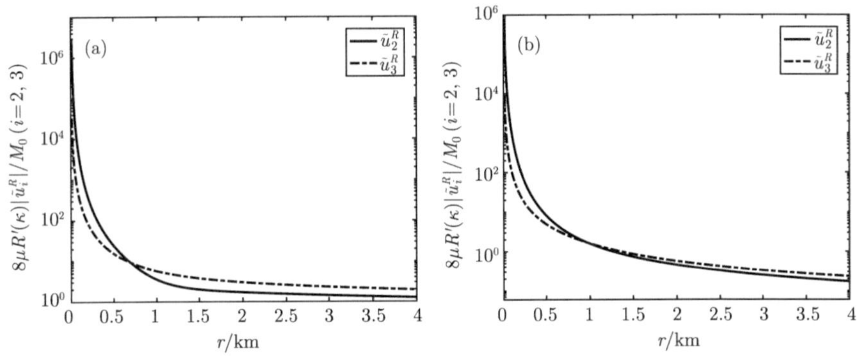

Fig. 7.99 Relative amplitude variation of $\tilde{u}_2^R$ and $\tilde{u}_3^R$ due to shear dislocation source with r. **a** Vertical left-lateral strike-slip fault; **b** Thrust fault. The relative amplitude is defined as $8\mu R'(\kappa)\left|\tilde{u}_i^R\right|/M_0$ $(i = 2, 3)$

displays the relative amplitudes of the Rayleigh wave displacement components for a vertical left-lateral strike-slip fault and a thrust fault as a function of horizontal distance r.

The similarity in the particle motion trajectories of far-field Rayleigh waves caused by a horizontal point force (Fig. 7.83), a vertical point force (Fig. 7.84), and the left-lateral strike-slip (Fig. 7.94) and reverse (Fig. 7.97) shear dislocation sources suggests that regardless of the specific form of the force, as long as the distance is sufficiently large, the single-frequency Rayleigh wave patterns are similar.

7.5.2.4 Rayleigh Wave in Time Domain Induced by Single Force Point Source

In the previous discussions, we analyzed the behavior and characteristics of Rayleigh wave motions generated by a single force and a shear dislocation source in the frequency domain, focusing on individual frequency components. We observed that, unlike the classic plane Rayleigh wave considered in traditional seismology, the near-field single-frequency Rayleigh wave motions induced by different forces exhibit a complex picture. However, in the far field, these Rayleigh wave motions simplify to resemble plane Rayleigh wave. A further question arises: given the expression for a single frequency component of Rayleigh wave, what characteristics would the Rayleigh wave in the time domain have if we performed an inverse Fourier transform (essentially a superposition of Rayleigh wave motions of different frequency components)?

Let's first consider the case of a single force point source. Recall that in Sect. 7.3, we validated our calculations by comparing them with the results from Johnson (1974) for the $\mathcal{LP}_2$. Figures 7.37 and 7.38 show the Green's function components for epicentral distances and source depth ratios of $\zeta = 5$ and 50, respectively. Using these as examples and employing the same computational parameters, we can perform an inverse Fourier transform using Eq. (6.5.16) to obtain the Rayleigh wave in the time domain.

Figure 7.100 compares the calculated results (black line) with the Green's function components (gray line). For easier analysis, the arrival times of various seismic phases are marked in the figure. It can be seen that the shape of the calculated Rayleigh wave around the Rayleigh wave arrival time "R" matches well with the complete Green's function components. At the arrival times of other seismic phases, there is no change in the Rayleigh wave curve. Notably, the G_{22}^{H} component, under the current source-field position combination, is primarily an SH wave component, so there is no significant Rayleigh wave present. A horizontal force along the x_1 axis or a vertical force along the x_3 axis only causes displacement components within the x_1x_3 plane. Therefore, particle motion trajectories can be plotted based on the horizontal and vertical displacement components in the x_1x_3 plane, which helps us intuitively understand particle motion in the elastic medium.

Figures 7.101 and 7.102 show the particle motion trajectories at the observation point (10, 0, 0) km (point A in the figure) caused by horizontal and vertical forces at point source depths of $x_3' = 2$ km and 0.2 km, respectively. The bold and fine dots represent the trajectories of only the Rayleigh wave and the complete motion including all seismic phases, respectively. Circles on each motion path indicate the paths near the Rayleigh wave arrival time. It is clear that the time-domain Rayleigh wave trajectories, calculated based on the Rayleigh wave excitation formula, accurately reflect the behavior around the Rayleigh wave arrival time on the complete motion path. Considering that the positive direction of the x_1 axis is the propagation direction of the seismic wave, all Rayleigh wave motions are retrograde.

One notable feature is that for a shallow point source, as seen in Fig. 7.102, most of the dynamic motion paths are traced around the Rayleigh wave arrival time, particularly for the vertical component. This indicates that for shallow earthquakes, the primary source of surface vibrations is the Rayleigh wave. This is a typical case where the free surface has a significant effect on seismic waves, and understanding the

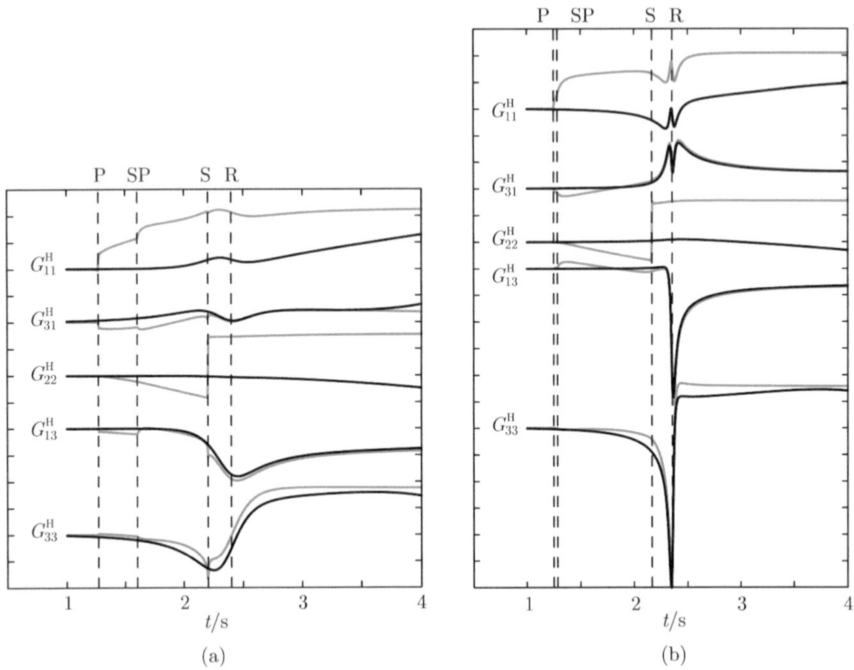

Fig. 7.100 Comparison of Rayleigh waves induced by point sources with Green's functions. **a** $x_3' = 2$ km; **b** $x_3' = 0.2$ km. The black line and gray line represent the Rayleigh wave and the complete Green's function components, respectively. The arrival times of different seismic phases are marked with dashed lines, where "P", "SP", "S", and "R" stand for the P-wave, SP-wave, S-wave, and Rayleigh wave, respectively. Refer to Figs. 7.37 and 7.38

movement and characteristics of Rayleigh waves is crucial for designing earthquake-resistant buildings. The excitation formula for Rayleigh waves provides a convenient way to conduct this research.

7.5.2.5 Rayleigh Wave in Time Domain Caused by Shear Dislocation Point Source

For the case of a shear dislocation point source, we still consider the two typical shear dislocation faults discussed in Sect. 7.5.2.3, as shown in Fig. 7.92. The point source location is the same as in the single force case of Sect. 7.5.2.4, with the only difference being that the source is now a shear dislocation point source instead of a single force point source. The observation point is located at (0, 10, 0) km, so for the strike-slip and thrust shear dislocation point sources considered in this section, the displacement component in the x_1 direction, u_1, is zero. To avoid significant Gibbs effects, a ramp function with a rise time $t_0 = 0.3$ s is used.

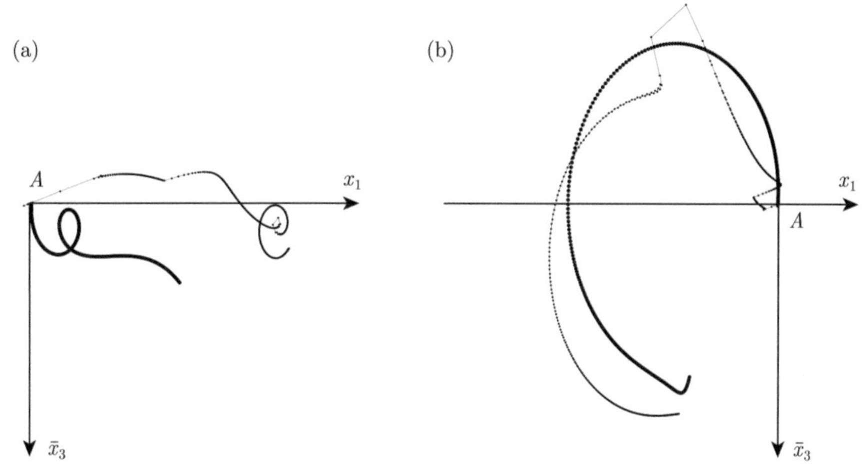

Fig. 7.101 Particle motion trajectory at point source depth of $x_3' = 2$ km. **a** Results of horizontal force; **b** Results of vertical force. The coordinates of point A are (10, 0, 0) km. Bold dots represent the Rayleigh wave trajectory, while thin dots represent the complete motion trajectory including all seismic phases. The $\bar{x}_3$ axis is obtained by translating the x_3 axis 10 km along the x_1 axis

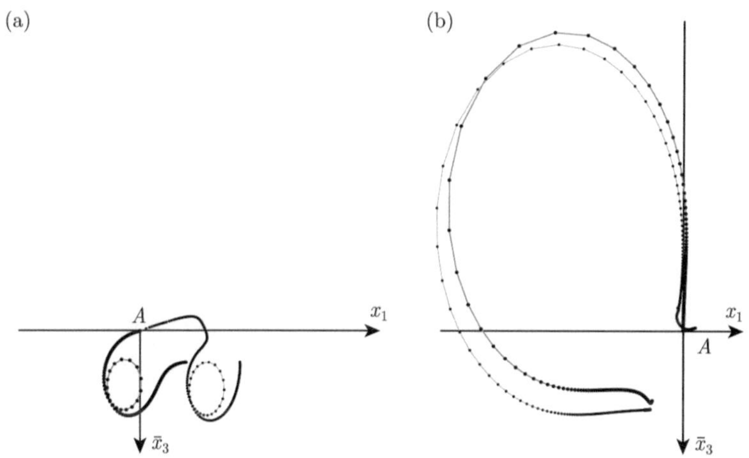

Fig. 7.102 Particle motion trajectory at point source depth of $x_3' = 0.2$ km. The same as in Fig. 7.101

For the vertical left-lateral strike-slip fault with $(\phi_s, \delta, \lambda) = (45°, 90°, 0°)$, Fig. 7.103 shows a comparison of the generated Rayleigh wave (black line) and the complete displacement waveform (gray line). Due to the use of the shear dislocation model and a more complex time function–the ramp function–the waveform is more complex than in the case of a single force with a step function time function. When comparing the waveform near the Rayleigh wave arrival time for a source depth of $x_3' = 0.2$ km to that for a depth of $x_3' = 2$ km, the waveform is not only more complex

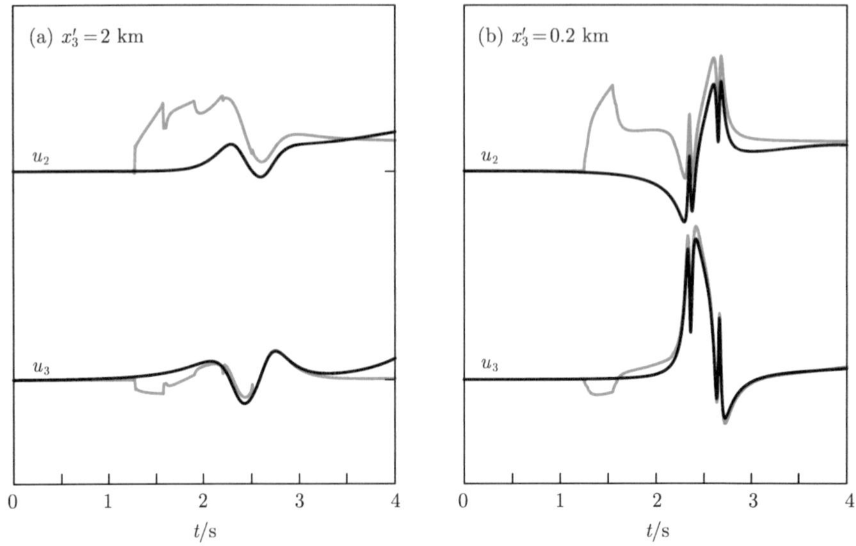

Fig. 7.103 Comparison of Rayleigh waves and complete displacements caused by strike-slip source. **a** $x_3' = 2$ km; **b** $x_3' = 0.2$ km. $(\phi_s,\ \delta,\ \lambda) = (45°,\ 90°,\ 0°)$. The black line and the gray line represent the Rayleigh wave and the complete displacement components, respectively. The time function is a ramp function with a rise time $t_0 = 0.3$ s

but also significantly larger in amplitude. Nevertheless, the Rayleigh waves in the time domain calculated using the Rayleigh wave excitation formula match well with the waveform near the Rayleigh wave arrival time.

Figure 7.104 shows the particle motion trajectories at different depths based on the displacement waveforms. The figure clearly illustrates that in the sections where the distance between points is relatively large (near the arrival of the Rayleigh wave), the results for the Rayleigh wave closely match those for the complete displacement components.

For the thrust fault with $(\phi_s,\ \delta,\ \lambda) = (90°,\ 45°,\ 90°)$, Figs. 7.105 and 7.106 show the comparisons of the corresponding Rayleigh wave (black line) and the complete displacement waveforms (gray line), as well as the particle motion trajectories caused by thrust point sources at different depths. Once again, we observe that the waveform characteristics near the arrival of the Rayleigh wave in the complete seismogram can be well depicted by the Rayleigh wave obtained using the Rayleigh wave excitation formula. The overall characteristics are similar to the strike-slip case, so they will not be elaborated further.[43]

[43] Of course, based on the results of the dislocation point source, Rayleigh waves generated by finite-scale shear dislocation sources can be obtained by superimposing point sources. However, we will not discuss this further here, and interested readers can explore this using the formulas provided in the book.

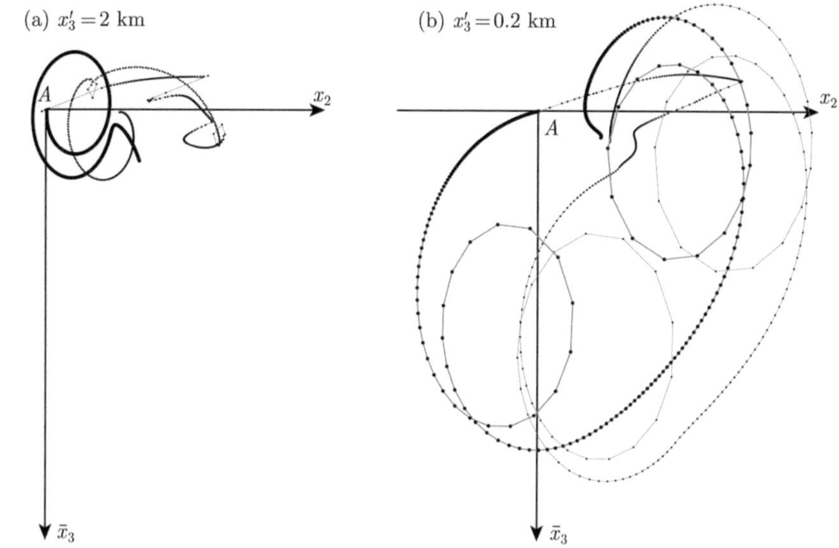

Fig. 7.104 Particle motion trajectories caused by strike-slip point source. **a** $x_3' = 2$ km; **b** $x_3' = 0.2$ km. The coordinates of point A are (0, 10, 0) km. Bold dots and thin dots correspond to the trajectories of the Rayleigh wave and the complete motion trajectory including all seismic phases, respectively. The $\bar{x}_3$ axis is obtained by translating the x_3 axis 10 km along the x_2 axis

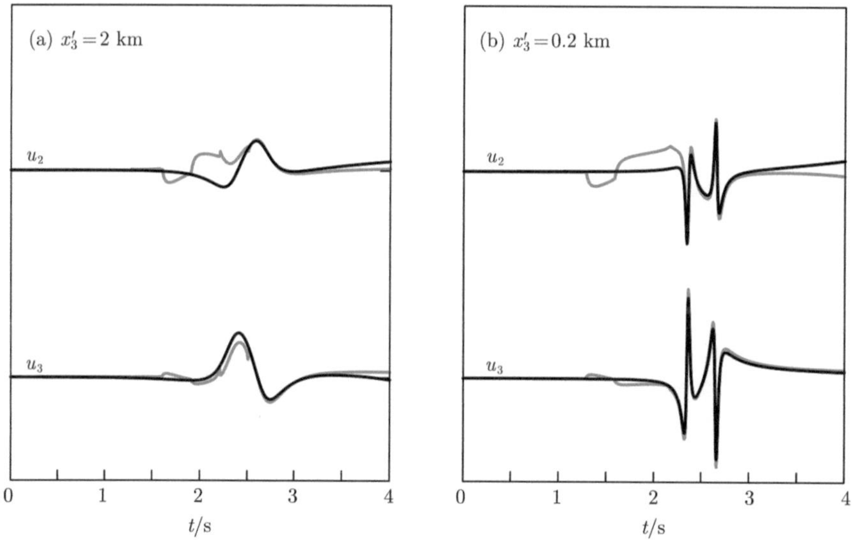

Fig. 7.105 Comparison of Rayleigh waves and complete displacements caused by thrust source. The same as in Fig. 7.103, except $(\phi_s,\ \delta,\ \lambda) = (90°,\ 45°,\ 90°)$

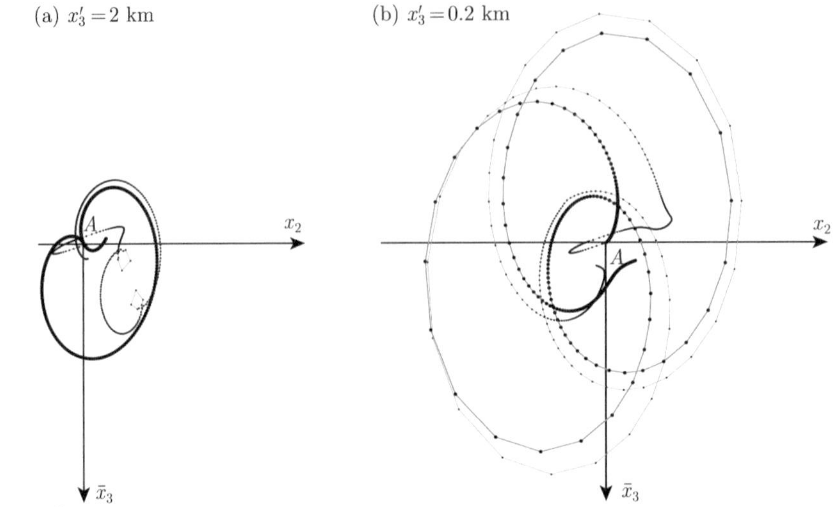

Fig. 7.106 Particle motion trajectories caused by thrust point source. The same as in Fig. 7.104

7.6 Summary

In this extensive chapter, building on the detailed discussion of the Lamb's problem in the frequency domain from the previous chapter, we first introduced numerical techniques related to wavenumber integration and discrete Fourier transform knowledge related to frequency integration. We then thoroughly validated the transition from formulas to program implementation. On this foundation, we examined the dynamic seismograms and static displacement fields generated by various sources, progressing from single force point sources and dislocation point sources to finite-scale sources. Finally, we explored the most prominent feature of the Lamb's problem compared to the previously considered infinite space model–the presence of a free surface–and examined the resulting special wave: Rayleigh surface wave. Based on the frequency-domain excitation formulas obtained in Chap. 6, we studied the characteristics of far-field and near-field Rayleigh waves generated by different sources, including point sources and dislocation point sources, and then obtained the Rayleigh waves in the time domain through inverse Fourier transform.

The various computational results presented in this chapter are aimed at revealing the properties of seismic waves implicit in the formulas obtained in the previous chapter from different perspectives. This is of great significance for a deeper understanding of the characteristics of ground motion presented in seismic records. The goal of seismology is to understand the internal structure of the Earth and information about seismic sources through seismic records, ultimately leading to the ability to predict earthquakes, mitigate earthquake disasters, and to some extent control earthquakes. Achieving this goal relies heavily on geophysical inversion, with a key component being forward modeling. It is not an exaggeration to say that the success of the inversion results largely depends on our understanding of the characteristics of seismic waves. It is hoped that the numerical results in this chapter will contribute to a deeper understanding of seismic waves.

The discussion of the Lamb's problem in the first volume comes to an end here. Reflecting on the entire book, several chapters were dedicated to setting the stage for the study of frequency-domain solutions to the Lamb's problem presented in the last two chapters. We began with a general review of theoretical seismology (Chap. 1) and then laid out the fundamentals of theoretical seismology–elastodynamics–in Chap. 2. We introduced the Green's function and the displacement representation theorem, which served as the foundation for establishing the source representation theorem and discussing equivalent body forces and the seismic moment tensor (Chap. 3). As the basis for studying the Lamb's problem, Chap. 4 addressed seismic wave problems in an infinite medium, deriving Green's functions in closed form and the seismic wave field generated by dislocation point sources. The infinite medium seismic wave problem captures the main characteristics of the elastic response of the Earth's medium to seismic sources. However, from a seismological perspective, this model fails to account for the surface's impact. Before diving into specifics, Chap. 5 reviewed the history of Lamb's problem studies, highlighting two main solution methods: the frequency domain solution based on Fourier transforms and the time domain solution based on Laplace transforms. Chapter 6 and this chapter then delved into the theoretical formulas and numerical implementation of the frequency domain solution.

In a sense, the introduction to the Lamb's problem is fairly complete. However, regarding the half-space model, a simpler and more specialized medium, there is a more elegant and brilliant solution–the direct time domain method. This is the best approach to solving the Lamb's problem, providing a single integral expression in the time domain. With careful study, this integral can even be expressed in terms of elementary functions and three types of standard elliptic integrals. Early on, results could be obtained by consulting elliptic integral tables, similar to trigonometric tables, and today they can be conveniently computed using standard functions in common mathematical software. Thus, the time domain method offers a generalized closed-form analytical solution to the Lamb's problem, which is valuable for both theoretical analysis and numerical implementation. The second volume of this book will thoroughly explore these related issues.

Appendix A
Branch Cut Construction of $f(z) = \sqrt{z^2 - z_0^2}$

A complex function of the form $f(z) = \sqrt{z^2 - z_0^2}$ ($z_0 = a - \mathrm{i}b,\ a, b \in \mathbb{R}^+$) involves the square root operation within the complex domain. We know that taking the square root in the complex domain results in a multi-valued function due to the multi-valued nature of the argument of the complex number. To ensure the analyticity of the complex function, a single-valued branch must be determined, and constructing a branch cut is an effective method for this. In this appendix, we will analyze in detail how to construct the branch cut for the above multi-valued function in the complex plane and discuss the argument on the branch cut.

First, it is clear that $z = \pm z_0 = \pm(a - \mathrm{i}b)$ are branch points.[1] Let $z = x + \mathrm{i}y$ ($x, y \in \mathbb{R}$), and $f_1 = X + \mathrm{i}Y$ ($X, Y \in \mathbb{R}$), we have

$$X + \mathrm{i}Y = \sqrt{(x + \mathrm{i}y)^2 - (a - \mathrm{i}b)^2}$$

thus

$$X^2 - Y^2 = x^2 - y^2 - (a^2 - b^2) \tag{A.1a}$$

$$XY = xy + ab \tag{A.1b}$$

[1] To determine whether a point is a branch point, the criterion is whether the function value returns to its original value after traversing a circle around the point in the complex plane. If it does not, then the point is a branch point.

H. Zhang, *Lamb's Problem in Seismology (Volume 1)*,
https://doi.org/10.1007/978-981-95-3170-7

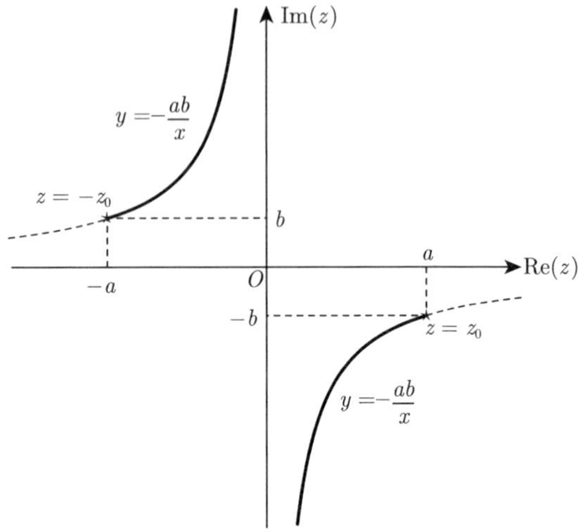

Fig. A.1 Branch cut of $f(z)=\sqrt{z^2-z_0^2}$ ($z_0 = a - \mathrm{i}b$). The horizontal and vertical axes represent the real and imaginary parts of z, respectively. The branch points at $z=\pm z_0$ are indicated by $\star$. The thick solid line represents the branch cut

To ensure the analyticity of $f(z)$, the usual approach is to constrain $X = \mathrm{Re}\{f(z)\} \geqslant 0$, where $X = 0$ actually indicates the location of the branch cut.[2] To find the branch cut equation, substitute $X = 0$ into Eq. (A.1b) to get

$$y = -\frac{ab}{x} \tag{A.2}$$

Substituting $X = 0$ into Eq. (A.1a), we have

$$-Y^2 = x^2 - y^2 - (a^2 - b^2) \leq 0$$

And further substituting Eq. (A.2) into this, we get

$$x^2 - \frac{a^2b^2}{x^2} - (a^2 - b^2) \leq 0 \quad \Rightarrow \quad \left(x^2 - a^2\right)\left(x^2 + b^2\right) \leq 0 \quad \Rightarrow \quad |x| \leq a. \tag{A.3}$$

Equations (A.2) together with the inequality give the branch cut equation, as shown in Fig. A.1.

As previously mentioned, the current method for drawing the branch cut ensures that the real part of $f(z)$ is zero. This means that any point on the branch cut in the complex plane is a purely imaginary number. However, the argument differs depending on whether the point is on the left or right edge of the branch cut.

[2] On the branch cut, the real part is zero, which means the numbers on the branch cut are purely imaginary. When $X = 0$, the square root of a negative number must be handled with care. For example, $\sqrt{-a^2}$ ($a > 0$) can be either $+\mathrm{i}a$ or $-\mathrm{i}a$, depending on whether the point is on the upper or lower edge of the branch cut. If the point is on the upper edge, where the argument is $\pi/2$, the value should be the former.

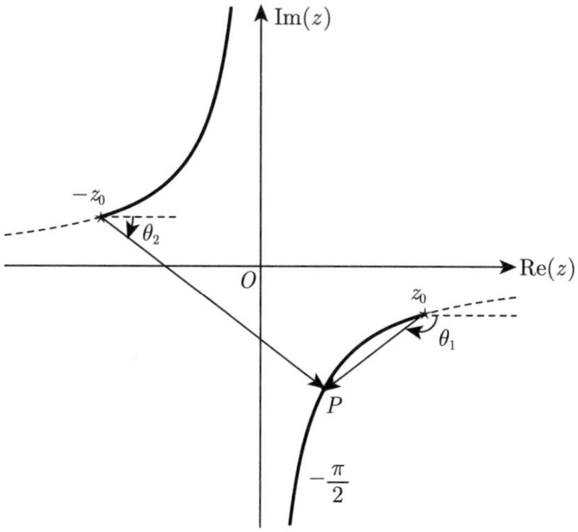

Fig. A.2 Diagram of argument on right side of branch cut in fourth quadrant of $f(z) = \sqrt{z^2 - z_0^2}$. The argument of $z - z_0$ is θ_1, and the argument of $z + z_0$ is θ_2. P is any point located to the right of the upper branch cut. The two arrowed line segments represent $z - z_0$ and $z + z_0$, respectively

Let's take the branch cut in the fourth quadrant as an example[3] to discuss the values of the argument of $f(z)$ when point P with coordinates (x_p, y_p) is on the right side, left side, or on the non-branch cut part of the hyperbola.

Express $f(z)$ as $\sqrt{z^2 - z_0^2} = \sqrt{(z - z_0)(z + z_0)}$, and let

$$z - z_0 = r_1 e^{i\theta_1}, \quad z + z_0 = r_2 e^{i\theta_2}$$

then,

$$f(z) = \sqrt{z^2 - z_0^2} = \sqrt{r_1 r_2} e^{i\frac{\theta_1 + \theta_2}{2}}$$

This means that the argument of $f(z)$, $\arg(f)$, is $(\theta_1 + \theta_2)/2$.

When point P is located to the right of the branch cut, as shown in Fig. A.2, both θ_1 and θ_2 are less than 0. Since

$$\tan(\pi + \theta_1) = -\frac{y_p + b}{a - x_p} \overset{(A.2)}{=\!=} \frac{b}{x_p}$$

$$\tan(\theta_2) = -\frac{b - y_p}{x_p + a} \overset{(A.2)}{=\!=} -\frac{b}{x_p}$$

[3] Readers can apply a similar analysis to the branch cut in the second quadrant.

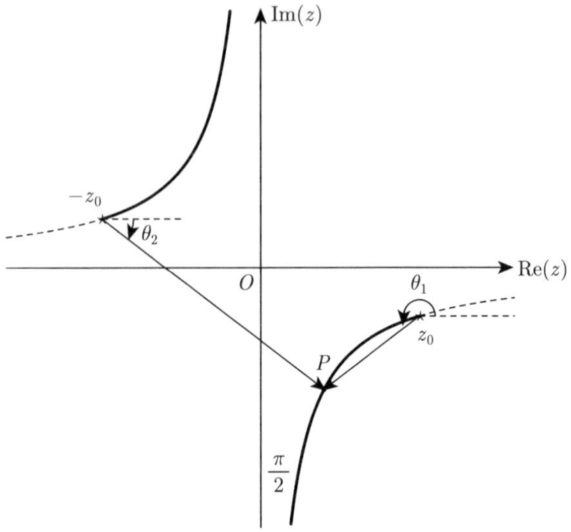

Fig. A.3 Diagram of argument on left side of branch cut in fourth quadrant of $f(z) = \sqrt{z^2 - z_0^2}$. The same as in Fig. A.2

we get $\theta_2 = -\pi - \theta_1$. Thus,

$$\arg(f) = \frac{\theta_1 + \theta_2}{2} = -\frac{\pi}{2}$$

This shows that on the right side of the branch cut, the argument of $f(z)$ is $-\pi/2$.

When point P is located to the left of the branch cut, as shown in Fig. A.3, $\theta_1 > 0$ and $\theta_2 < 0$. Since

$$\tan(\theta_1 - \pi) = -\frac{y_P + b}{a - x_P} \overset{\text{(A.2)}}{=\!=} \frac{b}{x_P}$$

we get $\theta_2 = -(\theta_1 - \pi)$. Thus, $\arg(f) = \pi/2$. This shows that on the left side of the branch cut, the argument of $f(z)$ is $\pi/2$.

When point P is located on the non-branch cut part of the hyperbola, as indicated by the dashed section in Fig. A.4, we have

$$\tan\theta_1 = \frac{y_P + b}{x_P - a} \overset{\text{(A.2)}}{=\!=} \frac{b}{x_P}$$

Therefore, $\theta_2 = -\theta_1$. Thus, $\arg(f) = 0$. This indicates that on the non-branch cut part of the hyperbola, the argument of $f(z)$ is 0.

Based on the above discussion, we can draw the following conclusions:

For $f(z) = \sqrt{z^2 - z_0^2}$, if we take its real part being zero as the branch cut, then in the complex plane of z, this corresponds to a part of a hyperbola, as shown by the

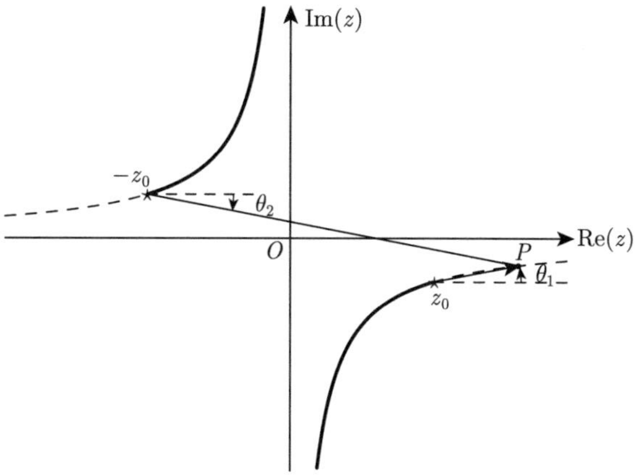

Fig. A.4 Argument of $f(z) = \sqrt{z^2 - z_0^2}$ on non-branch cut part of hyperbola in fourth quadrant. The same as in Fig. A.2

thick black line in Fig. A.1. The argument of $f(z)$ is distributed as follows: on the right side of the branch cut in the fourth quadrant, the argument is $-\pi/2$; on the left side, it is $\pi/2$. In the second quadrant, the argument is $\pi/2$ on the right side of the branch cut and $-\pi/2$ on the left. In the non-branch cut parts of the hyperbola in both quadrants, the argument is 0. Below the hyperbola in the fourth quadrant and above the hyperbola in the second quadrant, $-\pi/2 < \arg(f) < 0$. In the region between the two hyperbolas, $0 < \arg(f) < \pi/2$.

Appendix B
Zeros of Rayleigh Function

In this appendix, we will examine the roots of

$$R(x) = (1 - 2x)^2 + 4x\sqrt{m - x}\sqrt{1 - x} \quad \left(m = \frac{1 - 2\nu}{2(1 - \nu)}\right) \tag{B.1}$$

where m is a positive real number solely dependent on the Poisson ratio ν. Given that $0 < \nu < 1/2$,[1] it is easy to deduce that the range of m is $(0,\ 1/2)$. Our goal is to analyze and find the solutions of $R(x) = 0$.

Let

$$\overline{R}(x) = (1 - 2x)^2 - 4x\sqrt{m - x}\sqrt{1 - x},$$

then the product of $R(x)$ and $\overline{R}(x)$ yields a cubic equation in x^2:

$$f(x) = R(x)\overline{R}(x) = 16(1 - m)x^3 + 8(2m - 3)x^2 + 8x - 1 = 0 \quad \left(0 < m < \frac{1}{2}\right)$$

[1] According to the theory of elasticity, the theoretically predicted range for the Poisson ratio ν is $(-1,\ 1/2)$. However, negative Poisson ratios have not yet been observed in nature, making it practically meaningless to generally discuss negative Poisson ratios, and they are not considered here.

[2] The reason for considering $f(x) = 0$ is that directly solving $R(x) = 0$ is difficult. Transforming it into finding the roots of the cubic equation $f(x) = 0$ is much easier; however, we subsequently need to identify which of these roots correspond to $R(x) = 0$. In fact, because the second term in $R(x)$ contains radicals, directly solving $R(x) = 0$ requires rearranging and squaring, which is equivalent to solving $f(x) = 0$.

H. Zhang, *Lamb's Problem in Seismology (Volume 1)*,
https://doi.org/10.1007/978-981-95-3170-7

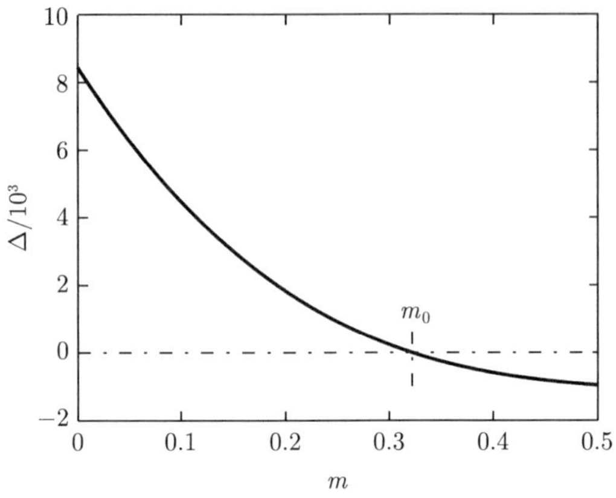

Fig. B.1 Variation of discriminant Δ with parameter m ($m_0 = 0.321498$ is the root of $\Delta = 0$)

According to the theory of roots for a general cubic equation,[3] we have $A = 256m^2 - 384m + 192$, $B = -16(k+3)$, and $C = 8(6k-1)$. Thus, the discriminant is

$$\Delta(m) = B^2 - 4AC = -49152m^3 + 82176m^2 - 47616m + 8448.$$

This is a cubic polynomial in m. Examining the properties of this discriminant theoretically is generally difficult, so the easiest approach is to directly plot it. Figure B.1 shows the variation of the discriminant Δ with the parameter m. The point where the discriminant is zero corresponds to $m = m_0 = 0.321498$.[4] The behavior of the discriminant is clearly illustrated in Fig. B.1:

[3] For a general cubic equation $ax^3 + bx^2 + cx + d = 0$, let $A = b^2 - 3ac$, $B = bc - 9ad$, $C = c^2 - 3bd$, and $\Delta = B^2 - 4AC$. Then,

$$\begin{cases} \text{There is a triple real root,} & \text{if} \quad A = B = 0 \\ \text{There is one real root and a pair of complex conjugate roots,} & \text{if} \quad \Delta > 0 \\ \text{There are three real roots, one of which is a double root,} & \text{if} \quad \Delta = 0 \\ \text{There are three distinct real roots,} & \text{if} \quad \Delta < 0 \end{cases}.$$

[4] This value can be accurately solved using Maple software. The exact value given by the `solve` command is

$$-\frac{1}{192}\left(77293 + 7296\sqrt{114}\right)^{1/3} + \frac{445}{192\left(77293 + 7296\sqrt{114}\right)^{1/3}} + \frac{107}{292}$$

Rounded to six decimal places, it is 0.321498.

(1) When $0 < m < m_0$, $\Delta > 0$, and $f(x)$ has one real root and a pair of complex conjugate roots.
(2) When $m = m_0$, $\Delta = 0$, and $f(x)$ has three real roots, with one being a double root.
(3) When $m_0 < m < 1/2$, $\Delta < 0$, and $f(x)$ has three distinct real roots.

Now let's examine the distribution of the roots of $f(x) = 0$ for different values of m. Noting that $f(x) = 8x(x-1)[2(1-m)x-1]-1$, we have $f(0) = f(1) = f(1/2(1-m)) = -1$. From the graph of $f(x)$ shown in Fig. B.2, it is evident that $f(x)$ always has one real root greater than 1. If there are two other roots, whether they are equal (when $m = m_0$) or distinct (when $m_0 < m < 1/2$), they must lie within the interval $(0, 1/2(1-m))$. For the case shown in Fig. B.2, where $m_1 < m_0$, there are no intersections between the curve and the x axis within the interval $(0, 1/2(1-m))$, indicating that there are two complex conjugate roots.

Now we need to address an important question: since $f(x) = R(x)\overline{R}(x)$ is the product of two functions, which roots belong to $R(x)$ that we are interested in, and which belong to $\overline{R}(x)$ that we are not? One convenient way is to verify each of the three roots for different values of m to see which satisfy $R(x) = 0$; those will be the roots of $R(x)$. Otherwise, they are roots of $\overline{R}(x)$. Alternatively, we can use Maple (with the `solve` command) to solve $R(x) = 0$ and $\overline{R}(x) = 0$ for different values of m and see the results.

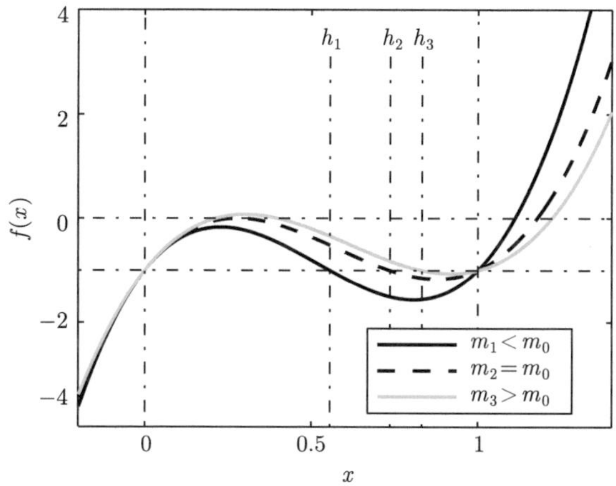

Fig. B.2 Behavior of $f(x)$ for different values of m. $m_1 = 0.1 < m_0$, $m_2 = m_0$, $m_3 = 0.4 > m_0$. The two horizontal dashed lines represent $f = 0$ and $f = -1$. The five vertical dashed lines from left to right represent $x = 0$, $x = h_1 = 1/2(1-m_1)$, $x = h_2 = 1/2(1-m_2)$, $x = h_3 = 1/2(1-m_3)$, and $x = 1$. It can be observed that for the three values of m, the curves each intersect the $f = 0$ line within the range $x > 1$, indicating a real root greater than 1

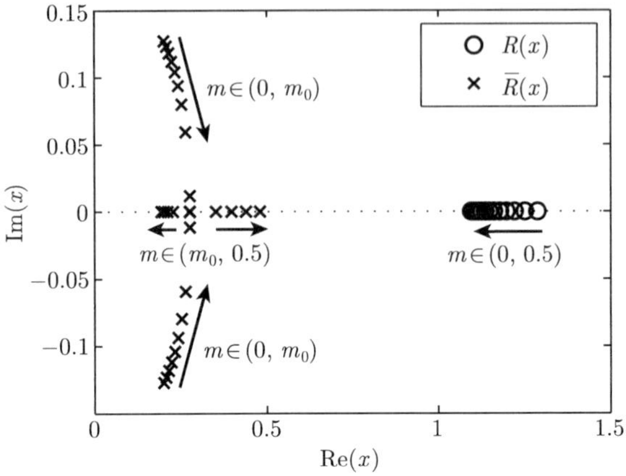

Fig. B.3 Distribution of roots of $R(x)$ and $\overline{R}(x)$ in complex plane of x for different values of m. The horizontal and vertical axes represent the real and imaginary parts of x, respectively. "∘" and "×" represent the roots of $R(x)$ and $\overline{R}(x)$, respectively. The dashed line represents the real values of x. Modified based on Fig. 4 of Liu et al. (2016)

For example, using the second approach for several m values from Fig. B.2:

(1) When $m = m_1 = 0.1$, the root of $R(x) = 0$ is 1.112177, and the roots of $\overline{R}(x) = 0$ are $0.221689 \pm 0.115299i$, which are two complex conjugate roots.
(2) When $m = m_0 = 0.321498$, the root of $R(x) = 0$ is 1.177537, and the root of $\overline{R}(x) = 0$ is 0.279690, indicating two equal real roots.[5]
(3) When $m = m_3 = 0.4$, the root of $R(x) = 0$ is 1.220466, and the roots of $\overline{R}(x) = 0$ are 0.398908 and 0.213958, which are two distinct real roots.

Figure B.3 illustrates the distribution and attribution of roots for different values of m. It is evident that for the entire range of m, $R(x)$ consistently has a single real root greater than 1. When $m \in (0, m_0)$, $\overline{R}(x)$ has two complex conjugate roots. When $m \in (m_0, 0.5)$, $\overline{R}(x)$ has two real roots, which are equal if $m = m_0$.

[5] How do we determine this is a double root, and the root of $R(x) = 0$ is not? Since $m_0 = 0.321498$ is an approximate value, the calculation results may have errors. For instance, the actual calculated roots for $\overline{R}(x) = 0$ are $0.279690 \pm 0.000192i$, with an imaginary part small enough to be considered negligible, thus confirming that the roots of $\overline{R}(x) = 0$ are actually two equal real roots.

By examining $f(x)$ in detail, we can refine our understanding: within the given range of m, $R(x) = 0$ always has one real root greater than 1. The other two roots, whether they are complex conjugate roots (when $0 < m < m_0$), double real roots (when $m = m_0$), or two distinct real roots (when $m_0 < m < 1/2$), are all roots of $\overline{R}(x) = 0$.

To summarize our conclusion in one sentence:

For the Rayleigh function defined in Eq. (B.1), within the range $m \in (0, 1/2)$, there is exactly one real root greater than 1.[6]

---

[6] For further reflection: when $x > 1$, both radicals in Eq. (B.1) have negative terms underneath them, which can result in multivalued outcomes for the square root operation. How should these be interpreted? Readers can refer to the discussion in Appendix A for analysis. Here's a brief analysis for reference: in the complex plane of x, the branch points are at $x = m$ and $x = 1$, with branch cuts extending from these points towards the positive real axis. When x is a real number greater than 1, it lies on the branch cut. Let $\arg(x - m) = \theta_1$ and $\arg(x - 1) = \theta_2$, then $\arg\left(\sqrt{m - x}\sqrt{1 - x}\right) = (\theta_1 + \theta_2)/2 + \pi$. When x is on the upper edge of the branch cut, $\theta_1 = \theta_2 = 0$, so $\arg\left(\sqrt{m - x}\sqrt{1 - x}\right) = \pi$. When x is on the lower edge, $\theta_1 = \theta_2 = 2\pi$, so $\arg\left(\sqrt{m - x}\sqrt{1 - x}\right) = 3\pi$. Both scenarios yield the same result, leading to $R(x) = (1 - 2x)^2 - 4x\sqrt{x - m}\sqrt{x - 1}$.

References

Abramowitz M, Stegun I (1964) Handbook of mathematical functions with formulas, graphs, and mathematical tables. National Bureau of Standards, Gaithersburg

Agnew DC (2002) History of seismology. In: Lee WHK, Kanamori H, Jennings PC et al (eds) International handbook of earthquake and engineering seismology (81A). Academic, Amsterdam, pp 3–11

Aki K, Richards PG (2002) Quantitative seismology. University Science Books, Sausalito

Aochi H, Fukuyama E, Matsu'ura M (2000) Spontaneous rupture propagation on a non-planar fault in 3-D elastic medium. Pure Appl Geophys 157:2003–2027

Apsel RJ (1979) Dynamic Green's functions for layered media and applications to boundary-value problems. Ph.D. Thesis, University of California, San Diego, La Jolla, California

Backus G, Mulcahy M (1976) Moment tensors and other phenomenological descriptions of seismic sources. II. Discontinuous displacements. Geophys J R Astron Soc 47:301–329

Ben-Menahem A (1995) A concise history of mainstream seismology: origins, legacy, and perspectives. Bull Seismol Soc Am 85(4):1202–1225

Ben-Menahem A, Singh SJ (1981) Seismic waves and sources. Springer, New York

Bouchon M (1979) Discrete wavenumber representation of elastic wave fields in three-space dimensions. J Geophys Res 84:3609–3614

Bouchon M (1981) A simple method to calculate Green's functions in elastic layered media. Bull Seismol Soc Am 71:959–971

Bouchon M, Aki K (1977) Discrete wave number representation of seismic source wave fields. Bull Seismol Soc Am 67:259–277

Bromwich TJ (1898) On the influence of gravity on elastic waves, and in particular, on the vibrations of an elastic globe. Proc Lond Math Soc 30:98–165

Burridge R, Knopoff L (1964) Body force equivalents for seismic dislocations. Bull Seismol Soc Am 54(6):1875–1888

Cagniard L (1939) Réflexion et réfraction des ondes seismiques progressives. Gauthier-Villars, Paris

Cagniard L (1962) Reflection and refraction of progressive seismic waves. (Trans. by E. A. Flinn and C. H. Dix.) McGraw-Hill, New York

H. Zhang, *Lamb's Problem in Seismology (Volume 1)*,
https://doi.org/10.1007/978-981-95-3170-7

Chao CC (1960) Dynamical response of an elastic half-space to tangential surface loadings. J Appl Mech ASME 27:559–567
Chen XF (1999) Seismogram synthesis in multi-layered half-space Part I. Theor Formulations Earth Res China 13:149–174
Chen XF, Zhang HM (2001) An efficient method for computing Green's functions for a layered half-space at large epicentral distances. Bull Seismol Soc Am 91(4):858–869
Chou PC, Pagano NJ (1992) Elasticity: tensor, dyadic, and engineering approaches. Dover Publications Inc., New York
Dahlen FA, Tromp J (1998) Theoretical global seismology. Priceton University Press, Princeton
Dahlquist G, Björck Å (1974) Numerical methods in scientific computing, vol 1. Prentice-Hall Inc., Englewood Cliffs
Das S, Aki K (1977) A versatile earthquake model. J Geophys Res 82:5658–5670
de Hoop AT (1960) A modification of Cagniard's method for solving seismic pulse problems. Appl Sci Res B8:349–356
Dix CH (1954) The method of Cagniard in seismic pulse problems. Geophysics 19(4):722–738
Eason G (1966) The displacements produced in an elastic half-space by a suddenly applied surface force. IMA J Appl Math 2(4):299–326
Ewing WM, Jardetzky WS, Press F (1957) Elastic waves in layered media. McGraw-Hill, New York
Feng X, Zhang HM (2018) Exact closed-form solutions for Lamb's problem. Geophys J Int 214(1):444–459
Feng X, Zhang HM (2021) Exact closed-form solutions for Lamb's problem–III: the case for buried source and receiver. Geophys J Int 224(1):517–532
Garvin WW (1956) Exact transient solution of the buried line source problem. Proc R Soc Ser A 234(1199):528–541
Guo DR (1991) Methods of mathematical physics (in Chinese). Higher Education Press, Beijing
Haskell NA (1964) Radiation pattern of surface waves from point source in a multilayered in a multilayered medium. Bull Seismol Soc Am 54:377–393
Honda H (1962) Earthquake mechanism and seismic waves. J Phys Earth 10:1–98
Johnson LR (1974) Green's function for Lamb's problem. Geophys J R Astron Soc 37:99–131
Kanamori H, Stewart G (1978) Seismological aspects of the Guatemala earthquake of February 4, 1976. J Geophys Res 83:3427–3434
Kausel E (2012) Lamb's problem at its simplest. Proc R Soc Ser A 20120462
Kikuchi M, Kanamori H (1982) Inversion of complex body waves. Bull Seismol Soc Am 72:491–506
Kikuchi M, Kanamori H (1991) Inversion of complex body waves-III. Bull Seismol Soc Am 81:2335–2350
Knopoff L, Gilbert F (1959) Radiation from a strike slip fault. Bull Seismol Soc Am 49:168–178
Knopoff L, Gilbert F (1960) First motion from seismic sources. Bull Seismol Soc Am 50:117–134
Kostrov BV (1966) Unsteady propagation of longitudinal shear cracks. J Appl Math Mech 30:1241–1248
Lamb H (1904) On the propagation of tremors over the surface of an elastic solid. Philos Trans R Soc Lond Ser A 203:1–42
Lapwood ER (1949) The disturbance due to a line source in a semi-infinite elastic medium. Philos Trans R Soc Lond Ser A 242:63–100
Liu TS, Feng X, Zhang HM (2016) On Rayleigh wave in half-space: an asymptotic approach to study the Rayleigh function and its relation to the Rayleigh wave. Geophys J Int 206:1179–1193
Maruyama T (1963) On the force equivalents of dynamical elastic dislocations with reference to the earthquake mechanism. Bull Earthq Res Inst 41:467–486
Mooney HM (1974) Some numerical solutions for Lamb's problem. Bull Seismol Soc Am 64(2):473–491
Nakano H (1923) Notes on the nature of the forces which give rise to the earthquake motions. Seismol Bull Central Meteorol Observatory Jpn 1:92–120

Nakano H (1925) On Rayleigh waves. Japan J Astron Geophys 2:233–326
Pekeris CL (1955) The seismic surface pulse. Proc Natl Acad Sci US 41:469–480
Pekeris CL (1955) The seismic buried pulse. Proc Natl Acad Sci US 41:629–639
Pekeris CL, Lifson H (1957) Motion of the surface of a uniform elastic half-space produced by a buried pulse. J Acoust Soc Am 29(11):1233–1238
Pinney E (1954) Surface motion due to a point source in a semi-infinite elastic medium. Bull Seismol Soc Am 44(4):571–596
Prudnikov AP, Brychkov YA, Marichev OI (1986) Integrals and series (vol 2, Special functions). Gordon and Breach Sciences Publishers, London
Rayleigh L (1885) On waves propagated along the plane surface of an elastic solid. Proc Lond Math Soc 1:4–11
Richards PG (1979) Elementary solutions to Lamb's problem for a point source and their relevance to three-dimensional studies of spontaneous crack propagation. Bull Seismol Soc Am 69:947–956
Scholz CH (1987) Wear and gouge formation in brittle faulting. Geology 15:493–495
Steketee JA (1958) On Volterra's dislocations in a semi-infinite elastic medium. Can J Phys 36:192–205
Tada T, Yamashita T (1997) Non-hypersingular boundary integral equations for two-dimensional non-planar crack analysis. Geophys J Int 130:269–282
Wang MZ, Wang W, Wu JK (2002) Course in elastic mechanics (in Chinese). Peking University Press, Beijing
Watson GN (1922) A treatise on the theory of Bessel functions. Cambridge University Press, Cambridge
Wu CS (2003) Methods of mathematical physics (2nd Edition, in Chinese). Peking University Press, Beijing
Xie XB, Zheng TY, Yao ZX (1992) Methods for calculating theoretical seismograms (in Chinese). Chin J Geophys 35(6):790–801
Zhang HM (2004) Theoretical study of 3D spontaneous rupture propagation on a planar fault in a half space (in Chinese). PhD Dissertation, Peking University, Beijing
Zhang HM (2006) Theoretical study of dynamic rupture in complex fault systems and preliminary study of supershear rupture under surface effects (in Chinese). Postdoctoral Report, Peking University, Beijing
Zhang HM, Chen XF (2001) Self-adaptive Filon's integration method and its applications to computing synthetic seismograms. Chin Phys Lett 18(3):313–315
Zhang HM, Chen XF (2006) Dynamic rupture on a planar fault in three-dimensional half space–I. Theory Geophys J Int 164(3):633–652
Zhang HM, Chen XF (2009) Equivalence of the Green's function for a full-space to the direct-wave contributions for a half-space and a layered half-space. Bull Seismol Soc Am 99(1):454–461
Zhang HM, Chen XF, Chang SH (2001) Peak and trough averaging method and its application in calculating synthetic seismograms for shallow sources (in Chinese). Chin J Geophys 44(6):805–813
Zhang HM, Chen XF, Chang SH (2003) An efficient numerical method for computing synthetic seismograms for a layered half-space with source and receivers at closed or same depths. Pure Appl Geophys 160(3/4):467–486
Zhou J, Zhang HM (2020) Exact Rayleigh-wave solution from a point source in homogeneous elastic half-space. Bull Seismol Soc Am 110(2):783–792

MIX
Papier aus verantwortungsvollen Quellen
Paper from responsible sources
FSC® C105338

If you have any concerns about our products,
you can contact us on
ProductSafety@springernature.com

In case Publisher is established outside the EU,
the EU authorized representative is:
Springer Nature Customer Service Center GmbH
Europaplatz 3, 69115 Heidelberg, Germany

Printed by Libri Plureos GmbH
in Hamburg, Germany